# Light Metals 2008

## LIGHT METALS 2008

***Related Titles***

- *Light Metals 2007*, edited by Morten Sørlie
- *Light Metals 2006*, edited by Travis J. Galloway
  - *Volume 1: Alumina and Bauxite*
  - *Volume 2: Aluminum Reduction Technology*
  - *Volume 3: Carbon Technology*
  - *Volume 4: Cast Shop Tech & Recycling Aluminum*
- *Light Metals 2005*, edited by Halvor Kvande

Other TMS 2008 Annual Meeting Publications

- *TMS 2008 Annual Meeting Supplemental Proceedings,*
  *Volume 1: Materials Processing and Properties*
- *TMS 2008 Annual Meeting Supplemental Proceedings,*
  *Volume 2: Materials Characterization, Computation and Modeling*
- *TMS 2008 Annual Meeting Supplemental Proceedings,*
  *Volume 3: General Paper Selections*
- *Magnesium Technology 2008*
- *EPD Congress 2008*
- *Carbon Dioxide Reduction Technology*
- *Aluminum Alloys: Fabrication, Characterization and Applications*
- *Mughrabi Plasticity, Failure and Fatigue in Structural Materials*

**ELECTRODE TECHNOLOGY**


*ORGANIZERS*

**John A. Johnson**
United Company Rusal
Krasnoyarsk, Russia

**Carlos Eduardo Zangiacomi**
Alcoa Inc.
Pocos De Caldas, Brazil

A Publication of **The Minerals, Metals & Materials Society (TMS)**
184 Thorn Hill Road
Warrendale, Pennsylvania 15086-7528
(724) 776-9000

**Visit the TMS Web site at**
http://www.tms.org



ISBN 978-0-470-94339-7 **ISSN Number 109-9586**



If you are interested in purchasing a copy of this book, or if you would like to receive the latest TMS Knowledge Resource Center catalog, please telephone (724) 776-9000, ext. 270, or (800) 759-4TMS.

# TABLE OF CONTENTS
# Light Metals 2008



## Alumina and Bauxite

### HSEC



### Equipment

## Bauxite



## Additives

## Operations



## Precipitation

# TABLE OF CONTENTS
# Light Metals 2008



## Aluminum Reduction Technology

### Sustainability and Environment

## Cell Development Part I and Operations



## Process Control

## Aluminum Industry in Mid-East

## Reduction Cell Modelling



## Cell Development Part II

## Fundamentals, New Technologies, Inert Anodes

# TABLE OF CONTENTS
# Light Metals 2008



## Cast Shop Technology

### Sustainability in the Casthouse



### Casthouse Operation

## Melt Handling and Treatment

## Foundry Ingots and Alloys



## Casting Processes and Quality Analysis

## Modelling

# TABLE OF CONTENTS
# Light Metals 2008



## Electrode Technology Symposium (formerly Carbon Technology)

### Carbon Sustainability and Environment Aspects



### Anode Raw Materials and Properties

## Anode Manufacturing and Developments



## Cathodes Raw Materials and Properties

## Cathodes Manufacturing and Developments

## Inert Anode

# TABLE OF CONTENTS
# Light Metals 2008



## Hot and Cold Rolling Technology

### Session I

# TABLE OF CONTENTS
# Light Metals 2008



## Recycling

### Light Metals

# PREFACE

Dear Colleagues:

This *Light Metals* proceedings represents an enormous investment in intellectual resources on the part of the authors and their co-workers in conducting the research and development work, and in preparing the papers. I offer my thanks to them all for their efforts. *Light Metals* remains the pre-eminent repository for aluminum and magnesium process technical information, and it is the contributions of the authors and their co-workers that make it so.

As in past years, this edition continues coverage of the aluminum process including alumina and bauxite, electrodes, reduction, and cast shop. As a new initiative, this year we have organized a session on hot and cold rolling of aluminum. As with many interfaces between production centers, there are many technical issues that require an understanding of both sides to solve. It is our intent that *Light Metals* grows to become a major technical forum for fabrication process technology while maintaining our strength in the upstream processes. In doing so, we will strengthen the optimization of the entire aluminum process.

Social, political and economic trends throughout the world affect each of us as technologists in a significant way. The consolidation of our industries continues at an increasing pace. One major consolidation, Rusal-Sual, was completed during the past year. Another, Rio Tinto-Alcan, is in final stages of approval, and a third major acquisition has been recently proposed. At the same time, the past few years have been unprecedented in terms of demand and prices for most metals, driven by strong growth in the BRIC countries. We have not seen this situation in the metals industry for at least 25 years, so it is a new phenomenon for most of us. It is projected that the demand for primary aluminum will reach 60 million tons by the year 2020, double the production levels of just a few years ago. These trends result in constant pressures on technologists to reduce costs at the same time as increasing productivity.

With respect to the environment, the obvious social pressures are to reduce the impact of our processes. Light metals are in a somewhat unique position to be able to be part of the solution in terms of light-weighting for transportation applications and recyclability, while at the same time we have significant challenges to reduce our own environmental footprint. To respond to these challenges, we have put a special emphasis on sustainability and environmental issues in this conference, with the opening session in each subject area covering this topic.

I thank the subject chairs and co-chairs in each subject area who performed the bulk of the work in soliciting and reviewing papers and organizing the sessions. I also acknowledge the work of the session chairs in reviewing the papers. Lastly, I thank the TMS staff for their outstanding professional support in the preparation of this volume.

David H. DeYoung

# EDITOR'S BIOGRAPHY

**DAVID H. DeYOUNG**
**LIGHT METALS 2008 EDITOR**

**David H. DeYoung** is a technical consultant at the Alcoa Technical Center in Pennsylvania, United States. He currently works in the primary metals research and development group with a breakthrough technology team. With Alcoa Inc. for more than 25 years, Dr. DeYoung has been manager of molten metal and casting technology, worked in the ingot casting division where he was responsible for molten metal treatment and quality, and worked on various smelting projects, including inert anodes, carbothermic reduction, and primary lithium production. Dr. DeYoung has published papers on various aspects of metal treatment in previous TMS *Light Metals* proceedings and holds 11 U.S. patents. He earned his doctorate in metallurgy from the Massachusetts Institute of Technology with a thesis on the chemistry of iron blast furnace slags.

# PROGRAM ORGANIZERS

## ALUMINA and BAUXITE

**Sringeri Chandrashekar** is manager of the alumina business at Rio Tinto Alcan, formerly Comalco/Rio Tinto Aluminium, in Brisbane, Australia, where he oversees the operations of two refineries. His work has also included commercial support and business development. Dr. Chandrashekar recently led a task force to integrate operations of Rio and Alcan refineries following the takeover of Alcan by Rio Tinto. Before joining Rio in 1998, he worked at Alusuisse's Gove refinery in the Northern Territory of Australia as the process engineering manager. Prior to his alumina and bauxite work experience, Dr. Chandrashekar worked with Zambia Copper Mines in various production management roles. He is a Fellow of Aus IMM and was the technical program director and one of the chief organizers of the AQW 2002 in Brisbane. Dr. Chandrashekar obtained his Ph.D. in extractive metallurgy from Imperial College, London.

**Peter McIntosh** took over as director of alumina technology at Hatch Associates in 2005; he began with the company in 2002 and had responsibility for the Hatch Light Metals business (bauxite, alumina, aluminium and magnesium) in Australia, Asia and China. Prior to that, he earned 30 years of experience in the alumina and aluminium industry in a range of technical engineering, operating and business roles, including at Queensland Alumina at Gladstone in Queensland and at Kaiser Aluminum in the United States.

## ALUMINUM REDUCTION TECHNOLOGY

**Geoffrey Paul Bearne** has worked in the aluminium industry for more than 25 years. He is currently general manager of the research center at Rio Tinto Alcan, formerly Comalco, in Melbourne, Australia. Mr. Bearne is also a director of CAST CRC Ltd., a light metals collaborative research center. Prior to Rio Tinto Alcan, he worked for Dubai Aluminium Ltd. Mr. Bearne has contributed to TMS through the submission of technical papers since 1995, as a session chairperson, and as assistant reduction technology subject organizer. He graduated with a degree in mechanical engineering from the City University in London and completed a professional engineering traineeship with the Central Electricity Generating Board.

**Martin Iffert** is managing director of the TRIMET Primary Group in Essen, Germany, where he is responsible for both TRIMET smelters in Essen and Hamburg, including the reduction department, anode plant, casthouse, site services, maintenance and administration. Mr. Iffert was also responsible for the organization and re-start of the Hamburg smelter in 2007 and the thrilling dry start-up of the first cell in March 2007. He has worked for TRIMET ALUMINIUM AG for 13 years. Mr. Iffert has published more than 10 technical papers at TMS and other technical conferences. For his 2002 TMS paper, "Reduction of $CF_4$ Emissions from the Aluminium Smelter in Essen," his team earned the TMS Reduction Technology Award. Mr. Iffert has served as session chair at past TMS symposia. He holds a graduate degree in aluminium smelting technology from the University of New South Wales, Australia, and is working on his doctorate thesis.

## CAST SHOP TECHNOLOGY

**Hussain Hassan Al-Ali** is general manager of casthouse and engineering service at ALBA and an executive with the ALBA management team. His other roles at Alba include chairman of the Industrial Relations Committee and member of the Executive Tender Board Committee. Mr. Al-Ali has more then 35 years experience in the aluminium industry. He is a chartered engineer, registered with the Engineering Council of UK. Mr. Al-Ali is a member of both the Institution of Mechanical Engineers and the International Project Management Institute. He graduated with a bachelor's degree in mechanical engineering from UK in 1979.

**David H. DeYoung** is a technical consultant at the Alcoa Technical Center in Pennsylvania, United States. He currently works in the primary metals research and development group with a breakthrough technology team. With Alcoa Inc. for more than 25 years, Dr. DeYoung has been manager of molten metal and casting technology, worked in the ingot casting division where he was responsible for molten metal treatment and quality, and worked on various smelting projects, including inert anodes, carbothermic reduction, and primary lithium production. Dr. DeYoung has published papers on various aspects of metal treatment in previous TMS *Light Metals* proceedings and holds 11 U.S. patents. He earned his doctorate in metallurgy from the Massachusetts Institute of Technology with a thesis on the chemistry of iron blast furnace slags.

# ELECTRODE TECHNOLOGY

**John A. Johnson** is manager of new cell technology for RUSAL Engineering and Technical Centre in Krasnoyarsk, Russia. Since 2001 he has held various positions with RUSAL primarily in the areas of carbon and reduction technology. He has more than 30 years experience working with primary emphasis in reduction technology. With Martin Marietta Aluminum, Mr. Johnson held positions of laboratory manger and technical manager, responsible for implementation of Sumotomo and Mitsubishi Søderberg dry anode technology, and subsequently, became technical manager of Commonwealth Aluminum. In the mid-1980s, he left the industry for four years to become the engineering-production manager for IMMP where he was responsible for developing powder injection molding of near net shape casting technology. Mr. Johnson then joined Kaiser Aluminum International as a technical manager on Kaiser's project at the OAO Krasnoyarsk Aluminium Plant and later worked on Kaiser's prebake technology programs. He holds a bachelor's degree in chemistry from Eastern Washington University and an MSCE from the University of Idaho.

**Carlos Eduardo Zangiacomi** is a consultant engineer for smelter process technology with Alcoa Primary Metals, Pocos de Caldas Operations, Brazil. At Alcoa Aluminum in Brazil since 2000, he has worked on process control and development in potrooms, potlining (cathode design, heat balance modeling, cell preheating and operation) and paste plant. Mr. Zangiacomi is the recipient of the ABAL award from the Brazilian Aluminum Association for the most notable technical paper published in 2005. He has also authored and co-authored papers presented at TMS meetings. Mr. Zangiacomi received his master's degree in material science and engineering from Sao Carlos Federal University in 2006, focusing on the project and operation of Hall-Heroult cells for aluminum smelting.

# HOT AND COLD ROLLING TECHNOLOGY

**Jürgen Hirsch** teaches material science and physical metallurgy at the technical university, RWTH Aachen, in Bonn, Germany. For nearly the last 20 years, he has also worked as a senior scientist in the aluminium industry, first at Alcoa in the United States and now at the R&D Center of Hydro Aluminium Deutschland GmbH in Bonn. Dr. Hirsch has an international reputation in research on aluminium fundamentals, production, application and simulation issues. He is head of the International Conference of Aluminium Alloys and also organizes international research and development projects for the Education and Technical Committee of the European Aluminium Association. Dr. Hirsch has 160 scientific presentations and publications to his credit. In 2007, he was awarded the Tammann-Medal of the German Materials Society. Dr. Hirsch earned his doctorate at RWTH Aachen.

**Ming Li** is head of the Solidification and Deformation Processing Section at Alcoa Technical Center in Pennsylvania, USA. He is widely recognized for contributions in integration of fundamental research in mechanics and materials into large scale manufacturing. Dr. Li has also made scientific contributions to the areas of metal shearing and cutting process, micro-mechanisms of fracture and failure, material instability, dynamic plasticity and impact mechanics. The author of more than 40 peer-reviewed publications and five U.S. patents, he has been an invited speaker at more than 30 seminars and lectures at universities and at both national and industrial research laboratories. Dr. Li received the 2003 Carnegie Science Award for Excellence-Advanced Manufacturing and Materials. He was elected as Fellow of the American Society of Mechanical Engineers (ASME) in 2004, and is the past committee chair of the Material Processing and Manufacturing Committee in ASME's Applied Mechanics Division.

# RECYCLING

**Gregory K. Krumdick** has worked with Argonne National Laboratory in Illinois, USA, for 17 years as an engineer. He has been involved in designing and engineering numerous pilot plant systems for Argonne's Process Engineering Section. Starting with industrial control systems, his field of interest moved to metallurgical processes where he worked on many light metals projects as well as the development of copper-based gas cleanup processes. Mr. Krumdick obtained his master's degree in bioelectrical engineering from the University of Illinois at Chicago, focusing on complex control systems.

**Christina E.M. Meskers** is currently pursuing her doctorate degree at Delft University of Technology, the Netherlands. Her research is about the effect of coatings on the recyclability of coated magnesium scrap. Completion of the doctorate research and thesis defense are scheduled for 2008. Ms. Meskers began her studies in resource engineering and obtained her master's degree at Delft in 2004. Her master's thesis on the relation between iron ore characteristics and sintering behavior during pelletizing was done in cooperation with Corus.

# ALUMINUM COMMITTEE
# 2007-2008

**Chairperson**
Morten Sørlie
Elkem Aluminium ANS
Kristiansand, Norway

**Vice Chairperson**
David H. DeYoung
Alcoa Inc.
Alcoa Center, Pennsylvania, USA

**Past Chairperson**
Travis J. Galloway
Century Aluminum Co.
Hawesville, Kentucky, USA

**Light Metals Division Chairperson**
Wolfgang A. Schneider
Hydro Aluminium GmbH
Bonn, Germany

**JOM Advisor**
Pierre P. Homsi
Pechiney Group
Voreppe, France

**Secretary**
Geoffrey Paul Bearne
Rio Tinto Alcan
Thomastown, Victoria, Australia

# MEMBERS THROUGH 2008

**Thomas R. Alcorn**
Noranda Aluminum Inc.
New Madrid, Missouri, USA

**Corleen Chesonis**
Alcoa Inc.
Alcoa Center, Pennsylvania, USA

**James W. Evans**
University of California
Berkeley, California, USA

**Travis J. Galloway**
Century Aluminum Co.
Hawesville, Kentucky, USA

**Pierre P. Homsi**
Pechiney Group
Voreppe, France

**Markus W. Meier**
R&D Carbon Ltd.
Sierre, Switzerland

**David V. Neff**
Pyrotek Inc.
Solon, Ohio, USA

**Ray D. Peterson**
Aleris International
Rockwood, Tennessee, USA

**Alton T. Tabereaux**
Consultant
Muscle Shoals, Alabama, USA

## MEMBERS THROUGH 2009

**Jean-Paul M. Aussel**
Initiatives Et Projets Industriels
La Buisse, France

**Todd W. Dixon**
ConocoPhillips Inc.
Bouger, Texas, USA

**Dag Olsen**
Hydro Aluminium Primary Metals
Porsgrunn, Norway

## MEMBERS THROUGH 2010

**Jean Doucet**
Alcan Inc.
Montreal, Quebec, Canada

**Rene Kieft**
Corus Group
IJmuiden, Netherlands

**Stephen J. Lindsay**
Alcoa Inc.
Alcoa, Tennessee, USA

**Morten Sørlie**
Elkem Aluminium ANS
Kristiansand, Norway

## MEMBERS THROUGH 2011

**Geoffrey Paul Bearne**
Rio Tinto Alcan
Thomastown, Victoria, Australia

**David H. DeYoung**
Alcoa Inc.
Alcoa Center, Pennsylvania, USA

**John A. Johnson**
United Company Rusal
Krasnoyarsk, Russia

**Halvor Kvande**
Norsk Hydro
Oslo, Norway

**Wolfgang A. Schneider**
Hydro Aluminium GmbH
Bonn, Germany

# ELECTRODE TECHNOLOGY

## Carbon Sustainability and Environment Aspects

*SESSION CHAIRS*

**Neal R. Dando**
Alcoa Inc.
Alcoa Center, Pennsylvania, USA

**Trygve Foosnas**
Norwegian University of Science and Technology
Trondheim, Norway

**Light Metals 2008** *Edited by: David H. DeYoung*
*TMS (The Minerals, Metals & Materials Society), 2008*

# TRIPLE LOW – TRIPLE HIGH, CONCEPTS FOR THE ANODE PLANT OF THE FUTURE

Dr. Dragan Dojc[1]; Dr. Wolfgang Leisenberg[2]; Detlef Maiwald[2]
[1]Cavnin C. A. Apartado 7549 Caracas 1040 A, Venezuela
[2]Innovatherm Prof. Dr. Leisenberg GmbH + Co. KG; Am Hetgesborn 20, D-35510 Butzbach, Germany

Abstract

For the next generation of anode plants six main goals should be achieved. Three of them, summarized in Triple Low, represent the targets of the process technology: Low Energy, Low Emissions, Low (investment and running) Cost. The three others, summarized as Triple High, stand for the product: High Quality, High Quality Consistency and High Density. This should be governed under the umbrella of maximum operational safety and efficiency.

Different concepts of anode plants and specific technologies are on the market, which only cover parts of the defined goals. This paper deals with technologies to achieve all of these partly contradicting goals which are available today or in the near future. Especially technologies with environmental impact like minimization of energy consumption and emissions will be spotlighted.

## 1. Introduction

Each ton of beautiful white metal (aluminum) is reduced by use of a half ton of carbon anode, manufactured in the complex, dirty, noisy, hot, sensible and expensive anode plant.

The anode is the biggest source of $CO_2$, evolved from the smelter. But since more than 90% of world wide energy comes from fossil fuel, we should look to the best efficiency of this fuel. The anode in the electrolysis shows an efficiency of about 60 % which is higher than that of any thermal power station. The carbon anode might be a large source of green gases, but the over all greenhouse gas mass balance suggests, that there is no better electrolysis process than carbon anodes.

Nevertheless a more efficient use of fuel in the anode plant should be realized, and there a considerable potential is left.

These words are only the reflection of the enormous influence and importance of carbon anodes to aluminum production and the answer why are strive to better understand the past and improve the design of future of the anode plants.

Today of Anode Production

- Plant capacity 35-40 t/h lines – continuous mixing/kneading
- Main warehouse for raw materials close to the port facilities, plant warehouse with one week production capacity, intermediary silos with couple of shifts spare capacity, liquid pitch facility or pitch melting facility
- Mixing/kneading temperature up to 175 degree C
- Operating Mode 6-8 weeks continuous production with stop of 3-7 days after the production cycle to clean repair and maintain
- 14 days/year general overhaul
- Optimization of activities and strict management
- 5-8 % green scrap, up to 5 % baked scrap
- One week inventory of baked anodes
- Change of quality and availability of raw materials
- Strict pollution control

Tomorrow of Anode Production

- Mega Plant 2 x 50 t/h, 3 x 50 t/h or 2 x 75 t/h lines

  – total continuous carbon plant
- Main warehouse for raw materials close to the port facilities, plant warehouse with one week production capacity, no intermediary silos, liquid pitch facility or pitch melting facility
- Mixing/kneading temperature up to 190 degree C
- Operating Mode 12-14 week continuous production with stop of max. 3 days after the production cycle to clean repair and maintain
- 10 days/year general overhaul
- Strict working discipline and total management
- Max. 4 % green scrap, up to 3 % baked scrap
- Decrease off quality and availability of raw materials
- Total production pollution control

The Carbon Plant of tomorrow will be the loyal supplier of the basic raw material to the pot line, financially competitive, environmentally correct and active participant in the clean, productive and simple aluminum production.

The targets for the tomorrow anode plant can be summarized in a double trio consisting of triple low and triple high portions:

- Low Energy
- Low Emissions
- Low (investment and running) cost

- High Quality
- High Quality Consistency
- High Anode Density

## 2. Paste plant

The trend for the future design of the anode plants according to these goals appears clearly:

The Continuous Plant

To reduce handling, self-grinding and contamination problems, the plant will operate continuously – only with the quantity of the raw materials in the system required for continuous processing. This concept will bring the reduction of size of equipment (no spare capacity necessary, as the process is continuous), decrease the quantity of the material in the processing system, decrease the speed of the equipment (as the quantity of the material in the system is constant), eliminate the intermediary silos and tanks between the others benefits.

Continuous processing is by definition less aggressive and therefore the equipment is not exposed to high wearing and damages. Continuous and slower motion of grains is generating far less suspended solids.

Butts are thoroughly cleaned and the residual material is briquetted and prepared as fuel for steel (re-carburizer) or cement factories.

The Paste Reactor

The central, controlling equipment of the green anode plant is the paste reactor: this must be extremely robust equipment, designed for large anode production cycles. The reactor controls solid component feeding, liquid component feeding and the dispersion of the liquid component into the solid component, bringing the prepared mass to the process area for kneading/mixing treatment. The paste reactor is a smart generation of apparatus, which is dynamically controlling the quality of the paste through energy input into the process – resulting in higher throughput of the paste, constant quality of the paste and lower revolution rate (longer technical life).

The expert process software controlling the reactor detects changes in raw material parameters at an early state.

Green Anode Plant and Anode Baking Furnace

The anode baking furnace, as the unit with surplus energy should provide all thermal energy for the paste plant. It is important to emphasize, that the use of the waste heat of the furnace in the green anode plant will reduce the price of the anode in the actual petroleum market development and competitively benefit the price of primary metal units.

The Environment Impact

The anode plant of tomorrow is characterized with a steady improvement of environmental conditions.

The dust recovery is improved through the use of better designed filters and efficiency through the concept of various, small, locally mounted units. The recovered dust is consequently returned to the process. Pitch fumes are treated in efficient and standard satisfying manner – no residual is generated.

The symbiosis of the baking furnace and the paste plant is necessary to reduce thermal environment contamination.

## 3. Baking

The breakthrough to achieve high anode quality in open pit furnaces was due to a better flue design in combination with the impulse firing technology developed in the seventies. Because of the high temperature consistency and the elimination of hot spots in the flue the top temperature could be brought down to 1150 degree C. Besides lower fuel consumption this technology resulted in longer refractory lifetime.

Firing Technology

In order to achieve a most flexible firing system adapting the flame shape to the flue design and to the individual process conditions a multi-mode-burner-technology has been designed. This technology allows the operation of gas burners with various gas velocities at the burner mouth up to sonic speed. Any specific burner mode is adjustable electronically without any manual interactions at the burner. Depending on the location of the burner within the fire arrangement the most suitable mode will be applied automatically. This technology offers a maximum flexibility in order to assure homogeneity of the pit temperatures.

Of course, the precondition of excellent baking results is a good state of the flue wall refractory. In order to supervise the flue condition a special software module for monitoring the flue condition has been designed.

Complete Pitch Burn Technology

The biggest step to lower fuel consumption and emissions was the use of the condensable tar pitch volatiles of the anodes as a fuel instead of a pollutant [1]. As shown in Fig. 1 this portion of volatiles leaving the anode at low temperatures contribute one third of the fuel which is stored in the anode.

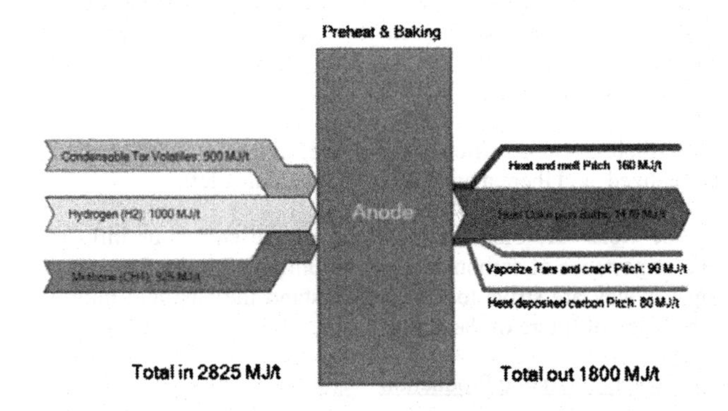


Fig. 1: Anode Baking Heat Balance

A self adapting software module is placed on top of the PID control loops modifying their set-points or limiting their outputs according to the temperature and oxygen demand of the pitch burn process.

Due to the higher heat transfer during the pitch burn phase, the baking level, i.e. the heat transfer to the anodes at identical firing profile in the flue is significantly higher and increases the anode quality without additional cost.

As shown in Fig. 2 the effective soaking time will be extended due to the earlier rise of the anode temperature caused by the pitch bum technology.

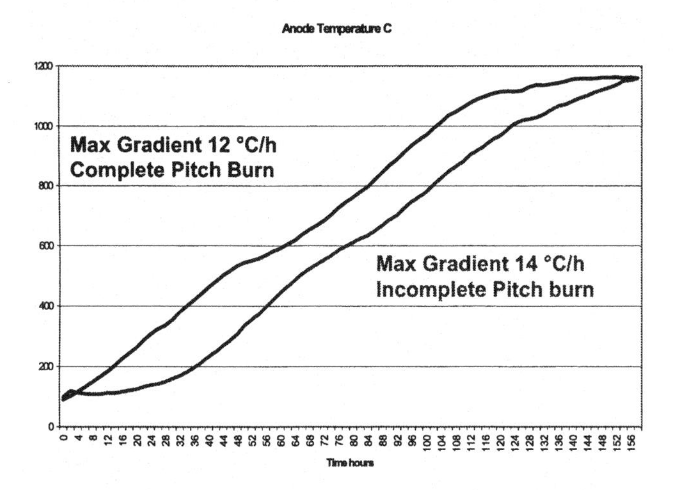


Fig. 2: How Pitch Burn influences Anode Temperature Profiles

A further positive effect of the pitch burn technology is the elimination of large deposits in the ducts. No cleaning of the ducts, no water cooling is required at the exhaust ramp and of course no danger of fires in the ducts which lowers significantly the maintenance cost.

High Density Anodes

In order to improve the efficiency of the electrolysis, most of the smelters are making efforts to increase the anode (real and apparent) density. A "High Density Module" has been created for that purpose, which controls the anode temperature gradient in the most critical phases of the preheat area.

With the pitch burn technology the heat transfer to the anode starts at a very early stage. This lowers the temperature gradient in the preheat area compared to a bad pitch-burn-situation (Fig. 2). This effect is very helpful for baking high density anodes which are sensitive to high gradients.

## 4. Fume treatment technology

Inorganic matter such as sulfur, fluorides, dust and condensed volatiles easily can be precipitated by dry adsorption while gaseous matter such as CO, methane and polycyclic aromatic hydrogen (PAH), especially benzene have to be oxidized.

The RTO Technology

Today the Regenerative Thermal Oxidation (RTO) technology is discussed to be the suitable fume treatment technology for anode plants.

The RTO plant consists of a double prefilter, a (4-chamber) RTO, an adsorption reactor for sulfur and fluorides and finally of a filter plant for removing the dust. In order to control the temperature in the RTO and for periodical cleaning of the prefilter the RTO technology needs additional fuel which is approximately 10 % of the furnace consumption. Since the RTO technology includes sensitive mechanical equipment permanent maintenance is required.

The Complete Pitch Burn as an Internal Oxidizing Process

As a side effect of pitch burn technology volatiles combustion of more than 99.8 % is achieved without any optimization of the oxidizing conditions. Therefore the pitch burn section of the furnace performs the thermal oxidation of the tar pitch volatiles.

It is evident, that compared with the complete-pitch-burn plus dry-adsorption technology the RTO technology must be higher in investment expenses and running cost. There is no question that the thermal oxidation of the volatiles in the pitch burn area is an ideal completion of a dry adsorption fume treatment plant (FTP).

Fume Treatment by the GTC

If we look to the plant arrangement of a smelter we wonder why the furnace needs its own FTP. Why can´t the fume gases of the furnace be directed to the Gas Treatment Centre (GTC) of the potline gases?

One reason why the fume gases of the baking furnace could not be merged with the potline gases was because of their tar content, building conglomerates of alumina in the filter plant. The other reason was that, as mentioned above, gaseous organic components are not removed well by dry adsorption.

Due to the pitch bum technology both of these reasons are eliminated. The future FTP of the baking furnace will consist of a dedicated to PAH and benzene removal firing technology, followed by a conditioning plant, which ensures constant low flue gas temperature and the scraping all condensed tar components. The conditioned tar-and-volatile-free fumes will be directed to the GTC.

Obviously this concept will lower the investment expenses and the running cost as well.

## 5. Heat recovery

Heat recovery is a goal which will become more and more important regarding fuel cost and also reduction of $CO_2$. As mentioned in chapter 2.3 a symbiosis between paste plant and furnace should be established. Fig. 1 shows that the anode baking process is a highly exothermal process [2]. Much more energy is fed to the process by tar pitch than what is needed to bake the anode.
If we look to the heat balance of a state-of-art baking process including the furnace structure and the packing material (Fig. 3), there is a lot of waste heat left, which is not used so far.

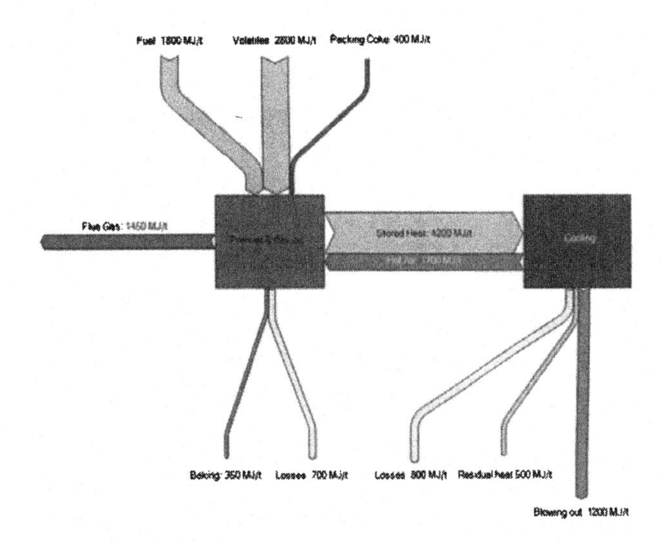


Fig. 3: Baking Process Heat Balance

The reason for this unsatisfying situation is firstly that not all of the energy of the pitch can be used in the baking process. A considerable portion of it leaves the furnace as a higher flue gas temperature.

Another big part of the heat disappears as heat loss due to radiation, convection and "out-blowing" or remains as residual heat in the anodes, the packing material and the refractory.

Secondly, a great portion of the heat stored in the cooling area is blown into the furnace hall, creating high ambient temperatures especially in hot countries. Nobody cared for the usage of this heat so far.

The reasons to waste lots of heat at the anode furnace are technical and historical as well. Low energy prices in the past, no engineering interfaces between paste plant, baking furnace and fume treatment plant and no technology nor concept to use the heat of the fume gases and of the cooling area. Last but not least no application for that superfluous energy has been found.

Heat Recovery for the Paste Plant

The only heat sink in the paste plant is the coke heating. The heat in the flue gas cannot be used effectively due to its low temperature. In order to meet the demand of this heat sink, higher temperatures are required if the paste plant technology should not be changed.

We see from the Sankey diagram, shown in Fig. 3, that the ring furnace shows relatively good fuel efficiency because of the recovery of heat from the cooling area. This heat at the end of the soaking phase is stored in the anodes, in the packing material and in the furnace structure.

A part of this heat is recovered to generate hot air for combustion in the firing zone and to heat the green anodes further downstream in the preheat area of the furnace.

The reason to use the energy from the cooling area is the high temperature level of more than 400 degree C. Therefore the temperature of 300 degree C required for the thermal transfer oil feeding the heating screw can be achieved easily.

In order to use the heat stored in the cooling area, a second duct besides the flue gas duct has to be installed, which collects the hot air from the cooling areas of all fires in a similar way as the flue gas exhaust ramp [3].

The hot air then is lead to a heat exchange unit (Fig. 4), transferring its energy to the thermal transfer oil for the paste plant. Since the hot air from the cooling area is not contaminated by corrosive nor acid matter the heat exchange unit is not affected by any corrosion.

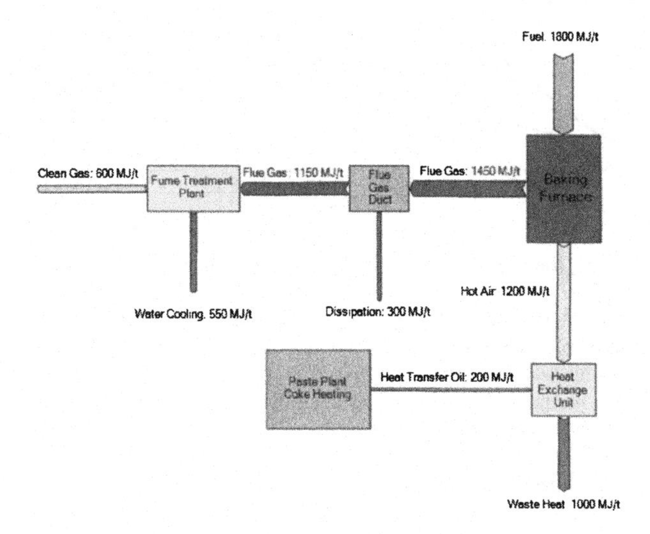


Fig. 4: Heat Recovery for the Paste Plant

This concept allows the complete supply of thermal energy for the paste plant without having to change the proven paste plant technology. The heat exchange unit just replaces the boiler for the heat transfer medium.

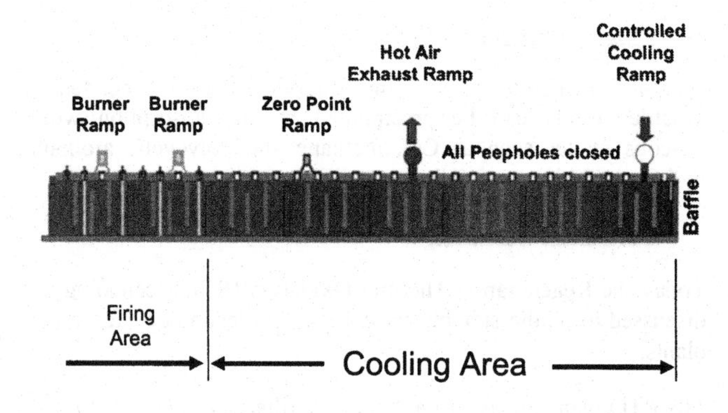


Fig. 4: Arrangement of Equipment in the Cooling Area

For startup operation or any lack of heat from the furnace a heat generator is placed in the hot air duct.

So neither the boiler for the thermal transfer oil in the paste plant nor any standby unit is required anymore. The investment cost for the second hot air duct including a fan and for the heat exchange unit is no higher than that for the boiler and its standby.

So as shown in Fig. 4 about 200 MJ/t of anode can be saved at roughly the same investment cost.

In order to realize this recovery technology, one of the two blower ramps in the cooling area is replaced by a second exhaust ramp. So the number of equipment on the furnace and the investment cost are kept the same (Fig. 4).

### The Total Heat Recovery Concept

A full heat recovery of the waste heat from the baking furnace can be realised if a thermal power station is located near the smelter.

Every thermal power station needs water for the boiler generating steam for the turbines. The amount of water required is so large and its temperature is so low, that this water performs an ideal heat sink to recover any remaining heat from the furnace and from any other heat source.

The energy fed to the water substitutes some fuel for the boiler in the power station for heating this water.

Since the firing efficiency of the boiler is about 90 %, every Joule of thermal energy fed to the boiler water saves 1.1 Joule of fuel at the power station.

Since the amount of water is huge even a portion of this water would suffice to recover all waste heat of the baking furnace.

As shown in Fig. 7, the remaining waste heat of the cooling area can be recovered using a second heat exchange unit in line with the high temperature unit for the paste plant.

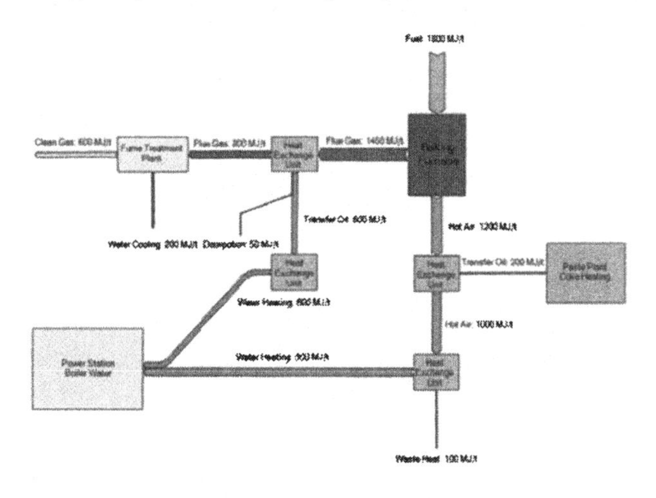


Fig. 5: Total Heat Recovery Concept

Not all waste heat of the flue gas as the second heat source of the baking furnace can be extracted because of the dew points of the acid components of the fumes. Therefore the fume temperature cannot be brought below about 130 degree C. But there is still a temperature difference of about 100 degree C left for heat recovery. This heat can be transferred to the water by a two-loop heat exchange plant.

As a side effect the heat recovery reduces the demand of cooling water in the conditioning tower considerably.

The overall heat balance in the Sankey diagram shown in Fig. 5 shows that the baking process delivers roughly the same amount of heat to the paste plant and to the power station as it needs for fuel. <u>With this Total Heat Recovery Concept the baking process is fed by tar pitch only and runs at zero fuel</u> [3].

### Extending the Concept to the Whole Smelter

Of course all waste heat sources can be treated in a similar way. Heat recovery of the potline gases would contribute four times more heat than that of the baking furnace. Also the casthouse can contribute considerably to the total heat recovery.

The great advantage of the boiler water as a heat sink is that it can use any energy, even if it is fed to the boiler water in a temporary manner or changing in quantity.

With this complementary concept using all major waste heat in the smelter complex, approximately 1.5 kWh per Kg of aluminium can be saved as fuel at the power station. The complete concept shown in Fig. 6 will set a new bench mark of fuel consumption in aluminium smelters.

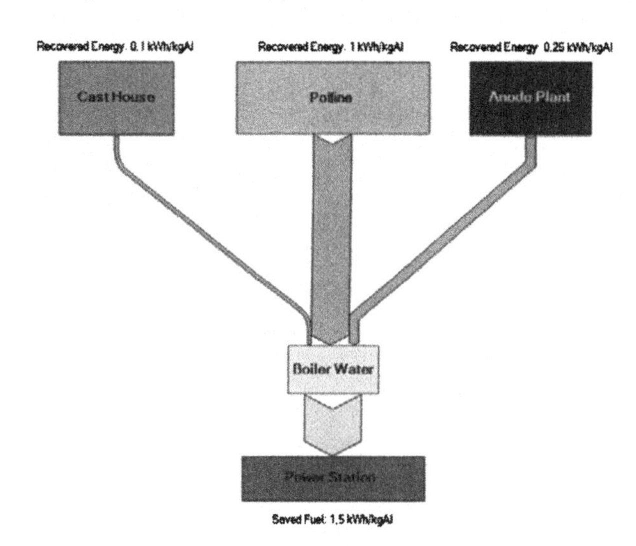


Fig. 6: Smelter-wide Total Heat Recovery Concept

## 6. Summary

With the technologies available and proven today and the concepts considered in this paper the target parameters can be brought very close to the values that can be achieved theoretically

| | |
|---|---|
| • High Anode Quality | Firing Technology |
| • High Quality Consistency | Sophisticated Control |
| • High Density Anodes | Controlled Gradients |

| | |
|---|---|
| • Very Low Energy | Zero Fuel Baking Process |
| • Low emissions | Lowest $CO_2$ Emissions |
| • Lowest cost | Lowest Energy cost |
| | No FTP nor RTO |

## 7. References


[1] W. Leisenberg, "Firing and Control Technology for Complete Pitch Burn and its Consequences for Anode Quality, Energy Efficiency and Fume Treatment Plant", 9th international Conference on non ferrous metals, July 8-9, 2005, Pune, India.

[2] B. J. Racunas, "Anode Baking Furnace Thermal Balöance" Light Metals 1980, 525 – 530

[3] W. Leisenberg "Zero Fuel Baking – The Total Heat Recovery Concept", to be published in Aluminium 1-2, 2008

**Light Metals 2008** *Edited by: David H. DeYoung*
***TMS (The Minerals, Metals & Materials Society), 2008***

# NEW ENVIRONMENTAL APPROACH AT BAKING FURNACE

Felippe Navarro[1], Francisco Figueiredo[1], Paulo Miotto[1], Hélio Truci[1], Hezio Oliveira[1], Marcos Silva[1], Raimundo Reis[1], Elisângela Moura[1], Fabiano Mendes[1]

[1]Consorcio de Alumínio do Maranhão (Alumar)
P.O. Box 661, 65.095-050 São Luis, Brazil

Keywords: Baking Furnace, biofuel, carbon credits

### Abstract

Consorcio de Aluminio do Maranhao (Alumar) is one of the largest aluminium smelters of Latin America producing 450'000 tons of primary metal. The plant has three baking furnaces with capacity to produce 700 anodes/day. The furnaces are operated with diesel oil due to the lack of natural gas distribution in the Northeast of Brazil.

Alcoa Inc. started in 2006 an effort to reduce green house gas emissions worldwide, increasing the sustainability of their business and allowing Alcoa to apply for carbon credits. Alumar's Baking Furnace started using in September a mix of light oil and biodiesel as standard fuel.

The development presented in this paper presents a new environmental approach where the main targets have been:

- Reduce $CO_2$ generation with lower cost impact;
- Implement an environmental fuel in the baking furnace with no impacts in anode quality, fuel efficiency and baking furnace operations.

### Introduction

The Kyoto Protocol was developed in 1997 with an objective to stabilize greenhouse gas concentrations in the atmosphere at a level that would prevent dangerous anthropogenic interference with the climate system. The countries that signed the protocol have promised to reduce their emissions by 5.2%, relative to 1990 levels, by the year 2012.

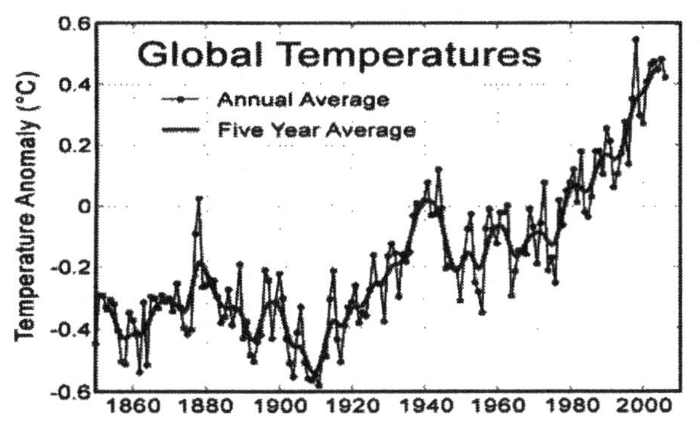


Figure 1 – Global mean surface temperature anomaly 1850 to 2006 relative to 1961–1990

Table I - List of the change in greenhouse gas emissions from 1990 to 2004

| Country | Change in greenhouse gas Emissions (1990-2004) |
|---|---|
| Germany | - 17% |
| Canada | + 27% |
| Australia | + 25% |
| Spain | + 49% |
| United States | + 16% |
| United Kingdom | - 14% |
| Japan | + 6.5% |
| China | + 47% |
| India | + 55% |

Alcoa has decided to engage in the Kyoto Protocol mission and has implemented a worldwide program to reduce greenhouse gases emissions through all of our businesses, worldwide. Alcoa Primary Metals Latin America decided to start on this mission by focusing on reducing the utilization of fossil fuels in its plants.

Alumar's baking furnaces operate with diesel oil, because there is no natural gas resource in Maranhão and the distribution system ends in Fortaleza (distance - 1000km). Fuel consumption is around 2 millions liters per month. Alumar is one of the biggest consumers of non-renewable fuel in Alcoa Latin America.

Several hypotheses were evaluated in an initial brainstorm session held during the end of 2005. The option taken to be developed was to substitute diesel with biodiesel. This option was aided when the Brazilian governor decided to add 2% of biodiesel to the diesel commercialized in the market. This decision motivated a huge expansion in the biofuel market in Brazil, In the last two years biodiesel production has increased allowing companies to replace fossil fuels per biofuel.

The main reason to focus on this topic was that the consumption of 1 ton of biodiesel (B2) can avoid the production and emission of 2.55 tons of carbon dioxide. This change would allow Alcoa to apply for carbon credits, each ton of eliminated carbon dioxide equivalent corresponds to 1 carbon credit.

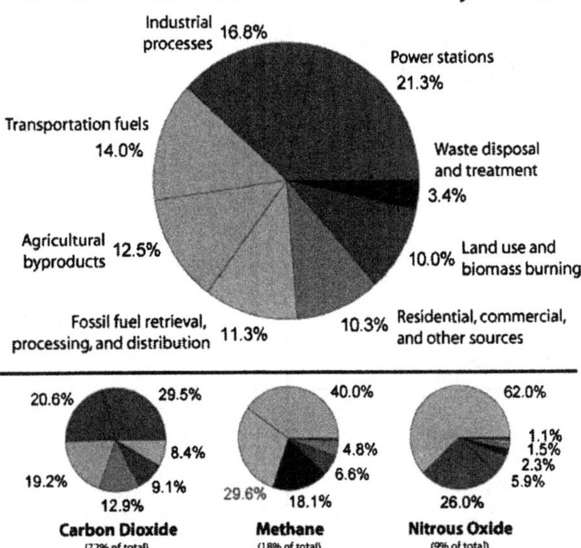


Figure 2 – Annual Greenhouse Gas Emissions by Sector (2006) Chart

Biodiesel can be produced from a wide range of biological sources; however production efficiency can vary with the raw material used. Generally, castor seed is the source for the production of biodiesel in Northeast Brazil.

Table II - Typical oil extraction from 100 kg. of oil seeds

| Crop | Oil/100kg |
|---|---|
| Castor Seed | 50 Kg |
| Copra | 62 Kg |
| Cotton Seed | 13 Kg |
| Mustard | 35 Kg |
| Sesame | 50 Kg |
| Soybean | 14 kg |
| Sunflower | 32 kg |

Independent of the differences in raw materials, standards have been developed to assure the properties of biofuel. Following the European Standard EN 14.214 and US Standard ASTM D 6751-02 the Brazilian National Petroleum Agency (ANP) has published preliminary specifications for pure biodiesel.

## Experimental

Alumar's smelter has 3 baking furnaces with a total of 9 fires (furnace 1 - 4 fires, furnace 2 - 3 fires and furnace 3 - 2 fires) that are responsible to produce 700 anodes per day. Furnaces 1 and 2 have 8 pits per section while furnace 3 has just six. Each fire has the following design:

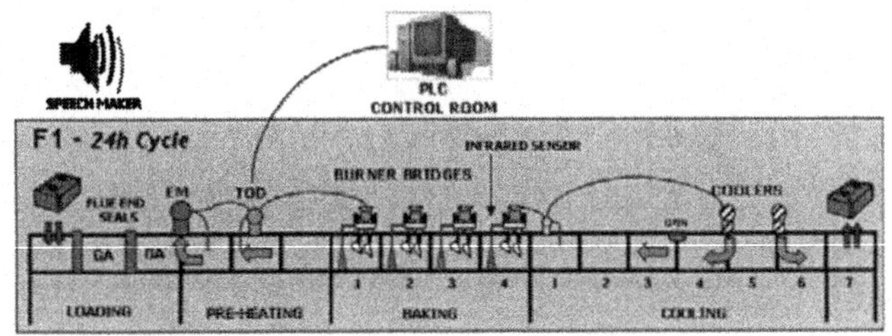


Figure 3 – Alumar's Fire sketch

Furnaces have two independent fuel feed systems allowing the independent supply of two different fuels. As furnace 3 is the smallest and it is responsible for only 17% of the production, it was chosen as a biodiesel test platform.

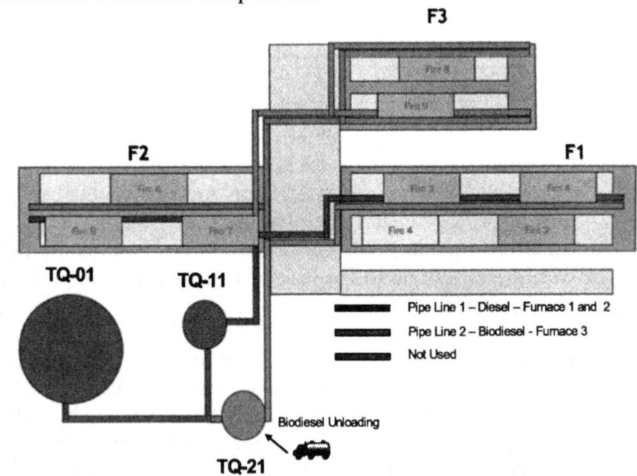


Figure 4 – Fuel Pipe Line Configuration

The initial trial was to test diesel mixed with 2% of biodiesel (commercially known as B2), the target was to achieve a mix of 20% Biodiesel (B20). Further evolutions would depend on first trial results and biodiesel availability in the market.

Before starting the test, furnace 3 was isolated in a single feed system, for a period of 20 days, to follow the process output results using diesel as fuel. Throughout this period, furnace 3 fuel consumption was measured daily, number of plugged burner occurrences was monitored, $SO_2$ and particulate total emission analysis in a 24 hours basis was performed and fuel properties analysis..

After the pre-test interval, TQ21 was fed with renewable fuel and the experiment was designed to burn 300,000 liters of B2 in about 26 days. The objective was to compare the results of the same parameters measured in the pre-test.

Results from both phases were used to evaluate the benefits and disadvantages of using the mix of diesel and biodiesel. If B2 is approved, the plan is to repeat the above procedure for each mix of both fuels (5%, 10%, 15% and 20%).

Fuel consumption was measured by the difference of heights of fuel inside tank 21 over two consecutive days.

Emissions test were performed using the isokinetic methodology, where the gases were collected from the base of the stack. An isokinetic sampling process means to collect a representative fraction of the gas flow with the same gas speed, avoiding a selective sample. Samples were collected according to Alcoa Method 4075A – TF and using the following equipment configuration shown in Figure 5

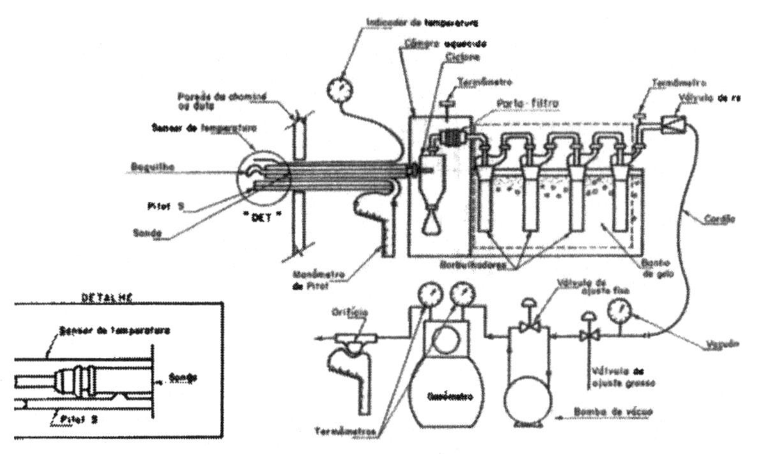


Figure 5 – SO2 and total particulate sampling equipment

Each test duration was 2 hours long and consisted of capturing particulate solids on a pre-filter and gaseous $SO_2$ in the solution scrubbing train, as shown above.

## Results

Physical analysis of diesel and biodiesel (B2) did not present any negative impact in any characteristic. As expected, merely the sulfur content of B2 can be considered as a substantial gain to the process.

Table III – Physical characteristics of Diesel and B2

| Properties | B2 | Diesel |
|---|---|---|
| Heating Power (BTU/lb) | 19355 | 19361 |
| Sulfur content (%) | 0.17 | 0.21 |
| Density (g/cm3) | 0.855 | 0.852 |
| Viscosity | 3.205 | 3.286 |

All process key parameters in baking furnace remained constant over the test period to allow a reliable assessment of the new fuel. Renewable Fuel Consumption was 1.05% higher than petrodiesel, as shown in Figure 6, which can be associated with the heating power difference.

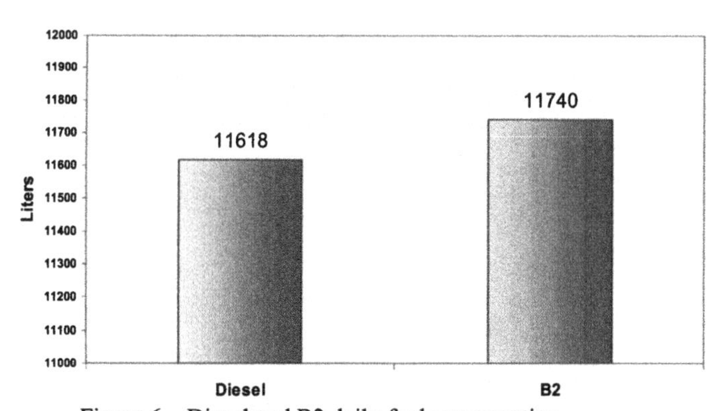


Figure 6 – Diesel and B2 daily fuel consumption

Theoretically biodiesel could reduce sulfur generation in anode baking process by 4 kg/day, as shown in the model below:

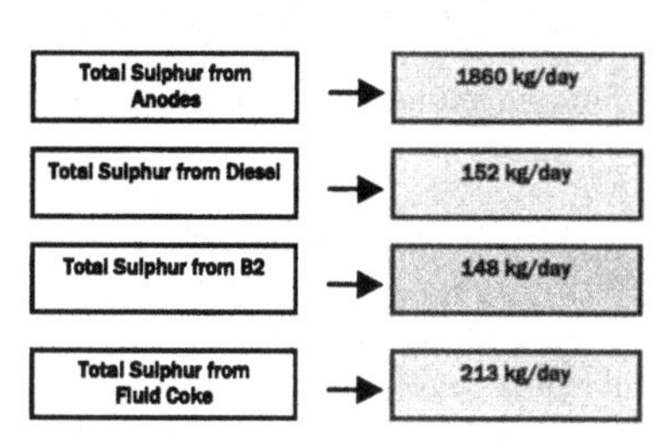


Nevertheless results of $SO_2$ emissions during B2 usage were higher than diesel. After full investigation an unexpected result connected to a new coke shipment with higher sulfur content was identified as the root cause.

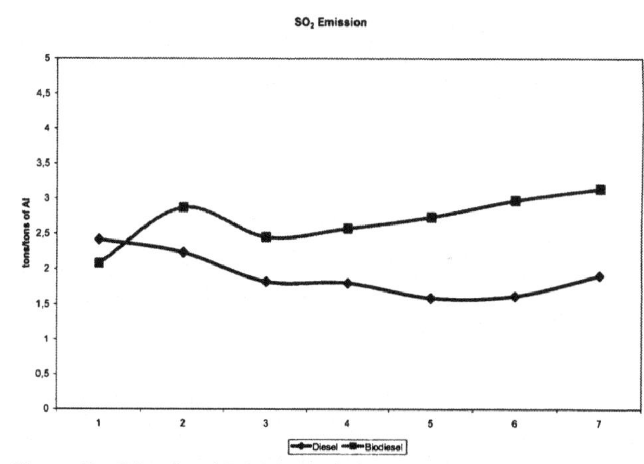


Figure 7 – Diesel and B2 $SO_2$ emission

Particulate emissions did not show any significant variation between either of these fuels.

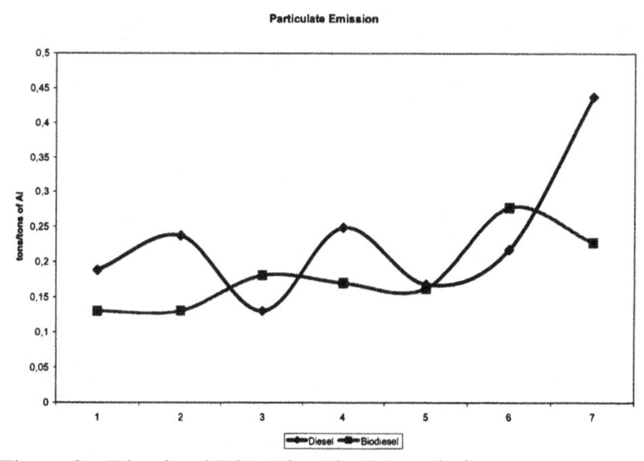


Figure 8 – Diesel and B2 total particulate emission

Biodiesel has a considerable solvent effect, so the number of plugged burner events was monitored to prevent any chaotic situation in the furnace due to line pipe cleanness. It was expected that the biodiesel would dissolve impregnated solid petro-materials from the internal area of the pipe system that could block the fuel injection inside the pits. An increase was observed, although it was not considered critical for the 2% mix. The number of plugged burner events increased from 0.6 to 1.4 per day.

## Conclusions

- All process aspects evaluated in this paper allowed Alumar to substitute petrol diesel by a mix of petrol diesel with 2% of biodiesel. Although fuel consumption increases, benefits from greenhouse gases emission reduction, especially $SO_2$ and $CO_2$, and carbon credits drove the company to decide for its implementation.

- Baking furnace reduced consumption of fossil fuel in 470.000 liters per year, consequently resulting in a reduction of 1018 tons of carbon dioxide emission.

- B2 results allow the process to move further. For the next step, a mix of 95% of diesel and 5% of biodiesel will be used. Subsequent steps will depend on biodiesel availability in Brazilian market.

## References


1. Expedido Jose de Sá Parente, *"Biodiesel – uma aventura Tecnológica num pais engraçado"* (Fortaleza,CE –Brazil: Tecbio, 2003) 66

2. Paulo Miotto, Marcos Aurelio Silva, Ciro Kato, Roberlito Silva, "New Method to Start Up Fires in Baking Furnaces", Light Metals 2005, 683 – 688

3. Jean Bigot, Magali Gendre and Jean-Christophe, *"Fuel Consumption: A Key Parameter in Anode Baking Furnaces"*, Light Metals 2007, 965 – 968.

4. Marco Antonio Conejero, and Marcos Fava Neves, "*Marketing de crédito carbono: um estudo multi-casos*", (Paper presented at the 30th EnANPAD Annual Meeting, Salvador - Brazil , 23 to 27 of September, 2006).

5. Tom M. L. Wigley, *"The Kyoto Protocol: CO2, CH4 and climate implication"*, Geophysical Research Letters (1998), vol. 25, pp. 2285–88.

6. Tom M. L. Wigley, *"The Climate Change Commitment"*, Science(2005), vol. 307, pp. 1766–69.

7. United Nations Framework Convention on Climate Change: Changes in GHG emissions from 1990 to 2004 for Annex I Parties.

8. Kyoto Protocol: Status of Ratification, 10 July 2006 UNFCC.

**Light Metals 2008** *Edited by: David H. DeYoung*
***TMS (The Minerals, Metals & Materials Society), 2008***

# PNEUMATIC TRANSFER OF FINE CARBON RESIDUES AT ALBRAS – A CASE STUDY AT THE CARBON PLANT

Paulo Douglas S. de Vasconcelos[1], André Luiz Amarante Mesquita[2]
[1]Albras Alumínio Brasileiro S/A; Barcarena, Pará - Brazil, Rod. Pa 483 Km 21 - 68447-000Phone: 0055 91 3754 6975 -
Fax: 55 91 3754 1515 – e-mail: pdouglas@albras.net
[2]Federal University of Pará; Phone/fax: 0055 91 3201 7960 – e-mail: andream@ufpa.br

Keywords: Pneumatic conveyor, dilute phase, pressure drop, gas-solid flow.

## ABSTRACT

Albras Alumínio Brasileiro S.A has upgraded the carbon plant to capture and transport the fugitive dust emissions from the operations of butt cleaning, crushing and grinding of butts and reject anodes, and also from the anode baking furnace tending cranes to reduce environmental pollution in the work place and the environment. This is described in the first part of the paper. Originally, the residual dusts, mostly carbon were conveyed by a screw conveyor for loading into a bulk solids truck for subsequent disposal. The rest of the paper describes the development, design, and implementation of a pneumatic conveying system developed as third part of a Masters dissertation to resolve the environmental issues, eliminating the coke contamination from the residual dusts while reducing annual maintenance and labor expenditures

## INTRODUCTION

This paper is the third part of a master degree dissertation developed by (Vasconcelos, 2005) to solve some process and environmental problems common to all aluminum smelters. Two of these papers were published and presented at the TMS meetings in Nashville 2000, 2003 in San Diego and TMS 2004 in Charlotte.

As plant engineers normally consider pneumatic conveyor system as a "black box", we will try to demystify this issue, discussing the routine of a project, from the computational simulation to the erection schedule.

## ORIGINAL PROJECT

In the original design, the bulk solids truck was loaded by a screw conveyor. Due to the short length of the screw conveyor, the bulk solids truck had to share the same access ramp with an open truck that supplies green coke to the paste plant – see photo 1 below.

This situation cause many problems such as: coke contamination by the carbon residues rich in sodium and iron, delay in time to load the bulk solids truck due to screw conveyor limitation and problems such as overload spillage of the carbon residues that, besides contaminating the coke, causes environmental pollution. The bulk solids truck typically uses one shift (eight hours) to load the 25 tonnes of carbon residues. These problems caused constant complaints from the truck's driver and the paste plant operators.

Photo 1 – Bulk solids truck, alongside the new pneumatic conveyor system during the erection phase

## CASE STUDY – ALBRAS' DESIGN

**Projects Criteria**: In the first two chapters of the master degree (Vasconcelos, 2005). Albras' operating conditions plus the literature concerning pneumatic conveyor in dilute phases were reviewed. Due to the lack of space we will make a resume of some calculations showing the theoretical foundation.

Projects developed by plant engineers cut the costs, because they enable better use of the plant facilities such as silos, belt conveyors and building's structures, and surplus materials from old projects (pipes, flanges, bolts and gaskets).

In a dilute phase pneumatic conveyor, the particles travel inside the pipe at high velocities (10 – 30 m/s), which is equivalent to a solids-to-air ratio less than 15 kg of solid by 1 kg of gas – see Figure 1 below.

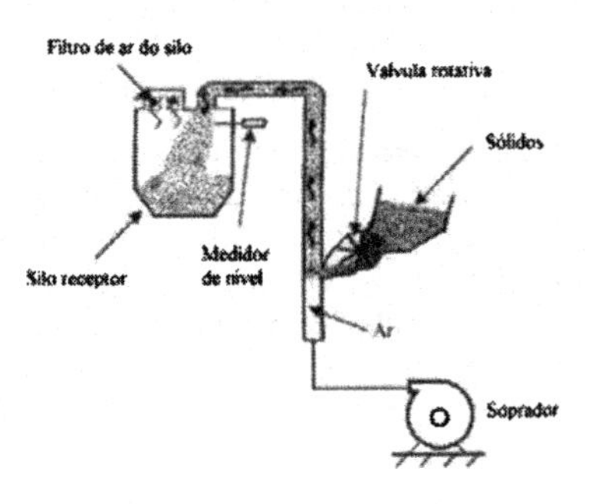


Figure 1 – Pneumatic conveyor of solids in the dilute phase

Before we begin the calculations, we will summarize some data of Albras' design and give some tips to prevent a variety of common problems that could occur with pneumatic conveyors.

- ✓ Maximize your installation layout – draw an isometric diagram of the circuit so that you can minimize the pressure drop
- ✓ Check physical chemistry data such as ambient temperature (33° C at Albras), solids grain size (105.2 $\mu m$ at Albras), real density of the solid grains (2200 kg/m3), bulk density of the solid grains (850 kg/m3), impurities (iron, sodium, sulfur, aluminum, etc), air density at 40°C – 1.127 kg/m3
- ✓ Optimize your project using a high level of automation (programmable logic controllers, pressure, velocity and level transducers, etc.) to reduce energy consumption, abrasion in bends, and a reliable operation and maintenance
- ✓ Choose among dilute phase or dense phase (dilute phase at Albras). This criterion can be adjusted depending on the level of automation – at Albras we reach a solids-to-air ratio of 18,95 – see Table 1.

| System | Solids-to-air ratio |
|---|---|
| Dilute phase | 0 – 15 |
| Dense phase | > 15 |

Table 1 – Systems' classification concerning solids-to-air ratio - source: Klinzing et al.

- ✓ Use software (Velarr) to simulate your project. Velarr is the mathematical model we developed in cooperation with our local university.

### THEORETICAL FORMULATION

After analysis we decided on a dilute phase system, but tried to maximize installation with a high level of automation.

**1. Flow range of the gas in the system**

The capacity of the screw conveyor is 30t/h, and we intend to reduce the time to load the bulk solids truck by the half, so we have to reach a solid flow of 60 t/h.

$W_{solid}/W_{gas} = 15$, then we have a minimum and a maximum of:

$$15 = \frac{30000t/h}{W_{gas}} \Rightarrow W_{gas} 2000t/h = \rho_{gas}.Q_{gas} \Rightarrow Q_{gas} = 2000/1.127 \approx 1800m^3/h$$

Therefore, if we double the feed rate we will double the volumetric flow of the gas: $Q_{gas} = 3600m^3/h$.

**2. Pipe specification for the system**

To reduce the abrasion and pressure drop for long systems we normally recommend a velocity around 15 m/s for the coke, but using the equation: $\phi = \sqrt{4Q_{gas}/\pi V_{gas}}$, this provides a minimum pipe's diameter ($\phi$) of 10". However, after analyzing the isometric diagram, we obtained an equivalent length for this system of 77 meters, therefore this is a short circuit, and then we can increase the gas velocity in order to use the pipe available in our store (8" schedule 40). Using this diameter we will work with velocities above 23 m/s. Even with a higher pressure drop, energy consumption and abrasion, this decision reduces the installation cost, and enables the project – see Figure 2.

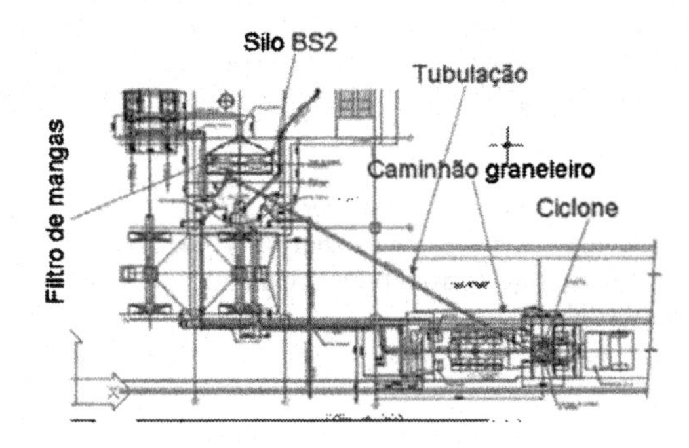


Figure 2 – Layout of the installation in the paste plant

**3. Dimensioning of the high efficient cyclone**

Figure 3 shows the closed circuit. The blowers are under the rotary valve of a storage silo of 150 tonnes in the paste plant. The pipe in red (~27m) is horizontal up to the cyclone inlet. After the cyclone outlet comes a vertical pipe (blue) until the top of the silo. Where, the bag house (red) is placed. According to (Macintyre, 1990), a high efficient cyclone can be calculated by the equation: $Q = 300d^2$ (3.1) (Q – volume flow in CFM and d - diameter in feet). Using the average value of the flow calculated in (3.1), and the model proposed by Rosin at al. (Vasconcelos, 2005) in Table 2, we find the cyclone dimensions:

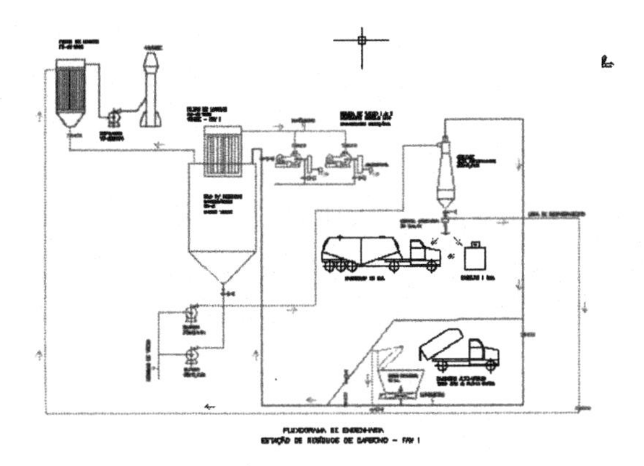

Figure 3 – Engineering flow chart for the proposed project

| Cyclone pattern A | | | | | | | |
|---|---|---|---|---|---|---|---|
| Dimension in (mm) | d | a | b | s | $d_s$ | h | H |
| Diameter multiply | 1 | 0,5 | 0,2 | 0,625 | 0,5 | 2 | 4 |
| factor (d) | 670 | 335 | 134 | 418 | 335 | 1340 | 2680 |

Table 2 – High efficient cyclone pattern A – H is the height of the cyclone.

In this cyclone the particles get 20 turns until they fall into the hopper, with a fractional efficiency greater than 99 % for a cut-off diameter of 2.26 $\mu m$. The cyclone pressure drop (102.3 mm water), was calculated using the equation (3.2) (Vasconcelos, 2005).

$$\Delta P = 8\rho_{gas}U^2_{cyclone} / 2g_{gravity} \tag{3.2}$$

### 4. High efficient bag house project

This is a very special design, due to the lack of space in the top of the silo, besides the high concentration of dust inside the silo (106 g/m$^3$), where for a standard bag house this concentration varies from 5 – 30g/m$^3$.

There are two import parameters concerning the bag house project, the air-to-cloth ratio and the velocities inside it. The standard bag houses admit air-to-cloth ratio of 1.5 m$^3$/min/m$^2$, and may reach a velocity greater than 1 m/s.

Due to the severity mentioned above, we will adopt an air-to-cloth ratio of 0.5m$^3$/min/m$^2$, and an ascending velocity of 0.5 m/s.

To calculate the number of bags of this filter, we divide the maximum volumetric flow entering the bag house by the air-to-cloth ratio – see calculations below. As this bag house can receive 60 m$^3$/min (3600 m$^3$/h calculated in (3.1)) plus 50 m$^3$/min from the pneumatic conveyor outside the paste plant, a total of 110 m$^3$/min, with a air-to-cloth ratio of 0.5m$^3$/min/m$^2$:

$$n = \frac{110m^3 / \min}{0.5m^3 / \min/ m^2} = 220m^2 \tag{4.1}$$

this is equivalent to 120 bags 12 ft long and 6" internal diameter – see Figure 4.

To calculate the internal velocities: Ascending velocity: divide the total volumetric flow by the open area of the bag house, and the interstitial velocity in the bags: divide the total volumetric flow by the liquid area (open area – total tube sheet area) – see Table 3.

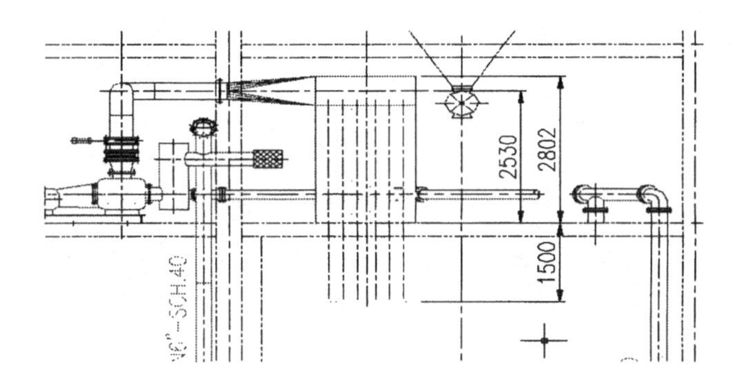


Figure 4 – Bag house inserted inside the silo (BS2) paste plant

| Velocities inside the bag house | |
|---|---|
| Width of the filter | 1.5 m |
| Length of the filter | 4.8 m |
| Transversal area | 7.20 m2 |
| Tube sheet area | 2.35 m2 |
| Filter open area | 7.20 m2 |
| Filter liquid area | 4.85 m2 |
| Ascending velocity | 0.37 m/s |
| Interstitial velocity | 0.78 m/s |

Table 3 – Velocities inside the bag house

With these values this bag house operates very well, even with the high dust concentration it will create a low pressure drop in the circuit. The pressure drop in a standard bag house in steady state condition is around 200 mm water. This pressure drop is proportional to the square of the air-to-cloth ratio, in other words:

$$\Delta P \propto \Delta P_0 (\frac{\Delta V}{\Delta V_0})^2 \equiv 200(\frac{0.5}{1.5})^2 \equiv 22mmWater \tag{4.2}$$,

but we have to consider that the concentration of dust is five times higher than in a standard bag house, so that the dust creates a thicker layer on the bag. The pressure drop is directly proportional to the cake thickness, so we hope that the pressure drop will be 22 x 5 = 110 mm water – set point value for a programmable logic controller.

### 5. Pressure drop through the pipe line (Klinzing et al, 1997)

The equations given here are based on the hypothesis that the gas-solid flow is in the dilute phase. Some assumptions such as: transients in the flow (Basset forces) are not considered nor the gradient pressure around the particles (this is considered negligible in relation to the drag, gravitational and friction forces). The pressure drop due to particle acceleration is not considered.

The flow is considering incompressible, omni dimensional and the concentration of solids particles is uniform. The physical properties of the two phases are temperature dependent.

The flow mass for each can be expressed as follows

$$m_g = \rho_g U_g (1 - \varepsilon_s) A \tag{5.1}$$

$$m_s = \rho_s U_s \varepsilon_s A \tag{5.2}$$

where the suffixes g and s denote gas and solid, respectively, U is the face mean velocity in relation to the tube cross section A, $\varepsilon_S$ is the volume fraction occupied by the solid as follows

$$\varepsilon_s = \frac{A_s}{A} = \frac{4m_s}{\rho_s \pi D^2 U_p} \tag{5.3}$$

D is the tube diameter and $U_p$ is the particle velocity. The void coefficient, in other words the fraction of volume occupied by the

$$\varepsilon_g = \frac{A_g}{A} = 1 - \varepsilon_s \tag{5.4}$$

The velocity and specific mass of the gas-solid mixture are expressed by:

$$U_m = U_{sg} + U_{ss} \tag{5.5}$$

$$\rho_m = \rho_s \frac{Q_s}{Q_s + Q_g} + \rho_g \frac{Q_g}{Q_s + Q_g} \tag{5.6}$$

Here, Q is the volumetric flow, $U_{sg}$ and $U_{ss}$ respectively the gas and solid superficial velocity, defined as the volume per unit of time of the phase that enters the tube divided by the tube cross section area.
Particle velocity calculation:
We consider in this paper the Yang models [1]

$$U_p = U_g - U_t \sqrt{\left(1 + \frac{2 f_s U_p^2}{gD} \varepsilon_g^{\ 4.7}\right)} \tag{5.7}$$

$$U_t = \frac{g D_p^2 (\rho_s - \rho_g)}{18 \mu_g}, \quad K < 3.3 \tag{5.8a}$$

$$U_t = \frac{0.153 g^{0.71} D_p^{1.14} (\rho_s - \rho_g)^{0.71}}{\rho_g^{0.29} \mu_g^{0.43}},$$

$$3.3 < K < 43.6 \tag{5.8b}$$

$$U_t = 1.74 \left[ \frac{g D_p (\rho_s - \rho_g)}{\mu_g} \right]^{1/2},$$

$$43.6 < K < 2360 \tag{5.8c}$$

where K is a factor that determines the range of validation for drag coefficient expressions, when the particle Reynolds number is unknown, and given by:

$$K = D_p \left[ \frac{g \rho_g (\rho_s - \rho_g)}{\mu_g^2} \right]^{1/3} \tag{5.9}$$

where $\mu_g$ is the gas viscosity, g gravity acceleration, $\rho$ the specific mass, $D_p$ the particle diameter, $f_s$ the solid friction loss, $\varepsilon_s$ fraction in volume occupied by the solids

**Pressure Drop Calculation**

The total pressure drop, $\Delta P_T$, for gas-solid flow is calculated from the contribution of the static, $P_E$ , and friction $P_F$, loss for both phases. In others words:

$$\Delta P_T = (\Delta P_E + \Delta P_F)_s + (\Delta P_E + \Delta P_F)_g \tag{5.10}$$

$$\Delta P_{Es} = \rho_s \varepsilon_s L g \tag{5.11a}$$

$$\Delta P_{Eg} = \rho_g \varepsilon_g L g \tag{5.11b}$$

and the contribution due to the friction given by the Darcy equation:

$$\Delta P_{Fs} = \frac{2 f_s \rho_s U_p^2 L}{D} \tag{5.12a}$$

$$\Delta P_{Fg} = \frac{2 f_g \rho_g U_g^2 L}{D} \tag{5.12b}$$

where, $f_s$ and $f_g$ are the friction coefficients for the solid and for the gas, respectively.

## Software (Velarr) to calculate the pressure drop

This is a computational routine developed to calculate the pressure drop in the gas-solid flow.

### Inputs

Equivalent circuit length (m)
Pipe roughness (mm)
Pipe diameter (mm)
Mean temperature inside de tube °C
Work pipe pressure (mWater, relative)
Gas Flow ( $m^3$ /min)
Flow with or without particle load
Particle diameter (mm)
Specific mass of the material conveyed (kg/ $m^3$ )
Flow of material to be conveyed (kg/h)
Pipe inclination (degrees)

Outputs

| | |
|---|---|
| Air velocity | m/s |
| Particle velocity | m/s |
| Mixture velocity | m/s |
| Flow Reynolds Number | $R_e \geq 10^5$ |
| Mixture specific mass | kg/ $m^3$ |
| Drag criterion | $\geq 1$ |
| Unit pressure drop | mWater/m |
| Total pressure drop | mWater |

Using the software (Velarr), we calculated the pressure drop in the closed circuit of the Figure 3. This circuit was divided into three paths or stretches to find the zero or atmospheric point.
As we can see in the Figure 3, the silo (BS2) is kept at a static pressure of – 600 mmWater, during the bulk solid track loading. Then we can classify this system as a push-pull, where the blower pushes the particles with a positive pressure and after the zero point the particles are pulled by the rotary piston blower installed after the bag house:

- The first stretch is vertical from the cyclone outlet to the top of the silo: 8" diameter with an equivalent length of 50m, 300 kg/h of ultra fine dust considering 99.5 % for the cyclone efficiency – the calculated pressure drop for this stretch is 302 mmWater. The liquid pressure for the next stretch will be: - 600 (static pressure in the silo) + 100 (pressure drop of the cyclone) + 302 (pressure drop in the first stretch) = -198 mmWater
- The second stretch is horizontal from the zero point to the cyclone inlet: 8" diameter with an equivalent length of 27 m, 60000 kg/h of dust. With the available static pressure of (-198 mmWater), we find the zero point at 6 m from the cyclone;
- The third stretch is horizontal from the zero point to the centrifugal blower outlet (rotary valve - feed point): 8" of diameter with an equivalent length of 27 m, 60000 kg/h of dust. The calculated pressure drop for this 21 m is 690 mmWater

The Figure 6 shows the pressure drop profile calculated for a 30t/h mass flow feed.

After this calculation we can specify the fan: flow rate from 40 – 60 m3/min at a static pressure of 1000 mmWater at 40 °C. The electric motor is 15 kw/440 Volts, three phases, 60 Hz, 3520 rpm. This motor will be controlled by a speed controller with PID, and square root functions available – see Figure 5 below

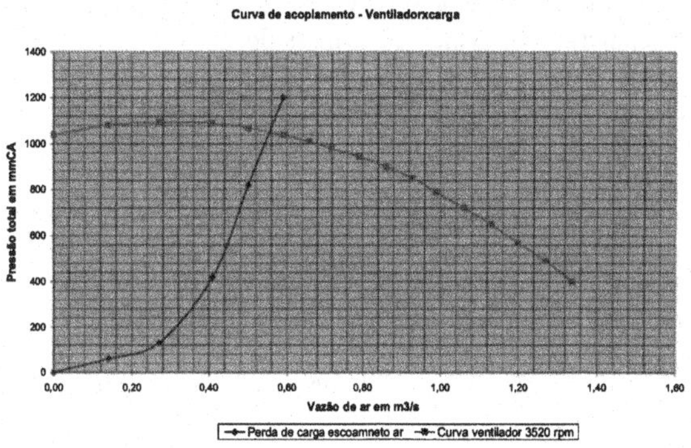


Figure 5 - fan curve-Static pressure versus flow rate at 40 °C

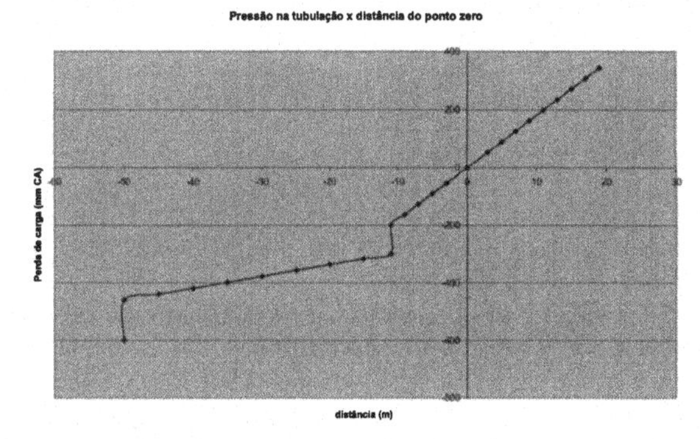


Figure 6 – Pressure drop profile for 30 t/h of dust
Pressure in mmWater versus distance in meter

The zero point is at 19 meters from the rotary valve, where the blower sees a positive pressure of 342 mmWater. The first sudden pressure drop occur at -12 meters and is due to the cyclone. The other sudden pressure drop occurs at -50 meter and is due the bag house. The linear pressure drop is due to the pipe (8" diameter and ~77 meter long). In the situation showed in this graph the fan controlled by a speed controller runs at 34 Hz or 2000 rpm and sees a static pressure of 342 mmWater. Therefore the total pressure drop in the closed circuit is 942 mmWater.
Now we can specify the rotary valve. To minimize the cost, we kept the existing lock Ducon valve of 12" with a rate capacity of 21 $dm^3$/rotation. This valve reaches the desired feed capacity in the range (30000–60000 t/h), or from (30–60 rpm) – see Figure 7.

Figure 7 – Standard lock ducon rotary valve

**Curves of a pneumatic conveying of solids to determine the optimum operation point (line in red) – Figure 8 below.**

Before starting the speed controller program, it is necessary to make a blower behavior survey, concerning the pressure drop of the pneumatic conveying versus mass flow feeding.
We divided the circuit in three stretches as was commented above. Fixing the blower speed at 2000 rpm, 2500 rpm, 3000 rpm and 3520 rpm (60 Hz), the mass flow varies from: 30t/h, 40t/h, 50t/h and 60 t/h.
Based on the fan/blower curve (Figure 5) shows the family of curves – Figure 8, for speeds varying from 2000 rpm (34 Hz) to 3520 rpm (60Hz). This is possible due to the similarity rule for blower and fans, where the flow rate is proportional to the speed, and the pressure is proportional to the second power of the speed. Using the software Velarr for the whole circuit, and fixing the mass flow, we vary the volumetric flow from 0 to 100 m3/min, for each speed mentioned above, and put the results (flow versus pressure drop) in an electronic chart, and then plotting the graphic of the Figure 8 below.

Figure 8 – Family of blower/fan curves

**Blower speed Control.**

Using the curves of Figure 8, it is possible to choose the optimum operation point, which in our system is at 30 $m^3$/min or 15 m/s for 8" pipe diameter – line in red.
In the beginning of our theoretical formulation, we chose an 8" diameter pipe in stead a 10" diameter pipe. In this option we will operate with a gas velocity of 24.3 m/s (45 $m^3$/min). These curves in Figure 8 have a very smooth slope near the minimum pressure drop, and even increasing the velocity by 50 % to operate in a more stable point, that corresponds to an increase of 50 mmWater in pressure drop, which is totally absorbed by the blower that was specified with 1000 mmWater static pressure.
To guarantee a safe operation, the system measures the gas velocity in the blower inlet and sends this signal to the speed controller that has a PID controller to keep the set point (velocity at 24.3 m/s – see line in red, Figure 8). This apparatus avoids pipe clogging, keeping the velocity at 24.3 m/s, in every type of feeding load by the rotary valve, from an absence of dust (only air) to a severe load of 60 t/h of dust.
The block diagram for this system is shown in the Figure 9.

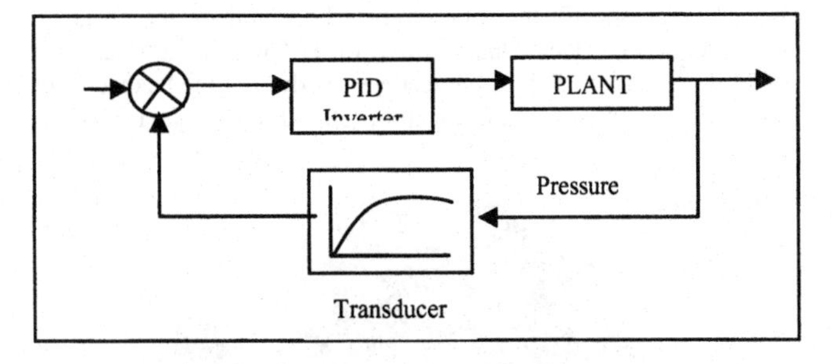


Figure 9 – Block diagram of the speed controller.

The pressure transducer sends a (4 – 20) mA signal to the speed controller. This signal is proportional to the dynamic differential pressure ($P_v$ - (0 – 2)" of water). The speed controller extracts the square root of this signal, compares with the set-point, and processes the error in the PID block and controls the motor speed to give the desired gas velocity (23.4 m/s).

Figure 10 – Supervisory screen plus the new truck load station

## CONCLUSION

Figure 10 summarizes our conclusion, showing the bulk load truck, in the process of loading, monitored via a computer screen. This is a successful project of science and technology, which solved an industrial problem and increased the academic knowledge.
This solution reduced the truck loading time by more than 70 %, and eliminated the following problems: coke contamination by sodium and iron, environmental pollution, in addition to operational and maintenance interferences.
Now the truck can be loaded every day of the week, and the silo (BS2) level is kept low, eliminating frequents rodding shop and baking furnace downtimes.

## REFERENCES

I. Klinzing, G. E., Gas-solid Transport, McGraw-Hill Book Company, 1981.
II. Macintyre, A. J. Ventilação Industrial, 2ª edição, 1990, Editora Guanabara Koogan S. A.
III. Vasconcelos, P. D, Improvements in the Albras Bake Furnaces Packing and Unpacking System – LIGHT METALS 2000, pp. 493 – 497.
IV. Vasconcelos, P. D, Exhaustion Pneumatic Conveyor and Storage of Carbonaceous Waste Materials – LIGHT METALS 2003, pp. 583-588.
V. Vasconcelos, P. D., Pneumatic Conveyor of Carbon Dust – Albras Case Studies, Master Degree Dissertation, Federal University of Pará, 2005.

**Light Metals 2008** *Edited by: David H. DeYoung*
***TMS (The Minerals, Metals & Materials Society), 2008***

# DEVELOPMENT IN SITU OF A SEMI-AUTOMATED SYSTEM IN THE CASTING STATION: A CASE STUDY AT ALBRAS

Emerson David Cavalcante Santos[1], Luis Carlos Carvalho Costa[1], Dourivaldo Sousa Dias[1], Antônio Ivan Arruda[1], Marco Antônio Reis[1], Edson Viégas Ribeiro[1], Antônio Leomar Dias Santos[1],

[1]Albras - Alumínio Brasileiro S/A; Barcarena, Pará - Brasil, Rodovia. Pa 483 km 21 - 68447-000. Phone: 0055 91 3754 6066 - Fax: 55 91 3754 1515 – e-mail: emerson.santos@albras.net

Keywords: Rodding anode process, Automation, Casting Station, Safety.

## Abstract

The work in the Casting Station of the rodding shop is very risky, because the flying sparks of molten cast iron are a safety hazard for the operators. The objective of this work is to present an automation project of the casting station developed by the Albras team. The semi-automated system consists of the static cabins with air-condition control and a large open view for maximum operator comfort; an automatic metal pouring device operated with an electronics-hydraulics drive, using proportional controller valves, controlled by electronics joysticks. Finally, results for casting quality, ergonomics and safety obtained by implementing the project are presented.

## Introduction

Albras Alumínio Brasileiro S.A. is a primary aluminum smelter, with a capacity of 450.000 tons of ingots per year located in the Pará state in the Amazon region of Brazil.

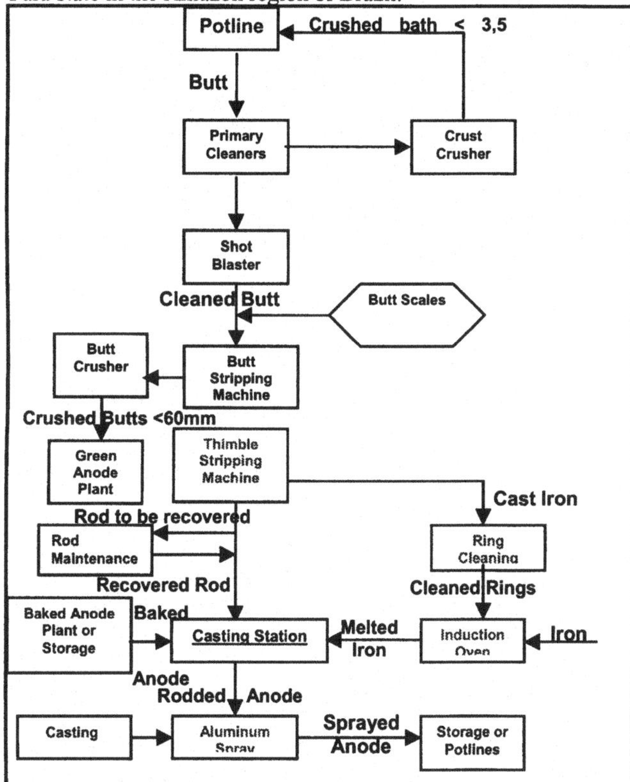


Figure 1 – Albras Rodding Process.

Albras has two complete pre-baked anode production lines on site, with two almost identical rod shop plants, phase I and II. They are operated as independent units, with a capacity of 45 anodes per hour and a productive time of 10 shifts per week. The rodding process is done in a conventional manner; the aluminum rod is connected to the yoke by four spider stubs using a bi-metallic transition joint that form a stem. The stem is mated to the block holes and fastened with cast iron in the Casting Station. At the end of the process all anodes are coated with an aluminum layer. Figure 1 illustrates a schematic diagram of the Albras rodding process.

## Casting Station

The original Casting Station was operated manually. This is the main workstation of the rodding process. It consists of three main sections:

1. A conveyor table that moves the anode block in the horizontal position that mate with the stem.
2. A stem alignment system that keeps the four stubs in the top of the block stub holes and ensures the perpendicularity between the carbon block and stem – see Figure 2.
3. A manually operated metal pouring device, called the "Rodding Cannon", consisting of a steel structure holding a vertical steel axis. This carried out rotational motions; one horizontal steel axis that carried out rotation, advance and return motions to empty the metal into the anode stub holes – see Figure 3.

Figure 2 – Illustrative photo of the Casting Station.

Figure 3 – Manual Rodding Cannon.

The operators carried out excessive and repetitive tasks, causing physical problems and occupational illnesses – see Figure 4. We did not have a precise dosing of the cast iron into the stub holes; the operators worked in risky conditions, because they stayed in a high temperature, due to molten iron. The operators worked next to the metal crucible. Working from inside a cabin would have been preferable as they frequently were subject to sparks of molten cast iron.

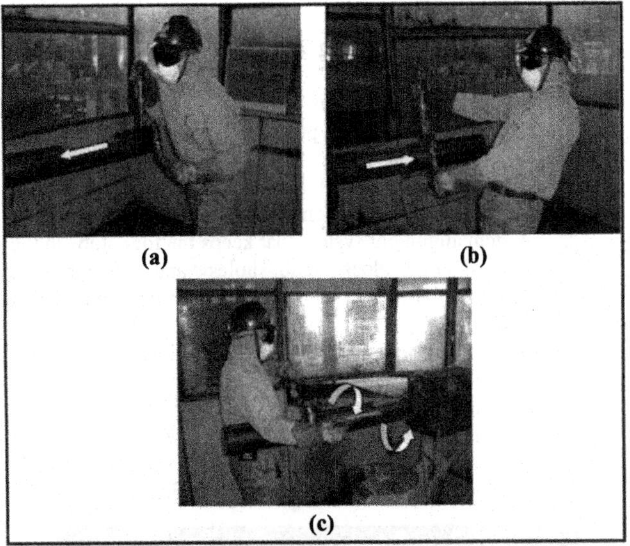


Figure 4 – Manual rodding cannon operation. In (a) and (b): Advance and return motions. In (c): Rotational motion for metal pouring in the stub holes.

The pouring crucible with the cast iron was located at the end of the horizontal axis (Figure 4c).

## Case study – The Automation Project

The project was developed by a teamwork composed for the operators, maintenance engineer, a technician, a mechanic, an electrician, three managers of the rodding plant and twelve contractor people – see Figure 5.

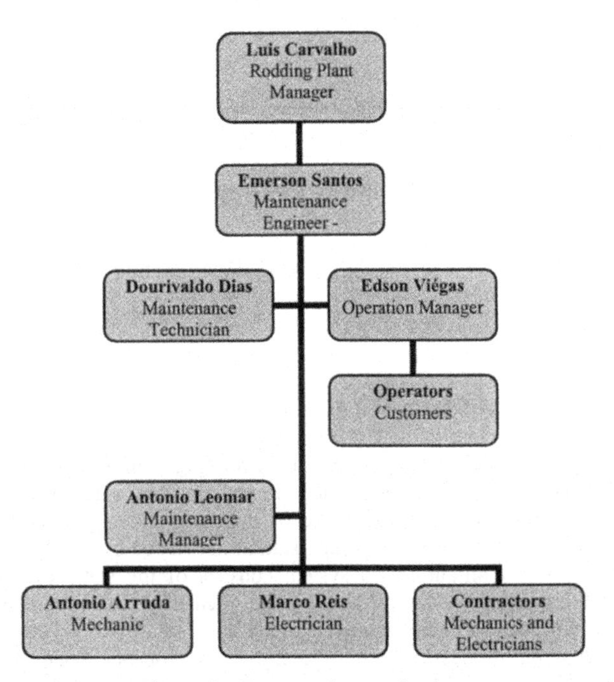


Figure 5 – Teamwork organization.

Before the beginning of the project, some tests in the equipments had been made for collection of data. The data of the tests were analyzed through of evaluation forms (Figure 6). In the forms, the operators (customers) gave the suggestions the teamwork analyzed and planned all the phases of the project, through of meetings. This stage lasted one year.

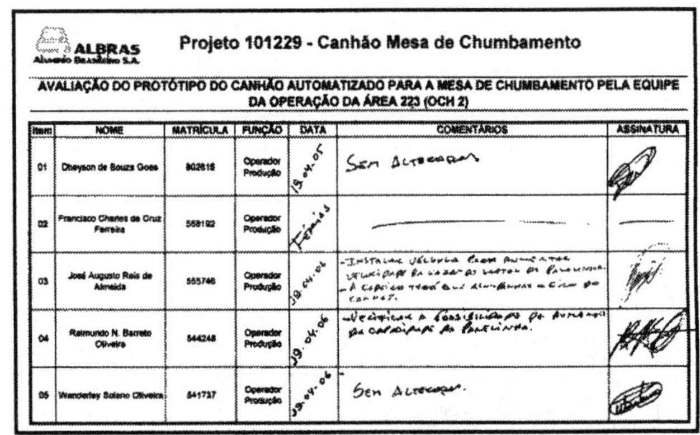

ALBRAS Alumínio Brasileiro S.A.

Projeto 101229 - Canhão Mesa de Chumbamento

AVALIAÇÃO DO PROTÓTIPO DO CANHÃO AUTOMATIZADO PARA A MESA DE CHUMBAMENTO PELA EQUIPE DA OPERAÇÃO DA ÁREA 223 (OCH 2)

| Item | NOME | MATRÍCULA | FUNÇÃO | DATA | COMENTÁRIOS | ASSINATURA |
|---|---|---|---|---|---|---|
| 01 | Dheyson de Souza Goes | 802616 | Operador Produção | 19.04.05 | [illegible] | [illegible] |
| 02 | Francisco Charles da Cruz Ferreira | 559192 | Operador Produção | Férias | — | — |
| 03 | José Augusto Reis de Almeida | 555746 | Operador Produção | 19.04.06 | [illegible] | [illegible] |
| 04 | Raimundo N. Barreto Oliveira | 544248 | Operador Produção | 19.04.06 | [illegible] | [illegible] |
| 05 | Wanderley Solano Oliveira | 541737 | Operador Produção | 19.04.06 | [illegible] | [illegible] |

Figure 6 – Evaluation form of the preliminary tests.

After of one year of data analysis, we started a challenging project, developed in three main phases:

1) Enlarge the static cabin and build an operational cabin to eliminate the burning risks and/or metal sparks.
2) Install air-conditioning control in the cabin for maximum operator comfort.
3) Improve the operators' ergonomic conditions and life quality, eliminating the excessive and repetitive tasks.

The team worked eighteen months in the development and execution of the cabin project, mechanic, hydraulic and electric projects.

Cabin Project

To improve the work conditions of the operators, a large open view cabin with an air-conditioning system was projected – see Figure 7.

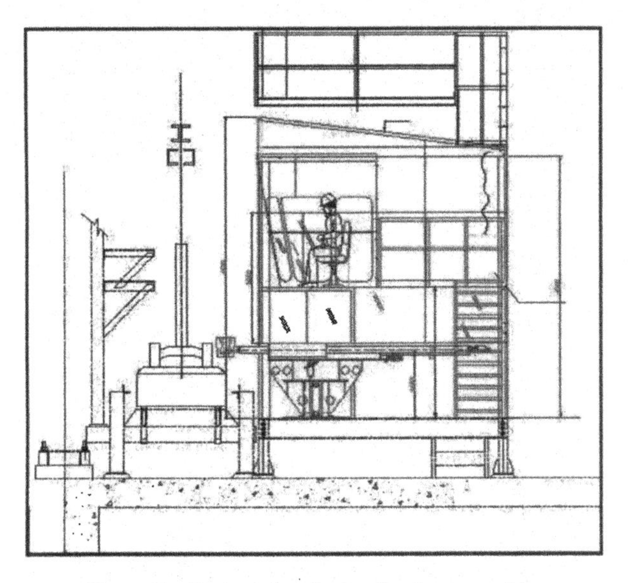

Figure 7 – Engineering design for the new cabin.

Mechanic Project

The new rodding cannon was projected to carry out all operator motions for pouring metal into the stub holes, controlled by an automatic system with electronics-hydraulics drive – see Figures 8 and 9.

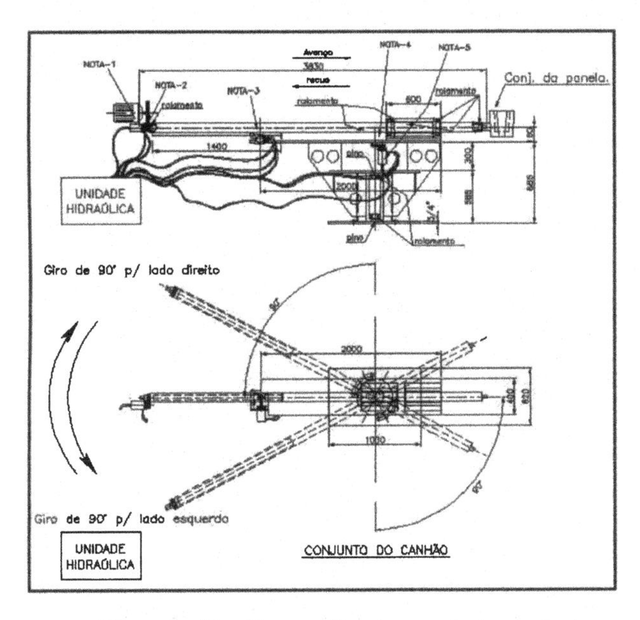


Figure 8 – The New rodding cannon design project.

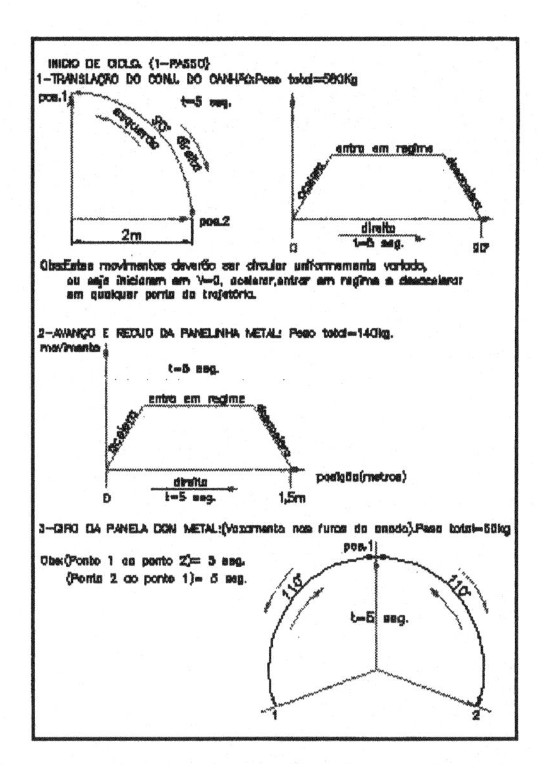


Figure 9 – Engineering chart detailing the operation cycle of the rodding cannon.

The motions were based on the operator's movements:

1 – Rodding cannon rotation to locate the pouring crucible between the anodes.

2 – Axis advance and return to make positioning of the pouring crucible in each stub hole.

3 – Pouring crucible's rotation with the left and right motions to pour the cast iron into the four stub holes.

Electric and Hydraulic Projects

The cannon's drive is made through of a proportional hydraulic control system. Proportional valves are used to control the velocity of the hydraulic motors, responsible for the rodding cannon's motions – see Figure 10.

Figure 10 – Rodding cannon and hydraulic system tests.

Two joysticks, manipulated by the operator (Figure 11) control the cannon's velocity and improve the cast iron dosing precision during pouring into the stub holes. The controls are located in an electric panel assembled in the operator's chair – see Figures 11 and 12.

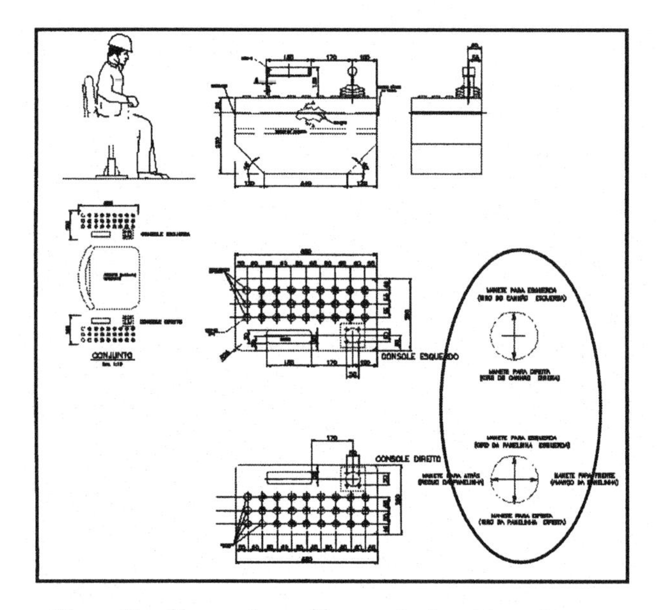

Figure 11 – Electro-electronic controls. Joystick motions are shown in details.

Figure 12 – Operator's chair during testing.

## Results

The main results obtained in the project were the following:

1. Improved safety work conditions

The operation static cabin was enlarged and built over two floors. The equipment operation is made from inside an air-conditioned and closed cabin, more distant from the metal crucible, eliminating the burning risks due to the high temperature and flying sparks of molten cast iron (hot metal projection) and the quantity of fire principles was minimized – see Figures 13 and 14.

Figure 13 – Photographs (a) and (b) showing the new operation cabin.

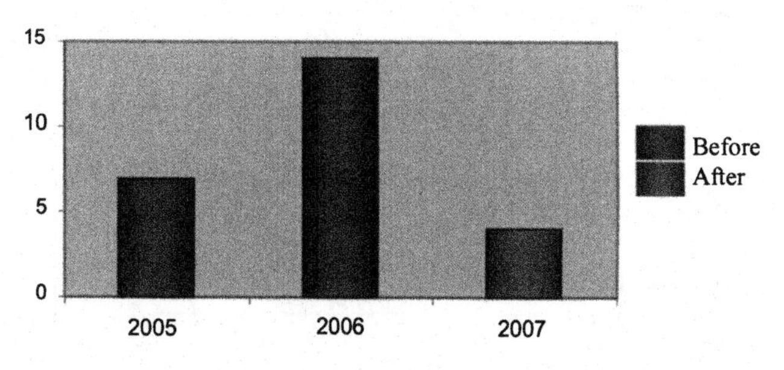


Figure 14

2. Improvement in the casting quality

The electronics-hydraulics control system it makes possible to minimize cast iron spilling (Figure 15), but improvement in the stub/carbon voltage and bath iron content was not identified in the reduction process.

Figure 15 – Rodded anode before of the cleanness process.

3. Improved ergonomic conditions

The control chair installation brought important improvements to the operators, because they do not work by carrying out excessive and repetitive tasks, which could cause severe physical problems and occupational illnesses. Today, the operators work seated in a chair.

Figure 16 – The operator working into the cabin.

The Figure 17 presents a graphic of operators' evaluation about the new project. In this graphic was evaluated four items, which the operators are displayed, before and after the project, as incorrect position, excessive and repetitive tasks, extreme heat and hot metal projection. The number varies in accordance with the exposition of the operator to the risk, *i.e.* , the numbers 0-3 represents little exposition, 4-7 represents average exposition and 8-10 represents high exposition.

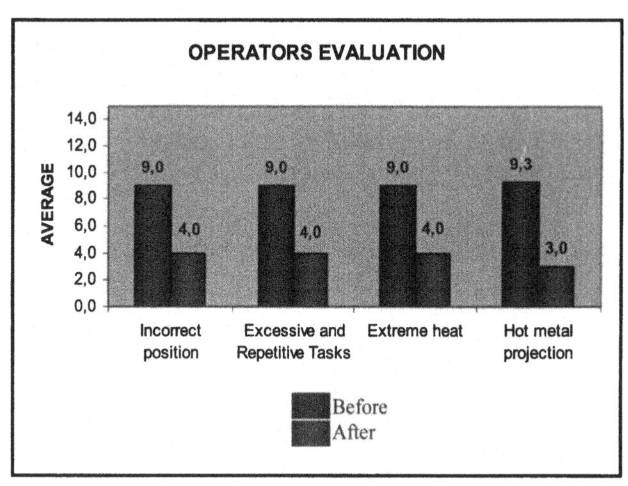


Figure 17

This graphic shows a significant improvement of the work and ergonomic conditions of the operators, after the implantation of the project. Therefore, the customers are more satisfied with the new workstation.

## Conclusions

This project has demonstrated that teamwork can give positive fruits. All the stages, *i.e.* , comment of the problems, survey of the causes, studies and project elaboration action planning and execution had the support of all the members of the team to reach the objectives.

The construction of a closed and air-conditioned cabin to isolate and to enlarge the operators' work area was the best solution to eliminate the risks of burning accidents and to prevent the exposition of the operator to the high temperatures.

Therefore, the casting station automation project was excellent to bring improvements in the life quality, casting quality and work safety.

## Acknowledgements

The authors wish to thank Albras for supporting projects that provide healthful, productive and safe work environments and for permission to publish the results.

## References

I. Gomes, A.S., "Carbon plant performance with blended coke" – Light Metals, 2005, pp. 659-660.
II. W. K. Fischer, et al., "Anodes for Aluminum Industry", R&D Carbon Ltd., 1995.

**Light Metals 2008** *Edited by: David H. DeYoung*
***TMS (The Minerals, Metals & Materials Society), 2008***

# CHEMICAL STABILITY OF FLUORIDES RELATED TO SPENT POTLINING

Wangxing Li[1], Xiping Chen[1,2]
[1]Zhengzhou Research Institute of Chalco, Zhengzhou, China, 450041
[2]Energy Science and Engineering School of Central South University, Changsha, China, 410083

Keywords: Aluminum electrolysis, Spent potlining, Fluoride; Chemical stability; Processing

## Abstract

Chemical stability of fluorides which are sodium fluoride, aluminum fluoride and cryolite related to spent potlining was studied. Experiments were carried out under conditions of exposure to air and exposure to calcium oxide or limestone. Results from exposure to air at temperature of 700~950 °C showed that most of aluminum fluoride reacted to alumina or volatilized when the temperature above 800 °C and the weight loss of aluminum fluoride surpassed 60 percent. Sodium fluoride did not volatilize or alter when heated. Cryolite and alumina formed when the mixture of aluminum fluoride and sodium fluoride was heated. When cryolite and sodium fluoride were heated up together, $NaAl_{11}O_{17}$ formed. When these fluorides related to spent potlining were exposed to calcium oxide or limestone and heated, Calcium fluoride formed easily.

## Introduction

Spent potlining is an unavoidable solid waste from aluminum smelters. In China, specific discharge of spent potlining is about thirty kilograms per ton Al excluding cathode rods [1, 4]. There are lots of fluorides which exist as $Na_3AlF_6$ and NaF in spent potlining [1~9]. Moreover, there are small amounts of cyanides which exist as NaCN and $Na_4[Fe(CN)_6]$ [1~9]. Cyanides and soluble fluorides are the main factors which lead to environmental pollution. Cleaning of spent potlining is becoming increasingly necessary in order to keep a sustainable development of the Chinese primary aluminum industry. Due to the solubility in water of sodium fluoride and cyanides, hydrometallurgy is not suitable for spent potlining processing. Thermal metallurgy processes are becoming leading methods in the world. Cyanides can be sufficiently decomposed by heat treatment, however fluorides are more troublesome than cyanides. Fluorides in spent potlining have high concentration, are highly dispersed and inlayed into other matters, so it is difficult to reduce the amount of soluble fluorides. Studies of the chemical stability of fluorides related to spent potlining can help to find some guidelines for spent potlining processing.

## Experimental

### Experimental Method

Tested samples are single fluorides or mixtures of different fluorides from the group of sodium fluoride, cryolite and aluminum fluoride. Fluorides are weighed and put into quadrate corundum crucibles. The crucibles are placed in the middle of the furnace hearth. During testing, the samples are exposed to air, calcium oxide or limestone. The testing temperature is 700~950 ℃ and the holding time is 120 minutes. When the heating program is over, the furnace is cooled to room temperature. The baked samples are collected and weighed, ready for the subsequent x-ray diffraction analysis. The percent weight loss of fluorides is calculated.

### Experimental Raw Materials and Equipment

Testing equipment is an auto-control electro-thermal furnace. The heating elements are silicon-carbon rods. The temperature range of the furnace is 0~1300 ℃. Before the test, turn on the controller, set the heating program, and run the furnace after samples are placed in the hearth.
All chemicals are industrial purity powders and their granule sizes are less than 0.074 mm.

## Results and Discussions

### Chemical Stability of Aluminum Fluoride

Chemical stability of aluminum fluoride is discussed according to the described experimental method. When exposed to air and heated, aluminum fluoride volatilizes or reacts with moisture in air. Calcium fluoride appeared when aluminum fluoride was heated up together with calcium oxide or limestone which was weighed in stoichiometric amounts and homogeneously mixed into aluminum fluoride. Results are shown in Table I and Figure 1.

Table I Weight loss of aluminum fluoride after baking, %

| Sample number | Ambience | 750 ℃ | 800 ℃ | 850 ℃ | 900 ℃ | 950 ℃ |
|---|---|---|---|---|---|---|
| FA-C | air | 36.7 | 60.7 | 65.4 | 68.3 | 71.9 |
| FA-Y | Calcium oxide | 21.5 | 21.7 | 21.8 | 23.7 | 24.7 |
| FA-T | limestone | 29.1 | 29.6 | 29.7 | 30.9 | 31.1 |

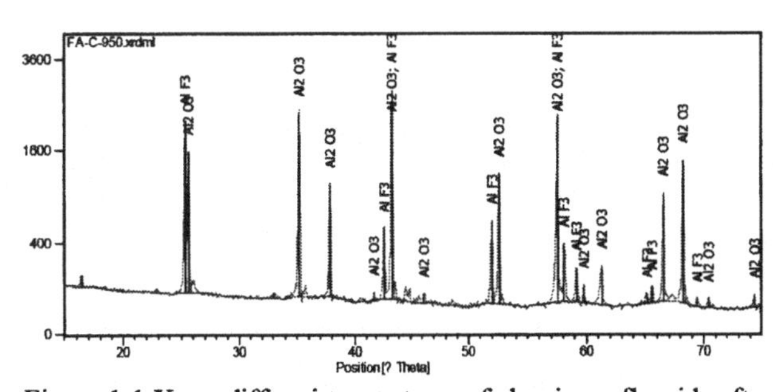


Figure 1-1 X-ray diffraction spectrum of aluminum fluoride after exposure to air at 950 ℃ for 120 minutes

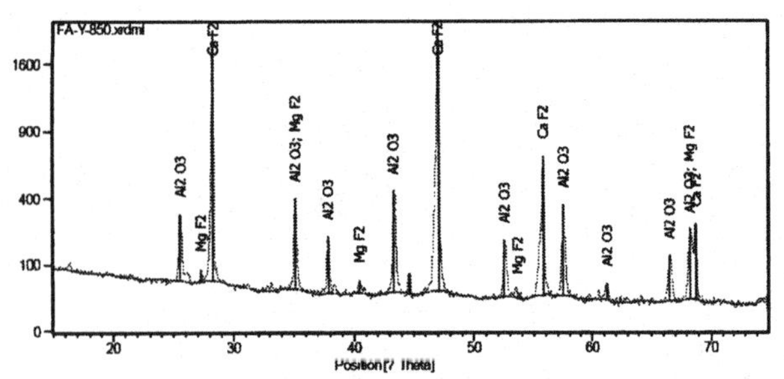

Figure 1-2 X-ray diffraction spectrum of aluminum fluoride after exposure to calcium oxide at 850 ℃ for 120 minutes

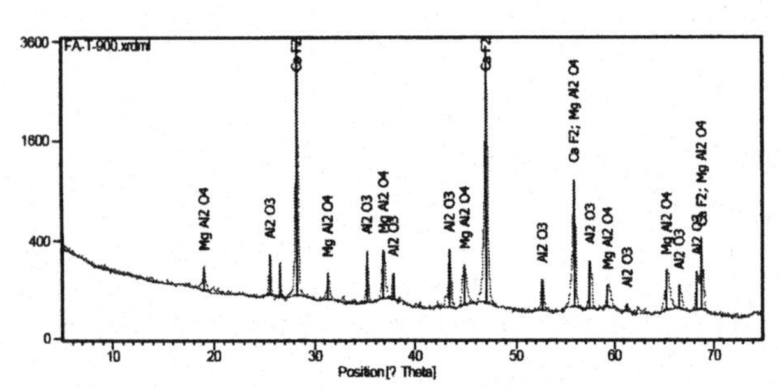

Figure 1-3 X-ray diffraction spectrum of aluminum fluoride after exposure to limestone at 900 ℃ for 120 minutes

It can be learned from Figure 1 and Table I that most of the aluminum fluoride changed into $Al_2O_3$ after exposure to air for 120 minutes at 750~950 ℃. Weight loss of aluminum fluoride increased with increasing heating temperature. Aluminum fluoride at 750 ℃ had less weight loss compared to the other. Calcium fluoride formed obviously when aluminum fluoride was exposed to calcium oxide at 750~950 ℃. In the case of limestone, the same phenomenon as for calcium oxide was observed. Moreover, from 750 ℃ to 950 ℃ FA-C, FA-Y and FA-T samples revealed similar behavior.

**Chemical Stability of Sodium Fluoride**

Chemical stability of sodium fluoride is investigated using the same experimental method as for aluminum fluoride. When exposed to air at 700~850 ℃, sodium fluoride was very stable, did not alter, and only partly volatilized. When above 850 ℃, sodium fluoride began to melt and penetrate slowly into the corundum crucibles. Sodium fluoride melted above 900 ℃. Calcium fluoride was formed when sodium fluoride was heated together with calcium oxide or limestone. Results are listed in Table II and Figure 2.

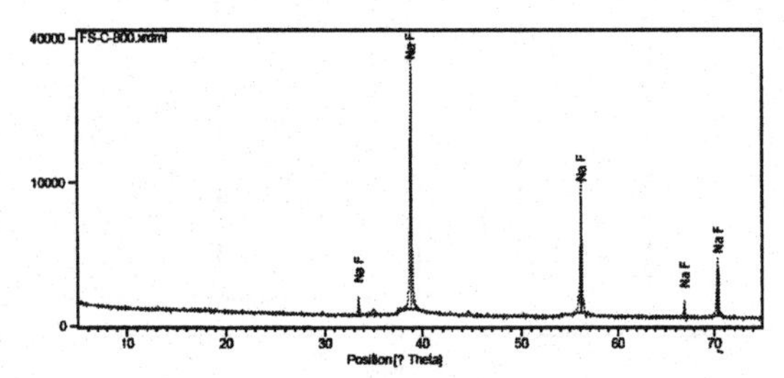

Figure 2-1 X-ray diffraction spectrum of sodium fluoride after exposure to air at 800 ℃ for 120 minutes

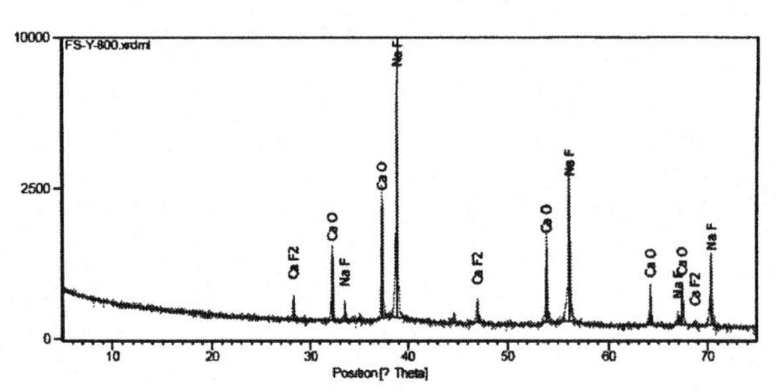

Figure 2-2 X-ray diffraction spectrum of sodium fluoride after exposure to active calcium oxide at 800 ℃ for 120 minutes

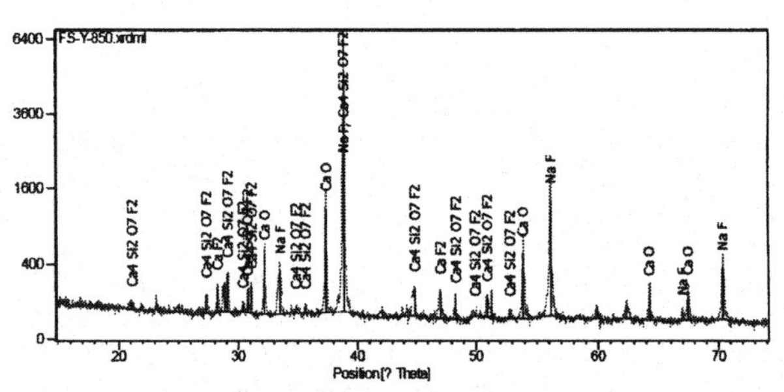

Figure 2-3 X-ray diffraction spectrum of sodium fluoride after exposure to active calcium oxide at 850 ℃ for 120 minutes

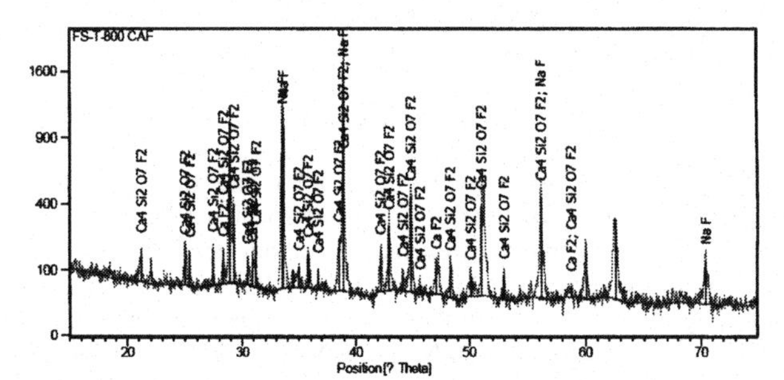

Figure 2-4 X-ray diffraction spectrum of sodium fluoride exposure to limestone at 800 ℃ for 120 minutes

Table II Weight loss of sodium fluoride after baking, %

| Sample number | Ambience | 700 ℃ | 750 ℃ | 800 ℃ | 850 ℃ | 900 ℃ |
|---|---|---|---|---|---|---|
| FS-C | air | 0.59 | 0.77 | 1.07 | 2.30 | / |
| FS-Y | Calcium oxide | 9.07 | 9.13 | 10.4 | 11.1 | 11.9 |
| FS-T | limestone | 25.6 | 26.2 | 26.4 | 26.6 | 27.2 |

It can be seen from Figure 2 and Table II that only small amounts of sodium fluoride volatilized after exposure to air for 120 minutes at 700~850 ℃. The reaction to form $CaF_2$ was slow when sodium fluoride was heated together with calcium oxide under 850 ℃. At 850 ℃ $Ca_4Si_2O_7F_2$ was formed. It was easier to form $CaF_2$ when limestone was used instead of calcium oxide. Sodium fluoride showed similar behavior whether it was exposed to calcium oxide or to limestone. Moreover, from 700 ℃ to 900 ℃ FS-C, FS-Y and FS-T samples showed similar behavior. The composition of baked sodium fluoride became increasingly complex with increasing heat treatment temperature.

**Chemical Stability of Mixed Fluoride #1**

The chemical stability of mixed fluoride #1 was studied. Mixed fluoride #1 was homogeneously mixed by sodium fluoride and aluminum fluoride according to the stoichiometry to form cryolite. Experimental conditions were the same as in the foregoing experiments. When the mixture was exposed to air at 700~900 ℃, $Na_3AlF_6$ and $Al_2O_3$ were formed. When the mixture was exposed to calcium oxide at 700~900 ℃, $CaF_2$ and $Al_2O_3$ were formed. When the mixture was exposed to limestone at 700~900 ℃, $CaF_2$ and $NaAlSiO_4$ could be observed. Results are described in Table III and Figure 3.

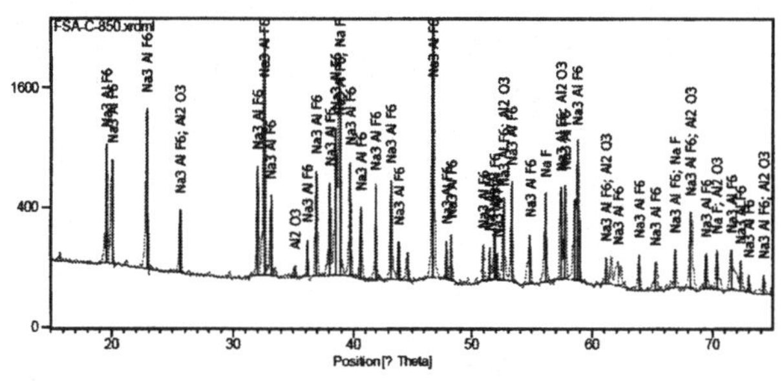

Figure 3-1 X-ray diffraction spectrum of mixed fluoride #1 after exposure to air at 850 ℃ for 120 minutes

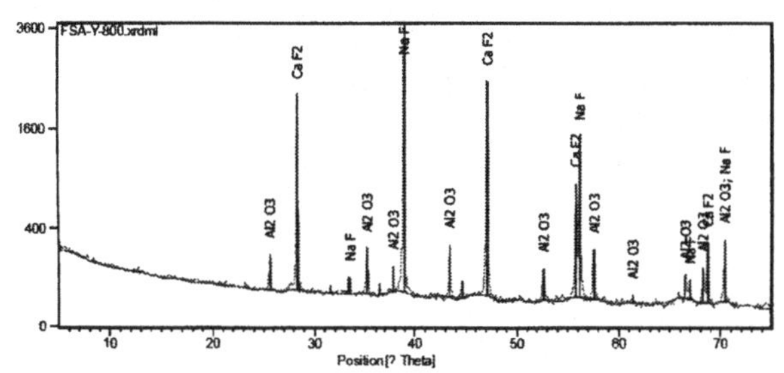

Figure 3-2 X-ray diffraction spectrum of mixed fluoride #1 after exposure to calcium oxide at 800 ℃ for 120 minutes

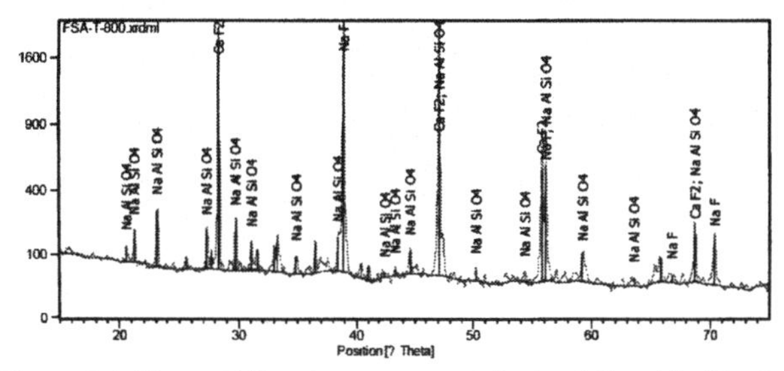

Figure 3-3 X-ray diffraction spectrum of mixed fluoride #1 after exposure to limestone at 800 ℃ for 120 minutes

Table III Weight loss of mixed fluoride #1 after baking, %

| Sample number | Ambience | 700 ℃ | 750 ℃ | 800 ℃ | 850 ℃ | 900 ℃ |
|---|---|---|---|---|---|---|
| FSA-C | air | 2.25 | 2.36 | 2.78 | 3.51 | 5.92 |
| FSA-Y | Calcium oxide | 7.30 | 7.64 | 8.92 | 9.63 | 10.8 |
| FSA-T | limestone | 21.4 | 21.7 | 21.7 | 21.9 | 22.3 |

It can be summarized from Figure 3 and Table III that the weight loss of the mixed fluoride #1 slowly increased with increasing heating temperature in exposure to air, calcium oxide or limestone at 700~900 ℃. Nepheline formed when mixed fluoride #1 was exposed to limestone. This was different compared to the case of exposure to calcium oxide. Furthermore, from 700 ℃ to 900 ℃ FSA-C, FSA-Y and FSA-T samples showed similar behavior and formed the same new compounds.

## Chemical Stability of Mixed Fluoride #2

Chemical stability of mixed fluoride #2 was discussed. The sample was made up of equal molecular sodium fluoride and cryolite. Experimental conditions were the same as for mixed fluoride #1. When mixed fluoride #2 was exposed to air at 700~900 ℃, $NaAl_{11}O_{17}$ was formed. When exposed to calcium oxide or limestone at 700~900 ℃, $CaF_2$ was formed. Results are shown in Table IV and Figure 4.

Table IV Weight loss of mixed fluoride #2 after baking, %

| Sample number | Ambience | 700 ℃ | 750 ℃ | 800 ℃ | 850 ℃ | 900 ℃ |
|---|---|---|---|---|---|---|
| FSC-C | air | 1.64 | 1.83 | 2.18 | 3.10 | 5.41 |
| FSC-Y | Calcium oxide | 10.6 | 10.6 | 11.1 | 11.6 | 11.7 |
| FSC-T | limestone | 18.7 | 20.1 | 21.6 | 22.7 | 23.0 |

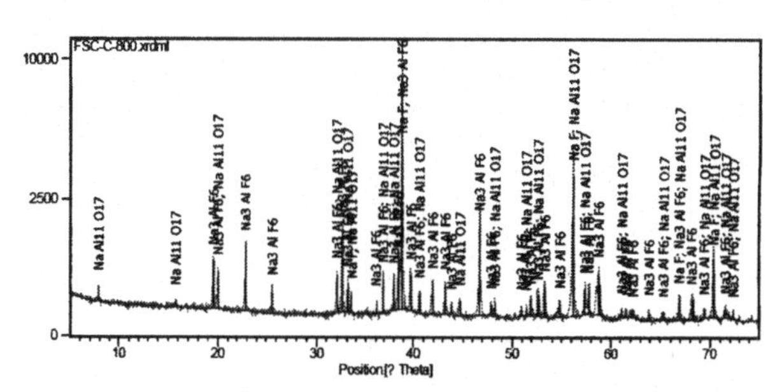

Figure 4-1 X-ray diffraction spectrum of mixed fluoride #2 after exposure to air at 800 ℃ for 120 minutes

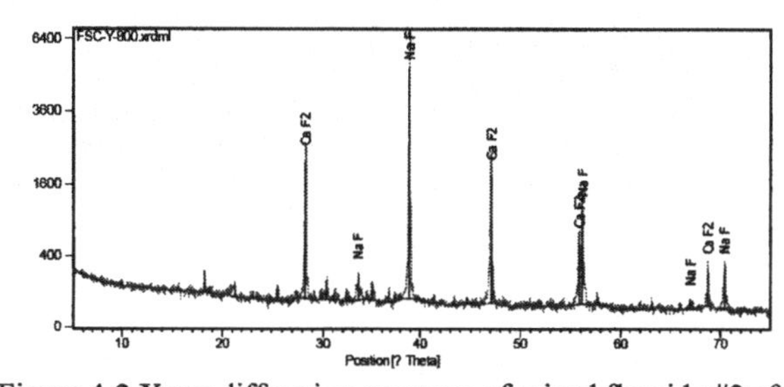

Figure 4-2 X-ray diffraction spectrum of mixed fluoride #2 after exposure to calcium oxide at 800 ℃ for 120 minutes

It can be learned from Figure 4 and Table IV that the weight loss of mixed fluoride #2 slowly increased with increasing heating temperature whether exposed to air, calcium oxide or limestone. Mixed fluoride #2 and #1 had similar chemical stability. Furthermore, from 700 ℃ to 900 ℃ FSC-C, FSC-Y and FSC-T samples revealed similar behavior and formed the same new compounds.

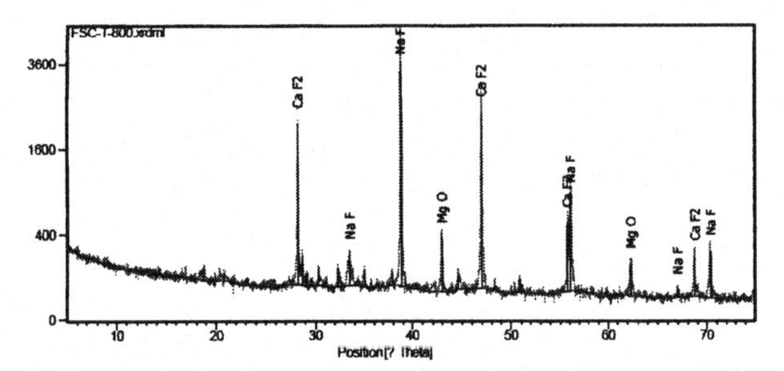

Figure 4-3 X-ray diffraction spectrum of mixed fluoride #2 after exposure to limestone at 800 ℃ for 120 minutes

## Conclusions

Aluminum fluoride has low chemical stability during baking in presence of oxygen and the calcium compounds. Almost all of the aluminum fluoride reacted to $Al_2O_3$ and $CaF_2$ when baking together with calcium oxide or limestone in air above 800°C. Only a small part of the sodium fluoride reacted to $Ca_4Si_2O_7F_2$ and $CaF_2$ when exposed to calcium oxide or limestone above 800°C. Mixed fluoride #1 has better chemical stability than aluminum fluoride when heat treated in the presence of oxygen and the calcium compounds. All aluminum fluoride in mixed fluoride #1 reacted to $Al_2O_3$ and $CaF_2$, but only small amounts of sodium fluoride reacted to $NaAlSiO_4$ and $CaF_2$ when exposed to calcium oxide or limestone. When mixed fluoride #1 is baked in air, $Na_3AlF_6$ and $Al_2O_3$ are observed. Mixed fluoride #1 has weaker chemical stability than sodium fluoride. Mixed fluoride #2 has weaker chemical stability than sodium fluoride during baking in presence of oxygen and the calcium compounds. All cryolite in mixed fluoride #2 formed $NaAl_{11}O_{17}$ and $CaF_2$; some sodium fluoride reacted to $CaF_2$ when exposed to calcium oxide or limestone. Results seem to hint that during processing of spent potlining, aluminum fluoride is the easiest to eliminate, but sodium fluoride is difficult to remove. The chemical stability of cryolite is between aluminum fluoride and sodium fluoride.

## Acknowledgement

All experiments were carried out at the National Aluminum Metallurgy Engineering and Technology Research Center. The work received great support from the Center. We give our sincere thanks to the Center and appreciate the help from our colleagues.

## References

1. Wangxing Li, Xiping Chen and Fengqin Liu, "Industrial Running Report of SPL Detoxifying Pilot Plant in CHALCO"(Report 2005-03, Zhengzhou Research Institute of CHALCO, 2005).
2. Wangxing Li, Xiping Chen, "Development of Detoxifying Process for Spent Potliner in CHALCO," *Light Metals 2005*, 515-517.
3. Wangxing Li, Xiping Chen, "Running Results of the SPL Detoxifying Pilot Plant in CHALCO," *Light Metals 2006*, 219~222.
4. Xiping Chen, Wangxing Li, "Studying on Properties of Spent Potliner", *Nonferrous Metals(extractive metallurgy) 2003(6)*, 35-37, China.
5. Morten, H. A. Øye. "Cathodes in Aluminum Electrolysis Cells", Norway, 1988, 151-201
6. David Belitskus, "Bench Scale Tests on Reuse of Spent Potlining in Cathode", *Light Metals 1991*, 299-303.
7. G.A. Wellwood, I.L. Kidd and R. Niven, "The COMTOR Process for Spent Potliner Detoxification", *Light Metals 1991*, 277-282.
8. I.L. Kidd, G.D. Dillett and D.P. Rodda, "Futher Development of the COMTOR Process for SPL Treatment", *Light Metals 1993*, 389-392.
9. Jean-Claude Bontron, Daniel Laronze and Pierre Personnet, "The SPLIT Process: Aluminum Pechiney Method for the Safe Disposal of Spent Potlining", *Light Metals 1993*, 393-397.

# ELECTRODE TECHNOLOGY

## Anode Raw Materials and Properties

*SESSION CHAIRS*
**Abdulmunim Binbrek**
Dubal Aluminum Co. Ltd.
Dubai, United Arab Emirates

**Charles Mark Read**
Bechtel Corporation
Montreal, Quebec, Canada

**Stephen L. Whelan**
Alcoa Inc.
Massena, New York, USA

**Light Metals 2008** *Edited by: David H. DeYoung*
***TMS (The Minerals, Metals & Materials Society), 2008***

# COKE BLENDING AND FINES CIRCUIT TARGETING AT THE ALCOA DESCHAMBAULT SMELTER

Michel Gendron[1], Stephen Whelan[2], Katie Cantin[1]
Alcoa, 1 Boul des Sources; Deschambault; Qc, Canada, G0A 1S0
Alcoa Massena, Park Avenue East; Massena, NY, US, 13662

Keywords: Coke, fines, sulfur, reactivity

**Abstract**

The continued increase of the demand for Aluminium metal combined with the fluctuations in the quality of aluminium grade coke makes it more challenging for the anode manufacturing plants to deliver steady quality anodes. The low sulfur coke material is becoming less available on the market and the price is steadily increasing. Environment regulations are aiming at reducing sulfur emissions, while the coke suppliers are offering higher sulfur material. The use of low sulfur material as feed stock increases anode $CO_2$ reactivity and therefore making the product less attractive for the downstream process. An attempt was made to use only the high sulfur material in the fines fraction, in order to optimize $CO_2$ reactivity while respecting sulfur emissions limit and coke supplier agreements. This paper presents the detailed results of this test, which later became a process flowsheet modification.

## Introduction

The idea of using a dedicated coke to manufacture the fine fraction of an anode has been discussed for a few years now. In this particular case, the Deschambault smelter used the technique to improve the anode's CO2 Reactivity. The timing was perfect since the potline operations were observing a lot of dusting in the pots.

The results presented include lab bench scale tests testwork and plant test results. This study required a lot of effort from plant personnel and the plant trial in the smelter got stretched over many months. In this study we used actual pots to benchmark the performance of the new formulated anodes against the normal ones. The benchmarking results will also be discussed.

Finally, this new approach has allowed the Deschambault smelter to target a dedicated coke to manufacture the fine coke fraction of the anodes. The final process diagram enables future coke trials at the plant.

## The problem – solution approach

The dusting cycle is characterized by anodes with a poor resistance to reactivity and hot pots operating at a lower current efficiency. Unstable pots can also lead to dusting in the pots and maintain the conditions that encourage the dust formation.

Anode plants need to produce anodes that perform better in the pots, thus reducing the tendency to generate dust in the pots. In order to do so, one approach is to dedicate the least reactive coke to the fines fraction.

Figure 1: Initial Process Diagram

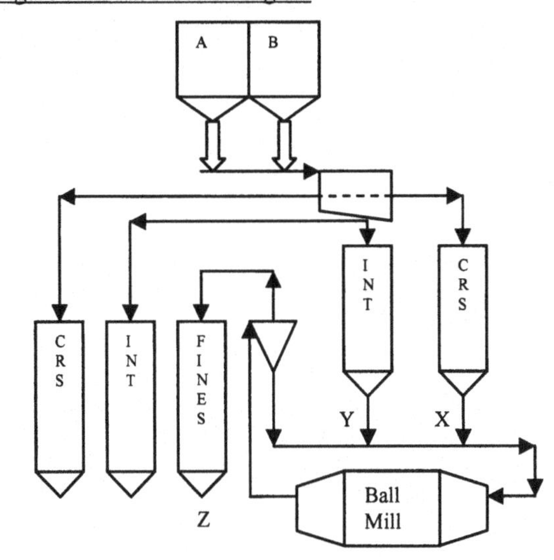


Figure 2: Modified Process Diagram

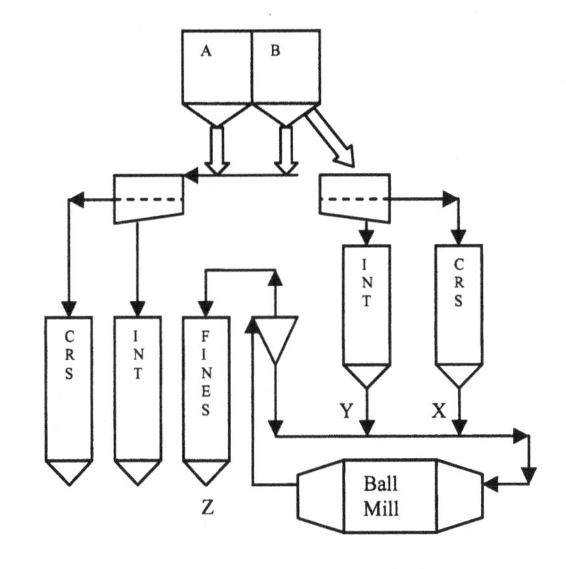


X : Coarse fraction
Y: Intermediate fraction
Z : Fines fraction

At the Deschambault smelter like many other recently built plants, the coke fractions used to make the anodes are characterized as coarse, intermediate and fines.

In the first step coke is received in the two main silos (figure 1 and 2: A,B ) for storage, these silos will keep enough inventory for over a month's consumption. In figure 1, it is indicated that the material is fed to the plant by a single conveyor from which the material is sieved and redirected to the appropriate size fraction. The two size fractions at this point are coarse and intermediate. A portion of each stream is directed to two other separate silos to feed the ball mill and manufacture the fines fraction to complete the anode recipe.

In figure 2, a dedicated stream is organized to feed the ball mill and the fines fractions stream. This improvement is the main topic of this paper.

Coke supply

Having two main silos allows the blending of two different coke sources to make an optimised mix. The purpose of the mix can lead to an improvement in many properties; in this particular case the objective was to significantly improve $CO_2$ reactivity.

The blending started with two different coke sources, one being higher in sulphur, while the second one is lower in sulphur (see Table 1). The higher sulphur coke has an obviously better $CO_2$ reactivity residue but produces high air reactivity. On the other hand the low sulphur coke produces poor $CO_2$ reactivity but excellent air reactivity. The mixture of the two coke sources make a good blend which presents properties that allow for a good anode performance in the pots while respecting environment regulations for $SO_2$ emissions.

Table 1: Coke specs

| | SULPHUR | CO2 REACTIVITY | AIR REACTIVITY |
|---|---|---|---|
| High sulphur coke | 3.3 % S | CRR + | ARR- |
| Low Sulphur Coke | 1.8% | CRR - | ARR+ |

Looking at it closer

When the anode is immerged in a bath of cryolite at 970°C the portion of the anode that is the most susceptible to $CO_2$ reactivity is the binder matrix which is composed of the fines and pitch. The most efficient way to improve anode performance in this case, is to eliminate the highly reactive coke from the binder matrix.

In the figure below (figure 3), the aim of the target composition is to fill the matrix with the fines fraction having the least reactive material. The higher sulphur coke being less reactive, this material is dedicated for that stream. The coarse and intermediate fractions are still being made with a mix of the lower sulphur coke and high sulphur coke.

Figure 3: Representation of an anode matrix

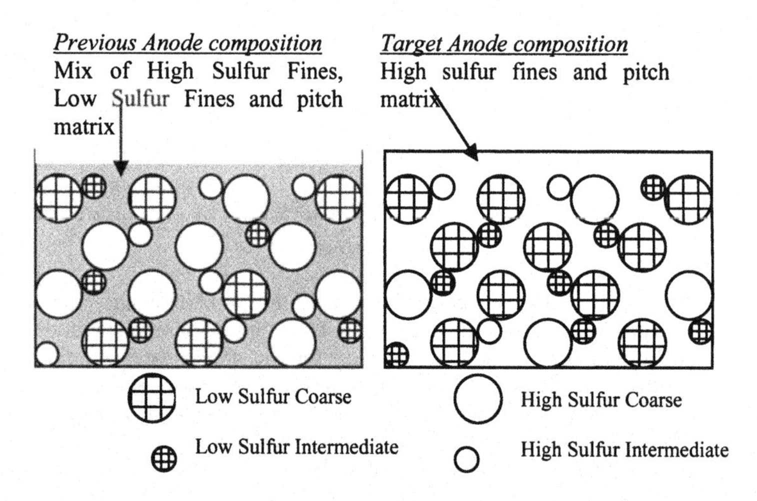


**Experimental Approach**

To test this concept a benchmarking exercise was organized in the potline. A test group (12%) was defined and its performance was measured on a weekly basis for 3 months. A batch of "special" anodes were manufactured every week to feed the test group and the "special" anodes were tagged to trace them along the process and to measure the efficiency of the placing system.

While laboratory analyses were done on the weekly samples from the "special" anodes production run, the usual quality control process was still taking place.

**Results and Discussion**

-Laboratory results:

There is a significant improvement with the special anodes formula. The most obvious result is about the $CO_2$ reactivity, as shown in figure 4 below, the net improvement is above 3% CRR. This clear improvement is confirmed with the dust portion from the $CO_2$ reactivity test which went from 2.9% to 0.8% as shown in Figure 5. Also, the standard deviation of the results is significantly improved.

Figure 4: Comparison of $CO_2$ Reactivity Residue

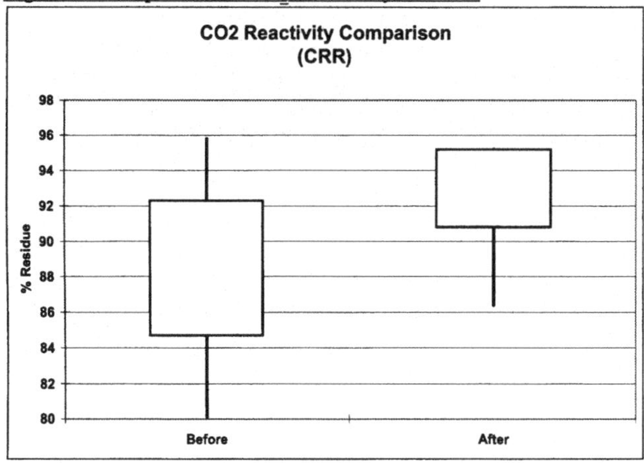


Figure 5 Comparison of $CO_2$ reactivity Dust

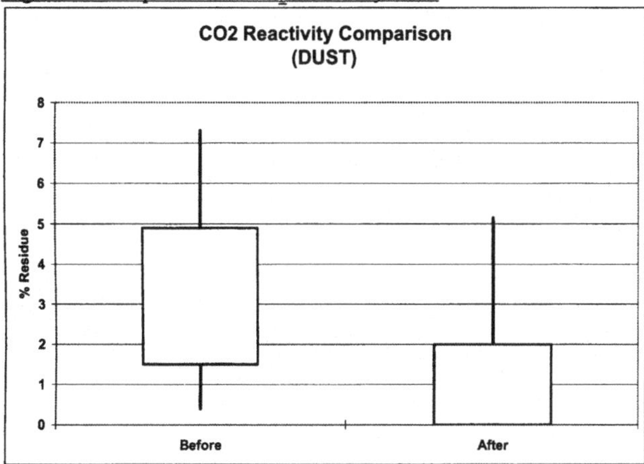


All other anode properties did not show significant changes and this can easily be explained since the total recipe remained in the same ratio for each coke. So the chemical results like: Ca, Fe, Na, S, V showed no significant change, while the pitch ratio was slightly reduced and baked density remained the same.

-Potline Benhmarking

The test group was chosen in the potline as being a section of pots having very intense dusting and unstable pots. A weekly audit was performed on all the pots in the potline, this way the test pots were monitored as closely as the benchmark group. The weekly audits involved observations at the tap hole and of the cover material. This operation is quite manpower intensive but allows a very good monitoring of the dusting evolution in the pots.

These observations were started early in 2006 to provide a good database and to have data to compare before and after the introduction of the "special" anodes.

The results of the audits are shown in figure 6. The investigation categorized dusting in three levels: code 0 being less than ½ inch of dust at the tap hole; code 1 is more than 1 inch of dust at the tap hole; code 2 being less than 1 inch and more than ½ inch of dust at the tap hole. Also, when observing the pot cover material, it is classified as being dusty or not. In figure 6, the pots with code 1 and 2 are reported as the dusty pots in percentage of pots in the potline.

Figure 6: Evolution of carbon dusting in the group

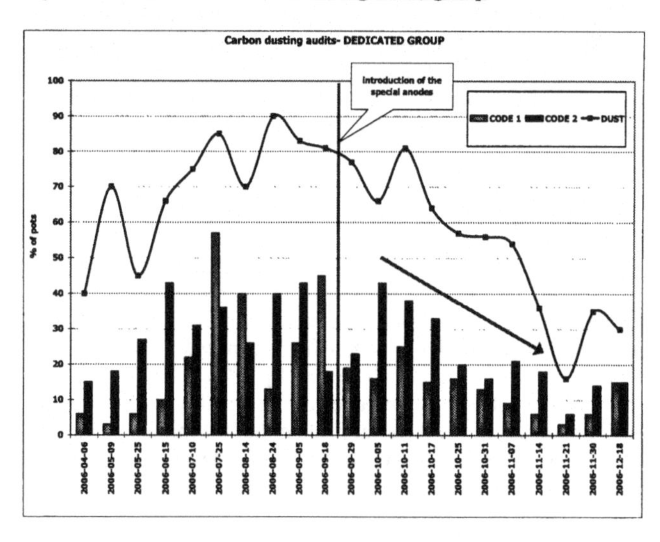


After the introduction of the "special" anodes in the test pots, one can observe the significant reduction in the frequency of dusty pots going from more as much as 90% to less than 30%. Also, the code 1 pots went down from about 30% on a steady slope to less then 5% at the best time and back to 15% at the end of the test. The trend for code 2 pots also evolved from an average 40% to less then 20% and went as low as 5% to finish the test at 15%.

Overall, the test group was significantly impacted by the introduction of the "special" anodes which confirms the new "special" anodes are of a less reactive formula.

Figure 7: Evolution of carbon dusting in the potline (excluding the test group)

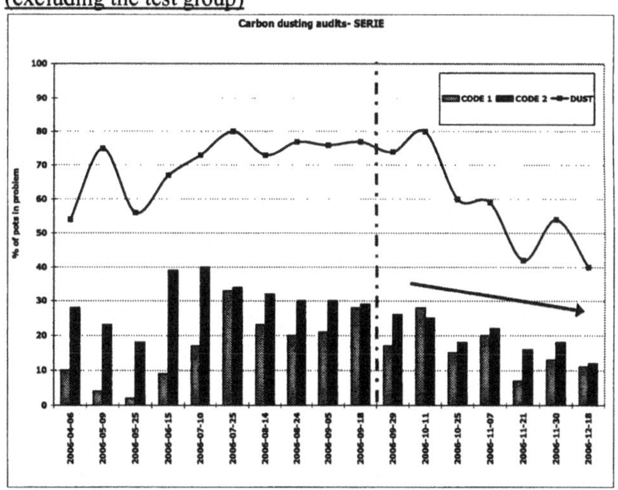


The results from the dust audits for the remainder of the potline (figure 7), show a more stable level of dusting in the pots and a reduction from approximately 75% to 45% at the end of the year. The code 1 pots evolved between 20% and 30% to complete the period of the test at 11%. The trend for code 2 continuously reduced on a very small slope from 30% to 20% and completed the test period at 12%.

These two graphs showed the significant impact the "special" anodes had on the test group compared to the rest of the potline. Obviously, the number of pots with dust presence was reduced significantly in the test group, even though the whole potline has experienced a reduction of the dusting issue, the pots in the test group went from being the worst pots to be the best pots.

-Process stability:
During the manufacturing of the fines fraction with a dedicated coke, it was observed that the overall sizing variations was significantly reduced. As shown in figure 8, the standard deviation of the sizing was compared and shows a very sudden reduction. The impact of this stability is confirmed with the slight reduction in pitch ratio which reduced by 0.3%.
One could explain this improvement in stability with the more stable hardness of the material feeding the ball mill since the two cokes used in the test showed a significant difference in hardness.

Figure 8: Sizing stability with dedicated coke for the fines

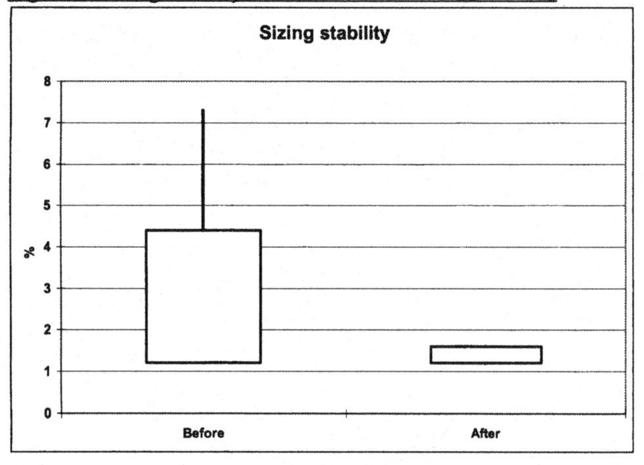


-Potential for a reduction in Sulphur mission :
In the present plant trial, the recipe used was stable and the ratio of the two coke sources used was unaffected as this was one premise of the test. With the present results, one could target to produce an anode of similar properties with a different ratio of coke. For instance, the high sulphur coke could be used to exclusively manufacture the fines fraction while the low sulphur coke to be used for the coarse and intermediate fractions. This would lead to a significant reduction of high sulphur coke consumption and the corresponding reduction in $SO_2$ emissions.

-Potential for other improvements
Coke targeting can lead to improvements in other fields than $CO_2$ reactivity. The same principles can be applied to improvements in density or air reactivity for instance. Once the systems are in place, it can be used to fulfill the need for improvement that coke targeting can deliver.
In some cases, the anode density might be the target issue. Using the lowest bulk density material to manufacture the fines to keep the highest bulk density for the coarse and intermediate fractions could lead to more dense anodes and thus reducing number of blocks produced to satisfy the needs of the customer.
Also, the same principle can be applied to manufacture anodes with better air reactivity. This would involve using the least reactive material in the fines fraction to avoid the presence of highly reactive material in the fines-pitch matrix.

## Coke Blending

Following this project of coke circuit targeting an additional step was reached in the coke blending path. Since the plant can receive coke in two main silos, and that one of the two silos is dedicated to the coarse and intermediate, this silo is now fed with a blend of two cokes.
In order to do so, an additional step of blending was realized at the port facility when loading the railcars as shown in figure 9.

Figure 9: Coke Blending at the port facility

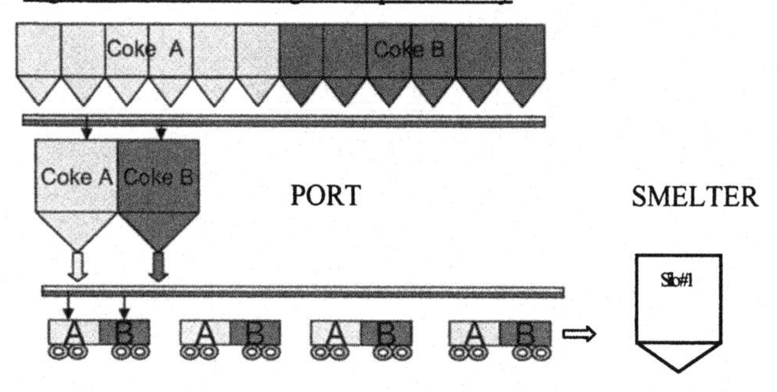


This coke blending diagram allows the introduction of 3 different coke sources in the plant at a controlled rate without increasing the process fluctuations while improving anode performance.

## Conclusion

In this paper we have discussed the coke targeting principle. It was demonstrated to be a proven application that the higher sulphur coke dedicated to a specific stream can positively affect anode properties. Many other opportunities in anode properties improvements are possible using the coke stream targeting approach.
Also, coke blending allows for a more flexible recipe once coke circuit targeting is implemented.


## Acknowledgements

The plant staff at Deschambault smelter were involved to make this project happen. The dedication of our manpower to continuously improve and remain a leader and benchmark in the industry is a key element to have this project realized today.
Thanks to all Alcoa technical staff who have directed us towards this great success, more specifically Steve Whelan for his sponsorship in this adventure.

**Light Metals 2008** *Edited by: David H. DeYoung*
***TMS (The Minerals, Metals & Materials Society), 2008***

# DESIGN CRITERIA FOR PETCOKE CALCINERS

Ravindra Narvekar[1], Arun K. Mathur[1], Jose Antonio Botelho[1]
[1]Goa Petcoke Consultancy Services, India

Keywords: Design, Process, Mechanical, Refractory

## Abstract

Goa Petcoke Consultancy Services provides comprehensive consultancy services and solutions in the field of petroleum coke calcination. During the course of working on various assignments, we probed into the fundamental principles of design of the rotary kiln systems for petcoke calcination, resulting into evolution of this paper.

The important process criteria in calciner design are retention time, heating up rate and operating drafts. These factors depend on kiln dimensions, slope, heat transfer rate and air injections. To understand this well, we have worked out mass and heat balance across kiln, cooler and incinerator.

The mechanical design should consider load factors, thermal expansions and strength of materials at elevated temperatures to ensure mechanical soundness and protection against flexing and distortion of kiln shell.

The refractory design should take into account differential thermal expansion coefficients, ability to withstand high temperatures, resistance to spalling, chemical attack , abrasion and good coatability.

## Introduction

A rotary kiln system for petcoke calcination consists of following major sections

- The **calcining** circuit consisting of the **Rotary Kiln, Rotary Cooler, Settling Chamber** and **Incinerator**

- The **material handling** circuit comprising of belt conveyors, vibratory feeders, trippers, mobile hoppers, crushers, magnetic separators, metal detectors, bag filters etc. for green coke unloading, storage, reclaiming, silo filling, blending and kiln feed as well as calcined coke storage and shipping

- **Utilities** section consisting of Electrical power High Tension (HT) / Low Tension (LT) , Power Control Centre, and Motor Control Centre; raw water, process water, cooling water; compressed air for bag filters, instruments etc.; fuel for burners; and mineral oil for dedusting calcined coke.

- **Environmental Protection** section consisting of Gas Conditioning Tower, Desulphuriser, Bag filters and Cyclones

- **Waste Heat Recovery** system consisting of Waste Heat Boiler, desalination plant (optional); and / or electricity generation plant.

Fig 1 below represents a typical petroleum coke rotary kiln calcination system.

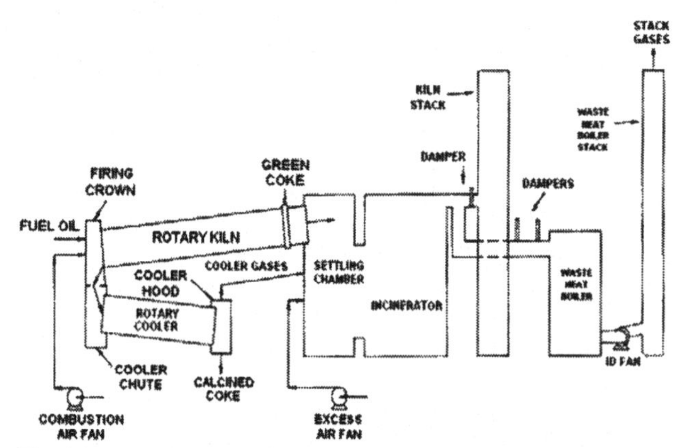


**Fig 1. Typical Rotary Kiln Calciner equipment flow diagram.**

This paper shall concern itself with the optimum design of the **Calcining Circuit** components.

**Rotary Kiln** - The rotary kiln is the heart of the calcining circuit as well as of the entire plant. Hence we have analyzed it in more detail. The following design aspects were studied

(1) Process (2) Mechanical (3) Refractory.

**Process** -The key parameters from the process point of view are :
(a) Retention time ( b ) Heating-up rate ( c ) Air & gas flow control

(a) **Retention time** is a function of kiln dimensions, slope, revolutions per minute (rpm) and the coke granulometry. It can be calculated by ***Boyuton's empirical formula*** $T = 1.77 \times \sqrt{\theta} \times L \times K / S \times D \times N$ , where T is travel time in minutes, L is length in feet , θ is the angle of repose in degrees for the coke , K is a factor that accounts for internal obstructions and changes in kiln diameter which can influence material movement (if no obstacles, then K = 1), S is the kiln slope in degrees, N is revolutions per minute, D is internal kiln diameter in feet. We have calculated retention times for various plant capacities to be 88 mins at 1.2 rpm.

Axial coke velocity and coke bed depth are inversely proportional to the retention time. Retention time is inversely proportional to the slope and rpm. A steeper slope would result into increased output for a given rpm or it would be possible to operate at lower rpm for a specific output. Lower rpm is preferred because of lesser wear and tear, with consequent lower maintenance costs. A steeper slope would involve higher construction costs.

The length / diameter (L / D) ratio is also critical. A higher diameter would cost more and lead to higher heat losses and a lower coke bed depth. A longer kiln will have better heat transfer but a higher construction cost, and higher entrainment of coke dust due to the higher flue gas velocity. The coke bed depth would depend on the L / D ratio, the kiln operating rpm, the kiln slope, coke feed rate etc.. The coke bed depth is crucial for heat transfer. With higher bed depths, a better heat transfer rate can still be achieved using lifters for turning the coke bed.

Based on above conclusions, optimum parameters should be -

(i) Length to internal diameter ratio: 20
(ii) Slope : 4.17 % or 2.388 degrees
(iii) Retention time : 88 minutes at 1.2 rpm and
(iv) Volume occupancy : 15 %.

Table I below indicates the application of these design parameters for different capacity rotary kilns, with diameters considered at 9 feet, 10 feet, 11 feet and 12 feet for the scope of this paper.

**Table I Major Design Parameters for Rotary Kiln**

| Sr.No. | Parameter | **90,000 tpa** | **132,000 tpa** | **180,000 tpa** | **240,000 tpa** |
|---|---|---|---|---|---|
| 1 | Kiln dia x length m | 2.74 x 46 | 3.04 x 52 | 3.35 x 58 | 3.65 x 64 |
| 2 | Shell thickness, mm | 18 / 32 / 45 | 20 / 32 / 50 | 22 / 32 / 60 | 25 / 32 / 65 |
| 3 | Total wt of rotary parts, MT | 405 | 546 | 625 | 776 |
| 4 | Power required BHP / KW | 95 / 71 | 170 / 128 | 207 / 155 | 314 / 236 |
| 5 | Green coke feed rate, TPH | 16 | 23.34 | 32 | 44 |
| 6 | Calcined coke discharge, TPH | 11.43 | 16.67 | 22.85 | 31.43 |
| 7 | Kiln volume occupancy % | 15.28 | 15.41 | 15.07 | 14.76 |
| 8 | Coke bed height, mm | 475 | 540 | 595 | 650 |

**Calculation of power requirement:**

BHP = N (4.75 d w + 0.1925 D W + 0.33 W ) / 100,000
N is the kiln rpm.
D is the diameter in feet of the riding ring while d is the diameter in feet of the kiln .Assume D = d + 2.
W is the total rotating load i.e. the dead load of kiln, refractory and kiln mountings plus the live load w, which stands for the weight of the material inside the kiln. The load is expressed in lbs. Once BHP is calculated, KW can be worked out as KW = 0.75 BHP.

**Calculation of kiln volume occupancy:**

The percentage of volume occupancy = v / V x 100 , where v is the volume occupied by the coke bed inside the kiln at any moment and V is the internal volume of the kiln.
The v can be calculated from the quantity of coke inside the kiln based on the coke feed and discharge rates, retention time and bulk density of the coke.
The V can be calculated by mathematical formula for the volume of a cylinder = $\prod/4 \, x D^2 x H$, where $\prod$ = 3.142, D is the internal diameter of the kiln, and H is the length

**Calculation of coke bed height:**

This can be calculated by applying a simple trigonometry function. In a circle, let r be the radius, H be the bed depth , then
Area of that segment = $r^2 \, Cos^{-1}(r–H/r)–(r–H) \, \sqrt{(2 \, x \, r \, x \, H–H^2)}$
Area of kiln face = $\prod x r^2$.

Assume a certain value for H and find out corresponding area of the segment. Then calculate the area of kiln surface and the percentage of the area of the segment compared to kiln surface area. This percentage should be equal to the volume occupancy percentage. For example for a 90,000 tpa capacity plant, H = 475 mm satisfies the segment area as 15.28 % of the kiln surface area, which is the volume occupancy for 9 feet diameter kiln.

(b) **Heating up rate** is the heat profile along the length of the kiln, a very important aspect of the process. Ideally, the kiln heating profile should match the VM evolution curve of the coke being calcined as shown in Figure 2. The VM evolution curve is the graph of the percentage of VM evolved v/s temperature

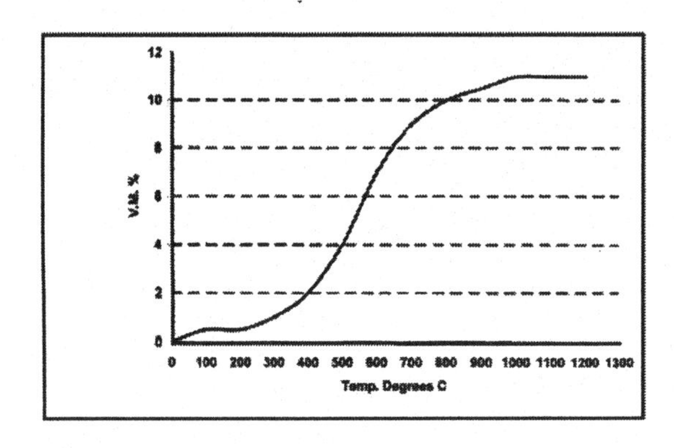


Figure 2 . VM evolution curve

The density or the porosity of the calcined coke depends to a large extent on the heating up rate. In the past, when external fuel was used as a source of energy, it was easier to control the heating up rate, but now with tertiary air injection, and the consequent released volatile combustible matter providing the major part of input energy, it is difficult and tricky to achieve the desired heating up rate. Tertiary air injection is introduced with the use of two blowers, mounted in opposing directions, with each blower injecting air through several nozzles, each angled in different directions – that is how heating up rate is sought to be controlled.

Heat transfer rate plays a crucial role in the design. Heat transfer is both by convection and radiation – by convection from the flowing gas heating the kiln brick and exposed bed surface, and by radiation from the red hot brick surface to the coke bed.

Another important aspect is thermal efficiency. In turn the kiln length is a major factor in determining thermal efficiency and kilns with a high L / D ratio exhibit greater thermal efficiency than those with a lower ratio. Thermal efficiency in the range of 65 – 75 % is considered adequate.

$$Us = 23.7x\ G^{0.67} \qquad \text{(Equation 1)}$$

Equation 1 is the empirical formula for determining heat transfer coefficient in which Us = is the heat transfer coefficient in J / ($m^2$.s.°k), where $m^2$ is the kiln surface area in $m^2$, s is time in seconds and temperature in degrees K; G is the mass flow-rate in kg/(s.$m^2$).

Therefore Q = Us.A.T, where Q is the quantum of heat to be transferred, A is the heat transfer area provided and T is the temperature difference.

(c) **Air and gas flow controls** inside the kiln are extremely important and have to be maintained so that the pressure inside the kiln is slightly negative. A marginally higher negative draft will result in the sucking of cooler steam inside the kiln, while a positive draft will pressurize the kiln resulting in puffing of fines, escape of volatile through air seals and back pressure at observation port, endangering the safety of the operation. This can be controlled by operating the damper located between the settling chamber and the incinerator, and controlling the air volume in the form of primary air, secondary air and tertiary air. Simple material and energy balance calculations across the kiln helps in deciding the fan design and operating capacities.

Table II below gives the values of mass and heat balance parameters calculated for four different capacity plants

**Table II Mass and Heat Balance Parameters for Kiln**

| Sr. No | Parameter | 90K tpa | 132K tpa | 180K tpa | 240K tpa |
|---|---|---|---|---|---|
| 1 | Primary air fan capacity,Nm³/hr | 900 | 1398 | 1804 | 1949 |
| 2 | Secondary air fan capacity, Nm³ /hr | 5000 | 8414 | 10,101 | 14,201 |
| 3 | Tertiary air fan capacity, Nm³ /hr | 2 x 12,000 | 2 x 19,939 | 2 x 23,926 | 2 x 33,646 |
| 4 | Kiln exit gases, Nm³ / hr | 26,956 | 44,152 | 53,920 | 74,723 |
| 5 | Heat transfer coeff. Kcal/hr.m².°k | 33.39 | 39.43 | 38.65 | 43.12 |
| 6 | Heat transfer area required, m² | 160 | 196 | 277 | 345 |
| 7 | Combustion Heat generation, Kcal/hr | 13.48 E+6 | 19.95 E+6 | 27.02 E+6 | 35.40 E+6 |

**Mass Balance across Rotary Kiln**

Inputs - Green coke, fuel oil / gas, primary air, secondary air, tertiary air.
Outputs - Calcined coke, carbon fines (both recovered and lost), flue gases.
Equating the inputs to outputs, we get a rough figure of flue gases generated.

**Chemical Balance across Rotary Kiln**

The composition of flue gases exiting the kiln is worked out as below:
(1) Assume a typical composition of volatile matter at - C-80 %, $H_2$-15%, $O_2$- 0.50%, $N_2$-0.25%, S-4 %
(2) The composition of fuel oil or gas with its carbon-hydrogen percentage is also known.
(3) It is assumed that 50 % VM burns inside the kiln and the balance 50 % in the incinerator.
(4) The coke burnt inside the kiln is calculated from the coke balance.

Then we have basic chemical equations (2) to (5) for combustion, such as

$$C + O_2 \text{ ------ } CO_2 \qquad \text{(Equation 2)}$$
$$2\,H_2 + O_2 \text{ ---} 2H_2O \qquad \text{(Equation 3)}$$
$$S + O_2 \text{ ------- } SO_2 \qquad \text{(Equation 4)}$$
$$N_2 + O_2 \text{ ----- } 2NO \qquad \text{(Equation 5)}$$

The moisture present in the green coke feed will be converted into steam content in the flue gases.
The quantity of input air is also known, and therefore $O_2$ and $N_2$ fed, consumed and the balance left after the reactions is known. This gives the composition of flue gases leaving the kiln.

**Heat Balance across Rotary Kiln**

(1) Heat inputs are - Enthalpy of green coke, enthalpy of fuel, enthalpy of primary air, enthalpy of secondary air and enthalpy of tertiary air.
(2) Heat outputs are- Enthalpy of red hot coke, enthalpy of coke fines, and enthalpy of flue gases.
(3) Heat generation - (a) Heat generated by combustion of fuel (b) Heat generated by combustion of VCM (c) Heat generated by combustion of coke.
(4) Heat loss – By applying Stefan Boltzman's equation, Q = SB x T E+4 x A, we calculate the radiation heat loss from the hot kiln shell.

Then the heat balance formula as below is applied -
Heat inputs + Heat generated = Heat outputs + Heat losses.

Typically, for a 180,000 tpa CPC plant, the calculation would be -

(0.88 x E + 6) + (27.02 x E + 6) = (27.14 x E+ 6) + ( 0.95 x E+ 6) Kcals/hr.

There can be a minor difference in balancing the two sides of the equation, because not all carbon is converted into $CO_2$ and instead some CO is generated, which is oxidized later in the incinerator.

(2) **Mechanical** - The dimensions and specifications of the main components of the kiln and cooler viz. (i) Shells; (ii) Supporting rollers; (iii) Riding rings; (iv) Thrust roller; (v) Girth gears; (vi) Pinion and drives – are all based on the load data, thermal expansion and mechanical strength.

Different **types of loads** have to be considered -

(1) Dead load of kiln / cooler along with refractory and all its mountings.
(2) Live load of coke bed and combustion gases.
(3) Unbalanced loads resulting from errors in balancing mounting accessories on the kiln such as tertiary air fans, air seals , slip rings , girth gears etc.
(4) Axial load resulting from friction between the supporting rollers and riding rings.
(5) Load resulting from the thrust exerted by riding rings on the thrust rollers.
(6) Abnormal loads resulting from kiln / cooler eccentricity.
(7) Eccentric loads resulting from malfunctions or from settling of foundations and seismic loads.

(i) **Shell** - The roundness of the shell is extremely critical and any ovality will result in refractory failures as well as unbalanced load on the drive.

(ii) **Supporting Rollers** - After determining the number of supporting rollers required based on the load data, the centre to centre distance between two adjacent rollers has to be determined to ensure uniform distribution of the load.

(iii) **Riding Rings** - The riding ring matched with the thrust roller called the "thrust" riding ring is different from the others plain riding rings. There are rings fitted on either side of the riding ring, called "wear" rings and they protect the retainer blocks. The diameter, face width and thickness with respect to the rollers, needs to be accurately designed to ensure easy floating of the kiln over the rollers when under thermal expansion. Due to differential thermal expansion resulting from temperature gradient between the shell and the riding ring, the clearance between the two is also critical . The normal clearance (in cold conditions) at discharge end riding ring should be 8 – 9 mm, so that under hot condition it would be approximately 3 mm , resulting in a 10 mm creep, assuming a temperature gradient is 100 to 120° C.

(iv) **Girth Gear and Pinion** - The dimensions of girth gear and pinion viz. its module, number of teeth, face width etc. are designed on the basis of the kiln dimensions, revolutions per minute (rpm) and the load data.

Table III below gives the details of mechanical properties of materials used in the construction of main components.

**Table III. Mechanical Design Data**

| Sr.No. | Machine component → Parameter ↓ | Shells | Supp.Rollers | Riding Rings | Girth Gear | Pinion shaft |
|---|---|---|---|---|---|---|
| 1 | Material of construction | IS 2002/62 Grade 2 B | ASTM A 148 Grade 115-95 | ASTM A 148 Grade 90-60 | IS 1570-61 Grade 35 Ni | EN 24 |
| | Carbon % | 0.22 | As applicable | As applicable | 0.3 – 0.4 | 0.35 - 0.45 |
| | Manganese % | Not specified | As applicable | As applicable | 0.6 – 0.9 | 0.45 - 0.70 |
| | Others % | Si : 1.35& S : 0.05 | S : 0.06 | S : 0.06 | Ni : 1.0 – 1.5 | Ni :1.3 - 1.8 |
| | Others % | P : 0.05 | P : 0.05 | P : 0.05 | Cr : 0.45 – 0.75 | Cr : 0.9 – 1.4 |
| 2 | Hardness BHN no. | 159 - 190 | 240 min | 180 - 220 | 200 min | 300 min |
| 3 | Yield strength, kg/mm2 | 50 | 95 | 60 | 55 | 53 |
| 4 | Tensile strength,kg/mm2 | 52 - 62 | 115 | 90 | 80 | 70 |
| 5 | Elongation % | 20 | 14 | 20 | 16 | 16.8 |

(3) **Refractory** – Although other types are available and also used, high alumina bricks are preferred for refractory lining of rotary kilns because of its following properties -
(1) High fusion point – 3200 to 3400 °C (able to withstand high temperatures).
(2) Acceptable deformation under hot loading (resistant to spalling).
(3) Low permeability (satisfactory coatability).
(4) Satisfactory hot strength.
(5) Good thermal shock resistance(resistant to spalling)
(6) Slight acidic nature (resistant to chemical attack)

Table IV below enumerates the typical specifications of alumina refractory bricks used in India

**Table IV Typical Specifications of Fire Clay and High Alumina Bricks**

| Sr.No. | Brand Name→ Parameter ↓ | **XXXX HD** | **XXXX 42D** | **XXXX 45 S** | **XXXX 62 S** | **XXXX 70 L** | **XXXX 80 S** |
|---|---|---|---|---|---|---|---|
| 1 | Alumina % min | 40 | 42 | 45 | 62 | 70 | 80 |
| 2 | Fe2O3 % max | 2.5 | 1.5 | 1.5 | 1.3 | 1.5 | 1.4 |
| 3 | B.D. gm/cc min | 2.1 | 2.25 | 2.3 | 2.35 | 2.5 | 2.6 |
| 4 | CCS, kg/cm² min | 250 | 500 | 350 | 600 | 420 | 750 |
| 5 | A.P. % max | 22 | 16 | 16 | 22 | 23 | 18 |
| 6 | P.C.E. min | 32 | 33 | 34 | 37 | 36 | 37 |
| 7 | RUL, deg C min | 1450 | 1500 | 1480 | 1500 | 1550 | 1540 |
| 8 | Applications | Hot Stack | Kiln Fire Hood | Settling Chamber | Kiln feed Zone | Kiln Calci-nation zone | Kiln Lifters |

**Rotary Cooler**

The design criteria for a Rotary Cooler are more or less similar to those for the kiln. However, the L /D ratio is normally taken as 10. The cooler revolutions per minute (rpm) is almost double that of the kiln, and the residence time can be calculated again using Boyuton's formula . A retention time of about 25 minutes is suitable for the cooler. The fraction of cooler volume occupied by the coke is usually 10 %.
The cooler exhaust fan is designed to handle the volume of steam generated in the quenching along with air. The steam generated can be worked out on the basis of 750 liters process water consumed per ton of coke quenched. The cooler exhaust fan has to draw this steam through a cyclone for trapping the coke fines and then to inject the same into the incinerator for either recovery or burning of carbon fines. The pressure drop across the steam duct is high and has to be considered while designing the fan capacity.

**Settling Chamber**

The settling chamber is a transition step between the kiln and the incinerator. It helps to accommodate the green coke feed pipe assembly and a damper at its outlet duct for controlling the exit of flue gases from the kiln.
The shape of the settling chamber is designed in the form of a hood at the top and a conical bottom so that the velocity of flue gases is reduced prior to their entering the incinerator, resulting in the dropping of entrained coke fines. The volume of settling chamber is so designed that based on the volumetric flow-rate of the kiln exit gases, a retention time of 3-4 seconds is available.

**Incinerator**

The main role of the incinerator is to -

- Burn the balance (approximately 50 %) of the volatile combustible matter;
- Burn the coke fines in the flue gases prior to discharge into the atmosphere;
- Complete the combustion and conversion of carbon monoxide into carbon dioxide.

The retention time is the crucial design criterion for the incinerator. From our experience, a retention time of about 10 seconds is necessary. Based on the volume of flue gases entering the incinerator from the kiln, the volume of the incinerator for a 10 seconds residence time can be calculated. Assuming a suitable L / D ratio, we can determine the incinerator diameter and length.

**Mass Balance across Incinerator**

Inputs - Exit flue gases from kiln, fuel burnt, burner air, Induced draft fan air, Cooler exhaust steam.
Outputs - Exit flue gases from the incinerator, dust recovered in the hoppers.

Equating the inputs and outputs, we get very rough figure of flue gases generated.

**Chemical Balance across Incinerator**

The composition of flue gases exiting incinerator is -

The 50 % unburnt VCM is completely burnt inside the incinerator.
The entrained coke particles are partially recovered and partially burnt.
The CO generated by partial combustion inside kiln is completely burnt to CO2.

We also know that the cooler exhaust gases take along with them steam generated from the quenching operation inside the cooler. The quantum of steam is known from the fact that 750 lts of water is consumed per ton of calcined coke during quenching.

Thus the composition of flue gases exiting the incinerator can be arrived at using the basic combustion equations and volumes of air used in various fans.

**Heat Balance across Incinerator**

Heat inputs are - Enthalpy of kiln exit gases, enthalpy of fuel , enthalpy of burner air , enthalpy of ID fan air , enthalpy of cooler exhaust gases.

Heat outputs are - Enthalpy of flue gases, enthalpy of coke fines recovered and lost.
Heat generated - (a) Heat generated by burning fuel; (b) Heat generated by burning balance unburnt VCM; (c) Heat generated by burning coke fines; and (d) Heat generated by complete conversion of carbon monoxide to dioxide.
Heat losses - Applying the Stefan Boltzman equation the heat loss by radiation from hot incinerator shell can be calculated.

Heat inputs + Heat generated = Heat outputs + Heat losses.
The typical values for a 240,000 tpa CPC plant are -

(28.73 x E+ 6) + (33.73 x E+ 6) = (58.95 x E+ 6) + (2.72 x E+ 6) Kcals / hr.

The two sides of the equation are found to balance out fairly well.

The mass and heat balance parameters for incinerators in different capacity plants are given in Table V

**Table V Mass and Heat Balance Parameters for Incinerator**

| Sr.No. | Parameter | **90,000 tpa** | **132,000 tpa** | **180,000 tpa** | **240,000 tpa** |
|---|---|---|---|---|---|
| 1 | Incinerator dia x length, m | 5.2 x 56 | 5.8 x 64 | 6.6 x 64 | 6.6 x 84 |
| 2 | Retention time , secs | 10.36 | 9.32 | 10.5 | 10 |
| 3 | Burner fan capacity , $Nm^3$ / hr | 8011 | 13,351 | 16,022 | 22,531 |
| 4 | Cooler exhaust fan capacity, $Nm^3$/hr | 18,107 | 30,179 | 36,215 | 50,927 |
| 5 | ID fan capacity , $Nm^3$ / hr | 35,889 | 55,931 | 62,456 | 82,769 |
| 6 | Exit flue gases , $Nm^3$ / hr | 73,649 | 118,804 | 139,951 | 191,510 |
| 7 | Combustion heat , kcal / hr | 12.85 E+6 | 18.75 E+6 | 25.76 E+6 | 33.73 E+6 |

**Summary**

The designing of a petcoke calciner is a blend of art, science and experience. The process key parameters viz retention time, heat-up rate and air & gas flow controls need to be addressesd correctly. Mechanical aspects such as load factors, thermal expansion and strength of material need to be accounted properly.Refractory design should consider differential thermal expansion coefficients,high temperature stability,resistance to spalling and chemical attack.

**Acknowledgements**


1. Perry's Chemical Engineering Handbook – Robert H. Perry and Don W. Green
2. Chemistry & Technology of Lime and Limestone - Boyuton
3. Steel Castings Handbook –Steel founders society of America revised by Davis Poweleit 1999.
4. Anodes for Aluminium Industry – R and D Carbon
5 Encyclopaedia of Chemical Technology ( 3rd ed. Wiley,New York , 1978 – 1980 ).

**Light Metals 2008** *Edited by: David H. DeYoung*
***TMS (The Minerals, Metals & Materials Society), 2008***

# UNDERSTANDING THE CALCINED COKE VBD- POROSITY PARADOX

Bernie Vitchus[1], Frank Cannova[1], Randall Bowers[2], Shridas Ningileri[2]

[1] BP Coke Marketing, 1990 West Crescent, Anaheim, CA USA 92801
[2] Secat Inc.,

Key Words: Calcined Coke Porosity, Vibrated Bulk Density, Packing Density

## Abstract

Carbon plants use either calcined coke bulk density or porosity analysis to predict pitch content and anode density. However, there is a calcined coke VBD-porosity paradox. That is, high bulk density calcined coke often has higher porosity while lower porosity coke tends to make more dense anodes. Understanding why this paradox exists should help lead to consistent high density anodes.

Mercury porosimetry and novel techniques were used to characterize the calcined coke's shape, internal porosity and packing characteristics. Calcined cokes sieved to a mesh size of -30+50usm from different coke sources were analyzed in the natural, jaw crushed, and roll crushed states. For particle shape analysis, calcined coke particles were suspended in a moving fluid and analyzed with a high speed camera to determine aspect ratios. Internal porosities were investigated by cross-sectioning particles, using optical microscopy techniques, and image analysis to determine area fraction per particle and pore morphology.

## Introduction

Obtaining consistently high density anodes is a goal of carbon plants. Most carbon plants use either calcined coke bulk density or porosity analysis to predict anode pitch content and/or anode density. In addition, these techniques are used to follow the affect calcined coke quality has on anode density. Internally BP Coke has successfully used a similar model to define how to optimize anode density via blending two calcined cokes of the same or different porosity. However, there is a calcined coke VBD-porosity paradox. That is, high bulk density calcined coke does not mean it has the lowest porosity while low porosity calcined coke tend to make more dense anodes. If the bulk density was primarily due to internal porosity, this paradox could not exist. Consequently, a coke particle's internal porosity is not the major factor affecting a coke's bulk density or anode density. However, since lower porosity cokes tend to make higher density anodes, some relationship exists. Studies were made to better understand what other factors affect a coke's bulk density and anode density. Understanding why this paradox exists should lead to a better understanding for achieving consistent high density anodes.

Studies were performed using standard mercury porosimetry as well as novel techniques to characterize the calcined coke's shape, internal porosity and packing characteristics. The initial studies consisted of determining the porosity and VBD of different particle sizes of single source calcined cokes. In addition, the larger sized particles were crushed to -30+50 usm and analyzed for porosity and VBD. The findings from this study lead to additional single source calcined coke samples being analyzed to expand the data base.

Literature data [1,2] has been presented that suggests method of crushing and crushing techniques also affect VBD. Consequently the expanded studies included crushing a high and low porosity coke via different techniques. These samples and a butt sample were sieved to obtain a natural occurring sample of -30+50 usm. In addition, larger particles from these two single source coke were crushed via roll crushing and/or jaw crushing followed by sieving to the same -30+50 usm particles. Besides analyzing these particles for porosity, RD and VBD these particles were analyzed for particle shape and between particle porosity. These results are reported in a companion paper [3].

After studying calcined coke particle properties such as bulk density, porosity, particle shape and relating these properties to density of commercial anodes, it appears that calcined coke porosity has a secondary influence on anode density due to its affect on crushed particle packing density.

## Density and Porosity

Calcined coke or anodes consists of solid particles and space. Knowing the real density (solid particles minus all pores) of the coke solids and the bulk density of the pile of coke or anodes allows the calculation of the porosity of the pile or anode. For example the following table shows the porosity of anodes with a known real density.

**Table 1 Anode porosity**

| If anode BAD is | 1.55 g/cc | 1.60 g/cc | 1.65 g/cc |
|---|---|---|---|
| At the indicated RD, | | | |
| It has the following porosity: | | | |
| 2.080 RD | 25.5% | 23.1% | 20.7% |
| 2.070 RD | 25.1% | 22.7% | 20.3% |

Where the porosity percentage is (1-(BAD/RD))*100

Notice that to increase anode BAD by 0.01 g/cc, the porosity needs to be reduced by 0.48%.

The same can be done for calculating the porosity of a coke pile knowing the coke's bulk density and real density. The initial study consisted of determining the porosity, real density and bulk density of -30+50 usm particles prepared by screening natural occurring particles or by crushing larger particles to -30+50 usm size for two different calcined cokes. The porosity curves of these particles are shown in the following plots Figures 1 and 2.

These plots show when the larger particles are crushed to the -30+50 usm size, the porosity decreases as some of the internal porosity of the bigger particles become external surface. Coupling the above data with real density data and vibrated bulk density data shown in Table 2 allows the calculation of the internal and between particle porosity.

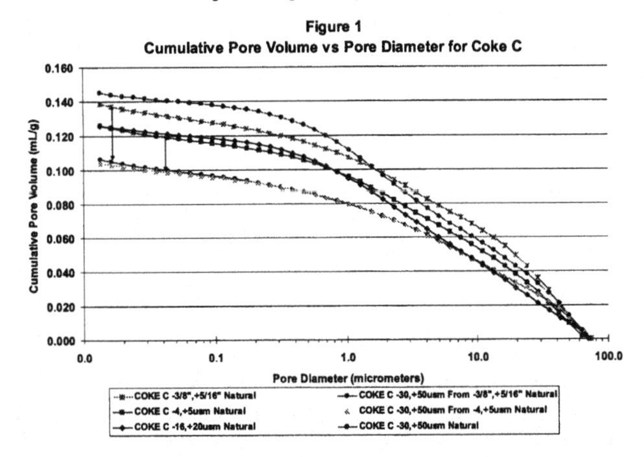


Figure 1
Cumulative Pore Volume vs Pore Diameter for Coke C

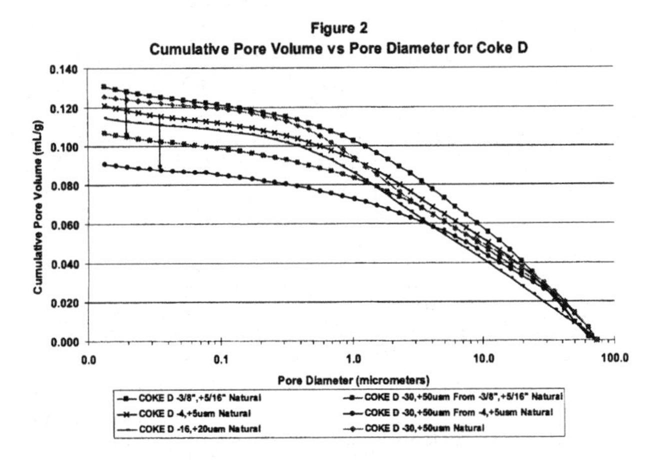


Figure 2
Cumulative Pore Volume vs Pore Diameter for Coke D

**Table 2 Calcined Coke Porosity Calculations for Coke C & D**

| Origin | Coke C | Coke C | Coke C | Coke C | Coke D | Coke D | Coke D | Coke D |
|---|---|---|---|---|---|---|---|---|
| Description | -3/8" +5/16" Natural | -30+50usm from minus 3/8"+5/16" | -30+50usm from - 4+5usm | -30+50 Natural | -3/8" +5/16" Natural | -30+50usm from minus 3/8"+5/16" | -30+50usm from - 4+5usm | -30+50 Natural |
| Hg Porosity @15K (cu. mm/g | 130.8 | 107.0 | 90.9 | 125.4 | 138.7 | 106.3 | 104.3 | 145.0 |
| VBD (g/cc) | 0.688 | 0.858 | 0.867 | 0.956 | 0.681 | 0.847 | 0.857 | 0.913 |
| Real Density (g/cc-He) | 2.084 | 2.081 | 2.082 | 2.072 | 2.085 | 2.086 | 2.082 | 2.071 |
| For 100 grams Of Material | | | | | | | | |
| VBD (volume, cc) | 145.3 | 116.6 | 115.3 | 104.6 | 146.9 | 118.1 | 116.7 | 109.6 |
| RD (volume, cc) | 48.0 | 48.1 | 48.0 | 48.3 | 48.0 | 47.9 | 48.0 | 48.3 |
| Porosity (volume, cc) | 13.1 | 10.7 | 9.1 | 12.5 | 13.9 | 10.6 | 10.4 | 14.5 |
| Inter Particle Volume VBD | 84.2 | 57.8 | 58.2 | 43.8 | 85.0 | 59.6 | 58.2 | 46.8 |
| % SOLID | 33% | 41% | 42% | 46% | 33% | 41% | 41% | 44% |
| INTERPARTICLE POROSITY | 9% | 9% | 8% | 12% | 9% | 9% | 9% | 13% |
| INTRAPARTICLE POROSITY | 58% | 50% | 50% | 42% | 58% | 50% | 50% | 43% |

The percentage of solid and porosity of a pile of particles can be calculated from the above data. The solid fraction is obtained by dividing the RD (solid without any pores) volume by the VBD volume; the within particle porosity is calculated by dividing the porosity by VBD volume and the between particle porosity is obtained by dividing the inter particle volume by the VBD volume. For the above samples of single source cokes the solid fraction ranges from 40% to 46%, the internal particle porosity ranges from 8% to 10 %, and the between particle porosity is from 47% to 57%.

The above type of analysis was repeated for two additional sets of single source calcined coke (high porosity and low porosity coke) samples (natural occurring, rolled crushed and jaw crushed) as well as a butt sample. These samples were also analyzed for particle shape and porosity by techniques discussed in the companion paper. The porosity, VBD, Real Density data as well as the porosity calculations are presented in Table 3. The between particle porosity is slightly lower for Coke A and Coke B. Coke A has a slightly different porosity in that the naturally occurring -30+50 usm particles have porosity very similar to the crushed particles. Because of this observation, additional samples from the same coker and calciner were analyzed which showed the normal trend found in Cokes C and D.

From the above data and calculations, it becomes apparent that the largest volume of space is between particle porosity. Consequently, minimizing this volume will result in higher density anodes or coke piles.

Notice the bulk density of the natural occurring -30+50 usm particles are significantly larger than -30+50 usm particles for the four different single source cokes obtained by crushing larger particles. However, the internal porosity of natural occurring coke particles is larger than the crushed particles. These results have been repeated many times on other single source calcined coke samples. What can explain this apparent paradox? Do the natural particles have a better shape to pack closer? Is there less between particle porosity as calculated above indicate?

Technical papers [1, 2] have addressed some of these issues, but the data in these papers was obtained on commercially blended coke samples. Blended commercial cokes consists of a large range of isotropy and porosity particles blended to meet metal and sulfur specifications. Using blended coke samples for technical studies can lead to conflicting results due to increased variability of particle properties. There is even difficulty in interpreting data obtained on single source coke samples due to isotropy and porosity variations.

Subsequent particle shape and porosity studies [3] and particle packing references [4] suggest that particle flowing friction affects particle packing. Using this logic, it is conceivable the crushed particles with lower roughness and compactness will generate less between particle friction and consequently higher packed density. Notice the natural occurring particles in Table 3I in the appendix from the companion paper have lower compactness and roughness than the crushed particles.

Lower porosity cokes normally result in higher density anodes. Initially it was thought that lower porosity cokes resulted in denser anodes because of their lower internal porosity. The previous presented porosity data indicates this is probably not the primary advantage to using lower porosity cokes. However, the higher packing density from lower porosity cokes could be due to lower roughness and compactness particles obtained when crushing these cokes.

**Conclusions**

1. Calcined coke bulk density is significantly affected by between particle porosity.
2. Crushed larger calcined coke particles have less internal porosity but lower bulk density than naturally occurring particles.
3. Naturally occurring particles appear to have a smoother surface than crushed calcined coke particles which could lead to denser particle packing [3].
4. Literature references indicate particle shape properties affect powder packing. Smoother particle surface properties may partially resolve the VBD-porosity paradox.

### Future efforts

This study as well as the subsequent particle shape and between particle porosity studies suggest that additional data for packing density is needed. In addition, between particle porosity studies are needed for different granulometry blends of calcined coke.

### References


1. D. Belitskus, "Standardization of a Calcined Coke Bulk Density Test", Light Metals 1982, pp. 673-689.
2. R.E. Gehlbach, et al, "Influence of Sample Preparation on Petroleum Coke Properties", Light Metals1995, pp 539-643.
3. Randall Bowers, et. Al., "New Analytical Methods to Determine Calcined Coke Porosity, Shape, and Size," Light Metals Presentation 2008.
4. Randall M. German, *Particle Packing Characteristics*, (Materials Engineering Department, Rensselaer Polytechnic Institute), pp.121-131.


## Appendix

**Table 3 Calcined coke porosity calculations for Coke A & B**

| Origin | Coke A | Coke A | Coke A |
|---|---|---|---|
| Description | -30+50 Natural | -30+50 Jaw Crushed | -30+50 Roll Crushed |
| Hg Porosity @15K (cu. mm/g) | 98.4 | 102.4 | 98.8 |
| VBD (g/cc) | 0.940 | 0.886 | 0.875 |
| Real Density (g/cc-He) | 2.077 | 2.077 | 2.077 |
| For 100 grams Of Material | | | |
| VBD (volume, cc) | 106.4 | 112.9 | 114.3 |
| RD (volume, cc) | 48.1 | 48.1 | 48.1 |
| Porosity (volume, cc) | 9.8 | 10.2 | 9.9 |
| Inter Particle Volume VBD | 48.4 | 54.5 | 56.3 |
| % SOLID | 45% | 43% | 42% |
| % INTERPARTICLE POROSITY | 9% | 9% | 9% |
| % INTRAPARTICLE POROSITY | 45% | 48% | 49% |

| Origin | Coke B | Coke B | Coke B | Butts |
|---|---|---|---|---|
| Description | -30+50 Natural | -30+50 Jaw Crushed | -30+50 Roll Crushed | -30+50 |
| Hg Porosity @15K (cu. mm/g) | 149.8 | 103.8 | 109.7 | 115.6 |
| VBD (g/cc) | 0.918 | 0.892 | 0.878 | 0.964 |
| Real Density (g/cc-He) | 2.115 | 2.115 | 2.115 | 2.104 |
| For 100 grams Of Material | | | | |
| VBD (volume, cc) | 108.9 | 112.1 | 113.9 | 103.7 |
| RD (volume, cc) | 47.3 | 47.3 | 47.3 | 47.5 |
| Porosity (volume, cc) | 15.0 | 10.4 | 11.0 | 11.6 |
| Inter Particle Volume VBD | 46.7 | 54.4 | 55.6 | 44.6 |
| % SOLID | 43% | 42% | 42% | 46% |
| % INTERPARTICLE POROSITY | 14% | 9% | 10% | 11% |
| % INTRAPARTICLE POROSITY | 43% | 49% | 49% | 43% |

Table 3I - Average Values of Measurements on Particles

| Measurement Parameters | Coke A | | | Coke B | | | Butt |
|---|---|---|---|---|---|---|---|
| | AN | AJC | ARC | BN | BJC | BRC | |
| ESD, μm* | 544 | 513 | 503 | 528 | 578 | 523 | 483 |
| Compactness* | 1.65 | 1.90 | 1.89 | 1.60 | 1.73 | 1.81 | 1.74 |
| Roughness* | 1.13 | 1.21 | 1.21 | 1.14 | 1.15 | 1.17 | 1.19 |
| Elongation* | 2.79 | 3.68 | 3.63 | 2.61 | 3.10 | 3.36 | 3.12 |
| Perimeter, μm* | 1936 | 1965 | 1923 | 1900 | 2094 | 1934 | 1816 |
| % B/W Particle Porosity | 45% | 48% | 49% | 43% | 49% | 49% | 43% |

* Data from Referrence 3

**Light Metals 2008** *Edited by: David H. DeYoung*
***TMS (The Minerals, Metals & Materials Society), 2008***

# "New Analytical Methods to Determine Calcined Coke Porosity, Shape, and Size"

Randall Bowers[1], Shridas Ningileri[1], David C Palmlund[2], Bernie Vitchus[3], and Frank Cannova[3]
[1]Secat, Inc
[2]Fluid Imaging Technologies
[3]BP Coke

## Abstract

Carbon Anode properties especially density are dependent on the structure and porosity of calcined coke used to fabricate the anode. Optimum anode density is produced from blending together natural, rolled, and crushed calcined coke with their inherent structure and porosity characteristics.

Two tests were developed to characterize the particle shape including internal and between particle porosity. The first method is a novel technique developed by Fluid Imaging of suspending particles in a moving fluid during which a high speed camera is used to obtain a silhouetted outline of the particle shape and size. Secondly, using traditional optical microscopy techniques and image analysis, internal porosity of the calcined coke was determined. The results of the two methods are presented and compared with evaluation techniques found in literature. These techniques were used to obtain a better understanding of the Calcined Coke VBD - Porosity Paradox presented in a companion paper.

## I. Introduction

With Primary Aluminum plants operating at higher amperage, maintaining or even increasing anode density has become a goal for most carbon plants. At the same time, calcined coke quality is becoming more volatile due to oil refinery economic pressures. Most carbon plants use empirical tests like vibratory bulk density (VBD) to define and monitor calcined coke's effect on anode density. A few other plants use porosity tests like Apparent Density. A more thorough analysis of a coke's porosity and bulk density leads to a paradox. That is, naturally occurring coke particles have a higher bulk density than crushed particles of the same size but the natural occurring coke particles have a higher internal porosity than crushed particles. At the same time experience has shown that lower porosity calcined coke generally leads to higher anode density. This apparent VBD-porosity paradox suggests that some other factor is having a significant affect on anode density. A more fundamental study of these calcined coke properties and factors affecting these properties could lead to producing more dense anodes.

Postulating what can explain the VBD-coke porosity paradox led to the development of the two tests. Two postulated theories (particle shape characteristics or in between particle porosity) were considered and test procedures developed to test these theories. Two measurement techniques were employed for the study of the calcined coke particles. First, a novel technique developed by Fluid Imaging Technologies was used to quantify particle shape. Second, traditional optical microscopy techniques were used to determine porosity between carbon particles and within carbon particles. These techniques and results on -30+50 USM particles are described in this paper.

## II. Experimental Procedures

Two commercially available calcined petroleum cokes and a crushed anode butt were analyzed in this study. Both of the commercially produced cokes were analyzed in their natural, jaw crushed, and roll crushed states. All of the cokes analyzed were from a sieved size of -30+50 USM. There were 7 different samples analyzed. Each sample was tested using a FlowCAM® (Flow Cytometer and Microscope) and using traditional optical microscopy techniques.

### A. FlowCAM® Apparatus

Fluid Imaging Technologies, Inc. has developed an instrument called a FlowCAM® which performs fluid particle analysis. The equipment consists of 3 components - a fluid component, an optical component, and an electronics component as shown in figures 1 and 2. The particles are mixed into a fluid capable of suspending the particles without sinking. The fluid is then inserted upstream of a flow chamber. A pump will draw the fluid through the flow chamber which sits in the optical path of a microscope. As the particles pass through the microscope path, a light source illuminates the particles and a camera attached to the microscope captures the images. The individual particle images are stored for quasi-real time processing of particle sizes, aspect ratios, equivalent spherical diameter, roughness, compactness, and many other parameters.

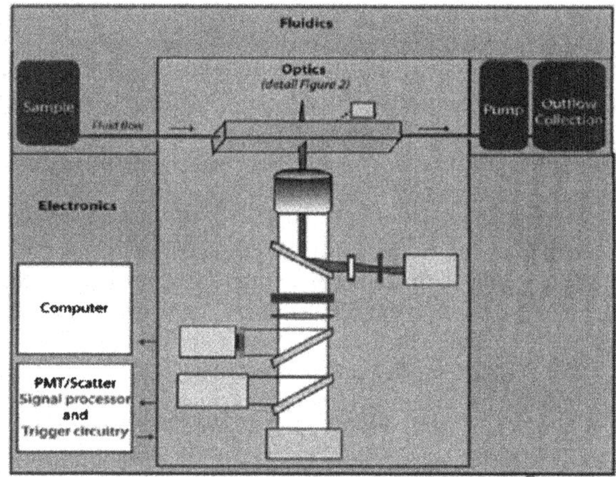


Figure 1 - General Schematic of FlowCAM®

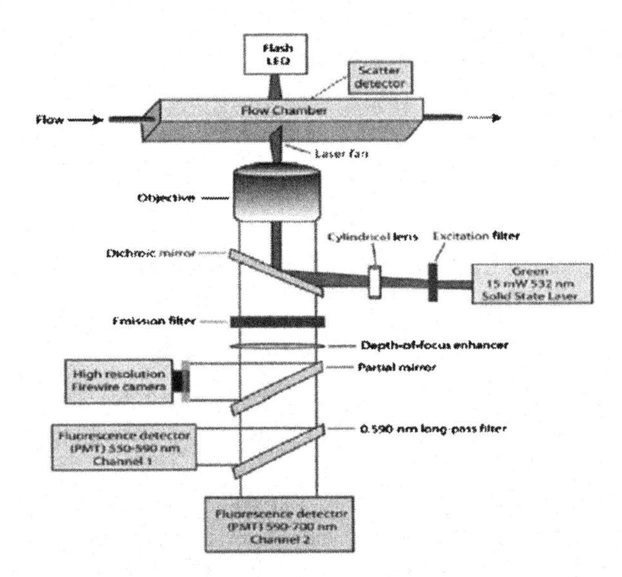


Figure 2 - Details of Optics in FlowCAM®

The FlowCAM® setup was run in an auto imaging mode with a 2X objective lens and a 10X camera. The total magnification of the particles was 20X and the minimum particle size was set to 50 μm. The fluid flow was set to 1000 μM. The suspension consisted of 12 mL of 5% polyvinylpyrrolidone (PVP) and water with 2 grams of calcined coke. Each sample had 100-700 particles imaged which took between 60 and 90 seconds. For each batch, several particle characterization parameters were calculated.

B. Traditional Optical Microscopy

To examine the internal porosity of the individual calcined coke particles, traditional optical microscopy techniques were employed. For each batch, the samples were vibrated in a sieve shaker and vacuum mounted with epoxy and polished using normal polishing procedures. After polishing, the samples were observed using an Olympus BH-2 optical microscope and analyzed using the Buhler Omnimet Software. Sample preparation details are given in tables V and VI of the Appendix.

For each sample, particles were observed using a 5X objective on the microscope and 5 images of the cross-section was captured. The whole image was used for calculation of the between particle porosity. From the 5 images for each sample, 10 individual particles were highlighted and extracted for calculation of the individual particle porosity and aspect ratios.

## III. Results and Discussions

A. FlowCAM® Results

The FlowCAM® unit is currently not capable of imaging the surface porosity of particles or direct measurement of packing of the particles. Therefore, focus is on the measurement of characteristics of the particles that will affect their packing behavior. Figures 3-6 are representative particle images from the FlowCAM® unit. The particles are of an irregular nature with rounded and jagged features. This characteristic of the particles makes prediction of particle packing behavior much more difficult. For each particle that was analyzed (>100 particles per sample), there is a variety of information obtained, some of which are equivalent spherical diameter (ESD), compactness, roughness, and elongation. These four parameters relate specifically to packing characteristics of particles due to packing arrangements and frictional effects on flow performance.

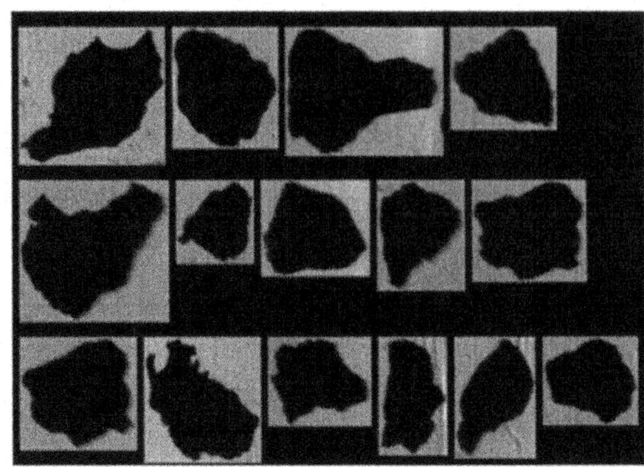

Figure 3 - Examples of Natural Particles

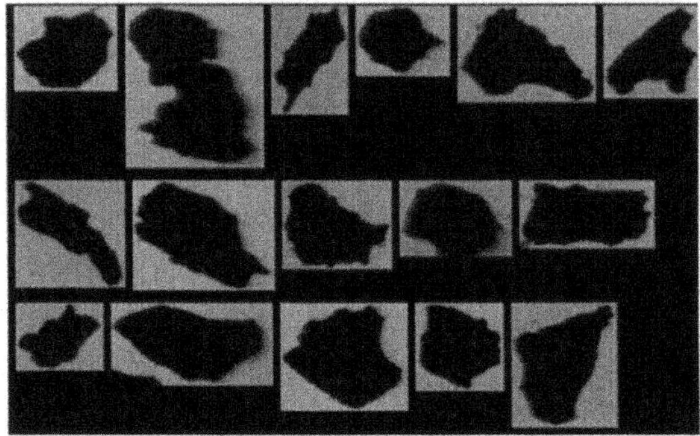

Figure 4 - Examples of Jaw Crushed Particles

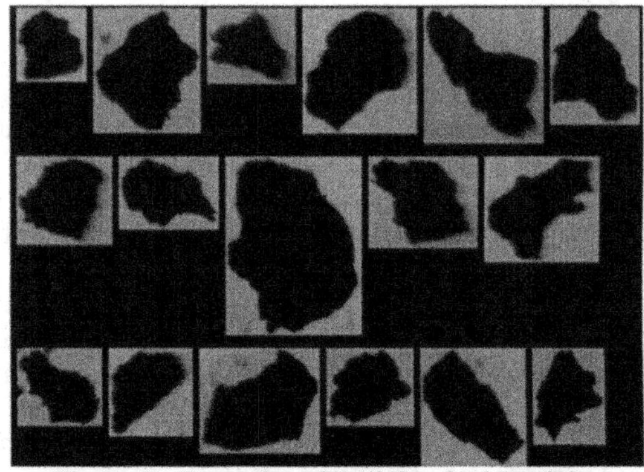

Figure 5 - Examples of Roll Crushed Particles

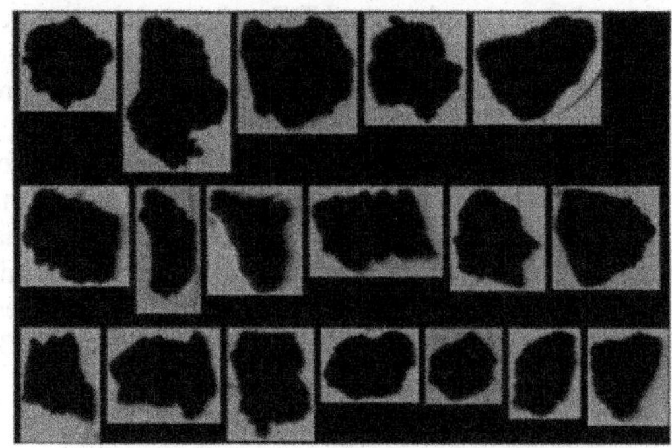

Figure 6 - Examples of Anode Butt Particles

In the powder metallurgy industry, the flow and packing of particulate matter has been studied in depth and many ideas of what factors affect packing behavior have been proposed. Although there is a large gap in size between many metal powders and the size of calcined coke, the concepts developed in the powder industry on particle packing can be extended into the coke industry. In packing of powders, the biggest factors affecting packing density are the sizes of the powder and the flow behavior of the powder [1]. In this study, the size of the calcined coke was sieved in order to approximately be considered equal. Therefore, the remaining analysis was focused on the three parameters compactness, roughness, and elongation which can be used to hypothesize about packing behavior of particles. The sample ID's used in the tables for the paper is given in table I.

Table I - Sample ID's for Tables

| Sample | ID Used |
|---|---|
| Coke A - Natural | AN |
| Coke A - Jaw Crushed | AJC |
| Coke A - Roll Crushed | ARC |
| Coke B - Natural | BN |
| Coke B - Jaw Crushed | BJC |
| Coke B - Roll Crushed | BRC |
| Crushed Anode Butt | Butt |

For the sets tested, the average values of the measurements are shown in table II along with a between particle porosity (B/W Particle Porosity) calculation based on VBD and Mercury Porosimetry tests [2]. The VBD and Mercury Porosimetry test results and analysis will be presented in reference 2 in more detail. The parameter ESD Diameter refers to the equivalent spherical diameter (ESD) and is used to compare sizes of irregular shaped entities. The sizes of particles follow a lognormal distribution. For Coke A, the largest average particle size was for the natural coke. For coke B, the largest was the jaw crushed coke. The anode butt sample had the smallest average particle size of all of the samples. The average particle size was approximately 0.5-0.6 mm. The remaining analysis of the FlowCAM® results are focused on the measurements which contribute to packing and flow of the particles assuming the sizes of the particles are the same.

Table II - Average Values of Particle Measurements

| Measurement Parameters | Coke A | | | Coke B | | | Butt |
|---|---|---|---|---|---|---|---|
| | AN | AJC | ARC | BN | BJC | BRC | |
| ESD, µm | 544 | 513 | 503 | 528 | 578 | 523 | 483 |
| Compactness | 1.65 | 1.90 | 1.89 | 1.60 | 1.73 | 1.81 | 1.74 |
| Roughness | 1.13 | 1.21 | 1.21 | 1.14 | 1.15 | 1.17 | 1.19 |
| Elongation | 2.79 | 3.68 | 3.63 | 2.61 | 3.10 | 3.36 | 3.12 |
| Perimeter, µm | 1936 | 1965 | 1923 | 1900 | 2094 | 1934 | 1816 |
| % B/W Particle Porosity* | 46% | 48% | 49% | 43% | 49% | 49% | 43% |

* Data from reference 2

The parameters compactness, roughness, and elongation are three parameters describing different aspects of a particle which could be correlated to particle packing and flow behavior. The affect on the behavior is inter-related between the three parameters. An attempt was made to analyze the characteristics separately but when comparing the between particle porosity results, the combined affect must be kept in mind.

The compactness is a measure of how far apart pixel groups are in the particle image. Therefore, for lower compactness numbers, the groups of pixels are closer together and would therefore have less appendages or smaller appendages protruding from the particle. The roughness of the particle is a measure of the unevenness or irregularity of a particle. Both parameters would relate directly to flow performance of particles during packing [1, 3]. The elongation is similar to an aspect ratio calculation but is a derived calculation which also includes information of area and perimeter of the particle. A 2 sample T-test was performed comparing the different distributions of particles and their characteristic measurements. The results are shown in table III. A brief description of the 2 sample T-test is in section B of the Appendix.

Table III - 2 Sample t-Test and VBD Results

| Parameter | Smallest << Largest |
|---|---|
| Compactness Order | BN << AN << BJC ≈ Butt << BRC << ARC ≈ AJC |
| Compactness Results | 1.60 1.65 1.73 1.74 1.81 1.89 1.90 |
| Roughness Order | AN << BN << BJC << BRC << Butt << AJC ≈ ARC |
| Roughness Results | 1.13 1.14 1.15 1.17 1.19 1.21 1.21 |
| Elongation Order | BN << AN << BJC ≈ Butt << BRC << ARC ≈ AJC |
| Elongation Results | 2.61 2.79 3.10 3.12 3.36 3.63 3.68 |
| B/W Particle Porosity Order | BN << Butt << AN << AJC << BJC << BRC << ARC |
| B/W Particle Porosity Results* | 42.6% 43.0% 45.5% 48.3% 48.6% 48.9% 49.2% |

* Data from reference 2

For the parameters listed in table III, the highest packing efficiency would occur for those with a minimum value for compactness, roughness, and elongation. A rougher and less compact particle would encounter more friction during flow which would reduce packing efficiency as would one with a more elongated structure. As each of the parameters are inter-related, the analysis becomes more difficult.

By looking at the compactness only, one can see the two natural cokes have the lowest compactness values of the group and that Coke A - Jaw Crushed (AJC) and Coke A - Roll Crushed (ARC) have the highest as well as being approximately equal when tested using a 2 Sample t-Test. The corresponding VBD analysis [2] shows the natural cokes have among the lowest between particle porosity while the Coke A - Roll Crushed and Jaw Crushed are among the highest with approximately 1% difference in porosity values. There is not a 1-to-1 correspondence between the compactness and porosity as can be seen for the Anode Butt (Butt) sample. The anode butt is around the middle value of compactness yet it actually has the second lowest porosity value. A corresponding analysis of the roughness and elongation show a similar trend as stated above for compactness.

To test how well the three parameters explain the VBD results, a regression analysis was performed with and without including the anode butt data. When the fit was tested with the anode butt data and using the three parameters compactness, roughness, and elongation, the $R^2$ value of the fit was 94.8%. When removing the anode butt data from the regression, the $R^2$ value improved by almost 4% to 98.3%. The regression is using a limited number of samples so must be verified with more data but does at least show a strong fit to the regression line. Furthermore, the F value of the second regression was 38.71. In an F distribution table [4], the critical F value for a 95% confidence is 9.55. As the F value is higher than the critical F value, the regression is supposed to be significant in that the fit is not based on chance.

B. Optical Microscopy Results

Figures 7-10 are example micrographs of Coke A in the natural, jaw crushed, and roll crushed states plus a crushed anode butt, respectively. Ten particles on each micrograph were extracted for further analysis of particle porosity values and aspect ratios. The particles extracted for analysis were approximately 0.5-1 mm in size.

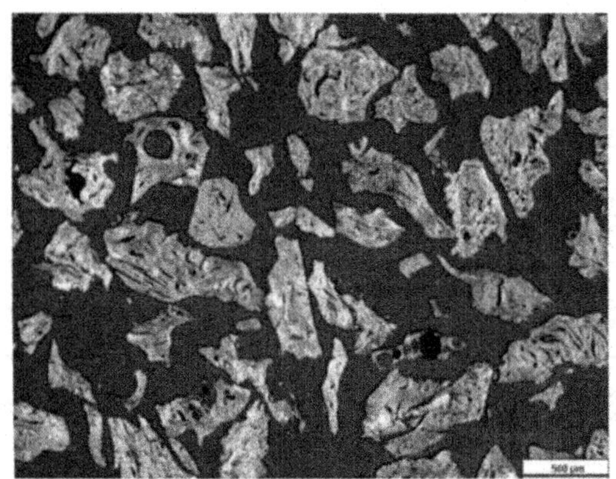

Figure 7 - Micrograph of Coke A - Natural

Figure 8 - Micrograph of Coke A - JC

Figure 9 - Micrograph of Coke A - RC

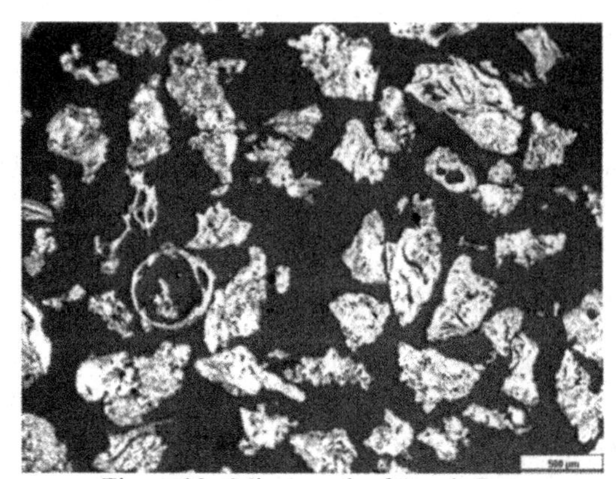

Figure 10 - Micrograph of Anode Butt

The between particle porosity for the 2 manufacturers with 3 process states (natural, jaw crushed and roll crushed) and an Anode Butt are shown in table IV and figure 11. The results indicate a large variation between each micrograph for the samples. The natural and anode butt materials had the highest variation in between particle porosity with coke B-natural having the highest between particle porosity.

Table IV - Between Particle Porosity Values, %

| Micrograph | AN | AJC | ARC | BN | BJC | BRC | Butt |
|---|---|---|---|---|---|---|---|
| 1 | 66.2 | 64 | 65.9 | 71.9 | 67.5 | 57.4 | 63.5 |
| 2 | 66.0 | 63.4 | 60.7 | 67.7 | 59.7 | 60.9 | 70 |
| 3 | 64.3 | 60 | 62.4 | 73.3 | 62.5 | 62.4 | 62 |
| 4 | 59.2 | 61.4 | 64.7 | 57.9 | 63.3 | 64.7 | 57.9 |
| 5 | 55.38 | 63.5 | 58.9 | 64.3 | 60.9 | 58.9 | 68.6 |

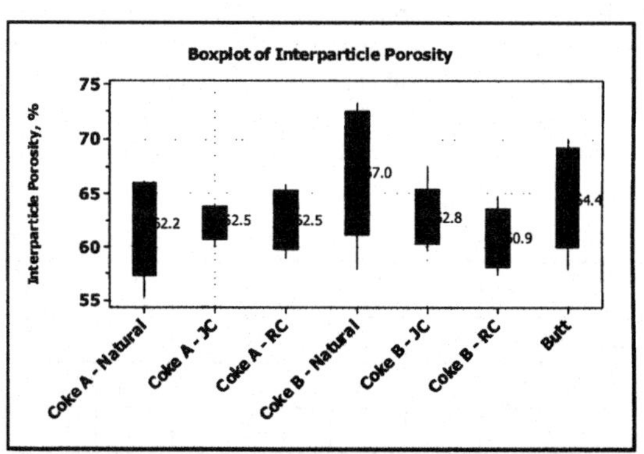


Figure 11 - Between Particle Porosity Values, %

In figure 12 below, the 50 extracted particles for each sample were analyzed for particle porosity and plotted. The natural coke particles had very high porosity with some calculated to be greater than 100%. Figure 13 shows a set of extracted particles with particle 16 having a porosity calculated to be 78%. The formula for the calculation of the particle porosity is the following.

$$\% \text{ Porosity} = \frac{\text{Area of Pores}}{(\text{Area of Particle - Area of Pores})} \quad (1)$$

The set of particles is from Coke B - Natural. Observed in the analysis was the tendency for natural particles to have an abundance of spherical pores whereas in jaw crushed or roll crushed particles, the pores were predominantly elongated or striated. A similar observation was noted in work done by R.E. Gehlbach et. al. [5] in which the authors observed that in natural coke, the porosity was more spherical in nature. As the coke was processed, the particles tended to become more jagged with striated porosity. They postulated the rounder natural cokes would pack better due to its better flow characteristics.

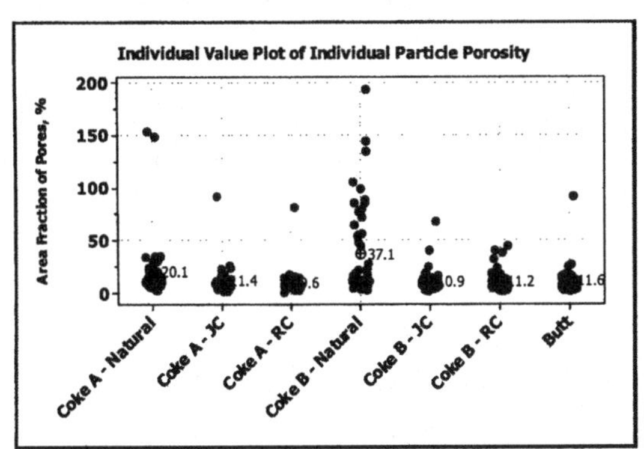

Figure 12 - Internal Porosity, %

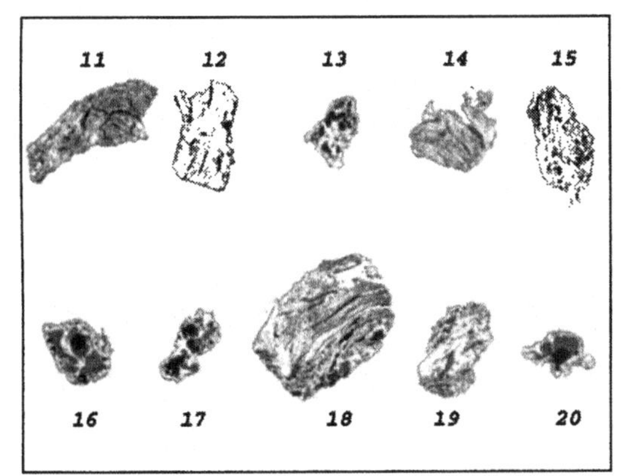

Figure 13 - Extracted Particles for Coke B, Natural

Figure 14 shows the individual values of aspect ratio calculations for the 50 extracted particles on each sample. There was no major difference observed in aspect ratios between the different lots. The aspect ratio calculation is,

$$\text{Aspect Ratio} = \frac{\text{Longest Dimension}}{\text{Shortest Dimension}} \quad (2)$$

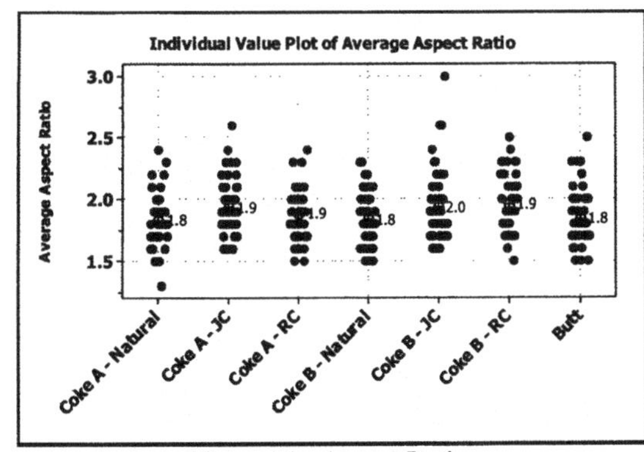

Figure 14 - Aspect Ratios

## IV. Summary and Conclusions

Analysis of the calcined coke particles reveal that the two methods of observations used have very different focus, but can be used in conjunction for studying the VBD-Porosity Paradox with some effectiveness. The particle analysis of size and shape is best suited to the FlowCAM instrument where large numbers of particles can be compared in a short amount of time. The optical microscopy method on the other hand is much better equipped to examine internal porosity differences in the coke particles themselves.

When trying to correlate bulk density to particle types, the measurements of the compactness, roughness, and elongation parameters show a similar trending to what was observed in the VBD tests. Although there was not a direct correlation in every case, obtaining a larger sample distribution for each particle type may improve the results. The use of more sample sets from more vendors may also prove useful for further study where a regression analysis can be used to determine the amount of correlation between the different parameters.

The use of optical microscopy in studying calcined coke should be restricted to work on examining differences in internal porosity of the coke particles themselves. Since the optical method requires a sectioning of the particles, the true shape is not necessarily being revealed in the cross-section but is rather the 2D projection of a 3D object. Furthermore, the use of a mounting media may tend to separate particles from one another as the particles may float or be moved as air escapes the packed powder when filled with the media. The optical observations did not provide more insight into packing arrangements of the particles themselves but did help to understand possible differences in the behavior of the particles as they are crushed during manufacture of the anode. Natural particles showed high amounts of internal porosity that were also predominantly spherical in shape. In contrast, the jaw crushed particles and roll crushed particles revealed striated porosities which may not breakdown during compaction.

Future work on deciphering the VBD-Porosity Paradox is suggested to focus on improving the flow performance of particles. Also, there may be a need to study particle sizing in more depth to find an optimum ratio of large to small particle sizes that can be mixed to improve pack density. Smaller particles can fill the spaces between larger particles and in conjunction with improving particle flow; a higher density anode may be achieved.

## V. References

1. Randall M. German, *Powder Metallurgy Science* (Princeton, NJ: Metal Powder Industries Federation, 1997), 62-64, 70-73, 163-171.

2. B. Vitchus et al., "Understanding the Calcined Coke VBD-Porosity Paradox" (Paper presented at TMS 2008).

3. E.C. Abdullah, D. Geldart, "The use of bulk density measurements as flowability indicators," *Powder Technology*, 102 (1999), 151-165.

4. Sam K. Kachigan, *Multivariate Statistical Analysis* (New York, NY: Radius Press, 1991), 292-294.

5. R.E. Gehlbach,et al, "Influence of Sample Preparation on Petroleum Coke Properties", Light Metals 1995, pp 539-643.

## VI. Appendix

### A. Sample Preparation Steps

Table V - Optical Microscopy Method Mounting Procedures

| Mounting Procedures | |
|---|---|
| 1. | Fill mounting cup half full of coke sample<br>(Mounting cup - 1.25" in diameter by 1.75" high) |
| 2. | Shake sample at the center of sieve shaker in bottom pan for 1 minute<br>(Sieve shaker - Gilson Company SS-15, ¼ HP shaker, orbital shaking motion) |
| 3. | Pour epoxy into mounting cup and apply vacuum for 5 minutes at 25" Hg |
| 4. | Bottom of mount was used for polishing and observations |

Table VI - Optical Microscopy Method Polishing Procedures

| Polishing Procedures | |
|---|---|
| 1. | Grind with 320 Grit SiC paper, complimentary rotation, water coolant, 300 rpm, and 4.5 lbf per sample for 1 minute or until plane surface obtained |
| 2. | Polish with 9 micron diamond suspension, complimentary rotation, diamond extender lubricant, 150 rpm, and 4.5 lbf. per sample for 5 minutes |

### B. 2 Sample T-Test

The 2 sample T-test is a statistical tool used for comparing the differences between means of two sample populations by way of hypothesis testing. The sample populations are assumed to follow a normal distribution. In the hypothesis test, there is a null-hypothesis, $H_0$, and an alternative hypothesis, $H_1$. During the test, the null-hypothesis is set to be that the differences in mean, $\mu$, is equal to some value. In this case the null-hypothesis is $\mu = 0$. Furthermore, the alternative hypothesis is that the $\mu \neq 0$ or in this case it is a one sided hypothesis of $\mu < or > 0$. If the probability, p value, is less than or equal to the $\alpha$-level, then the alternative hypothesis, $H_1$, is correct. Otherwise, the null-hypothesis, $H_0$, is correct. The $\alpha$-level is the value based on the confidence interval used such as for a 95% confidence, the $\alpha$-level would be 0.05.

$$H_0 = True \text{ if } p > \alpha - level, \text{ where } \mu = Value \quad (3)$$

$$H_1 = True \text{ if } p \leq \alpha - level, \text{ where } \begin{Bmatrix} \mu \neq Value, \text{ or} \\ \mu < Value, \text{ or} \\ \mu > Value \end{Bmatrix} \quad (4)$$

**Light Metals 2008** *Edited by: David H. DeYoung*
*TMS (The Minerals, Metals & Materials Society), 2008*

# COKE AND ANODE DESULFURIZATION STUDIES

Lorentz Petter Lossius[1], Keith J. Neyrey[2], Les Charles Edwards[2]
[1] Hydro Aluminium a.s Technology & Operational Support, P.O.Box 303, NO-6882 Øvre Årdal, Norway
[2]CII Carbon LLC, 2627 Chestnut Ridge Rd, Kingwood, TX 77339, USA

Keywords: Petroleum Coke, Anode, Sulfur, Baking

## Abstract

The sulfur level of high sulfur cokes used by the calcining industry for blending is increasing. During calcination, petroleum coke desulfurizes and the rate of sulfur loss is dependent on both the sulfur level and final temperature. Desulfurization negatively affects coke properties such as real density and porosity and additional desulfurization during anode baking can negatively affect anode properties. This paper is a follow-up to a 2007 TMS paper on desulfurization and presents the results of additional studies on anode and coke desulfurization versus equivalent baking level and discusses the effect of the level of calcination of high sulfur cokes on anode reactivity. The results indicate that the coke calcination levels must be set with coke sulfur levels in mind. It also shows that blending with high sulfur cokes need not be detrimental to anode properties.

## Symbols and Abbreviations

$T_{eq}$ Equivalent temperature in [°E] as a measure of anode baking level according to ISO 17499
$L_c$ Average carbon crystallite height in coke in [Å] according to ISO 20203 or ASTM D5187-02
S The sulfur concentration of the cokes in [wt%] determined by XRF according to ISO 12980 – 2000
HS High sulfur level petroleum coke, S > 3 wt%
NS Normal sulfur level petroleum coke, 2 < S < 3 wt%
LS Low sulfur level petroleum coke, S < 2 wt%

## Introduction

The loss of sulfur during calcination of petroleum coke was reviewed in detail in a paper published in 2007 [1]. At temperatures above 1250°C, thermal desulfurization of coke in rotary kilns and anode baking furnaces can significantly affect coke and anode properties. This is due to the creation of microporosity in the coke structure (typically <1µm) as carbon-sulfur bonds are broken and sulfur is driven out of the structure. Desulfurization has become an important issue for calciners and anode producers as the industry moves to blending higher sulfur cokes to achieve average coke sulfur levels.

The following paper presents the results of additional studies on coke and anode desulfurization and can be considered as Part 2 of the 2007 paper. The 2007 paper provides the necessary background for this work and should be referenced for additional context. This paper presents and discusses results in four areas:

- The variation in calcining level and desulfurization as a function of coke particle size from a typical rotary kiln.
- Sulfur-loss graphs and desulfurization rates for high and low sulfur cokes heated in an anode baking furnace.
- Desulfurization of anodes produced with high sulfur coke calcined at different levels.
- Modeling of sulfur emissions from a baking furnace section.

## 1. Variation in Calcining Level with Particle Size

Calcined petroleum coke contains a range of particle sizes from "fines" (<250µm) to larger particles up to 20mm in size. When green (raw) coke is calcined in a rotary kiln, the coke bed tumbles and coarser particles tend to "roll" on the surface of the coke bed. The coarse coke is subjected to higher radiant heat compared to the finer coke in the center of the bed. This is depicted in Figure 1.

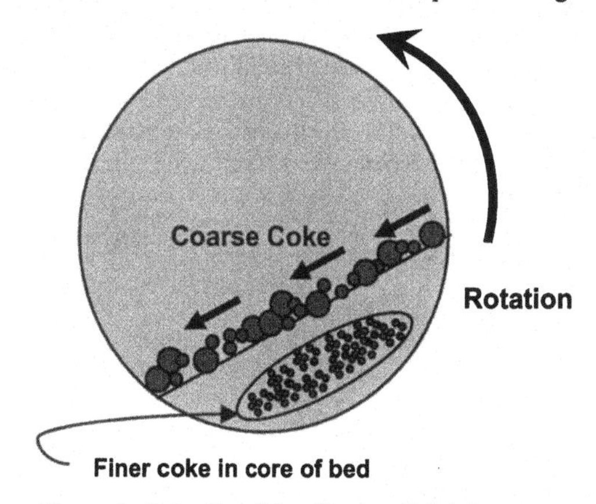


Figure 1: Coke Tumbling During Calcining

The ideal green coke feedstock for a rotary kiln is one with a good mix of coarse, medium and fine particle sizes. The coarse particles promote tumbling and mixing of the bed resulting in more uniform calcination. There will always be some variation in the degree of calcination between coarse and fine particles however, and this influences desulfurization. Coarser coke is always more highly calcined and therefore more susceptible to thermal desulfurization.

The data in Table I compares the properties of a typical calcined coke kiln product separated into three different size fractions: Coarse (+4.75mm), medium (1.70-4.75mm) and fine (<1.70mm). The properties are all measures of the degree of calcination or heat treatment where SER = specific electrical resistivity, $L_c$ = L sub C by X-ray diffraction and RD = real density. As calcination level or temperature increases, SER decreases, $L_c$ increases and RD decreases.

Table I. Coke Properties as a Function of Particle Size

| | SER (Ω.in) | $L_c$ (Å) | RD (g/cm³) |
|---|---|---|---|
| Coarse | 0.0360 | 34.3 | 2.070 |
| Medium | 0.0375 | 32.6 | 2.084 |
| Fine | 0.0400 | 30.1 | 2.078 |

Real density is the most common method used by calciners to control the degree of calcination and it is also the most common property specified by anode users. Calciners typically only measure average properties and the above shows clearly that the coarse coke is more highly calcined (as evidenced by lower SER and higher $L_c$) than the medium and finer size coke. The lower real density of the coarser fraction shows that some desulfurization has occurred due to the higher level of calcination.

Anode users need to be mindful of this sort of variability which occurs with all cokes produced in a rotary kiln. Cokes are typically delivered as a blend of high and low sulfur cokes with a range of particle sizes. The level of desulfurization of any given particle in a blend will therefore be a function of the starting sulfur level, the particle size and target level of calcination. Coarse, high sulfur particles are more susceptible to the negative effects of desulfurization described in Ref [1] at higher calcination levels, and during anode baking as described below.

## 2. Desulfurization of Cokes in a Baking Furnace

In the second part of this paper a series of calcined petroleum cokes with different sulfur levels were subjected to heat treatment in a commercially operating baking furnace. The series included single source, high sulfur (HS) coke, normal sulfur (NS) coke and low sulfur (LS) coke. Each coke is prefixed by a letter to distinguish the different coke sources with the full list shown in Table II. Most of the cokes were single source cokes. In addition a NS level blend containing a mixture of HS, NS and LS cokes was tested, along with a butts sample, also containing a mixture of cokes.

Test portions of 2.0-5.6 mm fractions of each coke were placed in crucibles under the packing coke of top layer anodes in Årdal Carbon Furnace #2, a Riedhammer type baking furnace. An equivalent temperature ($T_{eq}$) sample was sent through in parallel with each coke test portion to determine the baking level experienced by the coke.

Hydro Aluminium uses the equivalent temperature method to quantify the heat treatment anodes experience in the baking furnace. The method is described briefly in the next section of the paper with examples of use.

For each coke and each set of results, a linear curve was fitted to the data describing the desulfurization. Typical curves are shown in Figure 2. The horizontal line is the initial sulfur concentration before heat treatment. The thick line covers the measurement range. The dotted line is an extrapolation back to the initial sulfur concentration level, giving an estimate of the starting equivalent temperature of significant desulfurization.

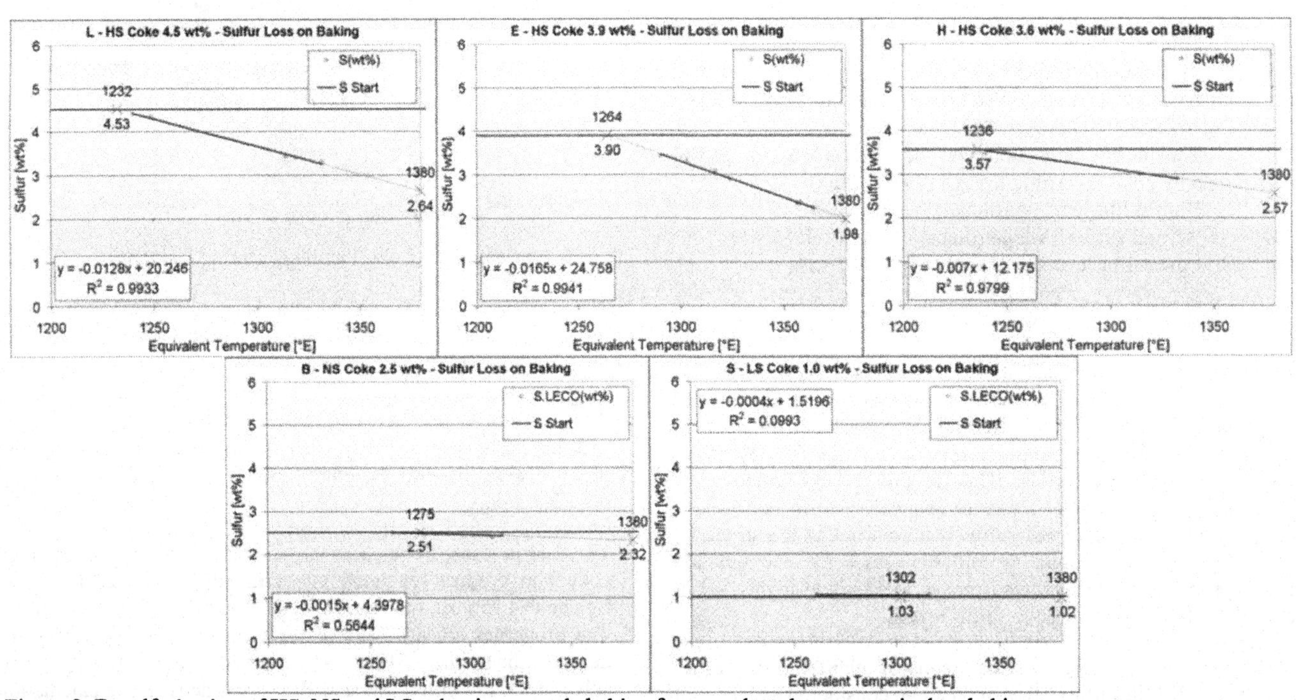


Figure 2. Desulfurization of HS, NS and LS cokes in an anode baking furnace plotted versus equivalent baking temperature.

The desulfurization rate varied significantly between the different sulfur level cokes where HS >>> NS >> LS ~ 0. It was observed that each single source HS coke and NS coke desulfurized linearly with equivalent baking temperature, and the linear curve fit was strong.

The important parameters of initial concentration, starting temperature for significant desulfurization and the rate of desulfurization are summarized in Table II. The single source cokes are sorted by decreasing sulfur level.

Table II: Petroleum coke and butts desulfurization. The starting temperature of significant desulfurization is measured by the equivalent baking level temperature.

| | Initial Sulfur Concentration [wt%] | Starting Temperature [°E] | Rate of Desulfurization [wt% / °E] |
|---|---|---|---|
| L-HS | 4.53 | 1232 | 0.0128 |
| E-HS | 3.90 | 1264 | 0.0165 |
| M-HS | 3.58 | 1233 | 0.0042 |
| H-HS | 3.57 | 1236 | 0.0070 |
| B-NS | 2.51 | 1275 | 0.0015 |
| N-LS | 1.37 | [a] | 0.0000 |
| S-LS | 1.03 | 1302[a] | 0.0004 |
| V-LS | 1.01 | [a] | 0.0000 |
| NS Blend | 2.23 | 1219 | 0.0019 |
| Butts | 1.75 | 1304 | 0.0040 |

[a] Uncertain starting temperature and little or no change in concentration.

Among the HS cokes, the rate varied considerably. The E-HS and M-HS cokes had comparable starting sulfur concentration, but the desulfurization rates were as different as 0.0165 wt%/°E and 0.0042 wt%/°E respectively, a factor of four. Except for E-HS, the HS cokes had a starting temperature for significant desulfurization around 1230°E.

The single source B-NS coke desulfurized linearly and showed a moderate desulfurization rate of 0.0015 wt%/°E, much lower than the HS cokes. The start of significant desulfurization was high, 1275°E.

The single source LS cokes showed little or no desulfurization in the observed range.

For all cokes with measurable desulfurization parameters the desulfurization ($\Delta S$) could be expressed by a linear relationship, e.g. for E-HS

$$\Delta S_E = 0.0165 \text{ wt\%/°E} * (T_{eq} - 1264\text{°E}) \quad \text{Eq. [1]}$$

For the NS blend and the butts, the linear curve fits are less well defined since they are based on only two data points. The NS blend was a blend of low sulfur (<1.5%) and higher sulfur coke (~3.2%). In general, desulfurization of blends will not be linear since blends are made up of several cokes which desulfurize at different rates, starting at different temperatures.

For the butts, the starting temperature for significant desulfurization was high, 1304 °E, since this material already had been baked at least once.

## The Equivalent Baking Level

Before continuing with the discussion on anodes an example of use of the equivalent baking level will be given to illustrate what equivalent temperatures correspond to normal baking levels of anodes.

The equivalent baking level is a measure of baking level using a temperature scale (°E) and equivalent temperature, $T_{eq}$. To establish the scale, a green reference coke is given a series of 2-hour heat treatments at different hold temperatures, and the resulting $L_c$ is measured. The series of ($L_c$, $T_{eq}$) pairs is used to establish a calibration curve, and from this, the equivalent temperature can be determined from the $L_c$ value after any heat treatment of the reference coke.

When a green test portion of the reference coke is sent with an anode through the baking furnace, the resulting $T_{eq}$-value is termed the equivalent baking level of the anode. A more detailed description is provided in ISO 17499 and Lossius et al. [2].

An important difference between the method described above and thermocouple measurements is that it includes the effect of both temperature and time. Longer soak times at a given temperature will give higher equivalent baking levels. At any given temperature, the coke $L_c$ will increase with increased soak time. Coke and anode desulfurization is also a function of time and temperature so the equivalent baking level method is very relevant.

The Hydro baking level target is 1230°E. Above 1320 °E, the anode is regarded as over-baked, and below 1150 °E, it is regarded as under-baked. The equivalent temperature is an important anode quality parameter and is part of Hydro Aluminium's routine analysis on anodes.

An example of equivalent temperature values observed in an anode baking furnace section from performance testing of a baking furnace are shown in Figure 3.

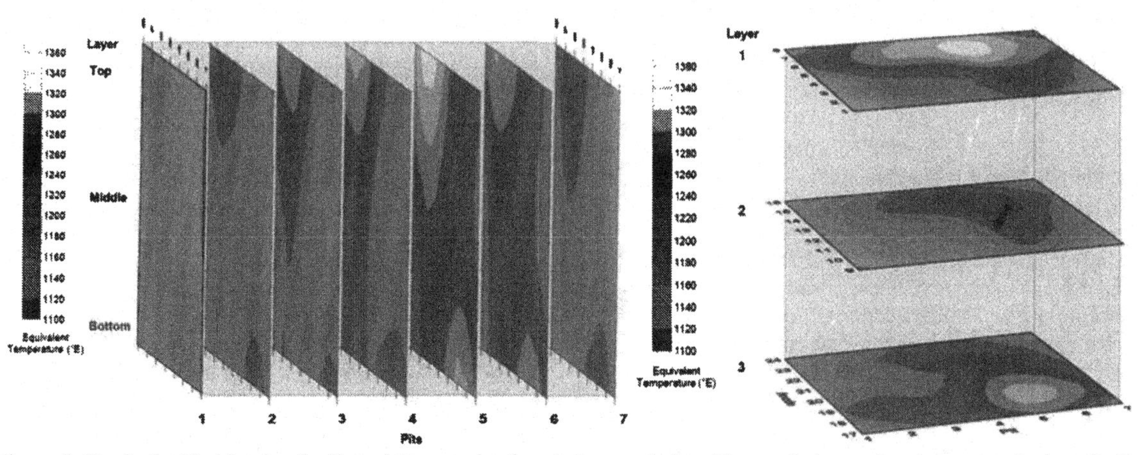


Figure 3. Equivalent baking level - Complete mapping by pit (upper plot) and by anode layer. Anodes are stacked vertically.

The distribution is obtained by determining the equivalent baking level of each anode in the section. The plots show one section of Årdal Carbon Furnace 3. Combined with Eq. [1] the sulfur emissions from an entire furnace section can be estimated.

## 3. Measuring Sulfur Loss from Anodes During Baking

A pilot scale anode testing program was run by Hydro to study effects of adding HS coke to anodes, especially how much HS coke can be added without negative effects. Measurement of the sulfur concentration and determining changes was part of the program.

Pilot scale anodes of this type can give valuable information if anode properties are realistic, and the pilot scale line in Årdal has been tuned to match the quality of production anodes at Årdal Carbon as shown in Table III. Most properties match well, although Young's Modulus is somewhat low and permeability somewhat high. The standard deviations were comparable.

The line consists of an Eirich batch mixer, a vacuum vibroformer (Wilkening design) and a laboratory furnace for controlled baking. Anodes are 3.9 kg with dimensions 130 mm diameter x 170 mm height.

Table III. Anode property comparison. Averages from 9 pilot anodes and 13 production anodes with the same coke fractions.

| Anode Property with Abbreviation and Unit | Pilot Scale Anodes (Avg) | Production Anodes (Avg) |
|---|---|---|
| Reactivity, $R.CO_2$ [$mg/cm^2h$] [a] | 14.9 | 15.3 |
| Reactivity, $Dust.CO_2$, [%] [a] | 3.2 | 2.0 |
| Reactivity, R.Air, [$mg/cm^2h$] [a] | 33.7 | 30.0 |
| Density, [$g/cm^3$] | 1.582 | 1.582 |
| Resistivity, SER, [$\mu\Omega m$] | 52.9 | 52.7 |
| Young's Modulus, YM [MPa] | 9394 | 10273 |
| Strength, CCS, [MPa] | 44.1 | 45.5 |
| Permeability, [nPm] [b] | 0.8 | 0.3 |
| Expansion, CTE, [$1/K*10^6$] | 4.5 | 4.7 |
| Thermal Shock, TSR | 19.6 | 18.1 |

[a] Reactivity by Hydro method (thermogravimetric method)
[b] Permeability by Hydro method (120 mm sample, 40% of RDC method)

### 3.1 Selection of a High Sulfur Coke for Use in Anode Tests

To determine which HS coke to use in the anode test program an investigation was made comparing several HS cokes to determine relative quality. The designations are the same as used in Table II. Rankings were (best first)

- Stability against loss of sulfur : M > H >> L > E
- Stability against shrinkage : H > M >> L

Based on this, coke L was judged to be a potentially difficult coke and was chosen for testing. The coke was calcined at two different levels (high and low) to compare the impact of coke calcination level. Specific electrical resistivity (SER) was used as the measure of degree of calcination. The high and low levels were selected to fall at the typical upper and lower levels of calcination in a commercial scale rotary kiln. It is not difficult to calcine coke to much higher and lower levels but this was not felt to be realistic, particularly for higher calcination levels. The SER targets selected were

- Low calcined (SER 0.040 Ω/inch)
- High calcined (0.036 Ω/inch)

### 3.2 Desulfurization of Anodes

The pilot anodes were made at the compositions shown in Table IV.

Table IV. Anode composition – $Y_{HS}$ is the percentage of L-HS coke which was set at 0, 25 or 50 %.

| Anode Data | S-level | Anode Composition |
|---|---|---|
| Butts | 1.83 wt% | 20 % of aggregate |
| NS blend coke | 2.23 wt% | 100 - $Y_{HS}$ % of coke |
| HS low calcined | 4.40 wt% | $Y_{HS}$ % of coke |
| HS high calcined | 4.10 wt% | $Y_{HS}$ % of coke |
| Pitch | 0.50 wt% | 14 % of green anode |
| Coke yield | | 66 % of pitch |

Part of the program consisted of small $2^3$ factorial designs to quantify the effect of treatments. Examples of treatments were

- Adding 25 % low-calcined or high-calcined L-HS coke
- Adding 50 % low-calcined or high-calcined L-HS coke
- Increasing baking level from 1200 to 1320°E
- Halving mixing energy or vibroforming energy input

Table V gives an example of a $2^3$ factorial. Only the effects relevant for desulfurization are described here.

Table V. Eight pilot scale anodes in 2^3 Factorial 1 with factor designation and sulfur concentration – HS coke was low calcined.

| A<br>L-HS low [%] | B<br>Baking [°E] | C<br>Mixing [min] | Sulfur<br>[wt%] |
|---|---|---|---|
| 0 | 1200 | 6 | 1.91 |
| 0 | 1320 | 6 | 1.94 |
| 0 | 1200 | 3 | 1.96 |
| 0 | 1320 | 3 | 1.90 |
| 50 | 1200 | 6 | 2.71 |
| 50 | 1320 | 6 | 2.51 |
| 50 | 1200 | 3 | 2.69 |
| 50 | 1320 | 3 | 2.36 |

Observations (percentages are averages from several factorials)

- At 1320 °E the desulfurization of the anode with 50% L-HS low calcined coke was -0.26 wt%, and the anode without HS coke 0.01 wt%. This meant the increase in desulfurization when adding 50% HS low calcined coke to the anode was 0.21 wt%.
- The corresponding increase in desulfurization for the 50% high calcined L-HS anode was 0.12 wt%.
- At 1260°E desulfurization of the low calcined L-HS anode was 0.02 wt%, and the high calcined L-HS anode was zero. The low calcined L-HS coke therefore started desulfurization at a lower baking level than the high calcined coke.

The change in sulfur concentration for the HS-coke added to the anode can be calculated using the information in Table IV. For the low calcined L-HS the change was 0.58 wt%, giving a sulfur concentration of 3.82 wt% at 1320 °E. For the high calcined, the change was 0.34 wt% giving a sulfur concentration of 3.76 wt% at 1320 °E.

This type of testing did not allow determination of a starting temperature for significant desulfurization, and therefore the rate of desulfurization cannot be determined accurately. Work is in progress to determine the starting temperature for significant desulfurization when a HS coke is used in anodes.

## 4. Estimating Sulfur Emissions in a Section

To model the sulfur emissions from a section in a baking furnace, one must know the desulfurization behavior of the components in the anodes, and the baking level of the anodes.

In Figure 4 the cumulative distribution is shown for two baking furnaces. Kiln 3 is a Hydro Aluminium vertical flue ring furnace with 105,000 ton annual capacity. Kiln 2 is an older Riedhammer vertical flue ring furnace with 25,000 ton capacity. The Kiln 3 data were used in the 3D representation of the baking level distribution in Figure 3.

Using the above furnace temperature distribution data, several cases were modeled to compare anode sulfur losses during baking.

Case 1: Anodes as in Table IV, but with a 50:50 NS:LS coke blend using the low sulfur S-LS coke in Table II. The sulfur content at the start was 13.4 kg/ton anode, and the desulfurization during baking was calculated to be:

Kiln 3 0.23 kg/ton anode
Kiln 2 0.60 kg/ton anode

Case 2: Exchanging the low sulfur S-LS coke with the high sulfur H-HS coke in Table II. This coke had an initial concentration of 3.57 wt%, starting temperature for significant desulfurization of 1236 °E and rate of desulfurization 0.0070 wt%/°E. The sulfur content at the start was 21.1 kg/ton anode, and the desulfurization during baking was estimated at:

Kiln 3 0.64 kg/ton anode
Kiln 2 1.71 kg/ton anode

Kiln 2, with the wider baking level distribution, would experience a higher increase in sulfur emissions when adding high sulfur coke to the aggregate.

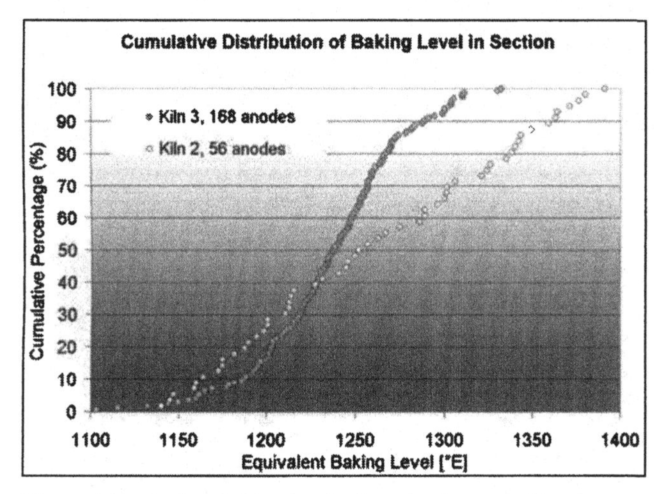


Figure 4. Equivalent baking level of anodes in two baking furnace sections, Årdal Carbon kiln 2 and 3.

Case 3: As in case 2 but with an average baking level 20 °E higher. The desulfurization during baking was estimated at:

Kiln 3 1.10 kg/ton anode
Kiln 2 2.19 kg/ton anode

For Kiln 3 the 20 °E increase in baking level would cause a significant number of anodes to reach the starting temperature for significant desulfurization, and to start desulfurizing, with a potential negative effect on anode properties.

To estimate the total sulfur emissions from a baking furnace, sulfur from fuel and from packing coke must also be included.

## Discussion

The results presented in this paper provide additional information on desulfurization of coke during calcining and anode baking. During calcining, coke is not calcined uniformly and coarser particles are calcined to higher levels than finer particles. The coarse particles in the high sulfur components of coke blends are more susceptible to desulfurization as a result of this. Excessive desulfurization of coarse coke must be avoided to reduce the potential negative impact on anode density.

Desulfurization results from anode baking furnaces have been shown for cokes where the heat treatment was quantified using the equivalent temperature. For each HS and NS coke, desulfurization was linear with the equivalent temperature, and it was possible to assign a starting temperature for significant desulfurization and a rate of desulfurization. With these parameters known, sulfur emissions during baking could be modeled. In the examples, the modeling was somewhat simplified by assuming linear desulfurization also for the NS-blend coke and the butts used.

Modeling sulfur emissions showed that increased use of HS cokes will demand stricter control of baking level for anode producers who operate under sulfur emission limits.

The interrelationship of coke calcining and anode baking on coke integrity is complex. Heavy calcining and further desulfurization during baking damage HS cokes. This will negatively affect anode quality. Observations made during the pilot anode study referred to above indicate that carboxy reactivity, and especially carboxy dusting, is sensitive to desulfurization damage and can be used as a measure. Two examples are given.

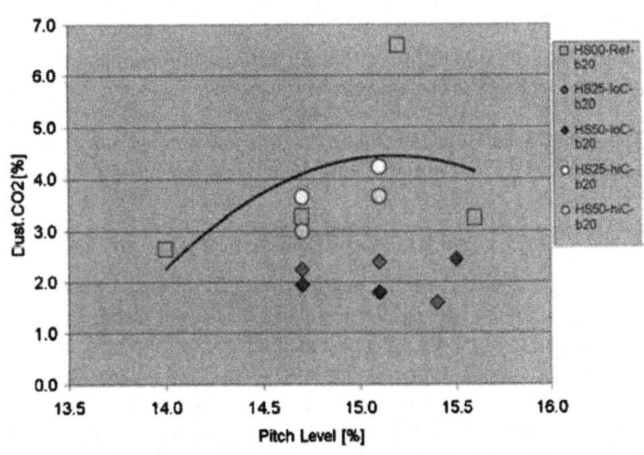


Figure 5. Carboxy dusting (Hydro dust index) for pilot scale anodes with 0, 25 and 50% addition of the high (hiC) and low calcined (loC) L-HS coke. b20=20% butts.

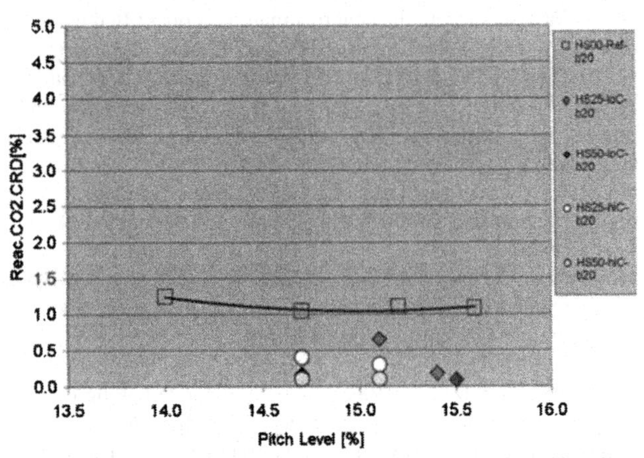


Figure 6. CRD Carboxy dusting (R&D Carbon method) for pilot scale anodes with 0, 25 and 50% addition of the high (hiC) and low calcined (loC) L-HS coke.

The carboxy dusting index by the Hydro Aluminium method is shown in Figure 5, and the carboxy dust percentage by the R&D Carbon method is shown in Figure 6. Generally, adding HS coke improves the carboxy reactivity properties due to the passivating effect of sulfur on sodium. This is seen in the plots, both for high calcined (circular legend) and low calcined coke results (diamond legend) relative to the blue line with no HS addition. However, the improvement is markedly better for anodes made with the low calcined HS coke (except outlier in Figure 6). Similar observations were made for the carboxy residue, R&D Carbon method, and the carboxy reactivity by the Hydro Aluminium method.

## Conclusions

High sulfur cokes undergo desulfurization during both calcining and anode baking. The rate of desulfurization depends on both the sulfur content and the level of calcination or heat treatment in the baking furnace. Desulfurization during calcining is also dependent on particle size due to the tumbling effect in the kiln where coarser particles tend to be more well calcined than finer particles.

The equivalent baking level is useful for quantifying the heat treatment of anodes in the baking furnace. It can also be used to quantify the desulfurization behavior of petroleum cokes and butts during baking. It was observed that desulfurization was linear with equivalent temperature and could be defined by a starting temperature for significant desulfurization and a rate of desulfurization. Linear expressions could be combined with the baking furnace baking level distribution to model average sulfur emissions per ton of anode produced.

The sulfur level of high sulfur cokes used in blends is increasing and will likely continue to increase. This work has shown that high sulfur cokes can be used successfully in blends without negative consequences but care must be taken to avoid over-calcining these cokes and over-baking anodes made with these cokes. Both have the potential to negatively affect anode properties.

## References

1. Les Charles Edwards, Keith J Neyrey, and Lorentz Petter Lossius, *A Review of Coke and Anode Desulfurization*, Light Metals 2007.
2. Lorentz Petter Lossius, Inge Holden, and Hogne Linga, *The Equivalent Temperature Method for Measuring the Baking Level of Anodes*, Light Metals 2006.
3. C. Dreyer, B. Samanos, F. Vogt, , "*Coke Calcination Levels and Aluminum Anode Quality*", Light Metals, 1996, pp535-542.

**Light Metals 2008** *Edited by: David H. DeYoung*
*TMS (The Minerals, Metals & Materials Society), 2008*

# STRUCTURAL EVALUATION OF COKE OF PETROLEUM AND COAL TAR PITCH FOR THE ELABORATION OF ANODES IN THE INDUSTRY OF THE ALUMINUM

Rafael J. Tosta M[1]., Evelyn Mercedes Inzunza[1]

[1]CVG Alcasa; Av. Fuerzas Armadas, Zona Industrial Matanzas; Puerto Ordaz, Estado Bolívar, 8050, Venezuela.

Keywords: Pitch, coke, anode

**Abstract**

**Abstract**

Petroleum coke is used as dry aggregated for production of electrodes in aluminum industry. This is a heterogeneous material that exhibits, at least, three predominant structures, which can be determined by microscopy techniques. Coal tar pitch, on the other hand, is the residual of the distillation or thermal treatment of the coal tar and it is also used in the conformation of anodes as the wet aggregated. Quality of anodes depends on coke structures and content and distribution of mesophases particles in pitch. Thus, this project is related with the structural evaluation of both components using different microscopy techniques.
The study involved the typical coke used in CVG- Alcasa and three types of pitches named A, B and C based on treatment temperature and soaking time.
Results indicated that coke shows different types of structures such as: elongated fibrous structures, fine and coarse mosaics, lamellar platelets, etc. Likewise it was observed planar ordering probably caused by temperature. Pores were also seen as a result of volatile matter. Related to pitch, results show that mesophase content depends on type of pitch analyzed. Type A presented smaller content in mesophase (0,77%) and better particle distribution, followed by B and C type pitch. As, the higher mesophase content the worse is the quality of pitch, type A pitch posses the best quality, and as a consequence a better anode will be obtained using it.

## Introduction

*Petroleum Coke*

Green coke from raw petroleum or green is the main component used in the production of anodes for aluminum industries. This is a solid form of coal, obtained by thermal decomposition and polymerization of the residuals derived from distillation of raw petroleum. It is a mixture of coal and heavy hydrocarbons, constituted by 88-95% carbon, 3-4% hydrogen, 1-2% nitrogen, 0,58-6% sulphur and 1-7% oxygen, chemically stable and inert under normal conditions. This initial coke depends on coking conditions; being its properties very affected for an incomplete treatment, high volatile content, low density apparent (bulk) and mechanical resistance. Coke of petroleum is a polycrystalline form of coal and it is considered as a random agglomerate of small glasses, ash and sludge. Its structure consists of arrangements of graphitics glasses with different sizes and orientations, Pores are also present and their dimensions vary according to purity, process conditions and the characteristics of petroleum residuals from which coke is obtained.

*Green Coke Quality*

The quality of raw coke is determined mainly as a function of sulphur, vanadium, iron and silicon contents. Classification of quality based on green coke purity is shown in table 1.

Table 1. Purity of green coke

| Quality / Purity | Good | Media | Bad |
|---|---|---|---|
| % S | < 2.5 | <5,5 | > 5,5 |
| Ppm de V | < 200 | < 400 | > 400 |
| Ppm Fe y de Si | <500 | <600 | > 600 |

*Coke types according to the production technology*

*Delayed Coke.*

This coke represents 90% of coke used in aluminum industry. Delay coking is a thermal process in which green coke dehydrogenation and polymerization occur. The main products are residuals and aromatic oils. The residuals contain asphaltic compounds that are mainly heterocyclic molecules. Aromatic oils comprise mainly aromatic polynuclear with rings coal -6 that produce the graphitic structures.
Several types of cokes are produced: needle, sponge and shot (pellet) cokes. The former product is obtained from raw petroleum with an important proportion of bonds with less than 6 coal atoms. In this case, the highly aromatic fractions continue plastic, even during carbonization allowing glasses form needle structures, with low coefficient of thermal expansion (CTE), and high conductivity. Isotropics or amorphous cokes are visibly very porous when coking and they are called sponge coke. The type shot (pellet) coke is a very rare type of coke that resembles small balls whose formation is favored with highly asphaltic residuals. The aromatic materials reduce the reaction of this coke type, although the fluidification in coke tanks can also produce it.

*Fluid Coke.*

Coking of fluid coke is a continuous process; where warm coke particles are feeding at 20-40 PSI and temperatures above 500°C and stay in contact with recirculated vapor, allowing coking process occurs. It is not used more than 10% of this coke in aluminum industry,, due to its high contents on sulphur and metals and also because it is available only in sub-millimetric sizes which hinder the mill and connection of anode.

*Coke types according to their origin*

According to their origin, coke is classified depending if the coke is derived from: aromatic residuals, asphaltic residuals or residuals paraphinic-naftenics.

*Anysotropic Coke.*

When coke is derived from aromatic residuals, coke is anysotropic, which is commonly called needle coke. It is obtained from condensation and polymerization of aromatic molecules, this coke type affects the properties of anode such as: resistance, reactivity and thermal expansion. Its properties are: Low level interleave, easy to become graphitic, different properties according to the direction, small quantity of sludge, small volumetric density, high porosity, relatively soft, small coefficient of thermal expansion, unidirectional and well interconnected pores and surrounded by walls of thickness. It can be used for anode production, after having calcined and, if necessary, after desulphurization and graphitization..

*Isotropic Coke.*

When coke is derived from asphaltic residuals it is called isotropic coke. This type of isotropic texture is produced due to a weak coking of highly volatil coal which is easily attacked by carbon dioxide.. A better coking produces thick circular and lenticular forms of the coal with lower reactivity. Typical properties of isotropic cokes are: High interleave level, difficult to graphitize, same properties in all directions, but pores closed, high sludge content, high volumetric density, high coefficient of thermal expansion, low crystallinity grade and small true density.

*Sponge Coke.*

When the coke is derived from Paraphinic-Naftenicos residuals it is denominated sponge coke. Its properties vary according to the formation mechanism and coke structures. Its hardness and density increase when asphalt content diminishes, In general terms it is the best coke used in the production of anodes for the industry of the aluminum. However, It is common the use of a mixture of anysotropic and isotropic cokes in order to take advantage of the individual characteristics of each one.

*Calcinations and process of graphitization of the coke.*

Coke calcination process involves a series of modifications in its crystalline structure, as a consequence of graphitization of coal. Carbonaceous mass of coke obtained, comprise different phases with different thermodynamic stabilities. At high temperatures, a number of transformations occur, from metastable to thermodynamically more stable structures.

Graphitization process can be considered as a process of multiple steps, where the resulting crystalline structure consists of some graphite layers parallelly ordered.. As the temperature of large coking is increased it is the structural classification of the same one. The purpose of coke calcination can be summarized as follow:

1. To increase the ratio carbon-hydrogen (C/H: 20>100) by elimination of great part of present hydrogen in the green coke.
2. To obtain grains strong enough to facilitate the handling and prosecution.
3. To minimize the shrink or grain volume reduction (<0.5%) by the end of anode thermal treatment.
4. To acquire enough thermal conductivity for an effective indirect heating.
5. To reduce electric resistivity.
6. To become less sensible to air oxidation
7. To obtain high purity.

*Measure of the optic texture.*

Petrography is the main tool to reveal both texture and microstructure of coke. The microstructure of coke is determined by porosity, the pores quantity and size, their wall thickness and quality of connecting. The texture and microstructure of coal are two important factors that affect the reactivity, and the stability of coke after reaction.

Optic texture is the appearance of coke surface. This can be examined using the microscope of reflected light, preferably with polarized light. Parallel poles are used as to the half of a wave slowed among the plate, the sample and the analyzer that allow to observe the interfaces of double color guided in the surface of the graphite sheets. An alternative method for the measurement of optic texture can be developed with pore radius. It is defined as the extension divided by the longitude of the diameter, where the extension (wide) is perpendicular to the diameter.

The nomenclature used to describe the characteristics has been developed by many years and they are now standardized by Marsh, 1989. The factor OTI (optical texture index) is a measure to describe the anisotropy of the structure of coal, where the dimensions vary according to resolution limits from 0.5 µm approximately to hundred of microns. According to Marsh, the definition of the optic texture anisotropy (OTI) is given by:

$$OTI = \Sigma f_i * OTI_i$$

Where:

$F_i$ = Fraction of each anysotropic domain.

$OTI_i$ = Factor for each anysotropic unit in the optic texture

In the Table 2 the different optic definitions of the cokes are shown.

Table 2. Optical definitions of the cokes.

| Characteristic | Limits | Factor OTI |
|---|---|---|
| Isotropic | Non optical activity | 0 |
| Fine Mosaic. | <0.8 µm diameter | 1 |
| medium Mosaic | >0.8, <2 µm diameter | 3 |
| Coarse Mosaic | >2.0, <10 µm diameter | 7 |
| Granular Flow | >2 µm long > 1µm width | 7 |
| Coarse fluid | >10 µm long > 2 µm width. | 20 |
| Laminar | >20 µm long > 10 µm width. | 30 |

*Coal Tar Pitch*

The tar is a highly aromatic hydrocarbon, derived after several successive distillations to high temperatures of the pitch of coal and of the residuals coming from the refinement the petroleum. This is used as agglutinative agent in the elaboration of anodes of coal.

The pitch of tar contains many solid particles of different origins formed during its production; such as: carbonaceas particles formed by the thermal craking of the vapors of tar, those dragged of coal and coke, and the insoluble particles formed in the phase it liquidates by means of the absorption of heat during the production of the pitch. The concentration of the particles accustomed to pyrolitics is a sensitive indicator of the grade of deshidrogenization of the vapors of tar happened during the carbonization. The particles accustomed to pyrolitics influence, in different grade, the determination of the point of softening and viscosity. When the pitch of tar absorbs heat, the particles accustomed to pyrolitics are united by the spherical bodies of the mesophase and they reinforce the liquidate plastic that doesn't penetrate easily in the porous structure of the filler. This phenomenon of the mesophase has a definitive influence on the properties (density in having cooked, forces and resistance) of the anode of coal.

*Quality and Physical-chemical Properties of the Pitch of Tar of Coal.*

The quality of the agglutinant is of supreme importance in the formulation of the anode of coal, because its composition complex singular and the thermal transformations of phases suffered by him during its production, they determine:

a. The interactions physique - chemical of the components of the anodic part ; b. The conditions of formation of a structure of coal; c. The mechanical, electric properties and speeds of consumption of the anode.

✓ The quality of the agglutinant is usually evaluated in function of the following properties physique - chemical:
✓ Good level of insoluble in quinolein (IQ) dear among (10 - 18%).
✓ Appropriate level of insoluble in toluene (IT) dear among (20 - 24%).
✓ Content of resins beta understood among (12 - 14%).
✓ Aromatic (fa) = aromatic C. / Total C..
✓ Mesophase.

*Mesophase*

The phenomenon mesophase discovered for JAMES D. BROOKS and GEOFFREY H. TAYLOR in the year 1965, they makes a decisive contribution to the improvement of the production of products of coal of high value (tar mesophase), such as coke premium for slowed coquification and fiber of coal for spinner. The mesophase is formed during the pyrolysis of residuals of petroleum (pyrolysis of pitch), or filtrate of pitch of tar of coal. During this process it is possible to observe under the microscope, using polarized light and to certain temperatures (300 to 500 °C), the formation of spheres anysotropics, which grow with long periods of time of reaction and increments of temperatures. Them coaleces around 500 to 600 °C, and they are transformed in semi-coke inside the phase, with marked anysotropics.

*Performance of the Mesophase with Temperature*

If it stays to 400 °C, during a time of 20 hours a sample of tar and with the use of a microscope with the polarized light one can observe the reactions that go happening, where the anysotropic prevails and the concentration begins in which the aromatic molecules go uniting, evolution of gases and superficial tensions that give to place to the formation of spheres denominated mesophase happening. These spheres begin to move and to grow culminating with the total cover of the womb that which can happen between spheres of different sizes where most is small and they flow around the biggest, for convection effect caused by the thermal treatment or the evolution of gases.

The mesophase behaves as a liquid glass, which is distorted by forces that happen inside the liquid tar, allowing increasing the viscosity as the mesophase loses its form esferoidal, until these are not visible. The figure 1 sample the behavior of the mesophase as it lapses the time to a temperature of 400 °C.

In the case of increasing the temperature to 460 °C one can observe a phase investment (solid-liquid) to (liquid - solid), being increased the evolution of gases progressively, which cannot escape due to the high viscosity of the tar, since in this phase the coking phenomenon begins. For this reason gradients of heating discharges should be avoided in the range of temperature among 350 to 600 °C during the cooking of the anode, because starting from 450 °C the coking of the tar begins, and the very violent exit of caught gases causes cracks and bigger porosity and it affects the mechanical resistance of the anode. Very slow heating gradients promote the mesophase formation.

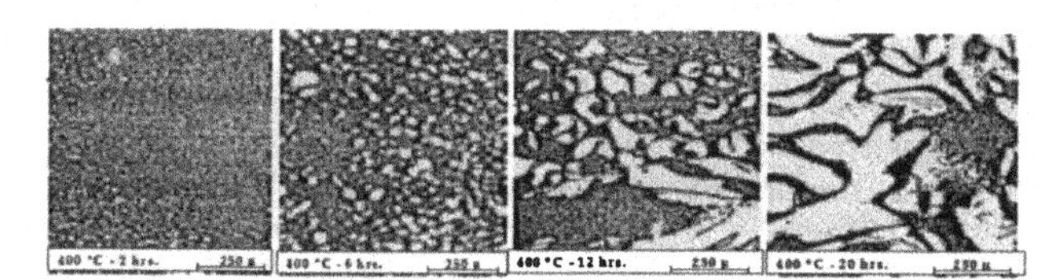


Figures 1. - Coalescent of the Spheres of Mesophase, as it lapses the time to oneself temperature.

*Effect of the Mesophase Over the Properties of the Anode.*

The references that describe the effect of the secondary IQ (mesophase) about the relative properties of the anode they are described to continue.

✓ In 1978, C. R. Mason, concludes that the agglutinant requirement for the anodic paste is increased with the increase of the concentration Mesophase.
✓ In 1981, D. Auguie, M. Oberlin, P. Hyvernat, informed on the highly harmful effect of the great quantity of spheres of secondary IQ in the tar it was found that the mesophase destroys the mixture, forming shells around the coke particles. Those shells reduce the wettability of the coke for the pitch of tar and consequently they degrade the properties of the anode.
✓ In 1986, T. Kagajo, K. Fujita, M. Died, in its work they conclude that the secondary IQ has a restrictive effect on the expansion of the anode during the cooking the IQ it increases the cooked apparent density.
✓ In 1989, A. Alscher, W. Gemmeke, F. Alsmier, it concludes that there is not negative effect on the properties of the anode, constituted with agglutinative tar that contains 1.9% of weight of secondary IQ.

*Cenosphere.*

It represents a smaller component of the pitch of tar of coal. This is formed for the quick pyrolysis of particles free of carbon. Microscopically they appear as hollow spheres or segments of them and their size typically is of around 10 to 500 microns. In polarized light, a cenosphere can be optically active, the size of the anysotropic pattern or mosaics depend on the line of the charred coal. The cenospheres is harder than the continuous phases and polished in relief. (to see figure 2).

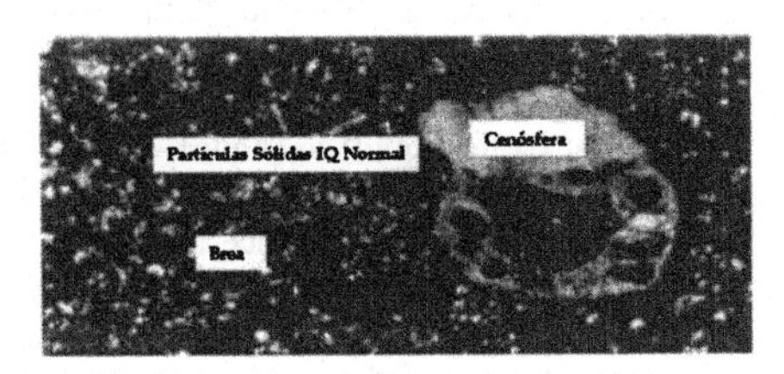


Figures 2. - Photomicrograph of the agglutinant showing the solid phase of IQ normal and solids of coal (Cenosphere) in the shown pitch .

**Experimental Procedure**

*Petroleum Coke*

The applied methodology is divided in two stages: The first one consists on the random selection of the sample and preparation of the same one and the second in the observation and analysis of the images.
The samples got ready applying the norm ASTM D3997, then they were refined for later on to be observed in an optic microscope (Olympus 1X70-S8F2) of invested light coupled an analyzer of images (Leco IA-3001 with camera nikon FX-350X) and an electronic microscope of sweeping (Philips XL 30) operated to a voltage of 10 KW in way of emission of secondary electrons, the samples were observed without cover

*Coal Tar Pitch*

They took samples of identified pitches as A, B and C For the determination of the content and distribution of size of the particles of Mesophase, were carried out a total of 105 samples of pitch of tar (TO, B and C), taking as range of temperature (110 - 550 ºC) and during a time of permanency of 1 at 3 hours.
For the production of the samples the mill technique was used for then to be configured granulometric and sifted in a Ro-Tap. Later on to the conformation granulometric, the samples of pitches of tar with grain <8 mesh was weighed and placed in small plastic bags. Of each sample you weight 50 gr which were transferred to a platinum hearth to subject them to the thermal treatments. Lapsed the lapses of time, each one of the fused pitches of tar was cooled
Once cold the samples of pitch of tar, were subjected to it roughdresses fine to be observed in the optic microscope coupled an analyzer of images. The samples were observed carefully jointly in the optic microscope with the analyzer of images. One carries out the respective program in the analyzer of images for the quantification and distribution of size of the mesophase particles for the samples of pitch of tar.

**Results and Discussion**

*Petroleum Coke*

The Table 5, it shows us the obtained results of the description present anysotropic in six types different from calcined petroleum cokes used in the industry of the aluminum. One can observe that the structure that prevails is the type needle and in smaller percentage the fine mosaic. This can also be observed in the micrograph shown in the figure 4, 5, 6 and 7. The studied cokes don't present porosity (> 10 m of long and> 2 m of wide) definite of the type thick flow with a value zero in all the samples what demonstrates that big lengthened pores and branching badges of the structure type needle exist.

Table 3. Description of the optic texture of the coke particles

| | CONTENT (%) | | | | | |
|---|---|---|---|---|---|---|
| TEXTURE | A | B | C | D | E | F |
| MF | 0.055 | 0 | 0.010 | 0.007 | 0 | 0 |
| M | 0.187 | 0.201 | 0.381 | 0.166 | 0.403 | 0.235 |
| MG | 2.423 | 3.295 | 6.826 | 4.154 | 8.380 | 3.642 |
| GF | 0.904 | 0.581 | 2.945 | 1.464 | 4.414 | 1.132 |
| CF | 0 | 0 | 0 | 0 | 0 | 0 |
| L | 96.40 | 95.92 | 89.83 7 | 94.21 | 88.80 3 | 95.00 3 |
| Ave | 99.97 | 99.99 | 100.0 | 99.99 | 100.0 0 | 100.0 1 |
| OTI | 19.519 | 19.46 | 18.66 | 19.24 | 18.52 | 19.34 |

The following images (Figures 3) in optic microscopy the classics structures present show in green cokes.

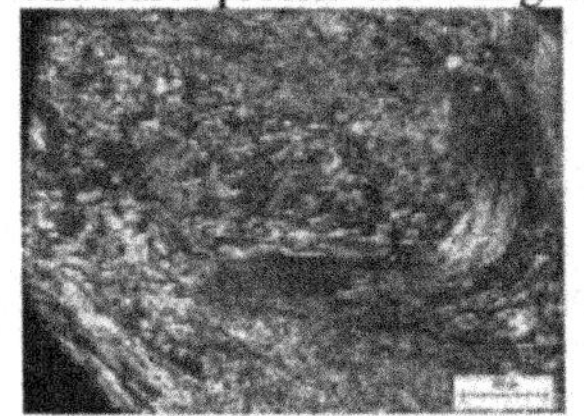
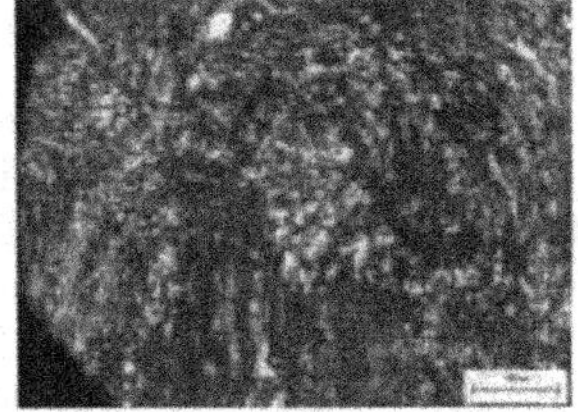

Figure 3. Images typical of green cokes of petroleum

In the following images we will show observed typical structures of calcined cokes of petroleum as much for optic microscopy as for electronic microscopy of sweeping.

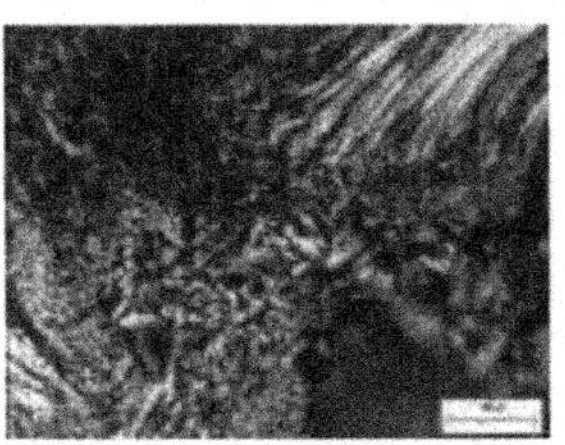

Figure 4. Coarse fluid anysotropic , small domains and coarse mosaics

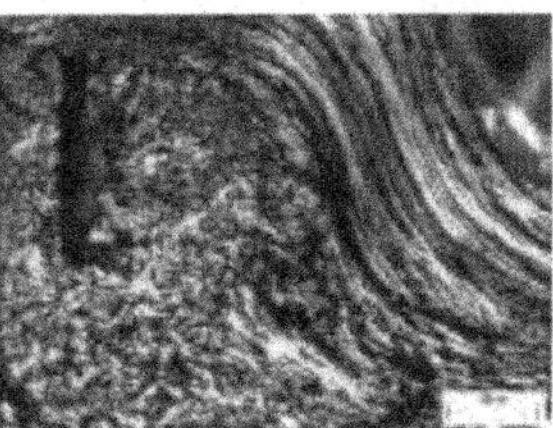

Figure 5. Medium Mosaic and fluid domains

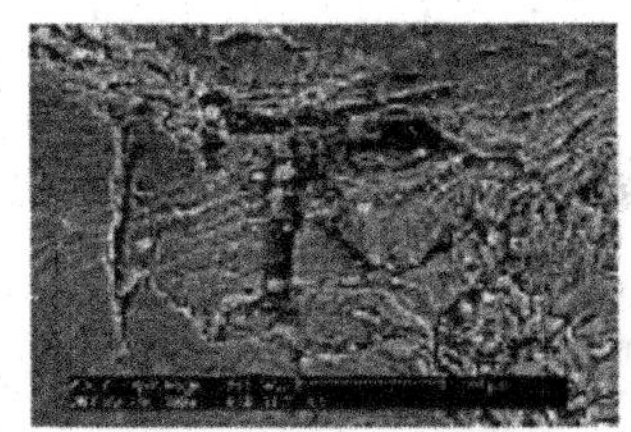

Figure 6. Elonged fiber structure (MEB)

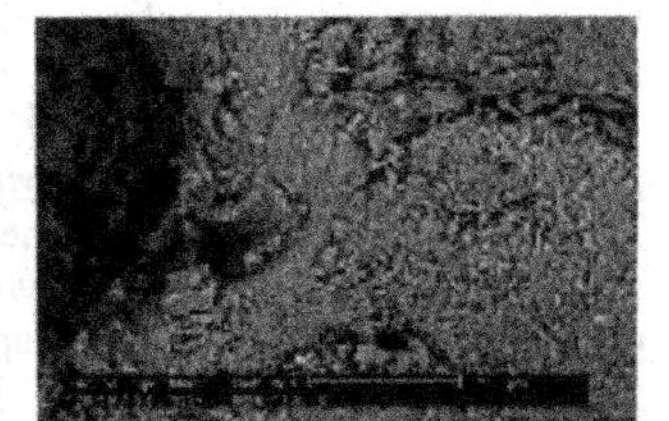

Figure 7. Lameliforme and elonged fiber structure (MEB

*Coal Tar Pitch*

*Mesophase evolution witch temperature*

The results shown in the Tables 4, 5 and 6 present the average of ten fields taken to each sample of each type of pitch, for each pitch some photomicrographs is also presented, showing the evolution of the mesophase.

Table 4. Average of the mesophase content in the pitch type A (for 10 fields each one)

| Content of the mesofase particles in the pitch type A | | | | | | | | | | | | |
|---|---|---|---|---|---|---|---|---|---|---|---|---|
| Temperature °C | 110 | | | 190 | | | 350 | | | 450 | | |
| Hour | 1 | 2 | 3 | 1 | 2 | 3 | 1 | 2 | 3 | 1 | 2 | 3 |
| Media 10 zoom | 0,4 | 0,5 | 0,7 | 0,8 | 1 | 1,1 | 0,3 | 0,8 | 0,9 | 0,4 | 0,5 | 0,5 |
| Media Total | 0,54 | | | 1 | | | 0,69 | | | 0.49 | | |

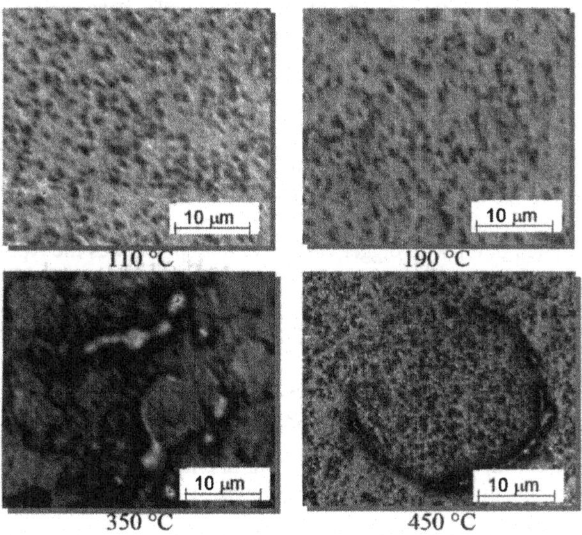


Figure 8. Photomicrographs of the evolution of the pitch type A with the temperature of thermal treatment

Table 5. Average of the mesophase content in the pitch type B (for 10 fields each one)

| Content of the mesofase particles in the pitch type B | | | | | | | | | | | | |
|---|---|---|---|---|---|---|---|---|---|---|---|---|
| Temperature °C | 110 | | | 190 | | | 350 | | | 450 | | |
| Hour | 1 | 2 | 3 | 1 | 2 | 3 | 1 | 2 | 3 | 1 | 2 | 3 |
| Media 10 zoom | 0,6 | 0,78 | 0,9 | 0,8 | 1,1 | 1,2 | 0,7 | 0,82 | 1 | 0,4 | 0,59 | 0,7 |
| Media Total | 0,75 | | | 1,04 | | | 0,85 | | | 0,56 | | |

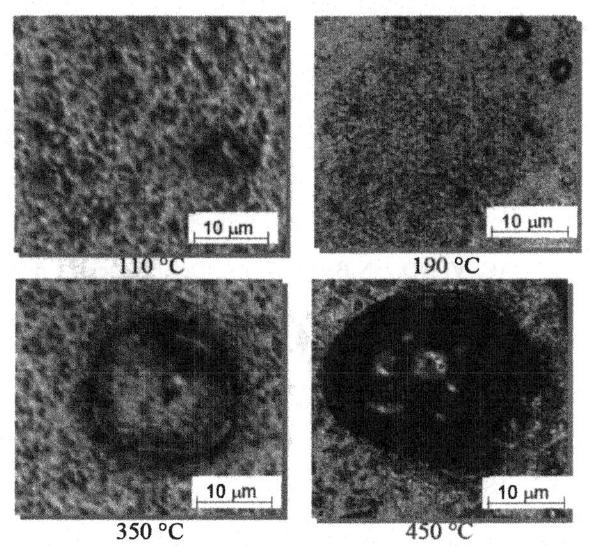


Figure 9. Photomicrographs of the evolution of the pitch type B with the temperature of thermal treatment

Table 6. Average of the mesophase content in the pitch type C (for 10 fields each one)

| Content of the mesofase particles in the pitch type C | | | | | | | | | | | | |
|---|---|---|---|---|---|---|---|---|---|---|---|---|
| Temperature °C | 110 | | | 190 | | | 350 | | | 450 | | |
| Hour | 1 | 2 | 3 | 1 | 2 | 3 | 1 | 2 | 3 | 1 | 2 | 3 |
| Media 10 zoom | 0,6 | 0,78 | 1 | 1 | 1,1 | 1,4 | 0,8 | 1,04 | 1,1 | 0,5 | 0,71 | 0,8 |
| Media Total | 0,79 | | | 1,15 | | | 1 | | | 0,66 | | |

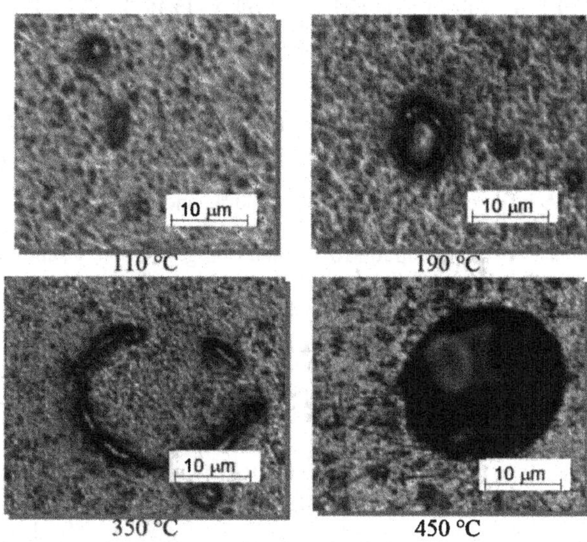


Figure 10. Photomicrographs of the evolution of the pitch type C with the temperature of thermal treatment

*Distribution and mesophase size found in the samples*

Table 7. Distribution and particle size for samples of pitch type A with the temperature and 2 hours of treatment.

| Temperature(°C) and 2 hours soaking time | | | | | |
|---|---|---|---|---|---|
| | Microns | 110 | 190 | 350 | 450 |
| Pitch A | 4-5 | | | | |
| | 5-6 | 5,09 | 5,64 | 5,74 | 5,36 |
| | 9-10 | | 9,56 | 9,82 | |
| | 13-14 | | | | |
| | 17-18 | | | | |

Table 8. Distribution and particle size for samples of pitch type B with the temperature and 2 hours of treatment.

| Temperature(°C) and 2 hours soaking time | | | | | |
|---|---|---|---|---|---|
| | Microns | 110 | 190 | 350 | 450 |
| Pitch B | 4-5 | 4,16 | 4,47 | | |
| | 5-6 | 5,15 | 5,58 | 5,86 | 5,51 |
| | 9-10 | 9,16 | 9,49 | 9,79 | 9,54 |
| | 13-14 | | | 13,52 | |
| | 17-18 | | | | |

Table 9. Distribution and particle size for samples of pitch type C with the temperature and 2 hours of treatment.

| | Temperature(°C) and 2 hours soaking time | | | | |
|---|---|---|---|---|---|
| | Microns | 110 | 190 | 350 | 450 |
| Pitch C | 4-5 | 4,12 | 4,46 | 4,92 | |
| | 5-6 | 5,16 | 5,44 | 5,83 | 5,57 |
| | 9-10 | 9,2 | 9,45 | 9,82 | 9,53 |
| | 13-14 | 13,15 | 13,15 | 13,46 | |
| | 17-18 | | | 17,45 | |

According to the total averages obtained in the Tables 7, 8 and 9, it indicates us that the content of the particles Mesophase in these three types of pitch of tar increases with the temperature until a certain value and then it diminishes; that is to say, the pitch of tar type A it increases of 110 °C up to 280 °C and then starting from this value it diminishes up to 550 °C., the pitch of tar type B increases of 110 °C up to 310 °C diminishing until 550 °C, and the pitch of tar type C it increases and it diminishes to the same temperatures that the pitch of tar type B. Like one can observe, the three types of pitches of tar spread to the same behavior, the difference resides that one increases and it diminishes in more proportion that another. This happens of equal it forms in the three types of pitch of tar as the sample remains more time exposed to the heat, that is to say, the content of the particles of mesophase increases in more proportion in the pitch of tar type C as this more time exposed to the heat remains, continuing in proportion the pitch of tar type B and lastly the pitch of tar type A that it is the one that presents smaller quantity of particles of Mesophase. This allows to settle down that in these three types of pitch of tar the particles of Mesophase experience their period of formation and growth to low temperatures until reaching an approximate value of 310 °C. The coalescens and volatilization of these particles of mesophase they are experienced starting from the 310 °C until the 550 °C.

The Tables 7, 8 and 9 present the distribution of size of the mesophase particle in each sample for certain temperatures and a time of thermal treatment of 2 hours, in these you can appreciate that the pitch of tar that presents bigger size of particles of Mesophase as it increases the temperature it is the pitch of tar type C, continued by the pitch of tar type B and lastly the pitch of tar type A. This behavior can be attributed to the fact that the pitch of tar type C possesses smaller quantity of insoluble in primary quinolein in comparison with the other two types of pitch of tar like one can observe when comparing the figures 8, 9 and 10 in which the evolution of the mesophase is observed and insoluble in quinolein of each type of pitch and which usually since with the mesophases the size of these they are in the order of 1 m, what originates that the little presence of insoluble in primary quinolein in this pitch of tar doesn't reduce the growth and coalescence of the spheres of Mesophase, since these like one knows they are not accepted by the particles of Mesophase and they remain in the surface of the same ones reducing I eat it was already said their growth and coalescence

In the figure 10 the stage of the 450 °C a hollow sphere is observed that is not more than a form of Mesophase already grown and they receive the name of cenospheres. These cenospheres represents a form of coalescence of particles of mesophase as it increases the temperature. After this it begins the period of volatilization, being generated a cenosphere semi-volatilizated leaving to the final pores that mean prints that there there was mesophase and also pores that come from the volatilization of lighter compounds.

**Conclusions**

*Petroleum Coke*

The use of petrographics technical, the optic microscopy and the electronic microscopy of sweeping, they increase the possibility to select a coke of petroleum of good quality for the elaboration of anodes in the industry of the aluminum.

The measure of the porosity of the coke samples determines that so compact that types and in that quantity is the morphologies definited of the coke of calcined petroleum.

*Coal Tar Pitch*

While higher is the content and distribution of size of the particles of worse mesophase is the quality of the pitch of tar, since, they increase the porosity in the anode of coal making it more reagent during its cooking process.

The pitch of tar with bigger content of mesophase was the type C with (0.96%), continued by the pitch of tar type B with (0.87%) and lastly the pitch of tar type A with (0.77%) respectively, being this last one that of better quality, since it possesses the necessary physical-chemical properties for form few particles of mesophase, smaller distribution of particle size, smaller porosity and sludges when it is subjected to thermal treatments.

According to the proportionate results at laboratory level, this study can be considered a method of alternative analysis for the evaluation of the pitches used in the industry of the aluminum.

**Bibliography**


1. Morten Sorlie, Harald Oye. *Cathodes in Aluminium Electrolysis*, (Aluminium-Verlag. Düsseldorf 1989).
2. P. Arredondo, Tesis de Maestría,"Determinación del porcentaje y distribución de tamaño de partículas de mesophase y evaluación de su influencia en la estabilidad del ánodo cocido", UNEXPO, 1984.
3. P.L. Lefrant, S.L. Hoff and J.J. Stefanelli, "Correlation of structural SEM data of cokes with graphite electrode performance", Carbon vol27, n°6, pp 945-949 1989.
4. Ross A. " The use of petrographic techniques for evaluation of raw material and process changes in an aluminium smelter", Light Metals 2000, pp 501-509.
5. ASTM D 3997 Standard Practice for Preparing Coke Samples for Microscopical Analysis by Reflected Light.
6. ASTM D 4616 Standard Test Method for Microscopical Analysis by Reflected Light and Determination of Mesophase in a Pitch.
7. ACEVEDO, N.B. (2000). "Determinación de la Mesophase en la Brea de Alquitrán Utilizada en la Fabricación de Ánodos de la Empresa C.V.G. ALCASA". Informe de Pasantía. IUTJAA.
8. ALSCHER, A.; GEMMEKE, W. ALMEI, F y BOENIGK, W. (1.989). "Evaluation of Electrode Binder Pitches for the Production of Prebaked Anodes Using a Bench scale Process". Light Metals.
9. ASTM. Designación: D 4616 - 95. Annual BooK of Estándar. "Microscopical Determinatión of Volume Porcent and the Maximum Spheroid Size of Mesophase in Pitch by the Point Count Method".

## ELECTRODE TECHNOLOGY

# Anode Manufacturing and Developments

*SESSION CHAIRS*
**Markus W. Meier**
R&D Carbon Ltd.
Sierre, Switzerland

**Alan David Tomsett**
Rio Tinto Alcan
Thomastown, Victoria, Australia

**Light Metals 2008** *Edited by: David H. DeYoung*
***TMS (The Minerals, Metals & Materials Society), 2008***

# ANODE BUTTS AUTOMATED VISUAL INSPECTION SYSTEM

Jean-Pierre Gagné[1], Marc-André Thibault[1], Gilles Dufour[2], Claude Gauthier[2], Michel Gendron[2], Marie-Claude Vaillancourt[2]
[1]STAS Inc. (Société des Technologies de l'Aluminium du Saguenay Inc.); 1846 Outarde; Chicoutimi (Que.), Canada G7K 1H1
[2]Alcoa Canada, Aluminerie Deschambault; 1 Blvd. des Sources; Deschambault (Que.), Canada G0A 1S0

Keywords: Anodes, Anode Butts, Anode Stubs, Inspection, Stub Cleaning

## Abstract

Anodes consumed in the smelting process for the production of primary aluminium need to be replaced regularly. Inspecting anode butts after they have been removed from the cells can be useful to optimize the electrolysis process, reduce the anode cycle and optimize anode fabrication. In most aluminium smelters, some manual measurements are taken on a few anode butts in order to obtain some feedback on anode performance.

STAS-Alcoa's R&D team has been working to develop an automated inspection system, with the objective of designing a low cost solution to automatically record a series of measures on all the anode butts processed at the rod shop. A plant prototype was implemented at "Alcoa, Aluminerie de Deschambault", Canada, on May 2006. The inspection system, based on artificial vision, is integrated within the existing conveying system of the rod plant. All anode butts are inspected in line while they are transported on the conveyor. This paper presents some of the results obtained.

## Introduction

The well-known Hall-Héroult process to produce aluminium uses carbon blocks as anodes. These anodes are consumed during the smelting process and need to be replaced on a regular basis. About two tonnes of alumina and 0.5 tonne of carbon are necessary to produce one tonne of aluminium. In a typical modern smelter, the useful life of an anode is about 25 days. Spent anodes are called "anode butts". Once removed from the cells, the anode butts are brought to the rod shop. After cooling, the butts are cleaned before the residual carbon and cast iron are crushed in order to be recycled. Figure 1 shows a carbon block with its anode rod and a typical anode butt after cleaning.

The Alcoa plant in Deschambault currently operates 264 electrolytic cells, and each cell is fitted with 40 anodes. Annually, the smelting process consumes about 150,000 anodes, that is, about 400 anodes per day.

While they are used in the electrolytic cells, the carbon blocks are consumed from contact with the oxygen ions that come from the dissolving alumina. The thickness of the anode butts on removal from the cells can vary depending on several parameters. Moreover, sometimes the oxidation is irregular and the anode butts may contain anomalies. These defects depend on many factors such as anode fabrication, raw materials and the electrolysis process. With a visual inspection system, it is then possible to optimise the anode fabrication process and/or the smelting process based on feedback related to the morphology of the anode butts processed at the rod shop.

Most aluminium smelters take weekly manual measurements on a few samples to obtain some feedback on the characteristics of the anode butts, but this approach provides only partial feedback from the available anode butts. Moreover, the results from manual measurements and evaluations can be subjective, inaccurate and not necessarily consistent. Until May 2006, Alcoa Deschambault was carrying out such manual measurements using a modified caliper. Only 30 anode butts were measured each week. The deformations were analysed in a qualitative manner by the operators, and since the measurements were not always performed by the same person, the analysis was somewhat unreliable.

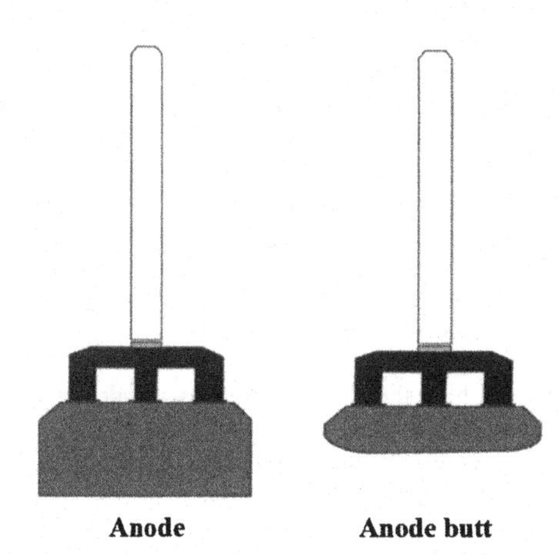


Figure 1. Typical anode as used at Alcoa Deschambault

In 2005-2006, the ALCOA-STAS Development Team developed an Anode Butts Automated Visual Inspection System to measure all the anode butts processed at the rod shop. Early on, it was decided that the inspection system would be based on artificial vision. In the preceding years, the same development team had successfully demonstrated in Deschambault the benefits of artificial vision applied to geometrical measurements through the installation of a system to set anodes within the potrooms.

The main objectives of the project were to develop a system with the following characteristics:

- Excellent repeatability;
- Good accuracy;
- Excellent robustness;
- No stop on the conveyer for measurements.

The Anode Butts Automated Visual Inspection System is located between the last cleaning step (of the anode butts) and the equipment used to remove the residual carbon.

This paper presents one type of the results obtained with the inspection system installed at Alcoa Deschambault (carbon thickness) as well as the team's work that led to the automatic classification of anode butt anomalies into standardized families of defects.

## Example of Measurements

About twenty parameters are automatically measured on each anode butt (angles, heights, surface, etc.). This section presents two of them: carbon thickness and deformation or anomalies. The method developed to classify the various butt anomalies into standardized families is also briefly described.

### Carbon Thickness

The thickness of the carbon remaining under the cast iron thimble is a very important parameter. Indeed, too thick anode butts means that they could have been used longer in the cells and the production cost has not been optimised. On the other hand, too thin carbon indicates a high risk of iron contamination in the electrolysis cell.

During anode casting, the anode rod is mechanically supported on its vertical axis, and the carbon block is raised. The longest stub is the first to stop against the bottom of the corresponding hole. Molten cast iron is poured into the holes, and the space under each shorter stub is filled with cast iron.

At Alcoa Deschambault, carbon thickness can now be precisely measured on every anode butt by combining the measurements made on the anode butts with the measurements of the lengths of the stubs, which are taken through the use of a second automated system that measures the anode stubs[1].

Figure 2 shows how the carbon thickness is calculated using the two measurement systems (butts and stubs). With these two systems, all measurements are taken in reference to the upper surface of the stubs horizontal traverse. Measures "A" (Figure 2) correspond to the distance from the top of the horizontal section to the underside of the butt. These measurements are taken by the anode butt analyser under each stub.

The second measure, "B", corresponds to the distance from the top of the horizontal section to the tip of the longer stub. This measurement is taken by the stub analyser[1] located downstream, after residual carbon and cast iron have been removed.

The carbon thickness, "C", is then calculated for each stub using measures "$A_1$, $A_2$, $A_3$" and "B" (Figure 2).

### Families of Defects

As indicated earlier, in practice, the shapes of the anode butts do vary. These variations depend on many factors such as anode fabrication, raw materials, electrolysis process, etc. It is thus possible to optimise the anode fabrication and/or the smelting process based on the morphology of the anode butts processed at the rod shop.

Given the multitude of shapes recorded initially, it became rapidly evident that some way of classifying the anode butt defects was needed. In order to develop an automatic classification system, a group of experts from Alcoa was involved. It was decided that the classification method would be based on the causes of the anode butt defects. Six families of defects were defined, each of them associated to their major causes. Figure 3 shows typical examples of each family, with the main causes of the defects.

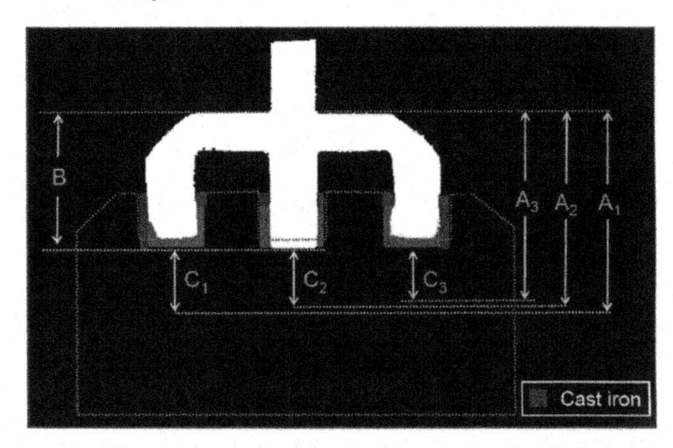


Figure 2. Carbon thickness under cast iron tumble

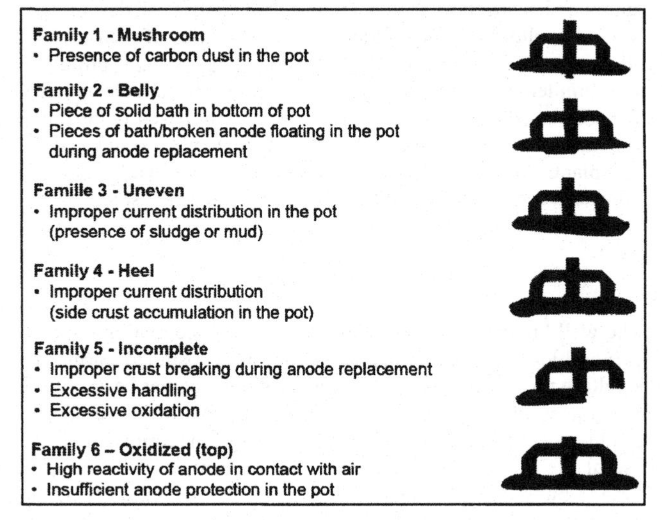


Figure 3. Families of defects

Figure 3 shows situations where the classification is obvious. However, in some cases (Figure 4), certain anode butts are more difficult to classify.

Figure 4. Example of anode butts difficult to categorise

A standardized classification method was thus required to obtain results that could be repetitive and objective. To this end, a reference group was defined. This group was made of about a hundred images of anode butts for which the deformations or anomalies were manually classified by the team of experts, using the chart shown on Figure 3.

Measurements of the main dimensional characteristics of the defects observed from the reference group were used as the basis to define a series of measurements to be made using the vision system. The image analysis software was programmed to automatically make these measurements on each anode butt. The following are examples of the measurements used to identify each type of anomaly.

Family 6, "Oxidized (top)", distinguishes itself from the others, because it makes reference to the upper portion of the butt.

The determination of a butt classified as "Oxidised" is made by superposing the image of the anode butt with that of a new anode (see Figure 5). The analysis software calculates the oxidation ratio of the anode butt.

Figure 5. Example of superposition for calculation of oxidation

An analysis of the various butts defined as "Oxidised" in the reference group allows to establish a lower threshold value of the oxidation ratio beyond which all butts exceeding this threshold value are automatically included in this category.

Family 5, "Incomplete Butt", includes all types of possible breaks of the anode butts. The analysis is made by determining the presence or absence of carbon nearby each stub. For that purpose, four control zones adjacent to each stub have been defined (see Figure 6). The presence of carbon in each of these four zones is evaluated and the algorithm determines if a butt can be classified within Family 5, depending upon the value measured compared with the threshold criteria determined for that family. See Figure 7 for two examples of butts included in this family.

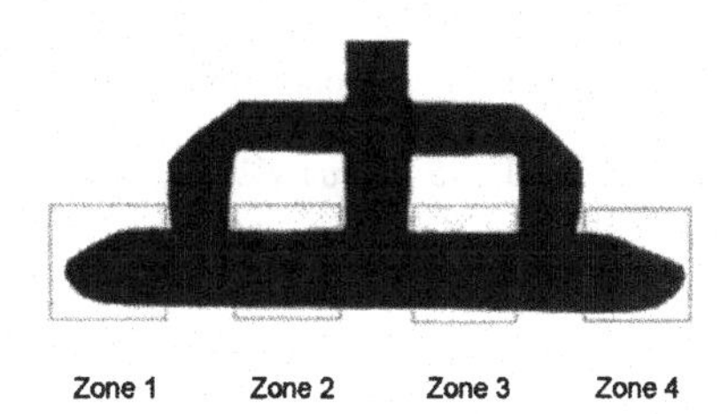


Figure 6. Location of control zones used for detection of broken butts

Figure 7. Example of butts belonging to Family 5 (incomplete butts).

For Family 3, "Uneven (uneven wear)", the system compares the angle between the lower surface of the butt and the horizontal traverse of the anode rod. Given the fact that all butts are not necessarily horizontal, the angles are measured with relation to the top of the image. First, the horizontal traverse that links the stubs is used to determine a reference angle (Angle A2 on Figure 8). Then, a second angle, "Angle A1", is determined by measuring the angle between the lower surface of the butt and the top of the image. Comparing angles A1 and A2 provides a value for classification analysis. When the anode wear is relatively straight, Angles "A1" are "A2" are more or less equal.

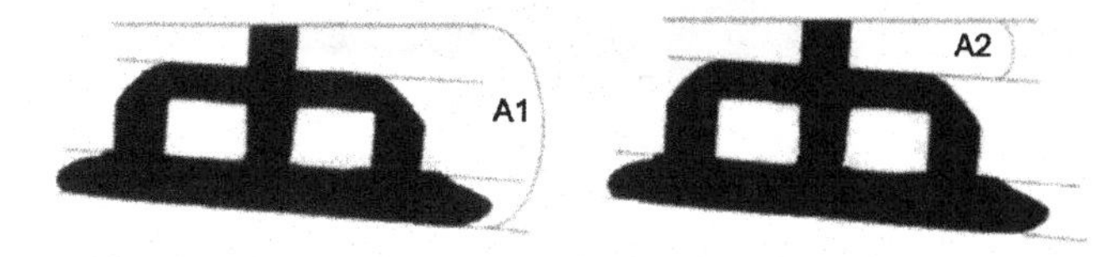


Figure 8. Measurements of wear angle

Families 1, 2 and 4, "Mushroom, Belly and Heel", are the families that are the most difficult to discriminate, because they are based on the same measurement parameters, that is, the characteristic measurements of the anomalies that are present on the butt lower surface. The system begins by locating anomalies. Then, more detailed measurements are taken in specific zones where the anomalies are detected, and data is collected about the different characteristics of these anomalies:

- Surface (S)
- Width (W)
- Height (H)
- Position (P)

The defects included in Family 1, "Mushroom", were characterised according to the low ratio of their width by their height (W/H) (Figure 9). Comparing this criterion with the threshold value determined with the reference group allows to differentiate the "Mushroom" butts from the "Belly" or "Heel" families.

Figure 9. Family 1 – Example

Further analysis is made by separating the butt section into seven specific zones and by associating each defect that has been identified with one or more of these zones. Figure 10 shows the zones that are defined for a typical butt. The defects of Family 4, "Heel", can be measured only in zones 1 and 7, even though these anomalies present geometrical characteristics similar to those of Family 2 (Belly).

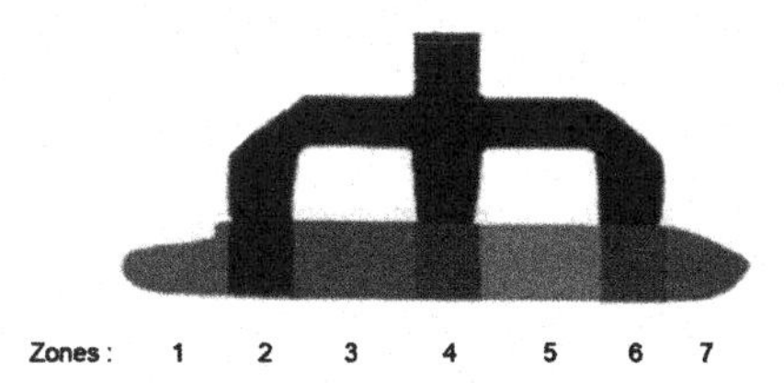


Figure 10. Selected zones of defects

The classification of defects within Family 2 (Belly) is made when anomalies are detected in zones 2 to 6 (Figure 11).

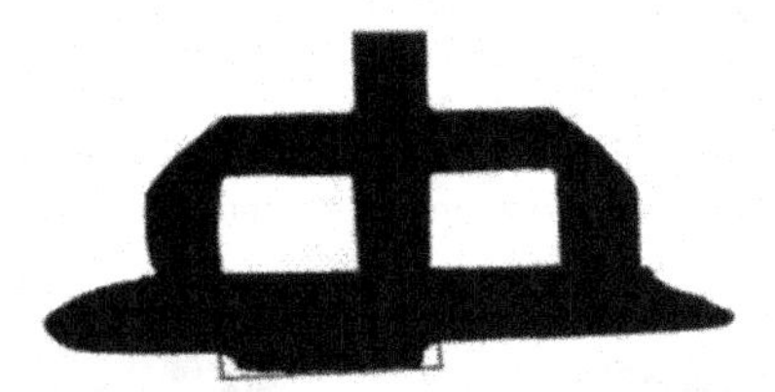

Figure 11. Family 2 – Example

The results from the different measurements taken on each butt are sent to the plant database. A specific software application has been designed to automate the classification process based on the data recorded in the plant data base (Figure 12). For that purpose, a series of classification criteria has been established based on each type of defects. These criteria are then used by the classification software to categorise each anode butt.

Results from the butt analyzer | Plant database | Software used to automate the classification process

Figure 12. Raw measurements are transmitted to the plant database, where they are automatically sorted and reported

As an example of the output available to the plant personnel, Figures 13 and 14 show the weekly percentage of butts that were classified into Families 1 and 5 during the first 37 weeks of 2007.

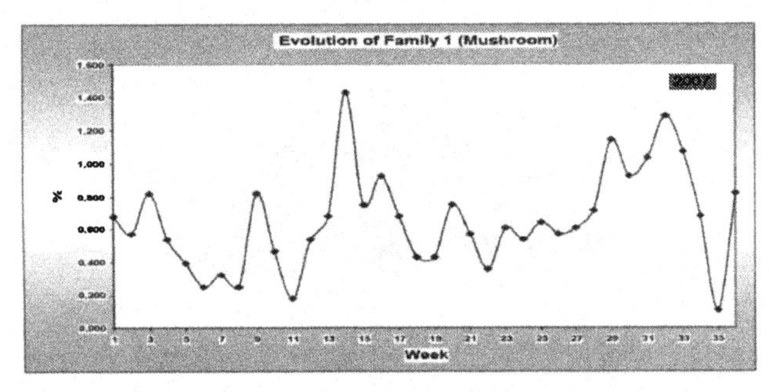


Figure 13. Evolution of Family 1 during weeks 1 to 37 in 2007

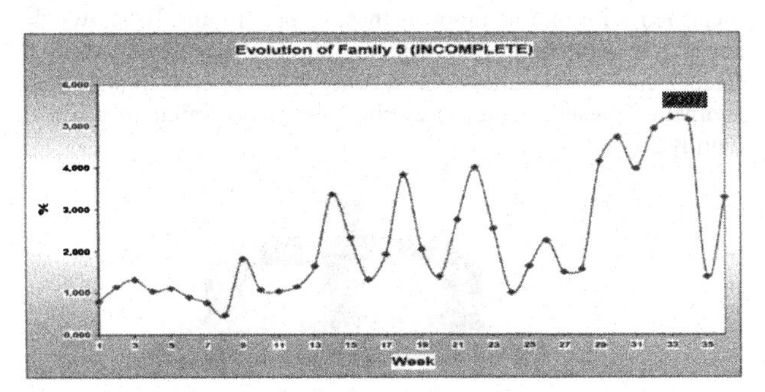


Figure 14. Evolution of Family 5 during weeks 1 to 37 in 2007

## Conclusion

To sum up, six families of anode butt defects were defined based on the major causes of the anomalies. To perform and automate the classification process, about a hundred images of butts showing geometrical anomalies were classified by a team of experts into their corresponding families. A detailed mathematical analysis of the anomalies was carried out to define the main characteristics of each type of defect. Then, the image analysis program was modified to perform these measurements. A second software was developed for the automatic classification of each anode butt based on the raw data recorded in the plant database.

As an outcome of this project, Alcoa Deschambault now owns an excellent system to follow-up the variations of the anode butt thickness. This system is based on the use of two automated artificial vision equipment: an "Anode Butts Automated Visual Inspection System" and an "Anode Stubs Inspection System"[1]. Both systems are designed with simple and proven components and are highly reliable. Since their installation in May 2006, no major repair has been required.

For all the anode butts that are removed from the pots, Alcoa Deschambault can easily follow-up the number of butts in each family of defects. In addition, the plant personnel can follow-up the evolution or variation of more than twenty additional parameters measured on each anode butt. The collected information is now used by Alcoa's personnel to react quickly when abnormal situations are detected. Further use of the available information is still being developed.

## Acknowledgements

Sincere thanks are due to the Alcoa Deschambault's personnel and the STAS design team, without whose dedication the success of this work would not have been possible.

## Reference

1. Jean-Pierre Gagné et al, "Anode Stubs Inspection System" (Paper presented at the TMS *(The Minerals, Metals & Materials Society)*, Orlando, Florida, 25 February 2007), 4.

**Light Metals 2008** *Edited by: David H. DeYoung*
*TMS (The Minerals, Metals & Materials Society), 2008*

# NEW RODDING SHOP SOLUTIONS

Nicolas Dupas
ECL, 100 rue Chalant, 59790 Ronchin, France

Keywords: Rodding Shop, anodes, carbon blocks, anode stems, bath cleaning, hot bath cleaning, loading station, shot blasting, butt stripping station, thimble press, caroussel, mating table, casting station, automation.

## Abstract

With more than 25 years of experience in rodding shop equipment design and commissioning, ECL is the leading supplier for modern smelters. All types of technology (2, 3, 4, or 6 stubs) are catered for, from turnkey solutions to individual machines, such as:
_ Overhead conveyors,
_ Loading/unloading stations with unique centering mechanism,
_ Bath breaking and removal stations,
_ Hot-bath cleaning and hot-bath treatment solutions,
_ Shot Blasting stations,
_ Butt stripping machines,
_ Thimble press, with new simultaneous breaking design,
_ New butt and thimble stripping combination,
_ Stem control and preparation stations,
_ Casting stations: standard carriage, articulated carriage, casting crane, mating tables and carrousels.

Centralized computer control of the rodding shop through "Level 2" solutions allows an increase in productivity and efficiency, and facilitates equipment maintenance.

As for all of the smelter's equipment range, environment, health and safety are the top priorities in the design and commissioning of the rodding shop.

## Introduction

The anode rodding shop plays a major role within prebake technology aluminium smelters and it is getting increasing attention when it comes to the efficiency of the smelter. Its performances in terms of production quality and productivity can dramatically influence the operation of the reduction lines, as well as the quality of the aluminium produced.

The role of the rodding shop is to recycle spent anodes and produce new ones. The aluminium anode stems, the carbon blocks, the cast iron used to mate them, and the frozen electrolysis bath are separated, and recycled in the aluminium production cycle. The stems are used for new anode assemblies (if they are physically sound), the carbon is crushed and sent to the paste plant, the cast iron is remelted, and the crushed frozen bath is re-used in the potrooms.

The reduction lines must be constantly supplied with new anodes as spent anodes are taken out of the pots, so the rodding shop has quality, productivity and reliability imperatives. The impact on the environment, mainly due to the anode cleaning process dust, also must be controlled and reduced.

To meet these requirements, ECL offers a global approach to the rodding shop design going further than mere equipment supply.

## Rodding shop design through flow management

The design phase of the rodding shop always starts with the smelters targets for production quality and productivity. From these imperatives, the type of equipment needed and the quantity of machines is determined, based on each machine's:
_ reliability rate,
_ availability rate,
_ cycle time.

Once the type and quantity of equipment is determined, the anode flow has to be optimized within the customer's given space. Physical location of the machines is critical, as is the size and availability of buffer stocks. These stocks are necessary so that a station down time (due to maintenance for example) is transparent for the production line. An example of anode flow simulation is shown in ***figure 1***.

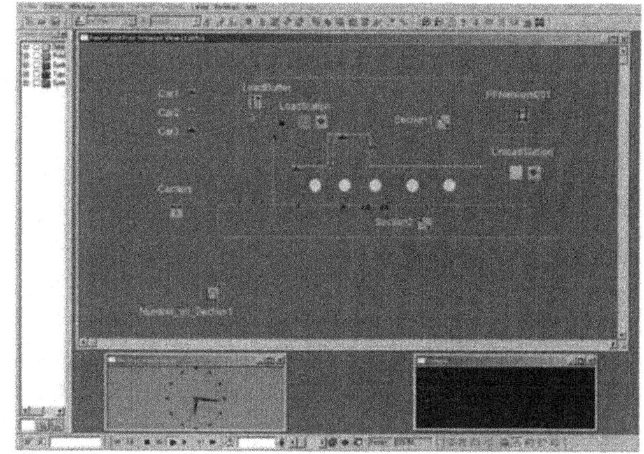

***figure 1: Anode flow simulation through the customer-specified rodding shop stations***

The material flows within the rodding room are organized through simulation using a commercial software (Witness®). The following variables are used: cycle times, availability rates, conveyor speed and machine locations. Various solutions are simulated to determine the best set of machines and rodding shop organisation for a given productivity target (***figure 2***).

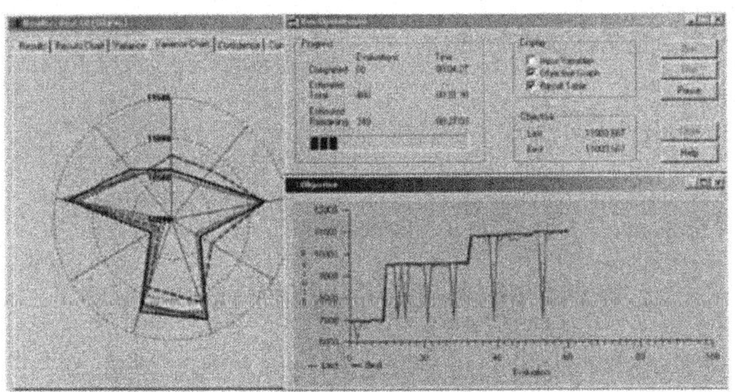

*figure 2: Rodding shop throughput simulation using Withness® software*

A full year of production is simulated, giving production curves for each piece of equipment. This allows ECL to guarantee an efficient solution with given production performances, before the production has started and rodding shop reorganisation is almost impossible.

### The rodding shop cycle

The rodding shop cycle is constituted of a succession of stations, each performing a specific duty for the cleaning of spent anodes or the manufacturing of new ones. Stations can sometimes be doubled-up to avoid bottlenecks and ensure the rodding shop's throughput.

The nature of the stations and their quantity are specified by the aluminium smelter. It is determined according to the desired performances of the rodding shop in terms of:

_ throughput: the quantity of spent anodes the rodding shop is able to clean, and the quantity of new anodes it is able to produce in a given amount of time,

_ cleaning efficiency: for example, maximum residual bath quantity can be specified at several steps of the spent anodes cleaning process. The quality of the cleaning process will have an impact on the bath and carbon materials recycling process efficiency.

_ impact on the environment: the spent anodes cleaning process creates dust and high noise levels that must be controlled to ensure a safe and clean working environment for the workers.

The rodding shop cycle follows a linear logic starting and ending with the loading/unloading station (***figure 3***):

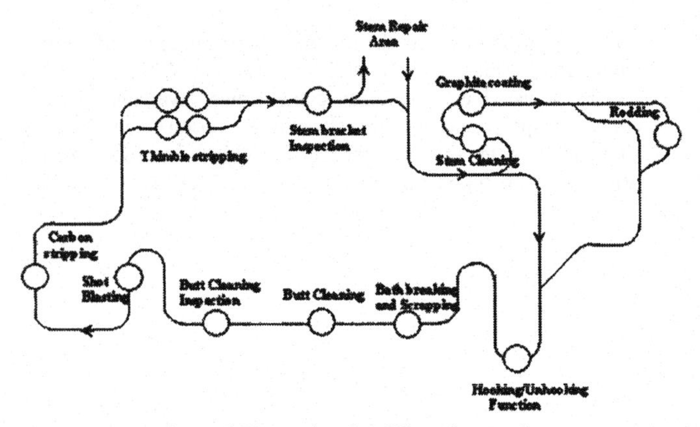


*figure 3 Example of rodding shop cycle*

Each station is assigned a specific task in the cleaning or rodding process. Its performance and cycle time have been calculated to maintain the desired performance of the entire rodding shop.

The technical solutions developed to ensure such performance are described in the following sections.

### Loading/unloading station

The role of the machine is to load spent anode pallets, centre anode stems and lock them onto the overhead conveyor, and to unload new anodes onto free pallets.

A new design of loading station was devised by ECL for the Hillside smelter rodding shop in 1995. This machine set a new standard for compactness and autonomy and is now a reference in the industry.

Traditional stem centering mechanisms use articulated arms and grips. These systems cannot guarantee the secure hooking of the anodes if they are at an angle (due to an uneven carbon block bottom for example). They also constitute additional mechanisms requiring critical space and specific attention from the maintenance teams.

By using centering pyramids (***figure 4***), ECL ensures the secure hooking of the stems, whatever their angle with regards to the carriage. This simple system is very compact and requires very limited maintenance. The hooking bell claws are also synchronised and self-tightening to ensure good and safe clamping.

*figure 4: Centering mechanism on a loading/unloading station*

To reduce the impact on the environment, the pallet cleaning process is automated and completed after each loading sequence in an enclosed area. While traditional pits often become safety hazards, the anode remains are collected into a dedicated receptacle and the rodding shop dedusting system collects the dust.

All movements are driven by VFDs (Variable Frequency Drives), as opposed to relay technology. This greatly reduces anode dropping risks due to the smoother movements allowed by this technology. Furthermore, rotary and laser encoders are used for high precision and repeatability of the positioning sequence. All of this leads to a cycle time down to 60 seconds per anode, and a higher reliability of the first rodding shop station.

### Bath breaking stations

As the first step in the spent anode cleaning process, this equipment breaks the frozen bath present on the top of the anodes. This material is very hard and resilient and its removal requires high forces. Here, it is mainly broken into pieces that will be removed at the next cleaning station.

Contrary to most designs, ECL's solution does not use vibratory motion for the breaking function. The use of vibration negatively affects equipment integrity and often results in a significantly reduced reliability. The ECL solution uses translatory motion tools which allows an efficient penetration into the hard material and reduces breaking force (***figure 5***).

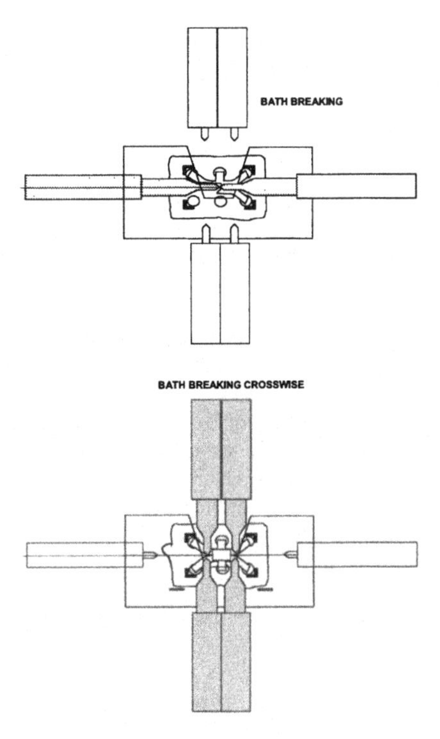


***figure 5: Breaking solution for hexapods using six breaking tools***

A variable number of tools are used: from one to six for hexapod type anodes. In this configuration, forces up to 80T are applied to the frozen bath. To avoid destructive forces on the overhead conveyors, the anode is clamped in a vertical position and its weight is carried by the machine (***figure 6***).

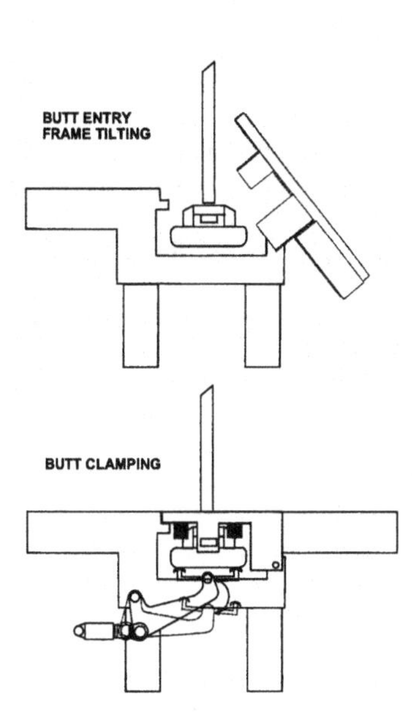


***figure 6: Anode clamping during the bath breaking phase***

During the breaking phase, the anode is kept inside a sealed enclosure: the dust is collected and the falling lumps of bath are collected under the machine. The actuators (hydraulic cylinders, gearboxes) are located outside of the enclosure so that they are not exposed to the bath particles, and to ensure easy access for the maintenance team.

For hexapod anodes (6 stubs), the cycle time can be brought down to 90s per anode (stem) as is the case at the Alouette smelter in Canada.

### Finishing stations

Once the frozen bath is broken, it is necessary to remove it completely from the top of the anodes. In the case of a hexapod anode assembly, a finishing station is always necessary for this task as lumps can be wedged between the two lines of stubs.

The anode is clamped and unhooked from the conveyor. It is then tilted at a 90° angle. Scrapers and rotating brushes (2 or 3) clean the surface of the carbon blocks, removing all but the smallest stuck bath elements from the anodes. This allows an average residual bath quantity of less than 4kg per anode.

As with the bath breaking station, the anode is inside a sealed enclosure to reduce the impact on the environment, and to keep the hydraulic cylinders and mechanical elements outside the enclosure for protection and easy maintenance. The cycle time can be as low as 90s per anode for a 6 stub stem.

### Hot bath cleaning stations

Most of the gas emissions from spent anodes take place during the cooling phase, just after they have been taken out of the pots.

Most smelters let the anodes cool down in specific areas before sending them to the rodding shop. In order to reduce gas emissions and minimise the impact on the environment, the anodes can be cleaned while the electrolysis bath is still hot.

To avoid material build up and for maximum heat resistance, specific profiles are used for the breaking tools. It is also necessary to use metallic rather than standard rubber conveyors to carry the hot bath pieces. These conveyors are hooded and equipped with specific fume collection systems to ensure the emissions of the rodding shop are effectively reduced.

While the breaking of hot bath requires significantly smaller forces than for cold bath, it is necessary to design machines that can work on the latter. The possibility that anodes reach the rodding shop after they have cooled down is high and underdesigned equipment would then be unable to carry out this duty.

Thanks to the lower resistance offered by the hot bath material, cycle times can be lower than the 90 seconds per anode reached by the standard cold anode cleaning stations.

**Shot blasting stations**

Depending on the global cycle time desired and the level of cleanliness required, this further anode cleaning station can be introduced in the rodding shop process.

Two technical solutions are available. The first one consists of a corridor where the anodes are shot at by up to 6 turbines facing each other. Although it is a simple method to implement, the station requires two complete shot recycling systems, and the recycling efficiency is reduced as more shot is broken. The second solution uses turbines on only one side blasting on a rotating anode. As a consequence, the recycling is simpler and more efficient, the cleaning of 6-stubs anodes is also possible, but the design is more complex and the cycle time is increased, starting at 50 seconds per anode. The choice is dictated by the type of anodes, the available space in the rodding shop building, as well as the cycle times required.

This solution brings the average residual bath quantity per anode below 400g (as opposed to 4kg). The efficiency of the materials recycling process is therefore greatly increased.

Flailing machines are sometimes used before the shot blasting stations to replace the visual bath inspection station. They allow removal of the bath residue stuck at the junction of the stubs and the carbon. It is often difficult for the upstream cleaning machines to reach these areas without risking collision with the anode stubs, and only these machines can totally remove the remaining particles of bath. Their design is simple but the chains have a limited lifetime.

**Butt stripping stations**

Once the bath particles are removed from the spent anodes, the carbon blocks can be detached from the stems and crushed to be recycled in the paste plant.

During the butt-stripping phase, the anode assembly stubs and yoke arms can be easily damaged or bent due to the forces exerted on the carbon blocks to break the carbon off the assembly. To avoid this, the base of the anode stem is supported while two retractable modules push on the carbon block to separate it from the stubs. This method also has the advantage of avoiding destructive strain on the overhead conveyor.

During this operation, a collecting table is placed at the bottom surface of the carbon block so that the broken pieces do not fall, but stay on the table. This greatly reduces dust and noise created by the machine. The collecting table is cleaned off at the end of the sequence and the carbon pieces fall into a noise insulated chute.

This design, as well as being efficient and environmentally friendly, has proven to be extremely reliable. As an example, the Butt Stripping Station commissioned at Hillside smelter (South Africa) in 1995 has never been opened or modified. Cycle times are also very low with an average of 60 seconds from anode entry to exit.

**Thimble press stations**

The thimble press is used to remove the cast iron mating the carbon and the stubs. The number of stations required depends on the type of anode and particularly on the number of stubs per anode.

For each stub, a pair of jaws clamp the stub while a breaking tool applies a breaking force up to 250 tonnes in a vertical manner (*figure 7*). The cast iron pieces then fall into a noise-insulated chute to be remelted for the rodding process. The jaw mechanisms are protected by a specific casing to avoid damage to the mechanism by broken pieces. Cycle time is as low as 28 seconds per stub. The total cycle time of the station depends on the type of anode and the number of presses installed.

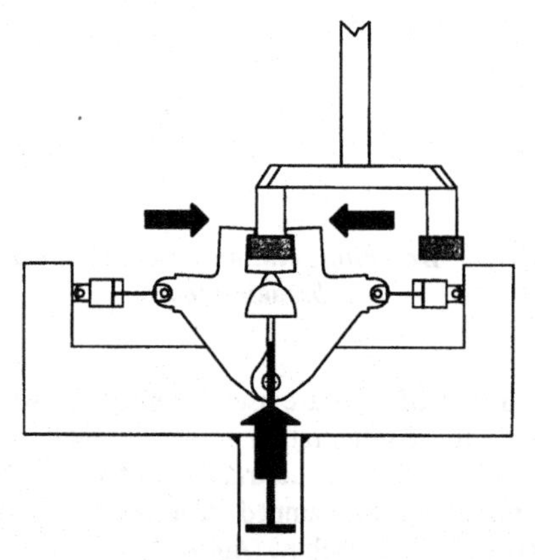

***figure 7: Upwards motion breaking tool of the thimble press***

In most applications, the thimble removal process is done in a sequential manner, using a limited number of thimble presses (usually two for a 6 stub stem).

It is also possible to break all thimbles at once, even for 6 stub anode assemblies. In this configuration, the use of floating jaws allows an alignment of each press. The cycle time can consequently be reduced to less than 60 seconds, allowing a gain of 25 seconds per stem on average.

### Butt and thimble combination press

For in-line stub anodes, it is possible to combine the butt stripping and thimble breaking functions bringing savings in capital investments and also reducing total cycle time.

Special jaws are placed around the anode stubs, protecting them from bending. They are equipped with knives on the underside that break the carbon block when it is pushed against them by the breaking tool coming from below. Once the carbon is removed, the same tool continues its upward movement to push the thimbles against the lower surface of the jaws, until the bottom of the anode stub reaches the level of the jaw. Both carbon and cast iron pieces fall on the same conveyor. They are separated for recycling by pushers on a sorting table.

The total cycle time is lowered down to 50 seconds per anode, as opposed to 88 seconds for separate stations.

### Control stations

Control stations for anode stem straightness and stubs physical integrity, traditionally use gauges. This type of design requires space and can lead to damaged anodes getting stuck in the gauges.

ECL has developed a solution using shape recognition technology. Three cameras are placed inside the perimeter of the station and are coupled to a computer that analyses the whole stem's shape. With this solution, no physical contact is needed and the control is performed over the whole stem assembly, rather than on specific points determined by the location of the gauges.

The shape recognition technology has many advantages over standard designs. Because the traditional rack and gauges are not necessary anymore, the shop floor space occupied by the control station is greatly minimized.

Also, the physical control can be performed while the anode stem is moving. As a consequence, the cycle time of this station is reduced to a few seconds as opposed to tens of seconds.

### Preparation stations

Before the rodding takes place, the anode stems can be brushed so that the quality of the electrical connection in the potroom is increased. The stems are easily covered by a fine layer of dust or grime that can be detrimental in the potroom where a very small voltage drop can mean a significant increase in power consumption.

The anode stubs can also be dipped in a graphite bath to improve the electrical connection between stub and cast iron and to reduce cast iron adherence so that the thimble stripping process requires smaller forces.

### Casting stations

The casting station is where the carbon blocks and the anode stems are mated by placing the stubs in the block holes and joining them with cast iron. This station is one of the most dangerous of the rodding shop, as molten metal and heavy loads are present. Flying sparks of molten cast iron also constitute a safety hazard and can impede the proper working of the station. To offer maximum safety for the operators, ECL supplies semi-automated designs with air-conditioned control cabins. For maximum operator comfort, the cabin is motionless with a large open view. No hydraulic equipment is used (only electro-mechanics and variable frequency drives) so that fire risks are minimized and movements are precise and smooth.

The first phase is the mating of the stem and the blocks. Conventional designs only ensure the vertical positioning of two elements. In the potroom, the positioning of the anodes in the pot is critical to the electrolysis process. An angle between the carbon block and the stem leads to a difficult vertical positioning of the anode with risk of reduced productivity of the pot.

ECL's mating station ensures the overall anode geometry from mating to rodding so that the bottom of the anode is at a precise and repeatable angle with respect to the stem.

Many rodding shops still use a casting trolley on rails. Cast iron spills and sparks often damage the wheels of these trolleys or can block them in their rails. The solution to these problems is offered by the use of a casting crane carrying a suspended ladle in which the tilting axis crosses the ladle's nozzle.

The semi-automated design allows reduced cycle time as well as a precise dosing of the cast iron. The ladle is pre-positioned in front of the anode stubs, the pouring starts and the operator validates the angle of the tilting at the end of the sequence. This system reduces cycle times (down to 120 seconds for 6 stubs) while minimizing cast iron spills that can contaminate the electrolysis pots.

### Level 2 solutions

High performance equipment alone doesn't make a high performance rodding shop. To ensure the productivity of this complex system, a global flow, production and maintenance management sytem has to be put in place.

To this end ECL provides level 2 solutions giving access to a large quantity of information regarding each station and the whole rodding shop operation. Such a system was first commissioned at Bécancour (ABI) smelter in Canada in 1984. Since then, it has evolved to offer operators and management a high level of information and analysis.

The centralized computer system receives for each station the following information:
_ input,

_ output,
_ alarms status,
_ fault status,
_ operator's shifts changes.

A typical maintenance screen is shown in ***figure 8***.

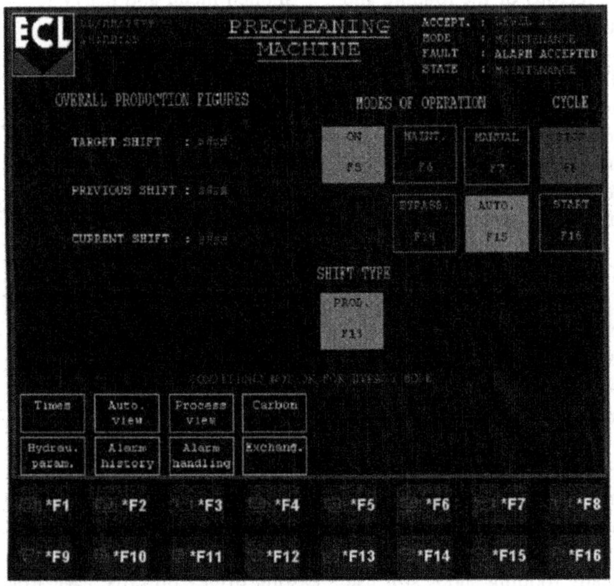

***figure 8: Example of a Level 2 screen in maintenance mode for a cleaning station***

Computer analysis helps determine for each machine the:
_ availability rate,
_ breakdown statistics and therefore its reliability,
_ productivity.

This information permits the optimization of the buffer stocks between each station so that maximum production is ensured while preserving a high level of finished anode availability. It can also assist in determining which station needs to be doubled or replaced for a possible smelter expansion.

Repeated faults due to failure or inappropriate settings are highlighted so that preventive maintenance or modifications can be carried out.

## Conclusion

By manufacturing and commissioning 24 anode rodding shops for smelters using a variety of reduction technologies, ECL has been able to fine-tune the design of each station to ensure maximum efficiency, reliability and productivity.

It is, however, the management of the rodding shop, assisted by the *level 2* solution, which guarantees the rodding shop's high performances. This is the case at Alba smelter where only 9 shifts per week are now necessary to produce the anode assemblies required for the 336 pots (AP35 technology), whereas 12 shifts were specified originally.

## Acknowledgements

The author would like to thank Franck Barrere, Didier Lescarcelle and Olivier Cousin for their help in writing this article.

**Light Metals 2008** *Edited by: David H. DeYoung*
***TMS (The Minerals, Metals & Materials Society), 2008***

# SAFE OPERATION OF ANODE BAKING FURNACES

Inge Holden[1], Olav Sæter[2], Frank Aune[3], Tormod Naterstad[4]
[1] Hydro Aluminium a.s Technology & Operational Support, P.O.Box. 303, NO-6882 Øvre Årdal, Norway
[2] Hydro Polymers a.s, Rafsnes. NO-3966 Stathelle, Norway
[3] Hydro Aluminium a.s Sunndal Carbon, P.O.Box 51, NO-6601 Sunndalsøra, Norway
[4] Hydro Aluminium a.s, Drammensveien 264, Vækerø, N-0240 Oslo

Keywords: Baking, Safety, Standards

## Abstract

The baking of anodes is a process in which combustible substances are released. Ring main fires do occur and even explosions in the fire zones, ring main or fume treatment plant have happened in carbon plants. The risks associated with different process deviations and the possible consequences of these, can be evaluated for both existing and new furnaces to be built.

European and IEC safety standards give useful guidelines for the design of process control and safety systems applicable to the baking process.

This paper will present and discuss methods for evaluating the risks and consequences, and give examples for how operational procedures and the design of safety systems can reduce the occurrence of unwanted events for open as well as closed top furnaces.

## Introduction

The total energy consumption in modern baking furnaces amounts to typically 4.8 – 5.2 GJ/t baked anodes, which is a combination of the following fuel sources:

- Oil or gas: 40 – 50 %
- Pitch volatiles: 40 %
- Packing coke: 10 – 20 %

The pitch volatiles may be classified in two main groups of hydrocarbons:

- Condensable hydrocarbons (tar), which are mainly poly-cyclic aromatic hydrocarbons (PAH).
- Non-condensable substances, which are mainly hydrogen and methane.

The condensable hydrocarbons appear during distillation of the lightest fractions in the pitch. The non–condensable substances appear in complex chemical reactions, polymerization and cracking at temperatures above 400 °C. In total, the volatilized components of the pitch coking are:

- Tar: Released at temperatures of 200 – 500 °C.
- Methane: Released at temperatures of 400 – 800 °C.
- Hydrogen: Released at temperatures of 400 – 1000 °C.

Particular safety aspects of baking furnace operations are related to the following specific features:

- Only 40 - 50 % of the total energy input is controllable by immediate actions.
- Strict control of the air to fuel ratio is practically hampered by false air ingress into the furnace atmosphere.

International safety recommendations, i.e. European and IEC safety standards, prescribe essential design and operational characteristics to be fulfilled for safe operation of similar furnaces.

A risk assessment of the anode baking furnace has been completed based on hazard analysis by explosion simulation. Explosion simulations are particularly useful to evaluate hazardous consequences caused by process disturbances, and to reduce the risks by design measures. This paper describes the furnace operation conditions used in simulations, the results of simulations and aspects of the safety systems required to safeguard against explosions and fires.

## Principal Risk Assessment

Flammability of Flue Gas Compositions of Baking Furnaces

The vaporization and pyrolysis of pitch generates a large number of components throughout the ordinary heat treatment range of anodes. The resulting fuel composition may vary [1,2]. For this study the following, typical average composition was chosen:

- Tar: 91.3 weight %
- Hydrogen: 6.5 weight %
- Methane: 2.2 weight %

Homogeneous, combustible gas-air mixtures are flammable within a limited range of compositions. In the flammable range a flame can propagate freely upon ignition. The flammable range is defined by the following limits:

- The Lower Explosion Limit (LEL).
- The Upper Explosion Limit (UEL).
- The Limiting Oxidant Concentration (LOC).

The relationship between combustible gas, air and inert gas and the location of the flammable range can be visualised by a ternary diagram as shown in Fig. 1 and Fig. 2. The stoichiometric air to fuel ratio $\lambda = 1$ is also shown.

Tars are known to include a large number of components (3000-4000). Analysis carried out by Charette et al. [3] found that the poly-cyclic aromatic hydrocarbon (PAH) components constitute approximate 70% of the total hydrocarbon loss. For this study, an average formula weight of approximately 200 g/mol with a C/H – ratio of 1.6, as for fluoranthene and pyrene ($C_{16}H_{10}$), was chosen as the average, representative characteristics of tar and is referred to as "Tar" in this paper.

Zabetakis [4] gives LEL of aromatic hydrocarbons as $(50\pm2)*10^{-3}$ g/l, giving the LEL of Tar as approximately 0.6 vol%, while the value of UEL of Tar can be derived as approximately 6,0 vol%. All values are normalised to 25 °C and atmospheric pressure.

In typical combustion regions, where the temperatures of the flue gas and the brickwork are higher than the Auto Ignition Temperature (AIT) of the combustible gases, the flammable region of the gas mixture is irrelevant and available oxygen will be consumed by oxidation. Process deviations may, however, increase the risk of forming gas-air mixtures within the flammable range downstream the combustion region of the fire zones:

1. Loss of draft situations followed by re-establishment of draft.
2. Sub-stoichiometric combustion. Too high fuel supply or too low draft, or a combination of the two conditions.

Oxygen deficit situations as described above may also be the root cause of formation of ignition sources. Cracking of heavier hydrocarbons forms soot/carbon particles. Light, glowing particles transported into the preheating sections facilitates ignition in pockets of flammable gas-air mixtures, or deposit in low velocity areas of the ring main system. High flue gas temperatures may also serve as ignition sources of soot/tar deposits in the ring main.

If the furnace is operated at oxygen deficit conditions caused by too low draft or excessive fuel, the resulting sub-stoichiometric concentration will be as shown in Fig. 1.

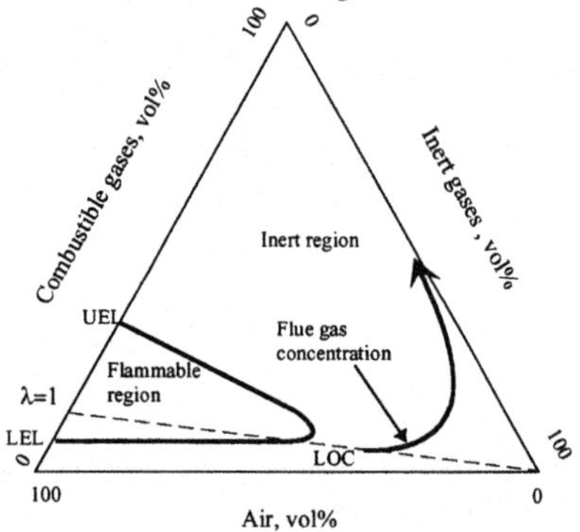


Figure 1. Idealized flue gas concentration in the combustion region by sub-stoichiometric combustion or lost draft.

Downstream of the flammable temperature range, dilution take place by air ingress through brickwork or access hatches, and the flue gas can enter the flammable composition range as principally shown in Fig. 2. The presence of an ignition source along the exhaust system will in that case initiate a hazardous fire or explosion, depending on the amount of fuel and the fuel/air mixture.

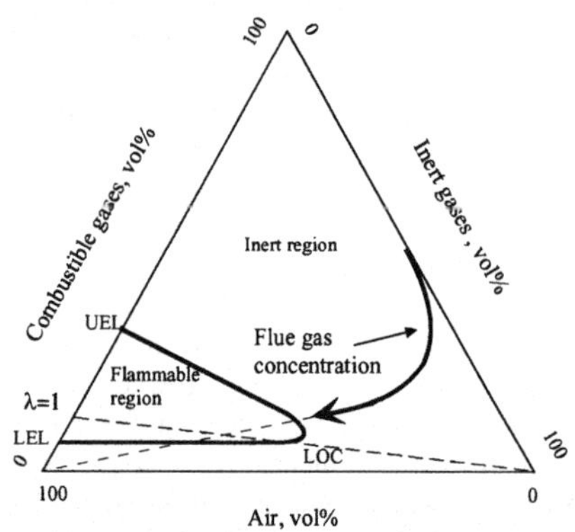


Figure 2. Idealized flue gas concentration in - and downstream the combustion region following a sub-stoichiometric combustion or lost draft situation.

Equation of Tar Combustion

The average values of reactants and products of combustion at normal operation measured in a closed top baking furnace are provided in Table 1.

The total of the combustion products is approximately 1300 $Nm^3/t$, while the total flue gas rate at the outlet of the combustion region is 1600 – 1900 $Nm^3/t$, since the furnace is operated at an oxygen surplus.

The flue gas entering the Tar combustion region contains the combustion products of the preceding fuel sources, e.g. propane, coke and the non-condensable substances from pitch pyrolysis. The principal equation of the stoichiometric Tar combustion ($\lambda$=1) in the fire can thus be described by:

$$C_{16}H_{10} + 18.5(O_2 + 3.76N_2) + \Sigma Inert \rightarrow 16CO_2 + 5H_2O + 69.6N_2 + \Sigma Inert \quad (1)$$

The average total of combustion products from energy sources other than Tar (*ΣInert*) at normal operation is approximately 193 [mol/mol Tar], as shown in Table 1.

Explosion simulations are carried out at stoichiometric air to fuel ratios. The stoichiometric air to fuel ratio of Tar combustion is 88:1 [$Nm^3/Nm^3$], as seen from Eq. 1, and is one of the essential factors considered by explosion simulation.

Table 1. Average composition of reactants and products of combustion

| | Unit | Coke | Propane | Pitch volatiles | | | Total | Total excl. Tar combustion |
|---|---|---|---|---|---|---|---|---|
| | | | | Tar | $H_2$ | $CH_4$ | | |
| **Reactants of combustion** | | | | | | | | |
| Fuel consumption | kg/t | 30,6 | 38,8 | 41,6 | 3,0 | 1,0 | | |
| Fuel consumption | $Nm^3/t$ | | 19,8 | 4,62 | 33,2 | 1,40 | | |
| Air for combustion | $Nm^3/t$ | 272 | 470 | 407 | 79 | 10 | 1238 | 831 |
| **Products of combustion** | | | | | | | | |
| $CO_2$ | $Nm^3/t$ | 57 | 59 | 74 | 0 | 1 | | |
| $H_2O$ | $Nm^3/t$ | 0 | 79 | 23 | 33 | 3 | | |
| $N_2$ | $Nm^3/t$ | 215 | 371 | 321 | 62 | 8 | | |
| Total products | $Nm^3/t$ | 272 | 510 | 418 | 96 | 12 | 1307 | 889 |
| Total products | $Nm^3/Nm^3$ Tar | 58,9 | 110,4 | 90,6 | 20,7 | 2,6 | 283 | 193 |
| Total products | mol/mol Tar | 58,9 | 110,4 | 90,6 | 20,7 | 2,6 | 283 | 193 |

* $Nm^3/t = Nm^3/(\text{tonne baked anode})$

Loss of Draft Situations

Following a loss of draft, a rapid increase in the concentration of combustible gases will take place as discussed above (see also Fig.1). For the actual Tar gas mixtures, the air access in the off-gas system (preheating sections, ring main, fume treatment plant) define the fuel compositions adequate for explosion simulations, see Case 1 - 3 in Table 2.

Methane and hydrogen are substantial parts of the pitch pyrolysis, and released at higher anode baking temperatures than Tar. False air ingress above the Auto Ignition Temperature will contribute to the combustion of methane and hydrogen, and thus limit the maximum concentration level at the outlet of the combustion region. An explosion simulation of a gas mixture with high hydrogen concentration is shown as in Table 2.

Table 2. Cases for explosion simulation, loss of draft situations

| | Oxygen concentration, [vol%] | Gas concentration at the outlet of the combustion region, [vol%] | |
|---|---|---|---|
| **Case 1** | 12.5 | Tar: | 1.7 |
| | | Inert: | 98.3 |
| **Case 2** | 16.0 | Tar: | 3.7 |
| | | Inert: | 96.3 |
| **Case 3** | 18.8 | Tar: | 10.3 |
| | | Inert: | 89.7 |
| **Case 4** | 12.9 | $H_2$: | 68.0 |
| | | Inert: | 32.0 |

Sub-Stoichiometric Combustion

Maximum $CO + H_2$ Concentrations. The primary effect of air deficit is sub-stoichiometric combustion of Tar. The principal equation of sub-stoichiometric Tar combustion is attributed to the ratio factor range of $0.432 < \varphi < 1$ in Eq. 2:

$$C_{16}H_{10} + \varphi 18.5(O_2 + 3.76N_2) + \Sigma Inert \rightarrow (16-x)CO_2 + (5-y)H_2O + xCO + yH_2 + \varphi 69.6N_2 + \Sigma Inert \quad (2)$$

A sub-stoichiometric air ratio of $\varphi = 0.432$ gives the theoretical maximum of $CO + H_2$:

$$C_{16}H_{10} + 8(O_2 + 3.76N_2) + \Sigma Inert \rightarrow 16CO + 5H_2 + 30.1N_2 + \Sigma Inert \quad (3)$$

$$Vol\%(CO + H_2)_{max} = \frac{16+5}{16+5+30.1+\Sigma Inert} * 100\% \quad (4)$$

The total products from the other energy sources ($\Sigma Inert$) will vary depending on the propane consumption and the flue gas rate.

Cases 5 - 8 in Table 3 refer to operational conditions, where air deficits provide conditions of maximum $CO + H_2$ concentration. Case 5 is sub-stoichiometric combustion of Tar. Case 6 – 8 refer to sub-stoichiometric combustion of Tar and propane. For all basic conditions included in Cases 5 – 9, maximum values of Tar release rates were assumed and estimated to be 25% above the average value.

Maximum Tar Concentration. At stoichiometric air to fuel ratios, the oxygen concentration in the off-gas system is the limiting factor in defining the maximum Tar concentration at low flue gas rates. The operating conditions for the explosion simulation are shown as Case 9, Table 3 with an oxygen content of 14.0 vol%.

Table 3. Cases for explosion simulations, sub-stoichiometric combustion

| | Process data of combustion region | Gas concentration at the outlet of the combustion region | |
|---|---|---|---|
| **Case 5** | Flue gas rate: 695 $Nm^3/t$ | $H_2$: | 4.5 vol% |
| | Propane supply: 0 | CO: | 14.3 vol% |
| | | Inert: | 81.2 vol% |
| **Case 6** | Flue gas rate: 1750 $Nm^3/t$ | $H_2$: | 7.7 vol% |
| | Propane supply: 108 kg/t | CO: | 10.0 vol% |
| | | Inert: | 82.3 vol% |
| **Case 7** | Flue gas rate: 1320 $Nm^3/t$ | $H_2$: | 17.5 vol% |
| | Propane supply: 108 kg/t | CO: | 18.8 vol% |
| | | Inert: | 63.6 vol% |
| **Case 8** | Flue gas rate: 1320 $Nm^3/t$ | $H_2$: | 7.6 vol% |
| | Propane supply: 66.5 kg/t | CO: | 11.4 vol% |
| | | Inert: | 81.0 vol% |
| **Case 9** | Flue gas rate: 260 $Nm^3/t$ | Tar: | 2.3 vol% |
| | Propane supply: 0 | Inert: | 97.7 vol% |

Explosion Simulations

Explosion simulations were carried out by use of the FLame ACcelleration Simulator (FLACS) Code [5]. All explosion simulations are performed at stoichiometric air to fuel ratios and at a flue gas temperature of 200 °C, assuming dilution of all mixtures to stoichiometric composition by air ingress

The explosion characteristics of Tar are not known. For the simulation, the explosion characteristics of propane were chosen. The Tar concentrations considered in Table 2 were scaled to the flammability limits of propane by maintaining the stoichiometric air to fuel ratio. Characteristics of propane adapted to Tar mixtures by explosion simulations are considered to be a conservative assumption.

The laminar flame velocity is an important characteristic for explosion simulations. Values for the reference cases in Tables 2 and 3 are shown in Fig. 3. Results derived by explosion simulations in the ring main are shown in Fig. 4 at compositions referred in Table 2 and 3 (Case 1-8).

As seen from Fig. 3 and 4, the highets safety risk is related to total loss of draft or very low flue gas rate (Case 1 - 4, 9). The main differences are attributed to the level of fuel enrichment reached prior to air access, i.e. to the period of lost draft. An extended duration of lost draft represents an additional risk by enrichment of non-condensable gases - $H_2$, $CH_4$ and CO at reduced inert gas concentration. Although Case 3 represents the highest explosion pressure, cases with high concentrations of CO and/or $H_2$ (as Case 4) represent a higher probability of occurrence due to a wider flammable region combined with a lower oxygen demand (lower LOC).

Sub-stoichiometric combustion of Tar and propane rendering high concentrations of CO + $H_2$ (Case 5 – 8, explosion pressure all below 0.05 bar) do not cause gas pressures damaging the ring main (Fig. 4). The reason is probably the high concentration of inert gases from the other energy sources, in combination with low volumetric energy content. However, the flame continuation and gas pressure breakthrough at weaker parts of the ring main design may still represent a high risk to the working environment.

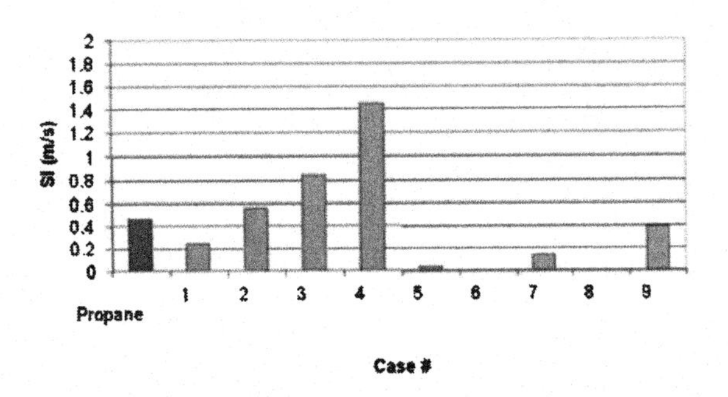


Figure 3. Laminar flame velocity of different gas mixtures.

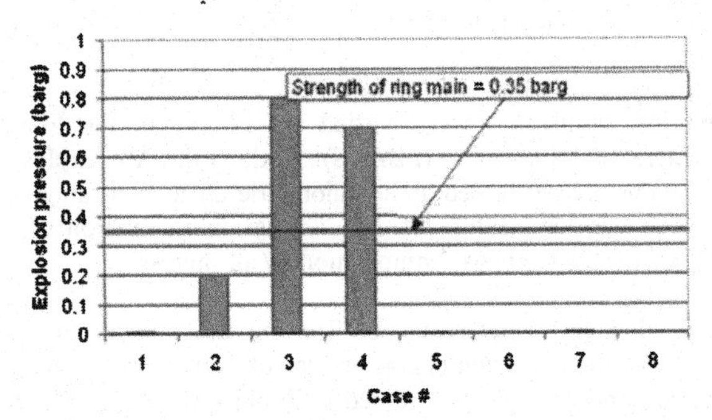


Figure 4. Explosion simulations of ring main pressures.

**Technical Solutions and Procedures for Risk Reduction**

European and International Electro-technical Commission (IEC) safety standards give useful guidelines for requirements and solutions for risk reduction. By definition, the Directive of Machinery (law in the European Union) applies to the Anode Baking System. The complete Machine includes the burner system, the baking furnace and the fume treatment plant/ emergency ventilation system. The Directive of Machinery outlines basic requirements, guidelines and standards related to safety, health and environmental requirements of baking furnace design and operation.

Safe Limits of Operation

Since no standards are specific to the anode baking process, the standard EN 1539 "Dryers and ovens, in which flammable substances are released – Safety requirements" gives useful guidelines for evaluation of the process.

Two main process requirements give guidelines to safeguard against fires or explosions in the ring main or fume treatment plant:

- An upper concentration of combustible substances as a percent of the LEL downstream of the combustion region (e.g. 25 %), dependent on to which level safety requirements are implemented.
- A Limiting Temperature for the flue gas, in practical terms defined to be the temperature at the exhaust manifold.

Basic process inputs and safety systems required to assure corrective action to unsafe conditions by Process Safety Supervision include the following functions:

Process Safety Function

The following are essential safety functions:

- The flue gas rate of from each combustion system.
- The fuel supply into each combustion system.
- The flue gas temperature in the exhaust manifold.
- Refractory temperature, interlocked to each burner group.

Safe limits of operation and corrective actions for each of the safety related functions must be defined for each furnace separately.

Safety Integrity Level

Operational safety levels are by international standards graded in terms of Safety Integrity Level (SIL) Figures, which are linked to the severity of hazardous consequences and the probability of occurrence.

Ring main fires do occasionally happen – and rare incidents of explosions as well. Hence the history provides strong arguments in favor of SIL 2 risk reduction measures.

The IEC standards of Safety Instrumented Systems [6, 7] cover aspects to be considered when electrical/electronic/programmable electronic systems are used to carry out safety functions. The requirements of SIL 2 risk reduction measures incorporates the following specific features:

The Process Shut Down Logic Solver must have redundant CPU and I/O on line fault diagnosis, and allow for online replacement of CPU and I/O cards. In case of failure of one CPU, the systems are allowed to operate as a single system for a limited period of time.

Safety Related Protective Systems

The protective systems are intended to take over when the control system fails, and bring the process to a safe condition. To achieve adequate safety, the protective systems must be separated from the control systems. The following functions can be defined as parts of the safety related protective system:

- Instrumentation necessary to supervise the process safety functions.
- Safety shut-off valves at the burner bridges.
- Safety shut-off valves at the main fuel circuit.
- Flue gas dampers necessary to bring the process to a safe position.
- An emergency alternative for the flue gas draft system.

The systems must be designed according to the fail-to-safe principle. For all functions a maximum reaction time needs to be defined.

Safety Related Utility Systems

The following function can be defined as a safety related utility system:

- Uninterrupted power supply to the safety related systems, including an emergency power system.

A maximum reaction time needs to be defined.

Procedures of Furnace Shut Down

Since only 40 – 50 % of the total fuel consumption can be controlled by immediate actions, the first priority during furnace operation is to maintain sufficient flue gas rate to provide oxygen surplus in each combustion system within acceptable time limits.

A total loss of furnace draft may occur with the ultimate risk previously discussed (Case 1 - 4). Hence, alternative ways of maintaining the flue gas in the non-flammable composition range below the flammability temperature must be considered, e.g. by flaring procedures. In the closed top furnace design, the temperature gradients along the fire zone have been proven to support transportation of flue gases towards high temperature sections in the fire zone. By reversing the flue gas direction in the sections with pitch pyrolysis, pitch volatiles are partly oxidized in the fire zone, and partly burned off by diffusion flames through hatches opened in the high temperature sections (T > 1000 °C).

When surplus air is established in all fires (visual observation of a clear transparent furnace atmosphere and no flame front inside or externally through openings on the furnace), normal operation can be re-established without any risk of forming flammable gas mixtures by restart of the draft. The risks involved with Case 1 – 4 are thus eliminated. The flaring procedure may take 3 – 12 hours for a closed top furnace.

Consequence Reduction Measures

The false air ingress along the fire zone makes a continuous, accurate, measurement of the air to fuel ratio challenging. Situations of sub-stoichiometric combustion can occur undetected by a safety supervision system and fire/explosions could occur in the ring main with minimal warning. By elimination of risks involved with loss of draft (Case 1 - 4), sub-stoichiometric combustion represents the highest remaining risk of furnace operation. Case 9, representing the highest explosion pressure, was used as the flue gas input concentration in the evaluation of pressure relief panels.

Pressure relief panels should be mounted at the ring main to direct any gas pressure away from working areas at a safe pressure level below the level of breakthrough for any other outlet hatches. The number of panels depends on the size of each panel. Fig. 5 shows the results from a FLACS simulation with different numbers of ø500 mm panels at each side of a 30 sections furnace. Eight panels where finally concluded to be sufficient. All hatches of the ring main shall include mechanical locking which is designed to withstand the explosion pressure.

Probability Reduction Measures

Steps can also be taken to reduce the probability of ignition sources in the ring main and ducting system. Ignition sources due to glowing of flammable substances can hibernate in the ring main and ducting system for long periods, e.g. due to embedded glowing particles in pitch/soot deposits with access to air through flanges/expansion joints.

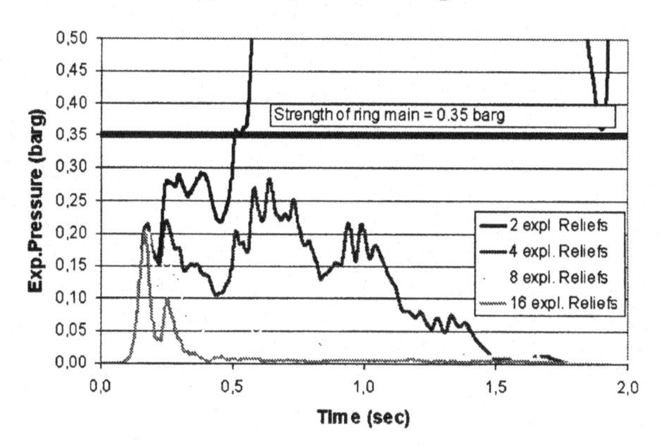


Figure 5. Explosion pressure in a ring main, Case 9.

The following counteractive measures reduce the existence of ignition sources:

- The intervals of cleaning of ring main and ducting system are chosen to reduce hibernation of ignition sources.
- The ring main and scrubber system is equipped with straps to provide effective earth potential along the whole flue gas system.

Another probability reduction measure is to ensure that air surplus for combustion of pitch volatiles are maintained by the procedures of energy and draft control at delayed fire step.

## Flue Gas Measurements

Flue gas analyses have been carried out to verify the concentration of combustible gases under normal operation and at enforced process deviations. The concentration measurements versus time at the outlet of the exhaust manifold required, include as a minimum: Tar, $H_2$, CO, $CH_4$, $O_2$ and $CO_2$, while some supplementary checks on $C_2H_6$ and $C_3H_8$ are recommended. Flue gas rate and temperature measurements were also measured.

The concentration of combustible substances [X] in volume% or [ppm] is determined by entries measured in the following equation:

$$X_{Mixture}[ppm] = X_{Tar}[ppm] + X_{CO}[ppm] + X_{H_2}[ppm] + X_{CH_4}[ppm] \quad (5)$$

The Lower Explosion Limit (LEL) of a gas mixture is found by the Lé Chatelier's principle [8].

$$LEL_{Mixture} = \frac{100\%}{\frac{Vol\%Tar}{LEL_{Tar}} + \frac{Vol\%CO}{LEL_{CO}} + \frac{Vol\%H_2}{LEL_{H_2}} + \frac{Vol\%CH_4}{LEL_{CH_4}}} \quad (6)$$

Where: Vol% Tar + Vol% CO + Vol% $H_2$ + Vol% $CH_4$ = 100%

The total combustible substances as % of LEL is calculated by:

$$\frac{X_{Mixture}}{LEL_{Mixture}} * 100\% \quad (7)$$

The Lower Explosion Limit decreases with increased temperatures [9] (t = temperature, °C):

$$LEL_t = LEL_{25^\circ C} * [1 - 0.000784(t - 25)] \quad (8)$$

Flue gas analyses were carried out on three closed top furnaces as part of the approach to establish safe limits of operation. Furnace A operates at a 26 hours fire step, Furnace B at a 36 h fire step and Furnace C at a 28 hour fire step. Measurements were carried out in the ordinary operation range (3000 – 4500 $Nm^3$/ton) and at stepwise, reduced flue gas rates. The results (Fig. 6) confirm the increased risk of fire or explosion under low draft conditions.

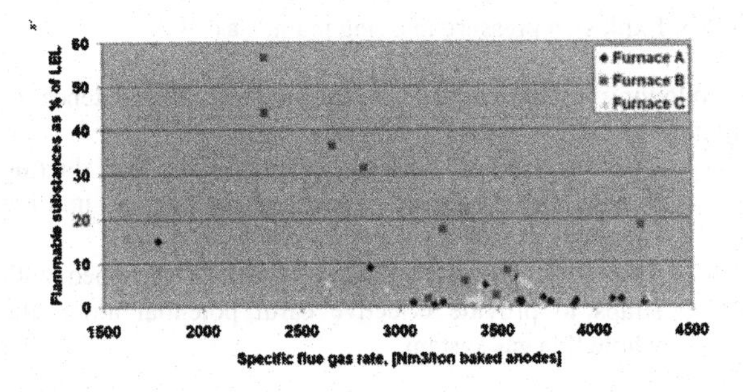


Figure 6. Calculated margins to LEL at normal operation and at enforced low flue gas rates. Measured at the exhaust manifold.

## Summary

European and IEC safety standards give useful guidelines in design of safety systems applicable to the baking process.

Two major process requirements are critical to safeguard against fires or explosions in the ring main or fume treatment plant:

- Maximum concentration of flammable substances as percent of LEL at the outlet of the combustion region.
- A limiting maximum flue gas temperature at the outlet of the fire zone.

A process safety supervision system, independent of the process control system, is considered to be essential to maintain a safe process condition.

In order to design a safety supervision system, the inherent safety functions need to be defined:

- Process safety functions to be supervised, e.g. flue gas rate, flue gas temperature and fuel supply.
- Safe limits of operation for each of the safety functions.
- Actions and equipment necessary to re-establish safe operation.
- Procedures to cater for loss of draft situations.

A process safety supervision system, which is designed to provide adequate corrective actions to safety offsets, basically needs to include:

- PLC according to IEC 61508-1. SIL 2.
- Safety related protective systems, intended to supervise the process and provide backup by any control system failure.
- Safety related utility systems.

Explosion simulations are useful to classify different kinds of process deviations and assign priorities to minimize the remaining risk level.

By flaring procedures, as proven for closed top furnaces, the risk involved with restart after loss of draft situations can be eliminated. Situations of sub-stoichiometric combustion are challenging to detect due to the false air ingress into the furnace, and represents the highest remaining risk of furnace operation. Corresponding flue gas concentrations are used for design of explosion relief panels.

Typical preventive measures to cater for the remaining risk level include:

- Routine schedule for cleaning of the ring main.
- Effective earth potential along the flue gas system.
- Pressure relief panels to direct any gas pressure away from working areas.

Some safety limits can be verified by direct measurement of combustible substances in the flue gas while others need to be estimated on the basis of measurements and calculations.

## References


1. Øyvind Gundersen, *Modelling of structure and properties of soft carbons with application to carbon anode baking* (Dr. ing. Thesis, Department of Engineering Cybernetics, Norwegian University of Science and Technology, 1998) 172-173.

2. Francois Tremblay and André Charette, *Cinetique de degagement des matieres volatiles lors de la pyrolyse d'electrodes de carbone industrielle* (The Canadian Journal of Chemical Engineering, Volume 66, February 1988), 93.

3. A. Charette at al., *Experimental and kinetic study of volatile evolution from impregnated electrodes* (Fuel, 1990, Vol 69 February), 194-202.

4. Michael G. Zabetakis, *Flammability characteristics of combustible gases and vapors*. (Bulletin 627, Bureau of Mines. 1965), 59, 60.

5. *Flacs User's Guide*, Flacs99, CMR-GexCon, Bergen, Norway

6. BS EN 61508. *Functional safety of electrical/electronic/ programmable electronic safety-related system,* (2002).

7. IEC 61511. *Functional safety – Safety instrumented systems for the process industry sector,* (2003).

8. National Fire Protection Association 69. *Standard on Explosion Prevention Systems*, (1997 Edition), 24.

9. National Fire Protection Association 86. *Standard for Ovens and Furnaces*. (1995 Edition), 31.

**Light Metals 2008** *Edited by: David H. DeYoung*
***TMS (The Minerals, Metals & Materials Society), 2008***

# Baking Furnace Optimisation

Amer Al Marzouqi, Tapan Kumar Sahu, Saleh Ahmad Rabba

Dubai Aluminium Company Limited,
P.O. Box 3627, Dubai, United Arab Emirates

Keywords: Productivity Increase, Baking Furnace Optimisation, Temperature Distribution


## Abstract

Dubai Aluminium Company Limited (Dubal) began its operation in 1979 as a 135,000 tons/annum of primary aluminium smelter. The electrode requirements for its three pot lines were supplied by two closed-top baking furnaces. Today, Dubal is a 890,000 tons/annum producer of primary aluminium operating four open top ring type Alcan-Alesa baking furnaces beyond their design capacity, to meet the ever expanding anode requirements. Recent increases in pot line amperage have demanded the use of larger anodes of varying configuration which has made the furnace operation extremely complicated.

For the past three years Dubal has optimised its baking furnace operations to improve baked anode quality and increase productivity, with the support of R&D Carbon.

This paper describes the improvement measures taken during these three years. Systematic implementation of various innovations in baking furnace practices have resulted in increased production while sustaining desired anode quality at Dubal.


## Introduction

Dubal has continuously increased aluminium production through improved cell technology and optimizing cell performance at higher amperages. The original Kaiser P-69 reduction cell technology was provided by National South Wire Aluminium, USA to operate cells at 150 kA using two stubhole anodes of size 1130 x 795 x 635 mm. By 1990, a series of continuous, innovative improvements to cell components and operational practices radically transformed the original cell technology to operate at 180 kA and is presently termed as the D18 design. Today, pot lines 1-4 operate with the D18 cell design at 185 kA using anodes of 1500 x 815 x 640 mm. Later, Dubal developed CD20 and D20 cell technologies which were used in future expansions. Pot lines 5 & 6 are constructed to the CD20 design and are operating at 225 kA using anodes of 1530 x 815 x 650 mm. Pot lines 7 & 9, with the D20 reduction cell design, use 1550 x 815 x 650 mm and 1530 x 815 x 650 mm anodes to operate cells at 230 kA and 227 kA respectively. In all the cell designs, anode sizes were increased in stages with the relocation of stub holes to boost metal production with higher line current. In order to accommodate increased anode dimensions in the furnace, the heights of all the furnaces were also increased. Baking anodes of different dimensions whilst balancing pit productivity is a constant challenge.

All these expansions increased the annual demand from 115,000 baked anodes required to 380,000 for eight pot lines in 2006. (Figure-1)

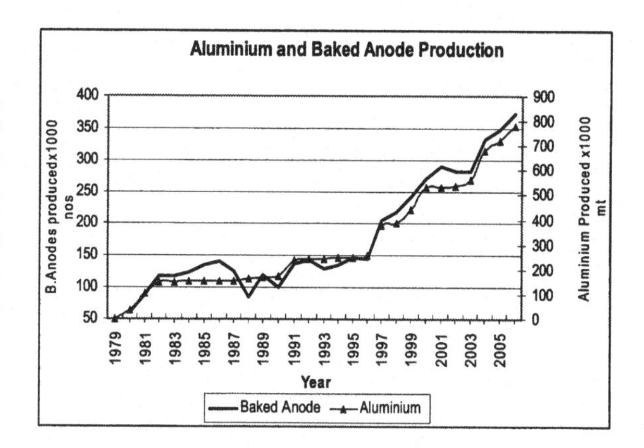


Figure-1: Aluminium and baked anode production.

The first concern of the carbon plant is to deliver anodes that must meet agreed quality standards to contribute to making potline operation as cost effective as possible. During the first smelter expansion, one furnace was retrofitted to Hydro Aluminium design with a 50% capacity gain to meet the anode requirement of 152,000 tpy. The third and fourth baking furnaces were added during each of the pot lines 5 & 6 expansion to fulfill the baked anode requirement. Retrofitting of old furnaces 1 and 2 to the open top design was completed in 1999 and 2003 respectively. In 2004, with the addition of 212 reduction cells, the 4$^{th}$ furnace was extended by 16 additional sections. In doing so, the total anode requirement of 350,000 was achieved through four furnaces operating at 12% more than the designed capacity [ 1 ]. Furnaces 1 & 2 consist of 32 sections each, with 7 pits to handle 3 layers of anodes per pit and 6 anodes in each layer. Furnace-3 consists of 32 sections with 8 pits to handle 3 layers of anodes per pit and 5 anodes in each layer. Furnace-4 consists of 48 sections and has similar pit dimensions as furnace-3. Furnaces 1 to 3 operate two fires each using R&D firing control system, whereas furnace-4 has three fires using Innovatherm firing control.

## Productivity Increase

The increase in metal production has not been proportional to the design capacity of all four baking furnaces. Therefore, the anode short fall was compensated by operating each furnace at a faster fire cycle to improve the production level by 17%.

The shorter fire cycles caused operational problems to a varying degree in terms of process control, emissions and quality variation of anodes.

The anode load per pit for a furnace dictates the pitch burn behavior and the availability of oxygen. Therefore, to maximize the production capacity of the furnaces, it was important to consider the productivity per pit and not the fire cycle time as the key driver for all process changes. Figure-2 illustrates the gradual increase in average productivity per pit for all furnaces.

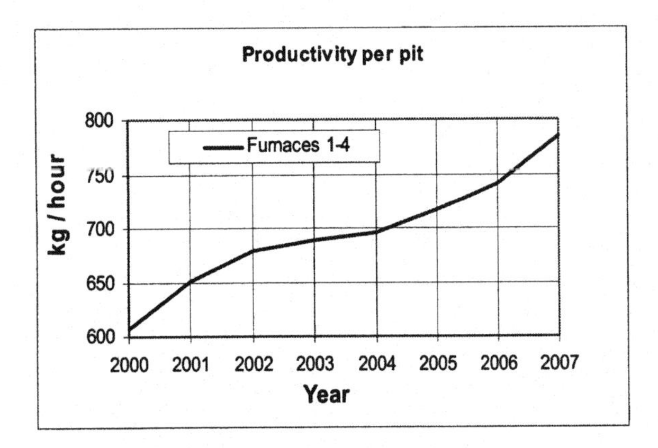


Figure-2: Increase in productivity per pit

Table-1 illustrates the relative increase in the productivity levels of all the furnaces in comparison to design capacity.

Table-I: Productivity per pit for furnaces 1 to 4

| Furnaces | 1 | 2 | 3 | 4 | Remarks |
|---|---|---|---|---|---|
| Productivity | 660 | 696 | 621 | 621 | Design |
| per pit, kg/hr | 830 | 814 / 847 | 735 | 706 | Q2, 2007 |
| Relative increase % | +26 | +17 / 22 | +18 | +14 | Avg. +19 |

Presently, Dubal is operating two furnaces with a very high productivity.

As this is a critical parameter, productivity is always balanced among the four furnaces whenever an anode dimension change is implemented.

## Process Optimisation

In the first quarter of 2005, the findings of a dynamic process optimization (DPO) were implemented in the two anode production lines in the greenmill [2]. A reduction of 0.7% pitch content without adversely affecting the anode properties was achieved through adjustments in the dry aggregate recipe and also in the preheating, mixing and forming process. Thereafter, it was thought appropriate to carry out process optimization of the baking furnace with the help of R&D Carbon.

### Correlation of Core Sample Results of Pilot Electrodes and Dubal Anodes

The optimization process was carried out in four steps:

1. Determination of optimum final baking temperature through pilot electrodes.
2. Mapping the temperature distribution across the section.
3. Comparison of anode core sample results with the above findings.
4. Implementation of results and establishment of a frame work for continuous monitoring of the furnace performance.

The optimum baking temperature depends on numerous variables such as raw material quality, green anode characteristics, furnace condition, process capability of firing system and finally, on the maximum allowance provided for environmental emissions levels. The first stage of the determination of optimum baking temperature was to develop calibration curves with the core results obtained from pilot electrodes [3]. Nearly 300 kg of green anode paste was collected after the paste cooler and sent to R&D Carbon for subsequent processing. A total of 45 pilot electrodes were produced in a pilot press using this paste, with a diameter of 146 mm and a height approximately 180 mm. Five sets of 6 green pilot electrodes were baked in an electrically heated laboratory furnace under controlled conditions to five different final baking temperatures of 1000°C, 1050°C, 1100°C, 1150°C and 1200°C. The cores from the pilot electrodes were analyzed for apparent density, thermal conductivity, air reactivity, $CO_2$ reactivity, density in xylene and elements. Based on the test results of pilot electrodes, calibration curves were drawn and they were utilized to establish the desired final baking temperature and the nature of the temperature distribution across the sections. A final anode baking temperature between 1100°C and 1150°C indicated optimum anode quality. An example of calibration curve is given in Figure-3.

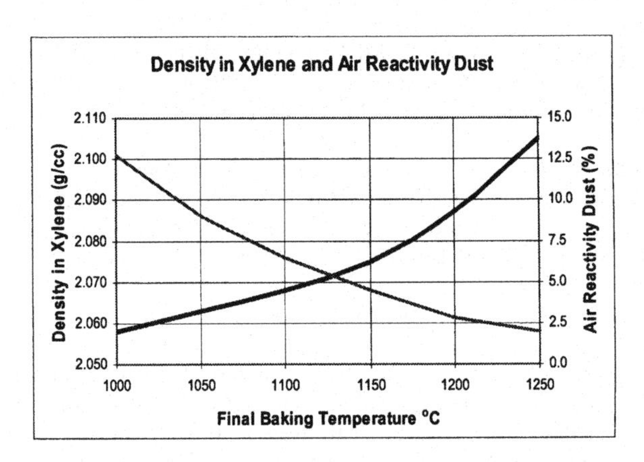


Figure-3: Calibration curve for key properties as a function of temperature

A total of 108 test anodes, produced from the same paste as pilot electrodes, were sampled as per a predefined sampling plan from all the four furnaces. They were also analyzed for same core characteristics as that of the pilot electrodes. In order to correlate the core sample results with the final baking temperature, calibration curves obtained from the pilot electrode test results were used (Figure-4). Three furnaces exhibited the same standard deviations (2 SD) of 50°C for the final baked anode temperature and this is a very good value.

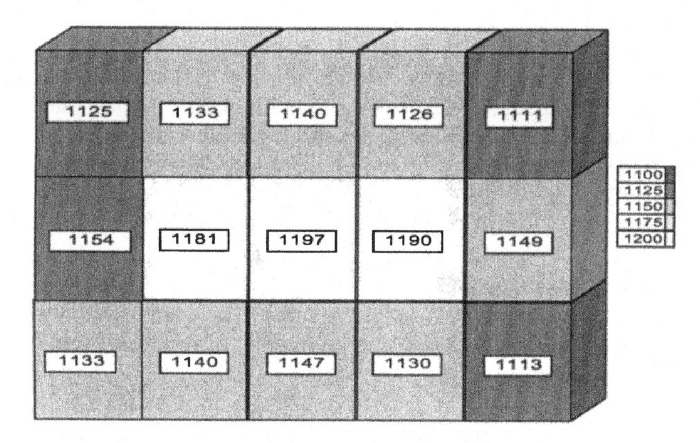


Figure-4: Illustration on the distribution of final baked anode temperature (°C) across a pit

In the fourth furnace, as seen on the above illustration, some of the anodes were baked close to 1200°C. This level of temperature is not beneficial for the anode quality as desulphurization was observed. Furthermore, $SO_2$ emission was identified as an issue.

Considering the higher final baking temperature, the density in xylene and thermal conductivity along with lower 'S' content of the anode, it was decided to reduce the final flue temperature for this furnace. This improved the temperature distribution and also reduced the $SO_2$ emission from the stack. Figure-5 illustrates the decreasing trend in $SO_2$ emission after the optimization measures were in place.

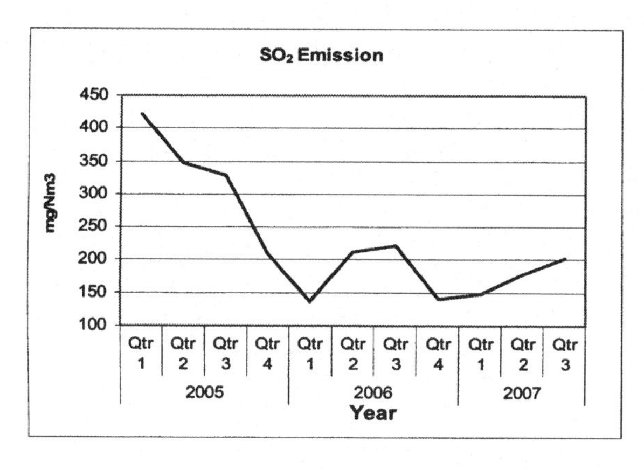


Figure-5: $SO_2$ emission from the furnace stack

## Equivalent Temperature Study

On a regular basis, Dubal uses thermocouples to observe the anode preheating rate and equivalent temperature (calculated from the Lc) method to determine the final baked anode temperature distribution in the baking kiln. Both the measurement practices are adopted because of procedural simplicity. The equivalent temperature of the baked anodes is determined by placing graphite crucibles, containing uncalcined coke, in the stub holes. After baking, an equivalent temperature (°E) is determined for the coke in each crucible and the data is used to map the baking level for entire section or pit. The results are interpreted taking into account the relative temperature variation across the pit. This measurement practice is established as an ISO standard, ISO 17499 [ 5 ].

It is generally observed that the resulting equivalent temperature is 80 to 120°C higher than the real temperature. Figure-6 illustrates the temperature mapping and improved temperature distribution pattern after the furnace optimization. The optimization measures eliminated the hot spots and improved temperatures lower than 1050°C.

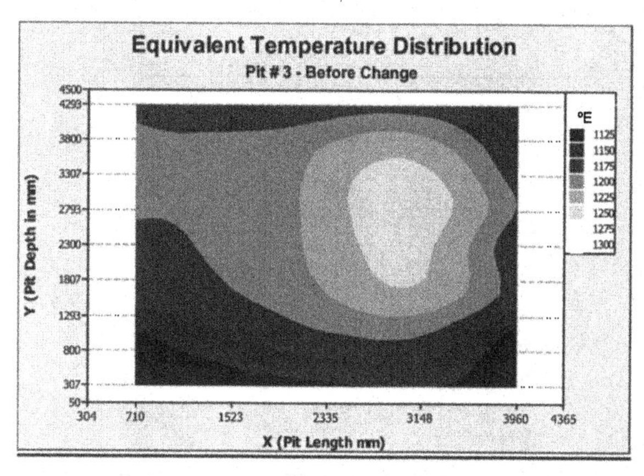


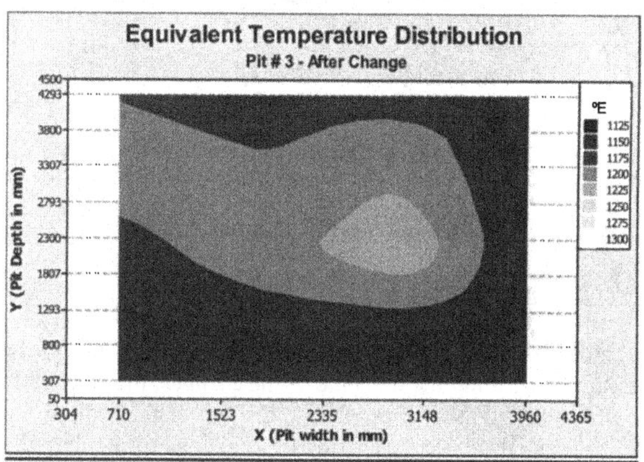


Figure-6: Mapping of final baking temperature of anodes

The elimination of hot spots was beneficial for the refractory life and helped in preventing desulphurization of anodes, which is detrimental to anode quality and stack emissions.

## Flue Gas Analysis

During the optimization process, the flue gas composition in front of the pitch burn zone was regularly measured. It gives a very good indication of the combustion efficiency with reference to availability of excess oxygen and the level of CO and $CO_2$. In order to eliminate soot formation, availability of excess oxygen is ensured through appropriate level of the draught. Figure-7 illustrates the improvement observed in excess oxygen availability after the process optimization.

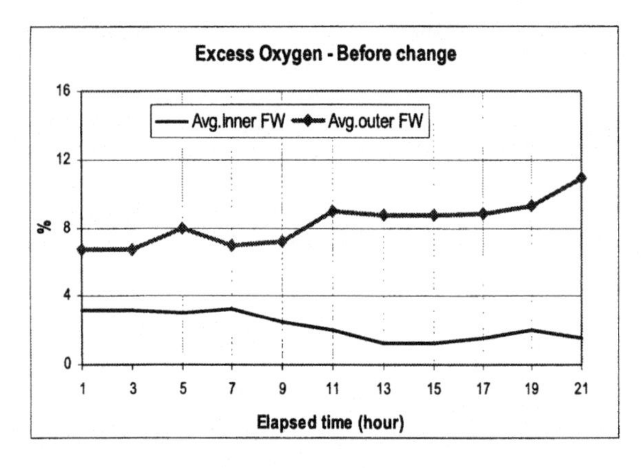


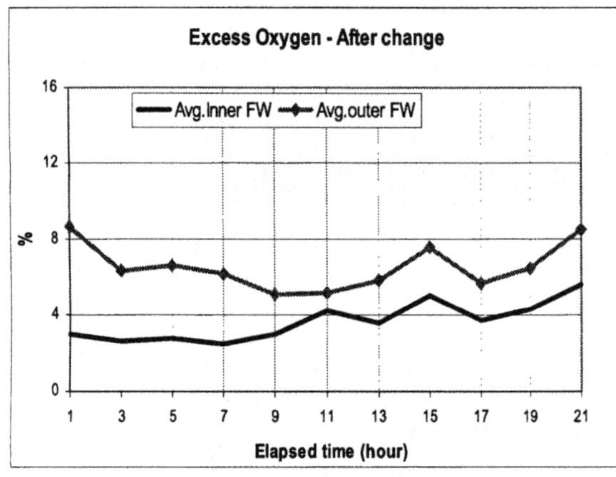


Figure-7: Improvement in excess oxygen availability.

Process Parameter Changes

During the optimization process, only one furnace was selected for trial and after confirming the outcome of the changes, the parameter changes were adapted to all fires. Due to the very high productivity in some furnaces i.e. > 800 kg/hr it was a challenge to regulate the fire in the preheating section with respect to pitch burn control and oxygen management.

During the process of optimization, main focus was given to the baking curve, burner operation and control parameters.

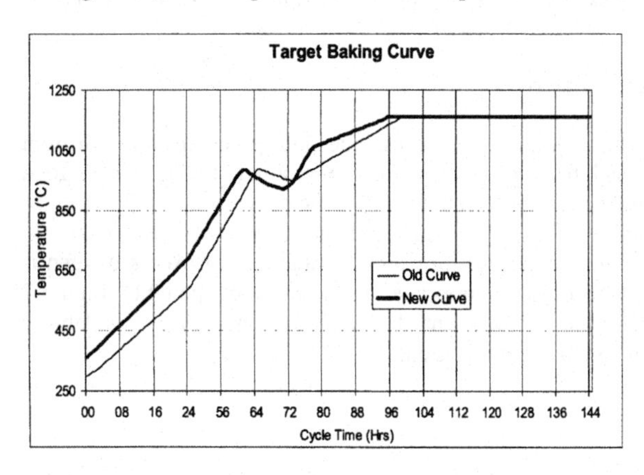


Figure-8: Target baking curve before and after optimization

The target baking curve was modified to advance the location of the pitch burn to gain additional soaking time (Figure-8). For this the heat-up rate, which is a critical parameter for anode quality, had to be increased. They were monitored by thermocouples embedded in the critical zones.

**Specific Energy Consumption**

Gas energy consumption for a given furnace is linked to anode composition, baking level, furnace condition, firing cycle, firing control and operation.

In order to improve the combustion efficiency of the flue gases, new venturi burners were introduced first in furnace-2 in 2003. Observing the satisfactory performance of these burners, furnaces 1 and 3 were also equipped with a similar type of new venturi burners.

Optimization of the fires along with the introduction of new burners helped in a gradual reduction of natural gas consumption from 2.1 GJ/t to 1.95 GJ/t of baked anodes in spite of the higher productivity level.

**Baked Anode Quality**

Baking is the most costly stage in the anode manufacturing process. The main factors of consideration when selecting the final baking temperature are the balance between the negative effect of anode dusting at the lower end and increasing anode reactivity at the upper end. The air reactivity residue is plotted as an example to illustrate the consistency in anode quality in spite of higher productivity and gradual increase of vanadium content in the coke to 230 ppm. (Figure-9)

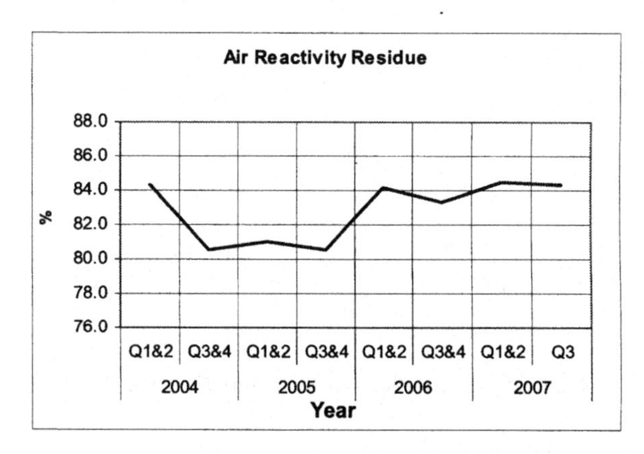


Figure-9: Air reactivity residue

**Refractory Life**

The deterioration of refractories is attributed to chemical attack and thermo-mechanical stresses. In a baking furnace, the following factors influence flue wall life.

- Mechanical damage during loading and unloading of anodes.
- Firing conditions and high temperature excursions.
- High flame temperature and heat intensity.
- Selection of refractory materials.

- Infiltration and reaction by alkali and aluminium fluoride from recycled anode butts.
- Hard carbon build-up on pit walls.
- Stresses resulting from inadequate expansion allowance.

In order to achieve desired baked anode quality with minimum refractory repair cost, regular refractory maintenance is implemented. In one furnace high flame temperature and heat intensity caused thermo-mechanical stresses.

The key activities undertaken to improve refractory life include:

- Regular flue wall cleaning on the anode side of flue wall to prevent coke sticking.
- Flue wall straightening if the deflection is > 50 mm.
- Restricting peak flue wall temperature to <1350ºC during firing.
- Reducing flame temperature and heat intensity by burner type and gas pressure.
- Improved butt cleaning practice.
- Maintaining the expansion gaps through regular maintenance.
- Trials with different fire clay brick types, instead of using conventional bricks with 42% alumina.
- Rebuilding the furnace with complete replacement of all flue walls while one fire is in operation.

With the above refractory repair practices we are able to achieve a bench mark level of more than 180 fires as average flue wall life.

## Conclusions

Baking furnace optimization process has been a successful journey for Dubal as it has established following gains:

- Productivity increase of ~17%.
- Improved baking homogeneity.
- Gradual reduction in furnace energy consumption.
- Reduced $CO_2$ emission through reduced energy consumption.
- Reduced $SO_2$ emission.
- Improved refractory life.

The use of pilot electrodes to assess the target baking temperature and subsequently optimize the entire baking process to achieve this target along with the desired core properties was found beneficial. In addition to this, periodic monitoring of the furnace process through anode preheating rate measurement, temperature distribution tests, flue gas composition monitoring and regular checks on refractory temperatures along with the suitable modifications of firing control parameters have proven to be the most effective method for continuous improvement.

## References


1. Masood Talib Al Ali, Raja Javed Akhtar, Saleh Ahmad Rabba, *Enhancement in Baking Furnace Operation*, Light Metals 2005.

2. R.J. Akhtar, S.A. Rabba, M.W. Meier, *Dynamic Process Optimization in Paste Plant*, Light Metals 2006.

3. V. Piffer, M.W. Meier et al, *Process Optimization in Bake Furnace,* Light Metals 2007.

4. A.J.M Kalban, K.V.Thomas, M.H.Vaz, T.Foosnaes, T.Naterstad, A.Werge-Olsen, *Baking Furnace Performance Optimization by Retrofitting and Operational Modifications*, Light Metals 1994.

5. Lorentz Petter Lossius, Inge Holden, Hogne Linga, *The Equivalent Temperature Method for Measuring the Baking Level of Anodes,* Light Metals 2006.

**Light Metals 2008** *Edited by: David H. DeYoung*
***TMS (The Minerals, Metals & Materials Society), 2008***

# NEW CONCEPT FOR A GREEN ANODE PLANT

Michael Kempkes, Werner Meier

BUSS ChemTech AG, Hohenrainstrasse 10, 4133 Pratteln 1, Switzerland

Keywords: Anode, Aluminium, Green Anode Plant, Carbon

## Abstract

Being the major element for primary aluminium production, the green anode plant is of highest importance.

Project realization-wise: as from start-up of the first pot-lines the green anode plant shall be operative to serve the electrolysis with own anodes.

Production-wise: as baking furnace and electrolysis would not excuse any interruption with anode supply or high fluctuation in product quality.

Cost-wise: as carbon consumption is one of the biggest cost factors besides alumina and power consumption.

Environmental-wise: as e.g. carbon emissions are or will be a cost factor and reduction of emissions is one of the major contributions to prevent further global warming, adding to the greenhouse effect respectively.

So, today's Green Anode Plant should:

- have a short construction period with straight forward start-up phase
- assure safe and stable production with high availability
- be cost efficient with strict adherence to budget
- be environmentally friendly

A combination of highly reliable equipment, avoidance of any unneeded components and strict simplification of process steps, where possible, leads to our new concept for the trendsetting Green Anode Plant with a throughput up to 80 tonnes per hour with one single production line.

## Introduction

Over the last 20 years there have been many proposed Green Carbon Plant design concepts and many built using these concepts. Common in all of the designs is the use of a continuous process with two to six calcined coke fractions (typically 3 to 5), liquid pitch and large in-process buffer silos or bins. The design of the Green Carbon Plant involves four primary functions, material handling (calcined coke crushing and screening, butt crushing and screening), weighing of the coke and butt fractions and liquid pitch; heating and mixing of the coke/butt with pitch; and anode forming.

However, there are two paradigms that are common to design and operation of all Green Carbon Plants.

The first paradigm is that the process is continuous when in fact, only the mixing process is really continuous.

The second paradigm is that as the throughput is increased, the size of the in-process buffer silos or bins shall also increase. The consequence is that with the increase in production, there has been a similar increase in size of the facility and little if any improvement in the production quality. This size increase of in-process buffers or bins has the negative effect of segregation of the before carefully screened fractions.

To give the dimension for the concept of a green anode plant it will be necessary to take a look at the current size of aluminium smelters.

The average size of today's aluminium smelter is in the range of 350 – 450000 tonnes per year. Required anode production accordingly is in a range of 200 - 270000 tonnes per year with green anode line capacity of about 35 tonnes per hour.

However, the next generation of aluminium smelters, presently in advanced planning phase, would either continue with the above capacities or its multiples (35 tonnes per hour line base), or go a step further by implementing a new single production line with a flexible capacity up to 80 tonnes per hour, from current standpoint, regarded as so called Mega Carbon Plant. The necessary key equipment is already available.

## Why a New Concept?

The projection of 80 tonnes per hour or higher capacities makes it necessary to change the traditional concept in terms of:

- physical dimensions of the plant building
- silo sizes for intermediate (daily or similar) storage
- more sturdy building structure increasing dynamic and static load
- raw material sourcing and quality fluctuations
- efficient pollution control
- rational resource consumption (energy, raw materials, scrap, waste)
- optimized logistic and material handling

## Why the Continuous Plant?

To be able to build and operate a mega plant, it is necessary to access to a totally continuous processing. Starting with the warehouse / storage (raw materials) and ending with the finished products (green anodes). Moreover, a totally continuous processing means no buffers' storages within the process stream. All processes are continuous, lean and easy controllable.

From green coke silos take-off, conveying, green coke processing (sieving, grinding etc.), butts recycling (crushing, grinding, sieving etc.), green scrap processing (grinding and feeding), fine fraction production (continuous ball mill and classification) proportioning of dry aggregate fractions (coke and butts), dry aggregate fraction preheating, liquid pitch handling and dosing, paste reactor, paste cooling to forming, the respective components are dimensioned to operate with chosen line capacity, where oversized equipment are not needed.

The continuous line design allows for both reduced and increased production without interruption to the continuous operation. This design approach is a tool for maintenance and production for higher flexibility with decreased risks.

Continuous processes allow easy and quick adjustable production, repeatable product quality and a high flexibility to changing raw material properties without the need for oversized facilities.

## Concept Milestones

A new concept does not mean re-inventing the wheel! The new concept for a green anode plant is the result of expertise from several first class suppliers combined with decades of knowledge from experts of this industry.

Butts

The butts are the main source for the coarse fraction and the butts fine fraction will be extracted from the process it is the main source of Sodium contamination. The extracted material is useful for other applications (e.g. steel industry, fuel).

Continuous, Vertical Ball Mill

The ball mill produces continuously the required amount of fines, no storage of fines is required and therefore no segregation and no bridging of the silo occurs.

Dry Aggregate Fractions

Three fractions (coarse, medium, fine) are the elements for totally controllable recipes for all operators' and end-users' needs.

Horizontal Paste Line Arrangement

The horizontal arrangement of coke pre-heater, paste reactor and paste cooler, one unit below the other, as well as a horizontal process flow within the units allow a reduced building height where all components have the minimum distance required for operation and maintenance. Only the continuous and horizontal flow of material allows a controllable and adjustable / adaptable process at high temperature processing profile and controlled outlet temperature (for vibro forming or for a press). Quick reaction on changing raw material qualities is possible!

Logistic and Material Handling

The best concept for a plant is meaningless if you do not consider general thoughts such as:

- guarantee of raw material supply
- proper and continuous logistic of it
- perfect layout integration of warehouse and appropriate location within the smelter complex (huge quantities of handled materials)

Emission Control

Best available technology to control and reduce emissions is a key contribution for a livable future. The concept is based upon the use of most advanced technologies like e.g. exhaust fume treatment by use of a RTO (Regenerative Thermal Oxidizer) or considerations about reduction of waste water consumption.

## Engineering

A concept and the ability to convey such concept to reality requires strong engineering competence for Basic and Detail engineering. The execution of world scale projects require competent partnering with experts in project execution. This can either be done with strong local partners of with global acting international companies.

## Key Equipment

The concept is based on a "Zero Weak Components" philosophy, meaning only dedicated equipment with proven high reliability and excellent performance results shall be considered. A second basic change is a reduced number of equipment in total. This allows the use of first class suppliers and most reliable equipment and still keeping the overall investment below industrial average.

Vertical Ball Mill [2] produces continuously the required quantity and quality of fines with high accuracy with a throughput regulation range from 25%-100%. An absolute dust-tight design and low-noise emissions allow an easy integration of the unit in the plant layout (no extra room in the plant building is required).

The compact design with integrated dynamic classifier reduces ducts and pipe work and due to the continuous production, also the required filter system is smaller than conventional ball mill. Instead of a large quantity of small balls, the vertical ball mill operates with five big milling balls rotating at low speed which results in drastically reduced wear of balls i.e. resulting in reduced metal impurities.

BUSS Horizontal Paste Line (HPL)

BUSS' Coke Pre-heater: The continuously operating horizontal coke pre-heater is a hollow flight type heat exchanger, heating up the dry fractions to 190 – 200 °C. The compact and sturdy design at low revolution makes this unit almost maintenance-free.

BUSS' Paste Reactor: The state-of-the-art design of the paste reactor operates with variable throughput of from 25 – 60 tonnes per hour (K 600 CPX) and 35 – 80 tonnes per hour (K 600 XL). The following paste reactor's phases are combined in one equipment as follows: filling, dry compression, liquid pitch injection, mixing and kneading, conditioning / homogenization, dynamic filling control.

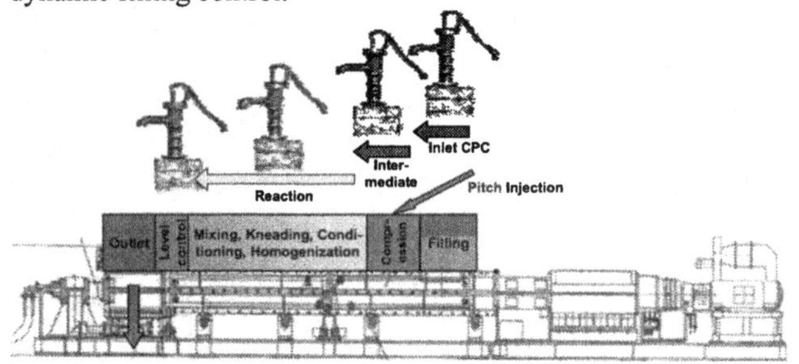


Figure 1: Pressure Progression within paste reactor

To achieve this complex operating ability and capacity it was not necessary to change the outside dimensions of the kneader; modification mainly took place at the internal parts of the machine.

This increase of 25% of throughput with the same good quality of paste (compared with the K 600 CP) was reached by several technical innovations which are basically:

- material feed of solid and liquid fraction is split. The coke will still be fed into the inlet opening of the kneader barrel but the liquid pitch is now fed through one of the first kneading teeth, directly into the process area resulting in the elimination of any material accumulation at the kneader inlet.

- during the kneading / mixing process, the density of paste varies from the beginning to the end of the barrel. In order to always achieve the same high process / reaction pressure and filling level within the different areas by adjusting the screw pitch in the different kneading zones to optimize both kneading effect and filling level.

- by using a dynamic filling control element at the end of the process zone, the mechanical instrument flap-die could be omitted. Instead of energy input increase by limiting the outlet opening dimension, this is now done with pressure increase by use of geometric means, dynamic and three-dimensional, with a change of the screw pitch of this last process element. The result is a full energy input into the paste instead of partial energy loss in the flap-die. In total, a reduced mechanical load increases availability and reliability of the machine.

- the number of kneading / mixing steps within the barrel could be increased, still keeping the overall dimension and the diameter-length ratio.

- due to the optimized material inlet, a higher throughput at lower revolution can be reached, and as a result the wear of the kneading elements (flight segments and stationary teeth) is noticeably reduced.

BUSS' Horizontal Paste Cooler: Desired adjustment of the paste temperature for the forming process is achieved in the continuous horizontal paste cooler. The cooling medium is water.

The paste temperature is adjustable according to the needs of forming equipment, either vibro compacting or press.

Anode Forming: As the decision for either vibro compacting or a hydraulic press [3] is questioned in this industry, both forming philosophies are possible to be integrated in our continuous concept.

RTO for pitch fume treatment [5] by using an RTO (Regenerative Thermal Oxidizer) 99% of the PAH (Polycyclic Aromatic Hydrocarbons) can be removed from the exhaust gas, this is today's best available technology without any negative cost effect.

Figure 2: Concept key components

## Conclusions

The New Smelter Generation requires Mega Carbon Plants. From an economical and technical standpoint, the Mega Carbon Plant has to be simplified (elimination of all unnecessary elements) and so be adjusted for the continuous operation.

Huge material flow requires very safe and stable production conditions. The operating costs of a mega plant oblige the plant to operate on very high efficiency standards. This is again only possible by producing continuously.

Considering these facts, the best available construction and production materials have to be chosen. Such a continuous process allows the implementation design of the highest environmental standards.

Technically and operationally, all apparatus and equipment are available to realize line capacities from 50 to 80 tonnes per hour.

As a last consequence to be mentioned, the Mega Carbon Plant is capable to adjust the quality of the anodes exactly as required by the electrolysis and not vice versa.

## References

[1] M. Kempkes, The future of anode manufacturing, Aluminium 83 (2007) 6, 70-73

[2] Thore Möller, Fines Production for Anode Manufacturing, Light Metals 2005, 653-658

[3] Alfred Kaiser, Hydraulic Pressing of Anodes, Aluminium 80 (2004) 3+4

[4] Kristine L. Hulse, Anode manufacture- raw materials, formulation and processing parameters, R&D Carbon Ltd. 2000

[5] Matthias Hagen, New requirements and solutions for the fume treatment at paste mixing and anode baking plants, Light Metals 2006, 615-619

**Light Metals 2008** *Edited by: David H. DeYoung*
***TMS (The Minerals, Metals & Materials Society), 2008***

# Modelling of Anode Thermal Cracking Behaviour

Odd Einar Frosta, Trygve Foosnæs*, Harald A. Øye*, Hogne Linga

Hydro Aluminium a.s Technology & Operational Support, P.O.Box 303, NO-6882 Øvre Årdal, Norway

* Department of Materials Science and Engineering, Faculty of Natural Sciences and Technology, Norwegian University of Science and Technology, Trondheim N-7941, Norway

Keywords: Carbon anode, Inhomogeneity, Thermal Shock Resistance, Anode Cracks, Modeling

## Abstract

Carbon anodes are used in the aluminium reduction process. As each anode is consumed it is replaced with a new one. The anode undergoes a severe thermal shock as it is submerged in the 960°C molten bath. As time passes, the temperature of the carbon anode rises due to the heat conduction from the bath, convection from the hot surroundings and the heat generation from the electrical current flow. Thermal stresses generated from this temperature rise may further alter the integrity of the carbon anode and sometimes cracking occurs.

*Figure 1. Three of the commonly observed types of crack configurations in carbon anodes are; A – corner cracking, B – vertical cracking and C – horizontal cracking [1].*

## Introduction

The mechanical performance of the anodes has a significant impact on the aluminium production, especially when anode raw material is degrading and cell current is increasing. Temperatures and stresses in a carbon anode have been determined using a three-dimensional finite element method in order to investigate how material properties are influencing the stress development when inserted into an electrolysis cell. The anode used in the model is 1590 * 700 * 605 mm and has four stub holes. The modelling is done using the FEM-program ANSYS. The study covers the first hour following an anode change.

The first modelling was based on a homogeneous anode. Some material properties are shown in Table 1.

Table 1. Some material properties used for modeling a homogeneous anode.

| Anode properties | Unit | Value |
|---|---|---|
| Baked Density | $kg/dm^3$ | 1,580 |
| Poisson ratio | - | 0,144 |
| Dynamic Youngs Modulus | GPa | 10 |
| Coefficient of Thermal Expansion | $10^{-6}/K$ | 4,6 |
| Cold Crushing Strength | MPa | 42 |
| Air Permability | nPm | 0,5 |

When the boundary conditions around the anode in the cell were established, the anode was stepwise divided into smaller parts. Each part has its own set of material properties. In total these parts resemble the inhomogeneity of the anode. Also the temperature dependence of some material properties was taken into account.

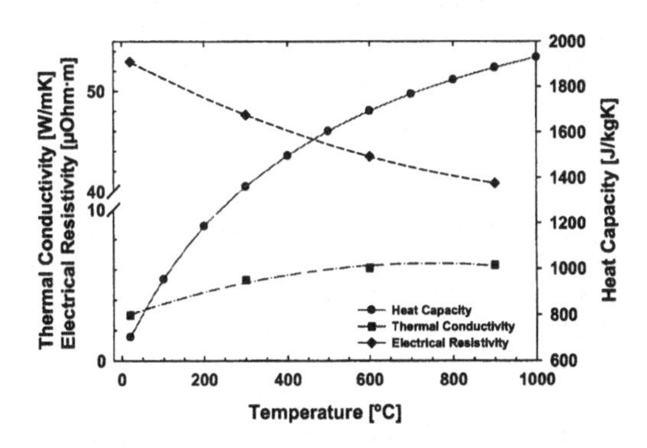


Figure 2. Electrical resistivity [2], Thermal conductivity [3] and Heat capacity[3] of anode shown as function of temperature.

## Modeling in ANSYS

All modelling is performed in ANSYS, included construction of the anode assembly. Due to symmetry effects parts of the modelling is only done on a quarter of an anode.

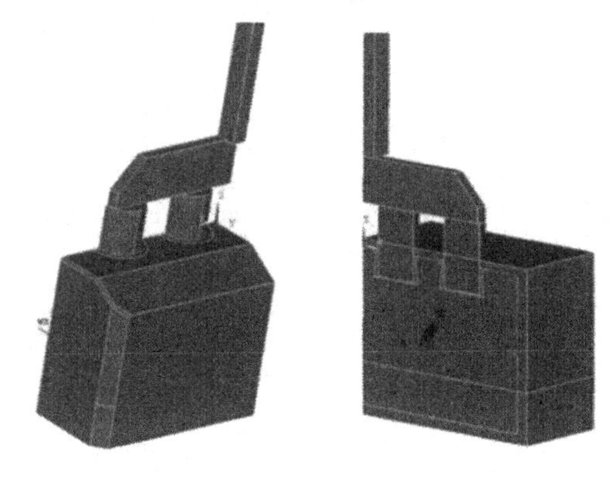

Figure 3. The figures show the geometry used in ANSYS. The total model includes carbon anode, cast iron, stubs, yoke, stem, frozen bath and cover material.

The analysis treats the problem as a case of transient state with some time-dependent boundary conditions. For each selected time increment, the thermal and mechanical differential equations are resolved in the three-dimensional space using the finite elements method. The anode is meshed by ten-node SOLID 98 elements. Once the temperatures at the nodes are obtained, stress and strain is calculated.

Due to only analysing temperature and stresses in the early age of the anode, current pick up, and thereby the Joule effect is not included.

Boundary temperature conditions;

| | | |
|---|---|---|
| Initial temperature of anode assembly | 20 | °C |
| Liquid bath temperature | 960 | °C |
| Bulk temperature of frozen bath | 625 | °C |
| Ambient temperature above anodes in cell | 115 | °C |

A layer of bath freezes immediately on to the immersed surface of the cold anode (lower 160 mm) and subsequently melts off. The bath crust which forms under the anode has a longer lifetime than that formed on the lateral faces. In the model different values of the heat transfer coefficient are used for different anode faces.

### Inhomogeneity of the Anode

Due to production parameters the anode is not homogeneous when it goes into the pot. The main reason for this is a result from the forming stage of the green block. The thick grey line in Figure 4 resembles the top of the paste before vibration forming started. The combination with high filling in the long sides, low filling in the middle and the chamfered top causes less energy to be transferred into the anode middle.

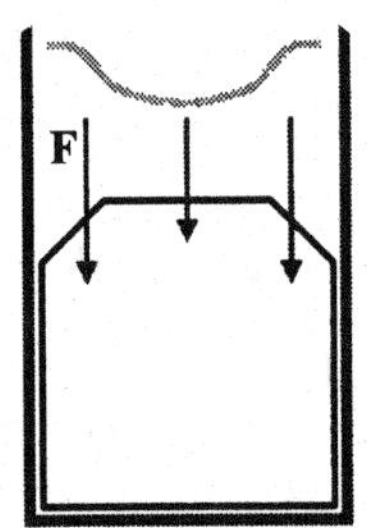


Shape of the paste top in mould before vibro compaction

Figure 4. Illustration of forming stage of a green anode. Due to uneven filling in form and the chamfered lid put on top, this will contribute making the anode body inhomogeneous.

Core samples have been drilled through baked anodes and analysed. By measuring material parameters in different positions it is possible to get en overview of gradients. An example of this is shown in figure 5, where density gradients through a horizontal cross section of an 1590 * 700 mm anode at 480 mm height (height of anode is 605 mm) are plotted.

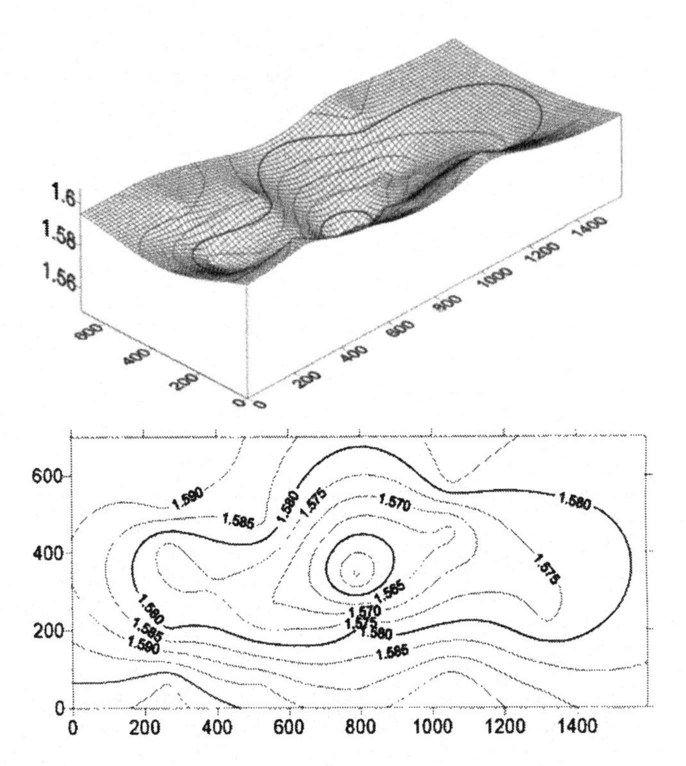


Figure 5. Contour plots of horizontal density distribution in the upper layer of a baked anode [4]. Density varies here from 1,550 to 1,605 kg/dm$^3$.

Other parameters measured were cold crushing strength (CCS), Young's Modulus, Thermal Shock Resistance (TSR)*, Coefficient of Thermal Expansion, Electrical Resistivity and Permability.

* Thermal Shock Resistance is calculated by the formula [5]

$$TSR = \frac{\sigma_{CCS}}{\rho \cdot \alpha \cdot E} \quad (1)$$

| | | |
|---|---|---|
| $\sigma_{CCS}$ | = | Cold Crushing Strength |
| $\rho$ | = | Electrical Resistivity |
| $\alpha$ | = | Thermal Expansion Coefficient |
| E | = | Youngs Modulus |

By interpolating the measured values, properties for each 10 mm through the anode are calculated. This means that the anode is divided into quadratic cubes with sides of 10 mm, each with its own set of material parameters. These values are stored in an Excel sheet.

The Excel sheet is exported to ANSYS and together with material parameters for the other elements included and boundary conditions, modelling temperatures and stresses in the anode are completed.

## Fracture Mechanics

From ANSYS we get results in colour plots with corresponding tension and stress scales. By combining these results with fracture mechanics, this will give us a good indication whether a crack might initiate or not, or if a latent crack will propagate.

In fracture mechanics there are three important variables, shown in Figure 6.

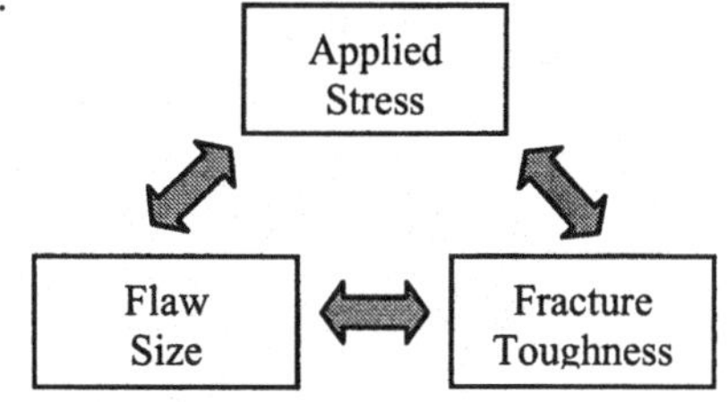


Figure 6. The three variables in fracture mechanics [6].

Many studies have dealt with initiation of cracks by thermal shock. However, resistance to crack propagation is at least as important for anode performance. Cracks can initiate by excessive stress early in the anode life or can be present already from the anode forming or baking stage. In either case, subsequent damage to the anode depends strongly on the resistance to further crack propagation. Baked anodes have therefore been investigated for mechanical properties that make us possible to better interpret the results presented from ANSYS.

Stress Intensity Factor

Fracture toughness was tested on an anode according to the standard BS 7448 [7], *Fracture mechanics toughness tests.* Results from this test can be used to calculate the stress intensity factor $K_I$. This parameter characterises stresses, strains and displacements close to a crack tip. The situation in Figure 7 gives the simplest form of $K_I$. For more complex crack situations this value is multiplied with a size factor *f*.

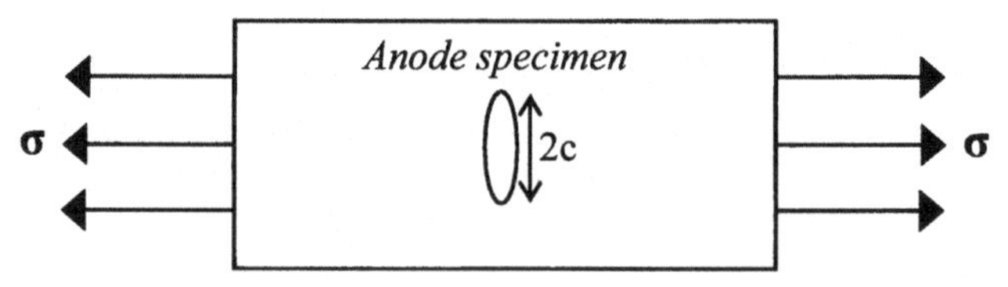


Figure 7. The figure illustrates a piece of an anode with an internal crack of length 2c. The sample is exposed to a tension stress σ.

$$K_I = f\sigma\sqrt{\pi \cdot c} \quad (2)$$

| | | | |
|---|---|---|---|
| $K_I$ | = | Stress Intensity Factor | $\left[MPa\sqrt{m}\right]$ |
| f | = | Factor describing crack geometry. | - |
| σ | = | Tension Stress | [MPa] |
| c | = | Half crack length | [m] |

The resistance against crack growth, the fracture toughness, is often characterised by the critical stress intensity factor $K_{Ic}$ of the crack. This is the stress intensity factor at the moment the crack starts to grow and fracture occurs. With a survival probability of 95 % results from 66 fracture toughness tests calculated for industrial scale anodes gives a

$K_{Ic} = 1{,}3 \left[MPa\sqrt{m}\right]$ [8].

In the calculation of the critical stress intensity factor, the size factor *f* was within the range of 0,45 to 0,55 and the maximum crack length was set at 15 mm. At crack lengths above ca 15 mm it is questionable whether formula (2) is still valid or not for carbon anodes [8].

Tension Test

Cores were drilled out from an anode and cut on a lathe to the dimensions of Figure 8. The largest grain size in the carbon anode tested was 14 mm. According to ASTM C 749 [9], the diameter of a tension test specimen should be at least three to five times the maximum particle size of the material. The geometry of the test samples is therefore within these specifications.

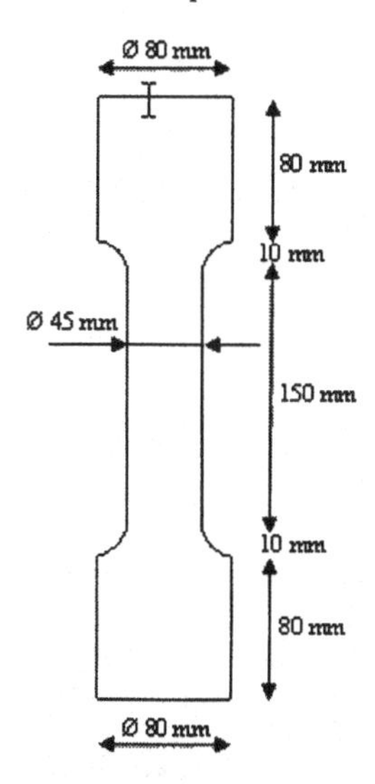


Figure 8. Dimension of tensile strength sample.

The test specimens were placed in a load train assembly. A load is applied to the specimen with means of measuring strain until fracture occurs.

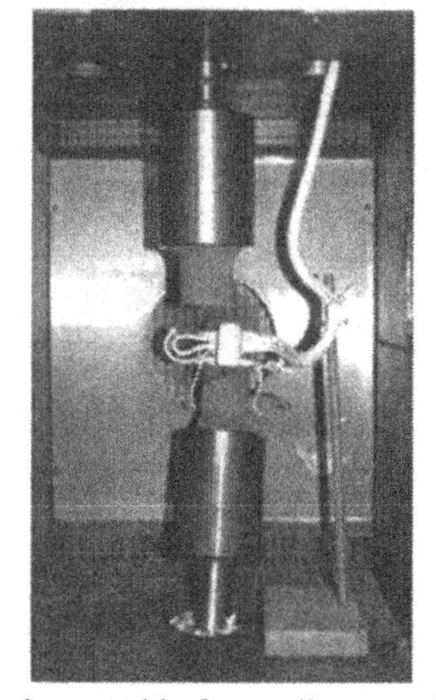

Figure 9. Load train assembly for tensile strength testing of anode samples.

Results from of 15 anode specimens gave a tensile strength of 5,60 ± 0,61 [MPa] [8].

Tests were executed on elevated temperatures as well. Figure 10 gives a plot of horizontal tensile strength at 20 different positions of core samples taken from 340 – 680 mm from one short end side. The temperature here is 400 °C and the average tensile strength is 7,9 MPa [8].

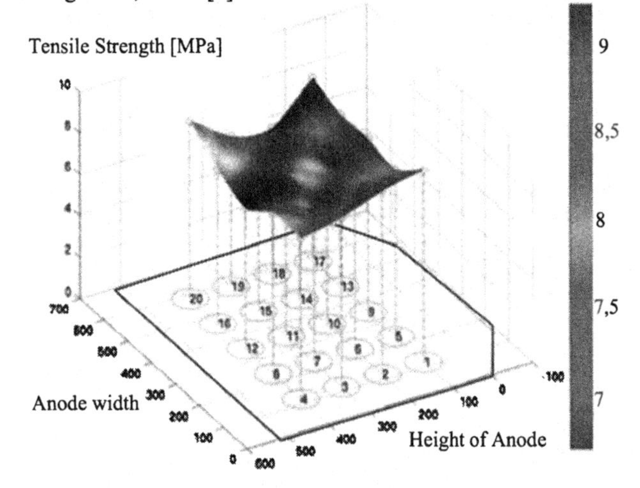


Figure 10. The plot shows tensile strength of anode samples at 400 °C.

**Modeling of tension stresses caused by gradients in the anode**

During the first 24 hours after the anode is set in cell, sometimes vertical cracking occur. Modeling this period, with cover material only at the top surface of the anode and frozen crust on the immersed part of the anode, a temperature development as shown in Figure 11 is presented.

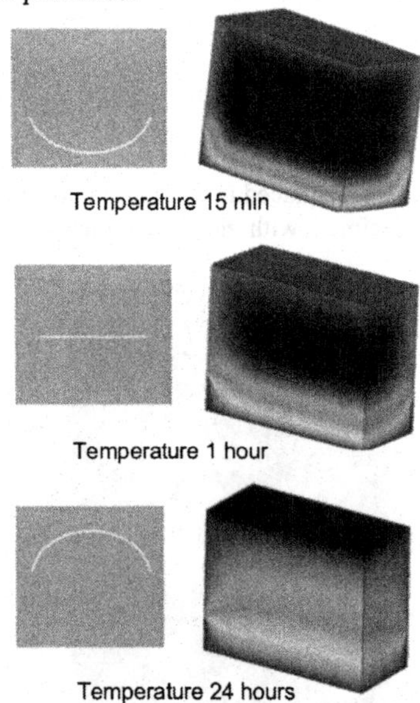


Figure 11. Temperature plots of a quarter of an anode seen from centre. After the first 15 minutes, a typical U-profile of temperature is shown in the lower part of the block. 45 minutes later the heat has moved further up into the body and the profile in the middle part is more leveled out. When the anode has been in the cell for approximately 24 hours, the temperatures in the upper part have an inverted U-profile.

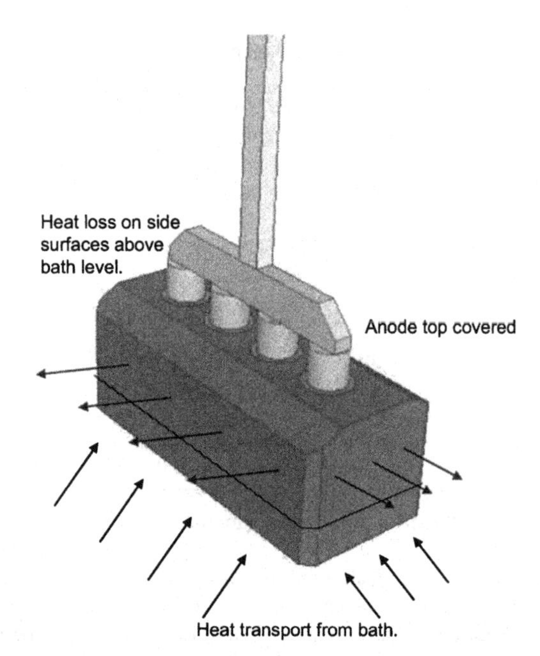


Figure 12. During the first period of the anode lifetime, heat from the frozen bath will be transferred to the anode. When this heat is conducted upwards in the anode, parts of it will dissipate through the side surfaces above bath level.

From extensive sample testing of anodes, a typical trough shape of several material property gradients is seen in the upper part of the anode. These gradients will influence the profile by increasing the temperature gradient, thereby also increasing the tension level in the upper part. When reaching the tensile strength these gradients can cause cracks similar to those yoke and stubs are responsible for. Given constant boundary conditions, larger material gradients with this profile will cause higher temperature gradients in the anode.

## Conclusions

Core samples taken through the whole anode gives a picture of the inhomogeneity of the block. Using these values in an ANSYS model, tension stresses after insertion in the cell can be calculated by finite element method.

By combining modeling and fracture mechanics, it is possible to predict whether an anode may crack or not.

Due to conditions in the vibration forming stage, often a trough pattern of material property gradients in the upper part of the anode is observed. This pattern is not advantageous regarding thermal shock resistance.

By alternating the measured values of material properties from the anode core samples and use these as input in ANSYS, know-how on where to do improvements in anode production in order to achieve more stress resistant anodes is obtained.

## References


1. M. W. Meier, *Cracking Behaviour of Anodes*, Sierre: R&D Carbon Ltd., 1996

2. Hydro Aluminium, Internal report

3. JANAF tables

4. L. P. Lossius, Internal Hydro Aluminium report, 2003

5. Formula used internal by Hydro Aluminium to calculate Thermal Shock Resistance

6. T. Anderson, *Fracture Mechanics Fundementals and Applications*, Boca Raton: Taylor and Francis Group, 2005

7. BS 7488, *Fracture mechanics toughness tests – Part 1: method for determination of $K_{Ic}$, critical CTOD and critical J values of metallic materials*, 1991. Incorporating Amendent No. 1 and corrigendum No 1.

8. L. Eliassen, *Characterisation of Mechanical Properties of Carbon Anode Materials*, Project report, Norwegian University of Science and Technology, 2006

9. ASTM C 749 – 92 (2002), *Standard Test Method for Tensile Stress-Strain of Carbon and Graphite*, 1992. Reapproved 2002

**Light Metals 2008** *Edited by: David H. DeYoung*
***TMS (The Minerals, Metals & Materials Society), 2008***

# CHARACTERISTIC AND DEVELOPMENT OF PRODUCTION TECHNOLOGY OF CARBON ANODE IN CHINA

Guanghui LANG Rui LIU Kangxing QIAN

SUNSTONE International 24F Building A TOPBOX No. 69 Beichen West Street, Beijing 100029, China

Keywords: Traditional technology, anode quality, improvement tred

## Abstract

Major technological advancements were achieved in China in a short time to support the rapid development of the domestic Aluminum industry.

The anode production plant of Sunstone is a good representation of the technological developments that took place in China to produce high quality anodes with a high level of efficiency.

The main design characteristics of the plant are described as follows:

• Shaft type calcining furnaces for petroleum coke are widely used in China producing a product of very consistent quality with low investment and long service life.

• Weigh hoppers are used for accurate control of the dry aggregate composition; Batch mixers are used for efficient paste mixing; green paste is then fed to vibroforming equipment to produce green anodes.

• The green anodes are baked in open top bake furnaces to obtain consistent anode properties that are within the worldwide typical range.

Sunstone dedicates major emphasis to the following factors:

• Quality and consistency of raw materials;
• Development of large granulometry in its shaft type coke kilns;
• Emphasis on environmental protection, maximum and best utilization of waste heat;
• Highest attention to consistency of anode quality.

## Introduction

Production technology of prebaked anode in China can be divided into two kinds of models generally: One is the traditional Chinese production model, i.e. shaft calciner → charge mixture in weighing storehouse → batch kneading → vibratory compaction → open-top baking furnace. The other model utilizes international technology: coke calcination by rotary kiln → continuous weighing of the dry aggregate and paste kneading → vibratory compaction → open-top baking furnace. With the rapid growth of the aluminum industry in China, anode production technology is quickly developing. The traditional Chinese anode production process is also rapidly evolving, from the simple original model to increasingly mechanized operations of large scale. The improved and more mechanized production model expanded very significantly in China. In 2006, this model manufactured 70% of anode production in China with 95% of it for international exports.

## Anode Production

### Desciption of the Sunstone Anode Plant

The anode plant of Sunstone is located in Linyi county, Shandong Province, with an annual output of 150,000 tons of prebaked anodes exported to the US, EU, Middle East. The plant includes the following processing areas: coke calciner, green mill, bake furnace and maintenance workshop. Major raw materials – green petroleum coke and coal tar pitch - are sourced from domestic suppliers.

### Coke Calcination

#### Features of Installation

Four calcination furnaces with 28 shafts each and one calcination furnace with 24 shafts:

• No added fuel, high energy efficiency, recycled heat of 600 kJ per hour and furnace that is used for paste mixing, pitch melting and other use

• Stable quality of calcined coke, high degree of calcination

• Low feed rate of materials during calcination leads to low degradation of granulometry

• Rate of carbon loss during calcination is less than 3 %

Production and Processing Parameters

• Output: 90 - 110 kg per hour and shaft

• Controlled parameters: temperature of the second flue 1250 - 1380°C

• Coke temperature: 1150 - 1200°C

• Negative pressure of the second flue: 15 - 30 Pa

• Heating cycle time: 40 hours

Calcined Coke Quality

The calcined coke shows the following typical properties, as summarized in table 1.

Table 1: Typical Properties of Calcined Coke

| Properties | Method | Unit | Average |
|---|---|---|---|
| Grain >4mm<br>>1mm | ISO12984 | % | 50%<br>>80% |
| Vibrated Bulk Density | ISO 10236 | g / cm³ | 0.800 |
| Real Density in Xylene | ISO 9088 | g / cm³ | 2.07-2.12 |
| $CO_2$ Reactivity | ISO 12988-1 | % | 25 |
| Air Reactivity | ISO 12989-1 | %/min | 0.05 |

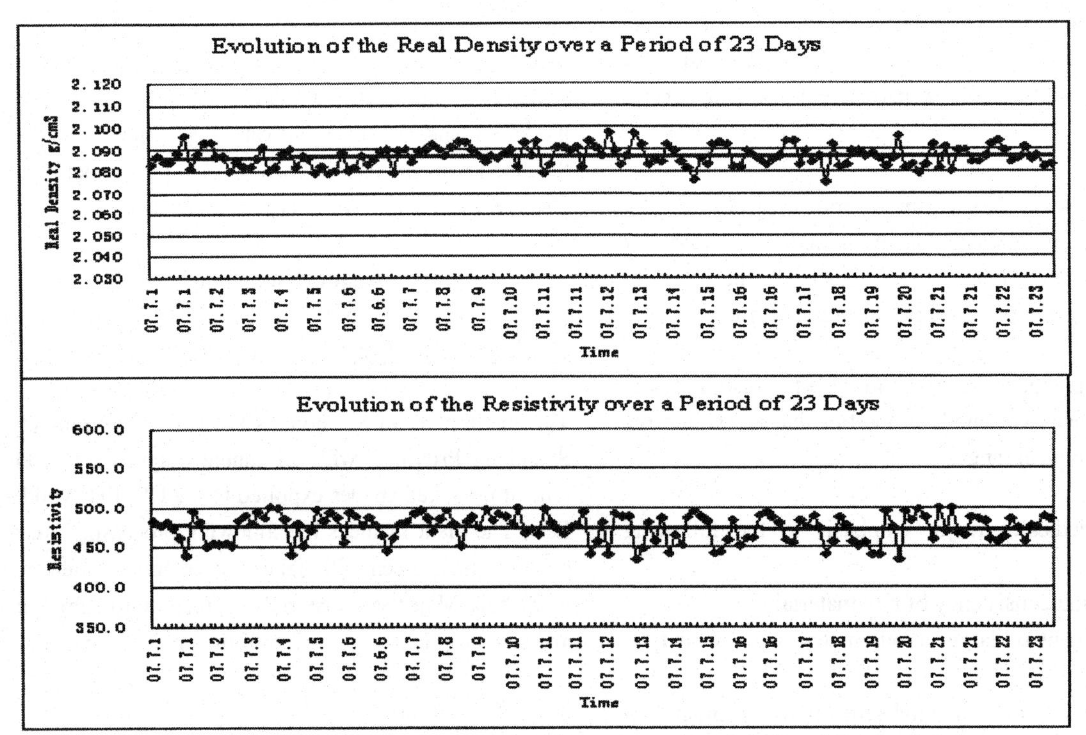


Figure 1. Effect of Time on Real Density and Resistivity of Calcined Coke

Production of Green Anodes

Crushing and Screening

- Major equipment: Impact crusher, three-layer vibrating screen, bucket elevator
- Production yield: 20 tons per hour
- Target granular purity: greater than 90 %

Milling

The fines are produced in a classical air swept ball mill, as illustrated in figure 2 with the following key features:

- Major equipment: Air swept ball mill (diameter 2.4 m; length 4.4 m), classifier, belt conveyor scale, electro-sound translator, cyclone, dust collector, fans, PLC control system
- Production yield: 5 tons per hour
- Target fineness of ball mill product: 4000 +/- 250 Blaine

Process Diagram of Petroleum Ccke in Air Swept Mill System

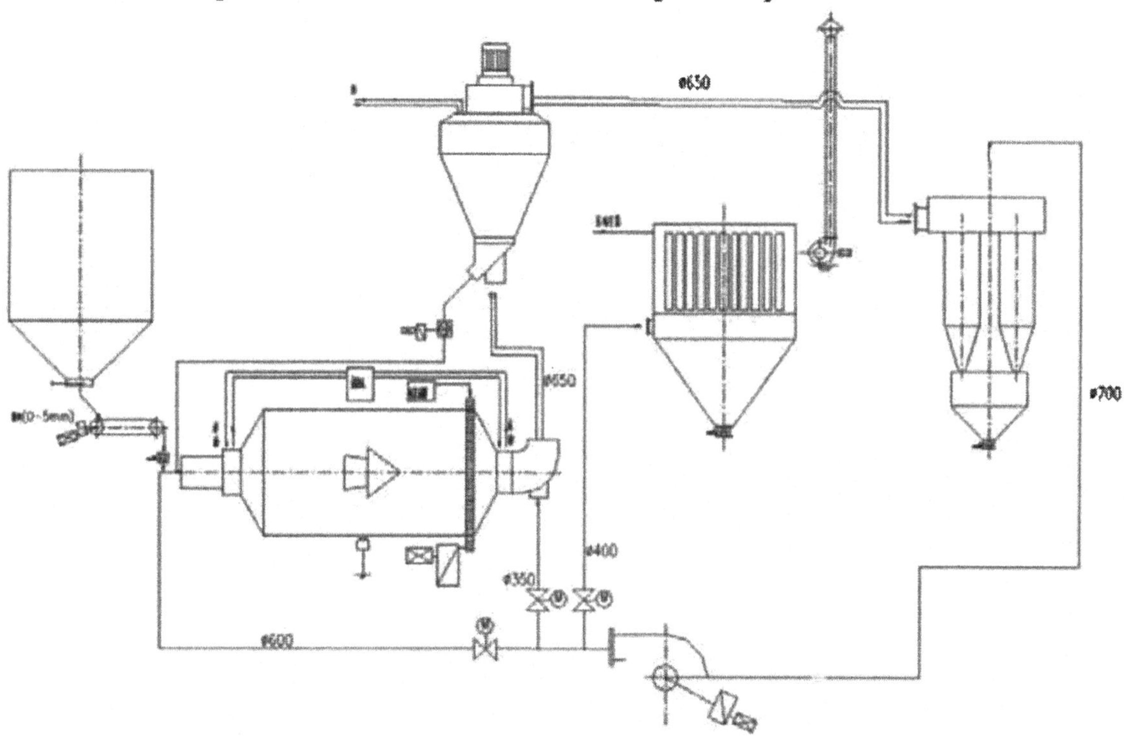


Figure 2. Air Swept Mill System

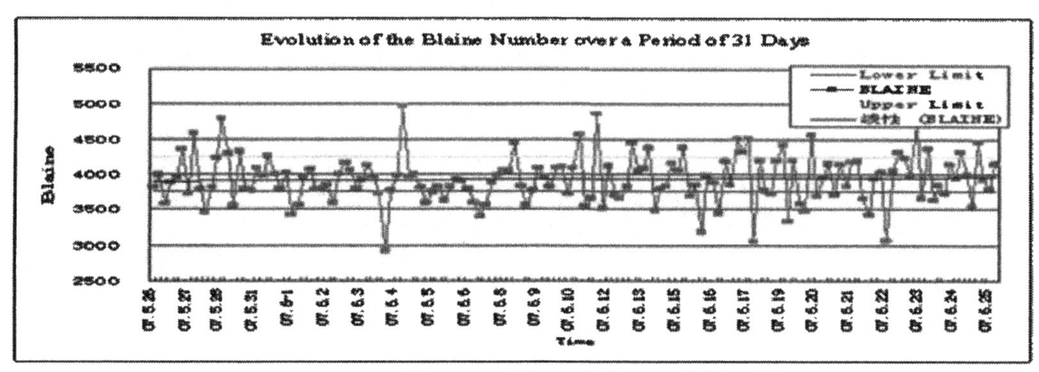


Figure 3. Effect of Time on Blaine Value

Butts Preparation

- Major equipment: 1000 x 900 jaw crusher; 600 x 400 jaw crusher; impact crusher 20 tons per hour for grains 0 - 20 mm; vibrating screens with two layers; elevator D250 of 37 meter
- Designed production yield: 20 tons per hour

- Target granular purity: greater than 90 %
- Actual production rate: 15 tons per hour
- Target grain size: less than 15 mm

Dry Aggregate Preparation

The layout of the dry aggregate preparation is illustrated in figure 4, with the following characteristics:

- Major equipment: Silos on load cells: 3 x 8 m3; screw conveyor with 630 mm diameter; two pitch batch vessels of 2 m3 each; PLC control system
- The dry aggregate consists of the following fractions: 4 - 2 mm, 2 - 1 mm, 1 - 0 mm. The fractions are individually weighed and collected in a batch system that is controlled by PLC.
- Control indicators: maximum moisture content in coke and solid pitch: 0.5 %
- Aggregate frequncy:     6-10-minute / times

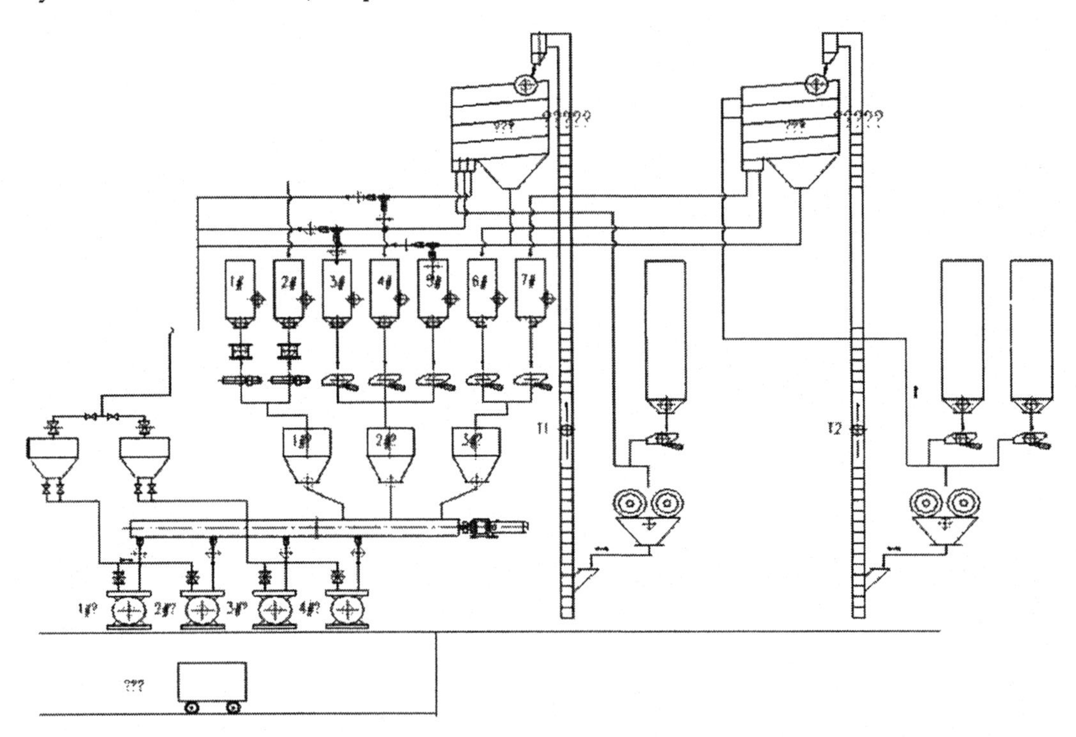


Figure 4. Dry Aggregate Preparation

It is full automation in the mixing process as shown in Figure 4.

Paste Mixing

- Major equipment: 8 batch mixers of 4000 liter each
- Mixing time: 60 minutes
- Temperature of heating oil: 230 +/- 3°C
- Paste temperature: 160 +/- 3°C

Anode Forming

- Major equipment: two single position vibro-compactors
- Production capacity: 22 anodes per hour
- Anode temperature: 150 +/- 3°C
- Excentric force: 600 kN
- Target green anode density: not less than 1.67 - 1.69 kg/dm3

Figure 5. Forming Station

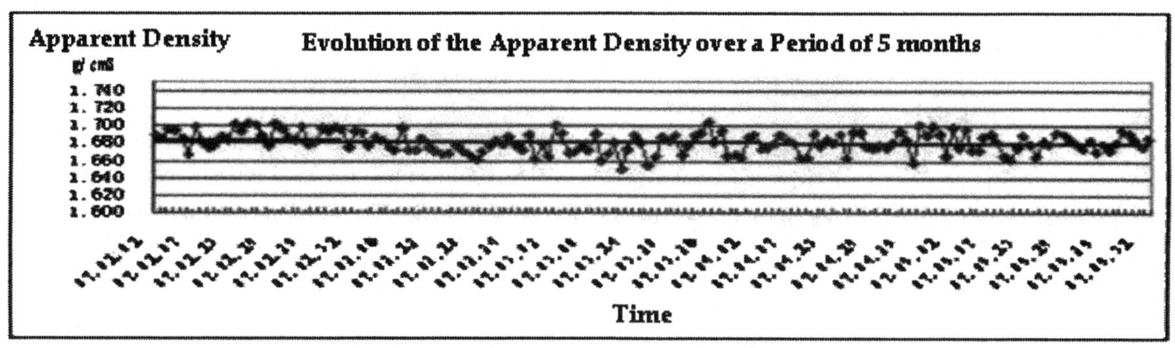


Figure 6. Effect of Time on Apparent Density of Green Anode

Anode Baking

Features of Bake Furnace and Baking Process

- Two open top bake furnaces of 38 sections each with 7 pits per section
- Size of pits (Length x Width x Depth): 5286 x 765 x 5500 mm and 5286 x 800 x 5640 mm
- Firing system: automatic control by PID
- Fuel: natural gas: 34 MJ / Nm3
- Target flue gas temperature: 1200 +/- 3°C
- Maximum temperature difference within pit: 50°C
- Suction capacity of crane: 45 m3 / hour
- Capacity of auxiliary crane: 5 x 3 tons
- Fire cycle time: 30 - 32 hours
- Maximum anode temperature: 1085 - 1125°C
- Fuel consumption: 2.65 GJ/t
- Exhaust fume treatment by cooling tower and electro-static precipitator

Quality of Baked Anodes

Table 2: Typical Quality of Baked Anodes

| Properties | | Method | Unit | Average |
|---|---|---|---|---|
| Apparent Density | | ISO12985-1 | kg/dm3 | 1.592 |
| Specific Electr.Resistance | | ISO11713 | μΩ•m | 54 |
| Compressive Strength | | ISO18515 | Mpa | 40 |
| Flexural Strength | | ISO 2986-1 | Mpa | 11.1 |
| Density in xylene | | ISO 9088 | kg/dm3 | 2.088 |
| Air Permeability | | ISO 15906 | nPm | 1.85 |
| CO2 Reactivity | Residue | ISO 12988-1 | % | 88.5 |
| | Dust | | % | 4.0 |
| Air Reactivity | Residue | ISO 12989-1 | % | 92.7 |
| | Dust | | % | 3.6 |
| Ash Content | | GB/T1429 | % | 0.32 |
| Elements | S | ISO 12980 | % | 1.27 |
| | V | | ppm | 123 |
| | Ni | | ppm | 216 |
| | Si | | ppm | 132 |
| | Fe | | ppm | 284 |
| | Na | | ppm | 116 |
| | Ca | | ppm | 231 |
| | Zn | | ppm | 9 |
| | Pb | | ppm | 10 |

## Characteristics of the Traditional Production Technology in China

### Coke Calcination in Shaft Calciners

Calcination of green petroleum coke takes place in the anode plant in shaft kilns. Calcined coke of a very stable quality is produced with high real density, low resistivity and low rate of burning loss.

### Green Anodes Manufactured by Batch Mixers and Vibroforming

The formulation of the dry aggregate and pitch dosing is controlled by a highly computerized automated batch system technology. As a result, our anode production is made with accurate ingredients, effective paste mixing and a high excentric force to achieve high green densities.

### Anode Baking Furnace

International advanced automatic baking technology has been introduced and applied in China. It applies to the anodes a very uniform temperature distribution and a very consistent anode baking leading to low anode consumption rates.

### Anode Manufacturing Technology in China

The anode manufacturing technology and the resulting anode quality has significantly improved and got international recognition.

### Capital Investment

The capital required to build Sunstone anode plant is very moderate and construction time is quite fast. Construction investment is about 2200-2500 Yuan / ton anode with a construction time of 18 months.

## Trends of Development in Technology

### Attention to the Quality of Raw Materials

We have known the importance of raw materials of high quality and of their impact to anode quality, and we look for a stable supply of petroleum coke including chemical elements. Suppliers of coal tar pitch are selected on the base of the technology and process requirement in the anode production plant.

### Large Development in Calcined Petroleum Coke

A shaft coke calcination plant of large scale was developed from 16 shafts to 28 and 32 shafts. The volume by shaft increased from 80 kg to 100 kg / shaft hour. The automation level of the discharge system was increased and the excess heat of the calciner is fully utilized.

### Improved Automation Level in Green Anode Plant

Plans are to continue to the level of automation and control of our anode plant, with expected improvement of operational uptime and product quality.

### Anode Baking

The trend in the baking furnace area is to increase pit capacity and to expand mechanized operations. More attention is payed to study ways to reduce fuel consumption.

### Attention to Environmental Protection

Our anode manufacturing plant will pay more attention to pollutant management, reduce the emission of pollutants, and particularly pitch fumes.

### Trend to Homogeneous Anode Quality

Anode production plant recognized that consistency of anode quality is more important than good quality of individual anode blocks. Anode quality optimization efforts at Sunstone focus in particular to proper selection of raw materials and operation technology.

## Summary

The Chinese traditional anode production technology improved dramatically in recent years as a result of major work and developmental efforts. Many weak aspects of the original design were re-engineered. The current upgraded version of the Chinese traditional technology is well proven to be capable to produce anodes of high quality that work efficiently in reduction cells of modern design. More work is in progress to achieve further technological upgrades, improved anode quality and lower production costs.

# ELECTRODE TECHNOLOGY

## Cathodes Raw Materials and Properties

*SESSION CHAIRS*
**Claude Gauthier**
Alcoa Inc.
Deschambault, Quebec, Canada

**Alexander Proshkin**
United Company Rusal
Krasnoyarsk, Russia

**Light Metals 2008** *Edited by: David H. DeYoung*
***TMS (The Minerals, Metals & Materials Society), 2008***

# ISO STANDARDS FOR TESTING OF CATHODE MATERIALS

Harald A. Øye
Department of Materials Science and Engineering
Norwegian University of Science and Technology
N-7491, Trondheim, Norway.

Keywords: Cathodes, ISO, Testing

## Abstract

ISO/TC 226 (Materials for the Production of Primary Aluminium) recently published a CD with 109 ISO Standards and 2 ISO Technical Specifications, covering the materials: Alumina, Pitch, Coke, Anodes, Cathodes and Ramming Paste. All standards are easily accessed from the CD. In addition to the test procedures for common room temperature properties, methods which characterize the cathode materials at operational conditions, have been developed. Examples are: Sodium expansion of cathodes with and without external pressure, rammability of paste and expansion and shrinkage of ramming paste. Additional studies of cathode property changes as function of temperature and time enable room temperature ISO standards to be extrapolated to operational conditions.

The standards are not only useful for material evaluation and quality control, but also important for development of more reliable mechanical-thermal-electrical models.

## Introduction

Aluminium production takes place at 960 °C in an extremely corrosive environment. About 60 % of the production cost is in materials: Alumina, Fluorides, Carbonaceous materials, Refractory and Insulation materials. The quality and consistency of materials are essential for a successful process. These materials are sold freely between suppliers and customers in different countries with test results. Before ISO started its work these test results were based on internal test methods or national standards that could give different results for the same properties dependent on details in the test methods.

The aim of ISO/TC 226 (Materials for the production of primary aluminium) (formerly ISO/TC47/SC7) is to obtain unified and complete standard test methods that suppliers as well as customers have agreed to. ISO/TC 226 is presently responsible for 109 ISO standards and 2 ISO technical specifications, see Appendix.

ISO/TC 226 is organized into 5 working groups each dealing with the following topics: WG 1 Pitch, WG 2 Solid carbonaceous materials, WG 3 Smelter grade alumina, WG 4 Smelter grade fluorides, WG 5 Cryolite resistance of dense refractories.

The test methods are useful for quality control, contracts and for development work. The ISO standards have to a large extent been based on modifications of internal methods and national standards. ISO/TCS 226 has, however, seen it as a special task also to develop international standards that characterize the materials at process conditions. This has been accomplished by involving test laboratories as well as universities in addition to suppliers and aluminium producers.

A clickable CD for all the 109 standards has been published. The standards are organized under the following subtitles: Alumina, Pitch, Coke, Electrodes, Ramming paste, Fluorides, Refractory (only cryolite resistance).

## Organization of Standardization Work and Choice of Methods

The development of the ISO standards for cathode materials has been a long struggle. During a 1.5 months sabbatical stay at RDC in Sierre in 1994 the existing standards for carbon materials of interest for the primary aluminium industry was reviewed, especially with respect to ASTM and DIN. At that time ISO had standards for alumina and fluorides, some for pitch and coke, but none for anodes, cathodes and ramming paste. An informal group was called together in Sierre after the review was completed. It consisted of participants from the 3 major cathode producers, 2 Norwegian smelters, RDC and Norwegian University of Science and Technology. This informal group was later incorporated into ISO/TC47/SC7 (later ISO/TC 226) as WG 2 (Solid carbonaceous material).

The carbon producers had most of the competence. They used different methods and wanted their specific method certified as an ISO standard. However, competence was also built up at Norwegian University of Science and Technology as well as in the Norwegian aluminium industry. The wish of the Norwegian aluminium industry had a high impact, as they were large buyers of cathodes and ramming paste material. The main standard test methods at room temperature were, to a large extent, modifications of ASTM, DIN or British standard. A major task was the choice of methods that simulates operating conditions. It was decided that the following methods had to be developed: Sodium expansion of cathode blocks, Rammability of paste and Expansion/Shrinkage of ramming paste. Sample preparation and baking procedures were also important issues.

Sodium expansion is heavily dependent on experimental conditions as bath composition, current density and electrolyte volume. It was necessary that the experimental conditions were described in great detail. As an example the sodium expansion as function of cryolite ratio is shown [1] (Fig. 1).

It was decided to make two standards, sodium expansion without external pressure and sodium expansion with 5 MPa external pressure. The bath composition, current density, temperatures, sample diameter and electrolysis time were the same for both standards.

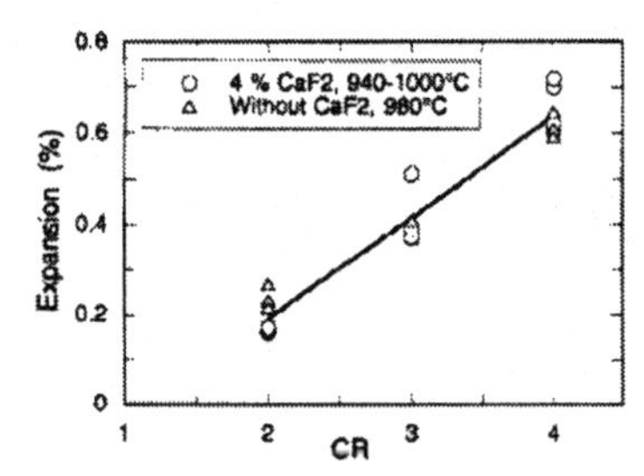


Figure 1: Sodium expansion test of semigraphitic cathode samples as a function of the cryolite ratio (mole NaF/mole $AlF_3$) [1].

Figure 2 shows the expansion/shrinkage of a ramming paste. A cylindrical ramming paste sample is heated at a rate of 90 °C/h to 950 °C (B), kept at 3 hours at 950 °C (C) and then cooled. The sample is viscous up to point A and the expansion to A is due to formation of vapour within the sample. A former test was to heat the sample to 950 °C, cool it and measure the dimension before and after the experiment. In the present case an expansion will be reported. The ISO/TC 226 opinion was, however, that the relevant property to measure is the shrinkage after the sample is solidified, *i.e.,* between A and C. In this temperature range shrinkage cracks can develop with possible penetration of aluminium.

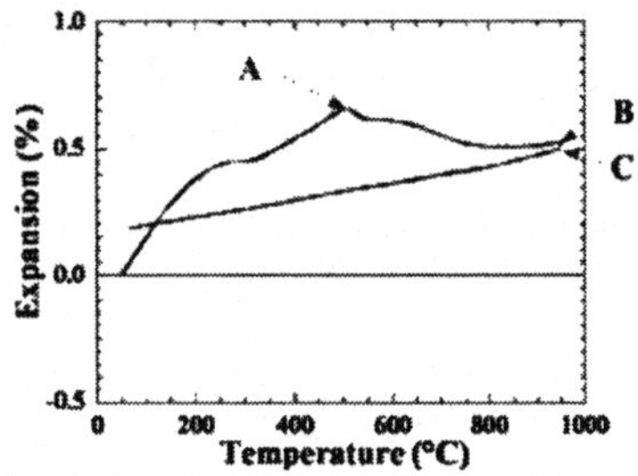


Figure 2: Expansion/shrinkage of ramming paste during continuous heating and cooling (ISO 14428).

An example of a test justification is the cryolite resistance for refractories. A very simple test was developed [2]. A 57 mm diameter hole with depth 40 mm was drilled into the refractory bricks. A salt mixture with 60 % cryolite and 40 % sodium fluoride was heated to 950 °C with a total time of 24 hours. The sample was cooled, cut diagonally through the hole and the penetration measured visually. At the same time the same materials were built into a trial cell which was stopped after 500 days. A very good correlation between test and plant results was found (Fig. 3) [2].

The cup test does not fully simulate the complex reactions taking place in the refractory. Siljan *et al.* [3] give a thorough discussion of the situation. Gaseous $SiF_4$ and especially Na will also be present, penetrate and react with the refractory. Allaire *et al.* [4] have proposed a test where bath and sodium penetration are determined separately. Nevertheless, the simple cup test is useful for ranking of cathode materials. Figure 4 shows that increased $SiO_2$ content in the refractory gives increased penetration resistance. This was also found by Brunk *et al.* [5]. These results had a large impact on the industries' choice of refractory materials. The test is being made into an ISO test (20292).

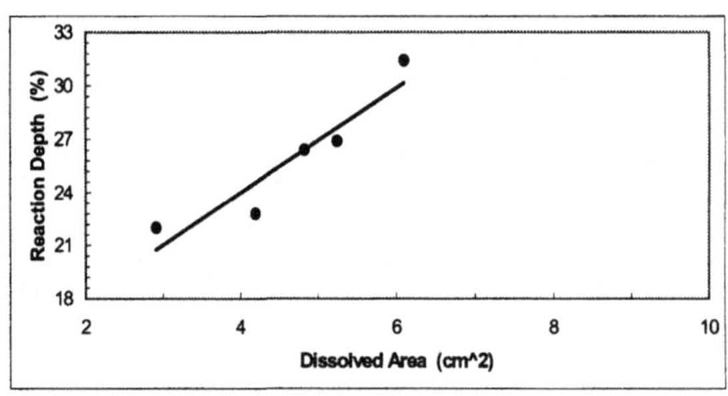


Figure 3: A comparison of laboratory investigation of measured dissolved area ($cm^2$) versus observed penetration depth in real-life cell linings. Experiments were performed in a 220 kA point-fed prebake cell with lifetime of 500 days. Reaction depth in the cell lining is given as percent of lining thickness reacted [2].

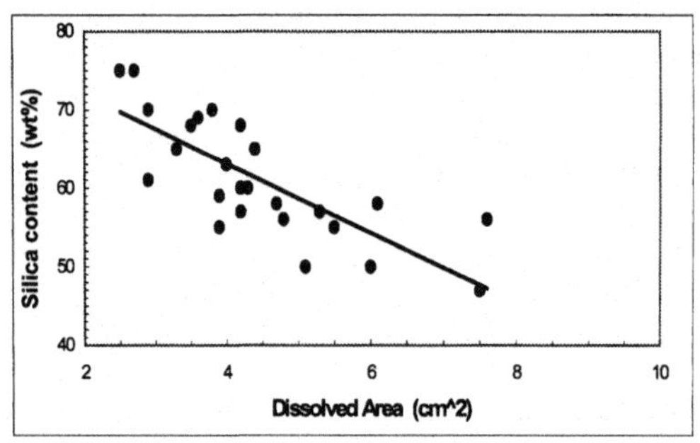


Figure 4: The effect of silica content on a measured dissolved area in alumino-silicate bricks. Linear regression trend-line determined by least-square analysis is included [2].

### Developed Standards for Ramming Paste and Cathodes

The developed ISO methods are given in Tables 1 and 2.

Table 1. Test methods for ramming paste.

| Method | ISO no. | Comments |
|---|---|---|
| UNBAKED PASTE | | |
| Binder and aggregate content | 14423 | General characterization. |
| Loss of volatile matter | 14425 | A measure of carbon yield. |
| Baking loss | 20202 | A measure of carbon yield. |
| Apparent green density | 14427 | Used to calculate density change during baking. |
| Rammability | 17544 | Important to establish temperature window for ramming. |
| Expansion/shrinkage during baking | 14428 | Very important property. Too high shrinkage after solidification may give crack opening in cathode. |
| BAKED PASTE | | |
| Electrical resistivity | 11713 | Not important. |
| Compressive strength | 18515 | Too high or too low strength is not wanted. |
| Open porosity | 12985-2 | Pore distribution more important than porosity. |
| Thermal expansion | 14420 | Not important. |
| Thermal conductivity | 12987 | Not important. |
| Baked density | 20202 | Should not be too low. |
| Ash content | 8005 | Not important but gives information about raw materials |

| Table 2. Test methods for carbon cathodes | | |
|---|---|---|
| Method | ISO no. | Comments |
| Bulk density, geometric | 12985-1 | Density is not directly relevant, but changes may indicate change of raw materials. |
| Bulk density, hydrostatic | 12985-2 | |
| Real density, xylene | 9088 | Indication of degree of graphitization. |
| Real density, helium | 21687 | Indication of degree of graphitization. |
| Open porosity | 12985-2 | Pore distribution more important than porosity. |
| Total porosity | 12985<br>9088<br>21687 | |
| Air permeability | 15906 | More important for anodes than for cathodes, but gives information about pore structure. |
| Compressive strength | 18515 | Less important than bending strength, but easy to measure. |
| Bending strength, 3 point | 12986-1 | The most important strength property. |
| Bending strength, 4 point | 12986-2 | The most important strength property. |
| Youngs (E) modulus | | Important property, the dynamic E-modulus is easiest to measure. |
| Electrical resistivity | 11713 | Very important property, but note temperature and time coefficient. |
| Thermal conductivity | 12987 | Important design parameter, but note temperature and time coefficient. |
| Thermal exp. | 14420 | Not important as quality parameter. |
| Ash content | 8005 | Unimportant, but changes may indicate change of raw materials. |
| Sodium exp. with pressure | 15379-1 | Very important property. Expansion decreases with graphitization |
| Sodium exp. without pressure | 15379-2 | |

A smelter's approval of a material is usually a major undertaking. In addition to evaluation of standard properties, extra testing is performed as well as performance test in trial cells. It is, however, important that the properties of an approved material stay constant, *i.e.*, no change in raw materials or production procedure. Such quality insurance is performed by using some easily performed tests which not necessarily is very relevant for the performance. Examples on such quality control procedures are given below.

Quality control of ramming paste (ISO number in parenthesis):

Apparent green density (20202)
Apparent baked density (14427)
Compressive strength of baked sample (18515)
Dilation / shrinkage (14428)

In addition storage durability should be checked by the rammability test (17544).

Quality control of cathode blocks (ISO number in parenthesis):

1. Real density (9088, 21687)
2. Apparent density (12988, 1-2).
3. Compressive strength (18515).
4. Electrical resistivity (both ⊥ and II) (11713)
5. Bending / shear strength (both ⊥ and II (12986, 1-2)
6. Ash content (8005).
7. Specification of slot cutting (for vibrated blocks).

**Additional Non-standardized Studies**

The ISO/TC 226 committee participants did also carry out additional tests which were not standardized, but which were of importance for thermo-electric-strain-stress modelling. Examples are Thermal expansion to 950 °C, Thermal and electrical conductivities to 950 °C, Static E-modulus, Sodium expansion as function of experimental parameters (pressure, bath composition, current density), Pore characterization by image analysis. Thermal and electrical conductivities are prime examples of such studies (Fig. 5) [6].

Note the very different temperature coefficients. The properties of non-graphitic materials also change with time due to graphitization (Fig. 6). With such data it is possible to extrapolate ISO standardized room temperature data.

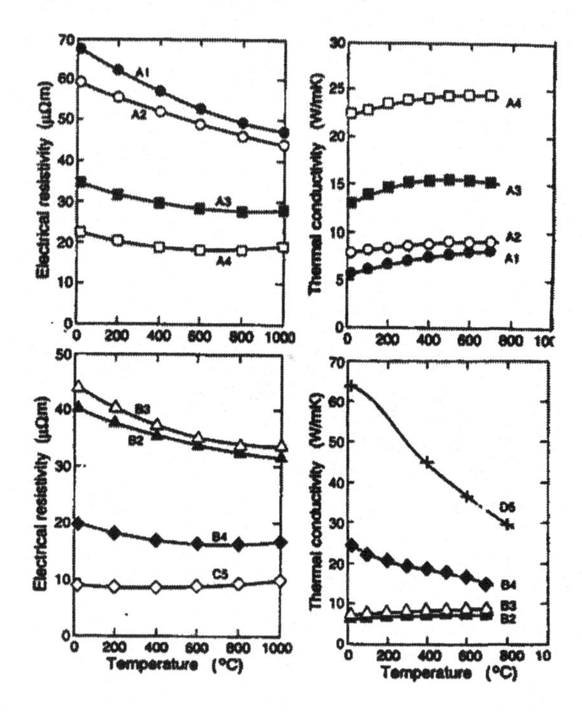


Figure 5: Electrical and thermal conductivity as function of temperature [6].

Figure 7 shows a strain stress modelling based on extrapolated room temperature ISO data. Sodium penetrates downward into the bottom of the cathode carbon block with time. The stress in the bottom of the cathode block increases while the stress on the top of the cathode block decreases due to the bottom expansion. The maximum stress on the surface of the cathode block is at arrow (12.2 MPa). The object of this modelling work was strengthening of the pot shell.

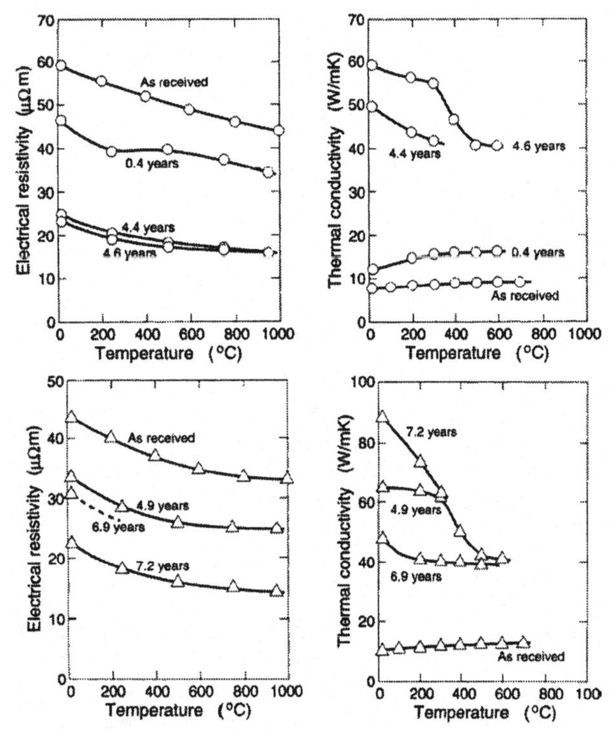


Figure 6: Electrical and thermal expansion as function of time. Changes with time are due to graphitization [6].

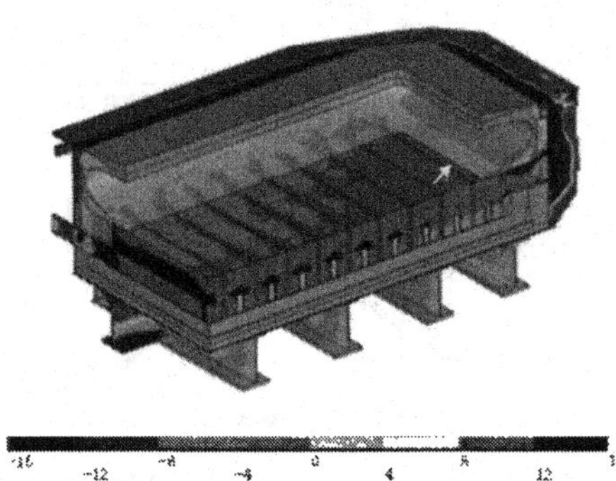

Figure 7: X-direction stress with thermal and sodium expansion, 400 days after start up (MPa)

### References


1. H.A.Øye, "Carbon Cathode Materials: Approval and Quality Control Procedure", JOM **47**, No. 2 (1995) 14-21.
2. O.J. Siljan, C. Schöning and T. Grande, "State-of-the-Art Alumino-Silicate Refractories for Al Electrolysis Cells", JOM **54**, No. 5 (2002) 46.
3. O.-J. Siljan, T. Grande and C. Schöning, "Refractories for Aluminium Electrolysis Cells. Part II", Aluminium **77**, 5, 385-390.
4. C. Allaire, R. Pelletier, O.-J. Siljan and A. Tabereaux: "An Improved Corrosion Test for Potlining Refractories", Light Metals 2001, 245-249.
5. F. Brunk, W. Becker and K. Lepére: "Cryolite Influence on Refractory Bricks. Influence of $SiO_2$ Content and Furnace Atmosphere". Light Metals 1993, 315-320.
6. M. Sørlie, H. Gran and H.A. Øye, "Property Changes of Cathode Lining Materials during Cell Operation", Light Metals 1995, 497-506.


**APPENDIX: The List of all the ISO/TC 226 Standards Included on the CD-Rom**

| **Alumina** | **Aluminium oxide primarily used for the production of aluminium** |
|---|---|
| ISO 802:1976 | Preparation and storage of test samples |
| ISO 804:1976 | Preparation of solution for analysis -- Method by alkaline fusion |
| ISO 805:1976 | Determination of iron content -- 1,10-Phenanthroline photometric method |
| ISO 806:2004 | Determination of loss of mass at 300 degrees C and 1 000 degrees C |
| ISO 900:1977 | Determination of titanium content -- Diantipyrylmethane photometric method |
| ISO 901:1976 | Determination of absolute density -- Pyknometer method |
| ISO 902:1976 | Measurement of the angle of repose |
| ISO 903:1976 | Determination of untamped density |
| ISO 1232:1976 | Determination of silica content -- Reduced molybdosilicate spectrophotometric method |
| ISO 1617:1976 | Determination of sodium content -- Flame emission spectrophotometric method |
| ISO 1618:1976 | Determination of vanadium content -- N-Benzoyl-N-phenylhydroxylamine photometric method |
| ISO 2069:1976 | Determination of calcium content -- Flame atomic absorption method |
| ISO 2070:1997 | Determination of calcium content |
| ISO 2071:1976 | Determination of zinc content -- Flame atomic absorption method |
| ISO 2072:1981 | Determination of zinc content -- PAN photometric method |
| ISO 2073:1976 | Preparation of solution for analysis -- Method by hydrochloric acid attack under pressure |
| ISO 2828:1973 | Determination of fluorine content -- Alizarin complexone and lanthanum chloride spectrophotometric method |
| ISO 2829:1973 | Determination of phosphorus content -- Reduced phosphomolybdate spectro-photometric method |
| ISO 2865:1973 | Determination of boron content -- Curcumin spectrophotometric method |
| ISO 2865:1973 / Cor 1:1991 | |
| ISO 2926:2005 | Particle size analysis for the range 45 µm to 150 µm -- Method using electroformed sieves |
| ISO 2927:1973 | Sampling |
| ISO 2961:1974 | Determination of an adsorption index |
| ISO 3390:1976 | Determination of manganese content -- Flame atomic absorption method |
| ISO 8008:2005 | Determination of specific surface area by nitrogen adsorption |
| ISO 8220:1986 | Determination of the fine particle size distribution (less than 60 µm) -- Method using electroformed sieves |
| ISO 17500:2006 | Determination of attrition index |
| ISO 23202:2006 | Determination of particles passing a 20 micrometre aperture sieve |

| Pitch | Carbonaceous materials for the production of aluminium |
|---|---|
| ISO 5939:1980 | Pitch for electrodes -- Determination of water content -- Azeotropic distillation (Dean and Stark) method |
| ISO 5940:1981 | Pitch for electrodes -- Determination of softening point by the ring-and-ball method |
| ISO 5940-2:2007 | Pitch for electrodes -- Part 2: Determination of the softening point (Mettler softening point method) |
| ISO 6257:2002 | Pitch for electrodes -- Sampling |
| ISO 6376:1980 | Pitch for electrodes -- Determination of content of toluene-insoluble material |
| ISO 6791:1981 | Pitch for electrodes -- Determination of contents of quinoline-insoluble material |
| ISO 6998:1997 | Pitch for electrodes -- Determination of coking value |
| ISO 6998:1997 / Cor 1:1999 | |
| ISO 6999:1983 | Pitch for electrodes -- Determination of density -- Pycnometric method |
| ISO 8003:1985 | Pitch for electrodes -- Measurement of dynamic viscosity |
| ISO 8006:1985 | Pitch for electrodes -- Determination of ash |
| ISO 9055:1988 | Pitch for electrodes -- Determination of sulfur content by the bomb method |
| ISO 10238:1999 | Pitch for electrodes -- Determination of sulfur content by an instrumental method |
| ISO 12977:1999 | Pitch for electrodes -- Determination of volatile matter content |
| ISO 12979:1999 | Pitch for electrodes -- Determination of C/H ratio in the quinoline-insoluble fraction |
| ISO 21687:2007 | Calcined coke – Determination of real density by helium pyknometry (can be used) |

| Coke | Carbonaceous materials used in the production of aluminium |
|---|---|
| ISO 5931:2000 | Calcined coke and calcined carbon products -- Determination of total sulfur by the Eschka method |
| ISO 6375:1980 | Coke for electrodes -- Sampling |
| ISO 6997:1985 | Calcined coke -- Determination of apparent oil content -- Heating method |
| ISO 8004:1985 | Calcined coke and calcined carbon products -- Determination of the density in xylene – Pyknometric method |
| ISO 8005:2005 | Green and calcined coke -- Determination of ash content |
| ISO 8658:1997 | Green and calcined coke -- Determination of trace elements by flame atomic absorption spectrometry (without precision) |
| ISO 8723:1986 | Calcined coke -- Determination of oil content -- Method by solvent extraction |
| ISO 9406:1995 | Green coke -- Determination of volatile matter content by gravimetric analysis |
| ISO 10142:1996 | Calcined coke -- Determination of grain stability using a laboratory vibration mill |
| ISO 10143:1995 | Calcined coke for electrodes -- Determination of the electrical resistivity of granules |
| ISO 10236:1995 | Green coke and calcined coke for electrodes -- Determination of bulk density (tapped) |
| ISO 10237:1997 | Calcined coke -- Determination of residual-hydrogen content |
| ISO 11412:1998 | Calcined coke -- Determination of water content |
| ISO 12980:2000 | Green coke and calcined coke for electrodes -- Analysis using an X-ray fluorescence method |
| ISO 12981-1:2000 | Calcined coke -- Determination of the reactivity to carbon dioxide -- Part 1: Loss in mass method |
| ISO 12982-1:2000 | Calcined coke -- Determination of the reactivity to air -- Part 1: Ignition temperature method |
| ISO 12984:2000 | Calcined coke -- Determination of particle size distribution |
| ISO 14435:2005 | Petroleum coke -- Determination of trace metals by inductively coupled plasma atomic emission spectrometry |
| ISO 20203:2005 | Calcined coke -- Determination of crystallite size of calcined petroleum coke by X-ray diffraction |
| ISO 21687:2007 | Calcined coke -- Determination of real density by helium pyknometry |

| Electrodes | Carbonaceous materials used in the production of aluminium |
|---|---|
| ISO 8007-1:1999 | Sampling plans and sampling from individual units -- Part 1: Cathode blocks |
| ISO 8007-2:1999 | Sampling plans and sampling from individual units -- Part 2: Prebaked anodes |
| ISO 8007-3:2003 | Sampling plans and sampling from individual units -- Part 3: Sidewall blocks |
| ISO 9088:1997 | Cathode blocks and prebaked anodes -- Determination of the density in xylene by a pyknometric method |
| ISO 11713:2000 | Cathode blocks and baked anodes -- Determination of electrical resistivity at ambient temperature |
| ISO 12985-1:2000 | Baked anodes and cathode blocks -- Part 1: Determination of apparent density using a dimensions method |
| ISO 12985-2:2000 | Baked anodes and cathode blocks -- Part 2: Determination of apparent density and of open porosity using a hydrostatic method |
| ISO 12986-1:2000 | Prebaked anodes and cathode blocks -- Part 1: Determination of bending/shear strength by a three-point method |
| ISO 12986-2:2005 | Prebaked anodes and cathode blocks -- Part 2: Determination of flexural strength by the four-point method (without precision) |
| ISO 12987:2004 | Anodes, cathodes blocks, sidewall blocks and baked ramming pastes -- Determination of the thermal conductivity using a comparative method |
| ISO 12988-1:2000 | Baked anodes -- Determination of the reactivity to carbon dioxide -- Part 1: Loss in mass method |

| | |
|---|---|
| ISO 12988-2:2004 | Baked anodes -- Determination of the reactivity to carbon dioxide -- Part 2: Thermogravimetric method |
| ISO 12989-1:2000 | Baked anodes and sidewall blocks -- Determination of the reactivity to air -- Part 1: Loss in mass method |
| ISO 12989-2:2004 | Baked anodes and sidewall blocks -- Determination of the reactivity to air -- Part 2: Thermogravimetric method |
| ISO 14420:2005 | Baked anodes and shaped carbon products -- Determination of the coefficient of linear thermal expansion |
| ISO 15379-1:2004 | Cathode block materials -- Part 1: Determination of the expansion due to sodium penetration with application of pressure |
| ISO 15379-2:2004 | Cathode block materials -- Part 2: Determination of the expansion due to sodium penetration without application of pressure |
| ISO 15906:2007 | Baked anodes -- Determination of the air permeability |
| ISO 17499:2006 | Baked anodes -- Determination of baking level expressed by equivalent temperature |
| ISO 18515:2007 | Cathode blocks and baked anodes -- Determination of compressive strength |
| ISO 21687:2007 | Calcined coke -- Determination of real density by helium pyknometry (can be used) |
| **Ramming paste** | **Carbonaceous materials used in the production of aluminium** |
| ISO 8005: 2005 | Green and calcined coke -- Determination of ash content (can be used) |
| ISO 11713:2000 | Cathode blocks and baked anodes -- Determination of electrical resistivity at ambient temperature (can be used) |
| ISO 12985-2:2000 | Baked anodes and cathode blocks -- Part 2: Determination of apparent density and of open porosity using a hydrostatic method (can be used) |
| ISO 12987: 2004 | Anodes, cathodes blocks, sidewall blocks and baked ramming pastes -- Determination of the thermal conductivity using a comparative method (can be used) |
| ISO 14420:2005 | Baked anodes and shaped carbon products -- Determination of the coefficient of linear thermal expansion |
| ISO 14422:1999 | Cold-ramming pastes -- Methods of sampling |
| ISO/TS 14423:1999 | Cold-ramming pastes -- Determination of effective binder content and aggregate content by extraction with quinoline, and determination of aggregate size distribution |
| ISO/TS 14425:1999 | Cold-ramming pastes -- Determination of volatile-matter content of unbaked pastes |
| ISO 14427:2004 | Cold and tepid ramming pastes -- Preparation of unbaked test specimens and determination of apparent density after compaction |
| ISO 14428:2005 | Cold and tepid ramming pastes -- Expansion/shrinkage during baking |
| ISO 17544:2004 | Cold and tepid ramming pastes -- Determination of rammability of unbaked pastes |
| ISO 18515:2007 | Cathode blocks and baked anodes -- Determination of compressive strength (can be used) |
| ISO 20202:2004 | Cold and tepid ramming pastes -- Preparation of baked test pieces and determination of loss on baking |
| **Fluorides** | **Cryolite, natural and artificial** |
| ISO 1619:1976 | Preparation and storage of test samples |
| ISO 1620:1976 | Determination of silica content -- Reduced molybdosilicate spectrophotometric method |
| ISO 1693:1976 | Determination of fluorine content -- Modified Willard-Winter method |
| ISO 1694:1976 | Determination of iron content -- 1,10-Phenanthroline photometric method |
| ISO 2366:1974 | Determination of sodium content -- Flame emission and atomic absorption spectrophotometric methods |
| ISO 2367:1972 | Determination of aluminium content -- 8-Hydroxyquinoline gravimetric method |
| ISO 3391:1976 | Determination of calcium content -- Flame atomic absorption method |
| ISO 3393:1976 | Determination of moisture content -- Gravimetric method |
| | **Sodium fluoride primarily used for the production of aluminium** |
| ISO 3429:1976 | Determination of iron content -- 1,10-Phenanthroline photometric method |
| ISO 3430:1976 | Determination of silica content -- Reduced molybdosilicate spectrophotometric method |
| ISO 3431:1976 | Determination of soluble sulphates content -- Turbidimetric method |
| ISO 3566:1976 | Determination of chlorides content -- Turbidimetric method |
| | **Anhydrous hydrogen fluoride for industrial use** |
| ISO 3699:1976 | Determination of water content -- Karl Fischer method |
| | **Cryolite, natural and artificial, and aluminium fluoride for industrial use** |
| ISO 2830:1973 | Determination of aluminium content -- Atomic absorption method |
| ISO 4280:1977 | Determination of sulphate content -- Barium sulphate gravimetric method |
| ISO 5930:1979 | Determination of phosphorus content -- Reduced molybdophosphate photometric method |
| ISO 5938:1979 | Determination of sulphur content -- X-ray fluorescence spectrometric method |
| ISO 6374:1981 | Determination of phosphorus content - Atomic absorption spectrometric method after extraction |
| **Refractory** | **Materials for the production of primary aluminium** |
| ISO CD 20292 | Dense refractory bricks –Determination of cryolite resistance |

**Light Metals 2008** *Edited by: David H. DeYoung*
***TMS (The Minerals, Metals & Materials Society), 2008***

# TEST AND ANALYSIS OF NITRIDE BONDED SiC SIDELINING MATERIALS: TYPICAL PROPERTIES ANALYSED 1997-2007

Egil Skybakmoen, Jannicke Kvello, Ove Darell, and Henrik Gudbrandsen
SINTEF Materials and Chemistry, NO-7465 Trondheim, Norway

Keywords: Aluminium Electrolysis, Sidelining, $Si_3N_4$-SiC

## Abstract

During the last ten years, SINTEF has tested a large number of different commercial nitride bonded SiC sidelining materials produced world-wide. The following properties will be summarized in the paper: Apparent porosity, density, mineral phase analyses ($\alpha$-$Si_3N_4$, $\beta$-$Si_3N_4$, $Si_2ON_2$, Si, and SiC), chemical analyses (level of Al, Fe, Ca and Ti), LECO analyses (total oxygen and total nitrogen), bending strength, hot modulus of rupture (HMOR), cold crushing strength, thermal expansion, thermal conductivity, oxidation resistance, and chemical resistance. Some comparisons between blocks with good and poor properties will be given, as well as some examples of variations in properties within the same block.

## Introduction

In modern high amperage aluminium cells, the use of $Si_3N_4$ bonded SiC refractories represents the state of the art as sidelining materials. It opens for use of thinner sidelining, larger anodes, higher current, and increased productivity.

$Si_3N_4$-SiC materials are produced by mixing SiC-particles (up to mm size) with fine grained silicon powder and a binder. SiC may contain some $SiO_2$, Si, Fe and C, coming from the production process (Acheson process). The bricks are shaped by pressing/vibration before drying. The bricks are sintered in nitrogen atmosphere at temperatures of about 1400 °C. During the nitration process Si reacts with nitrogen, and $Si_3N_4$ is formed as a binder between the SiC grains. The final product will then consist of large SiC grains with a microporous binder phase of $Si_3N_4$. Typical composition is 72-80 wt % SiC and 20-28 wt % $Si_3N_4$. Different types of oxides may also be present, especially $Si_2ON_2$, $Fe_2O_3$, $Al_2O_3$ and CaO. The thicknesses of the blocks used in aluminium cells varies from 40 – 100 mm, dependent on cell technology. SiC lining material can either be used only as a top lining or as a complete lining covering the entire steel shell.

There are many suppliers of such refractories world-wide, and there is a need for a quality evaluation process to be able to rank the different types on the market. For this reason we first developed a test for chemical resistance [1,2], and thereafter also included several other parameters to be analysed [3]. The experience obtained by testing and analysing commercial materials produced world-wide will be summarised in the following.

## Test methods and results

### Sampling

Due to natural variations of different properties across the block, it is important to record the sample positions within the block for each test to be performed. Our procedure when testing these types of materials is to take out samples in different positions, as indicated in Figure 1.

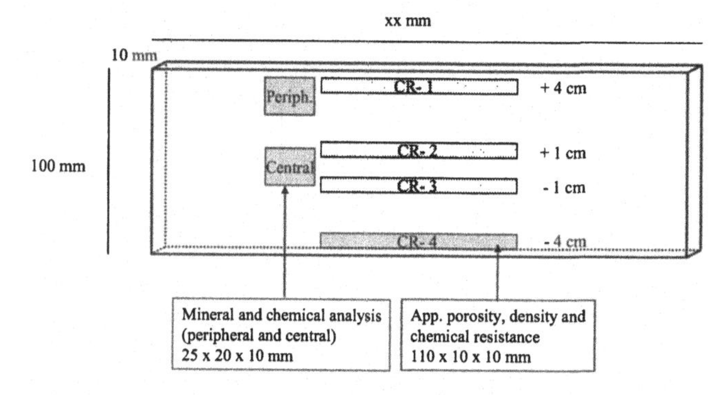


Figure 1. Samples from a 10 mm thick slice cut from the block. Sampling positions for chemical resistance (CR1..CR4), apparent porosity, density, mineral phase analysis, and chemical analysis are shown here.

### Apparent porosity and density

The apparent porosity and density are measured by the ISO 5017 standard method. Normally, several samples from each block are measured. The overall results for these tests are given in Table 1.

Table 1. Apparent porosity and density.

| Parameter | Range | Average | # samples | # types |
|---|---|---|---|---|
| App.porosity (%) | 11.2 – 20.5 | 15.5 ± 1.7 | 412 | 38 |
| Density (g/cm³) | 2.51 – 2.81 | 2.68 ± 0.05 | 412 | 38 |

### Mineral phase analysis and chemical analysis

The samples for chemical anlysis (both wet chemical analysis and XRF) and mineral phase analysis (XRD) were crushed to a fine powder (-100µm) in a tungsten carbide mortar. The samples for XRD were further milled down to < 45µm. Oxygen and nitrogen contents were analysed by a selective hot gas extraction method (LECO TC436DR).

The carbon content was determined according to the FEPA, 45-GB method by heating the samples to 1150°C. The $CO_2$ formed was adsorbed and gravimetric analysis was used to determine the amount of $CO_2$. Lately, we have also been using LECO analysis for C determination in order to estimate the SiC content.

Free Si was determined by treating the samples with NaOH and measuring the amount of $H_2$ formed in a colomn according to DIN 51075, but also quantitative XRD analysis was performed for the content of free Si. Elemental analysis of Fe, Al and Ca was performed by atomic absorption, according to NS-4770, after dissolving the sample in a mixture of HCl and HF.

An overview of the results obtained is given in Table 2. All values are given in wt %.

Table 2. Chemical analysis based on LECO-analysis and by atomic absorption analysis (wt %).

| Parameter | Range | Average | # samples | # types |
|---|---|---|---|---|
| SiC | 68.8 – 87 | 78.3 ± 3.2 | 53 | 27 |
| $Si_3N_4$* | 12.7 - 29.3 | 20.7 ± 3.5 | 79 | 33 |
| Si | 0.1 - 4.7 | 0.6 ± 0.8 | 53 | 27 |
| Fe | 0.07 – 0.8 | 0.3 ± 0.15 | 31 | 20 |
| Al | 0.02 – 1.17 | 0.2 ± 0.3 | 31 | 20 |
| Ca | 0.01 – 0.23 | 0.1 ± 0.1 | 31 | 20 |
| Nitrogen | 5.08 -11.68 | 8.26 ± 1.39 | 79 | 33 |
| Oxygen | 0.32 – 2.12 | 0.82 ± 0.38 | 82 | 33 |

*Based on total nitrogen content. May also include any $Si_2ON_2$.

Lately we have introduced quantitative mineral phase analysis performed by Ceram [4] where the amount of $Si_2ON_2$, $\alpha$-$Si_3N_4$, $\beta$-$Si_3N_4$ and Si are measured. This method is difficult and challenging (need very accurate calibration) and not very precise. Nevertheless, it gives very useful information regarding the binder phase. See also comments later in the paper regarding the homogeneity of the block. The overall results obtained by quantitative XRD for the peripheral and central part of the block are summarized in Table 3.

Table 3. Quantitative XRD analysis. 11 samples and 10 types.

| Phase | Central | | Peripheral | |
|---|---|---|---|---|
| | Range | Avg. | Range | Avg |
| Alfa-$Si_3N_4$ | 2 -13 | 7.7 ± 4.2 | 5.7 - 23 | 14.3 ± 5.8 |
| Beta- $Si_3N_4$ | 1 - 14 | 7.3 ± 4 | 1 - 15 | 4.6 ± 4.3 |
| Alfa/Beta | 0.2 -13 | 2.5 ± 4 | 0.4 - 20 | 7 ± 7 |
| $Si_2ON_2$ | 0 – 5.9 | 2.1 ± 1.8 | 0 - 4 | 0.7 ± 1.2 |

## Mechanical properties

*Cold crushing strength (CCS)* was measured by a slightly modified PRE R/14 standard method on adjacent samples cut from the block. Due to the high strength of this type of material, the sample dimensions were reduced from the dimensions specified in the standard (Ø50 mm and height 50 mm) to a square cross section of 30 x 30 mm. The height was kept at 50 mm.

*The modulus of rupture (MOR)* was measured in a 3-point bending test according to ISO 5014. The sample size was 25 x 25 x 150 mm.

*The hot modulus of rupture (HMOR)* was measured according to the PRE/R 18,78 standard method . The heating rate was 300 °C / h up to 950 °C. The sample was kept at this temperature for 30 minutes before the load was applied. The sample size was 25 x 25 x 150 mm.

The results obtained for CCS, MOR and HMOR are summarized in Table 4.

Table 4. Cold crushing strength, modulus of rupture and hot modulus of rupture.

| Parameter | Range | Average | # samples | # types |
|---|---|---|---|---|
| CCS (MPa) | 124 - 401 | 249 ± 65 | 65 | 19 |
| MOR (MPa) | 21 - 68 | 45 ± 12 | 48 | 12 |
| HMOR (MPa) | 38 - 78 | 54 ± 9 | 34 | 10 |

## Physical properties

*Thermal expansion* was measured according to the ASTM C372 standard method. The sample size was 50 x 10 x 10 mm. The temperature range was from 20 – 1200 °C in an atmosphere of synthetic air with a heating rate of 2 °C/min.

An example of the thermal expansion (given in % expansion) during heating and cooling versus temperature is shown in Figure 2. The overall results with calculated coefficient of linear thermal expansion are given in Table 5.

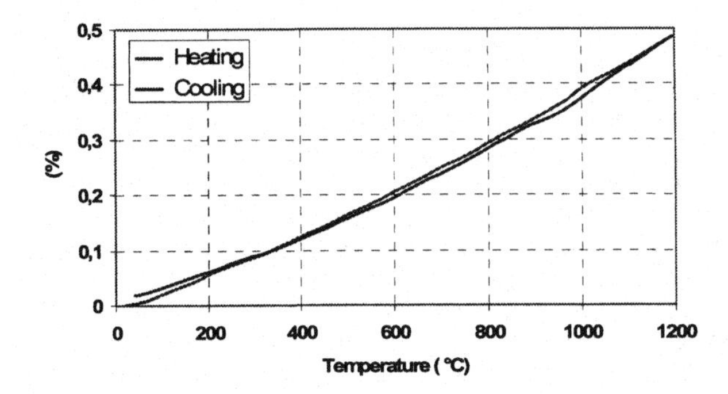


Figure 2. Example of thermal expansion up to 1200 °C.

Table 5. Thermal expansion 20 – 1200 °C and linear thermal expansion coefficient.

| Parameter | Range | Average | # samples | # types |
|---|---|---|---|---|
| Th.exp (%) | 0.47 – 0.53 | 0.50 ± 0.02 | 13 | 10 |
| Coefficient ($10^{-6}$ $K^{-1}$) | 4.00 – 4.55 | 4.3 ± 0.17 | 13 | 10 |

*Thermal conductivity* was measured by using the transient hot strip method (THS). The principle of this method was first described by Gustafsson et al. [5]. The technique has since then been modified [6]. A planar metal foil is used both as an internal heat source and as a temperature sensor (Figure 3). A constant, known current is passed through the strip (A-A) for a short period of time in order to heat it a few degrees, and the resistance increase over a defined strip length (B-B) is measured as a function of time.

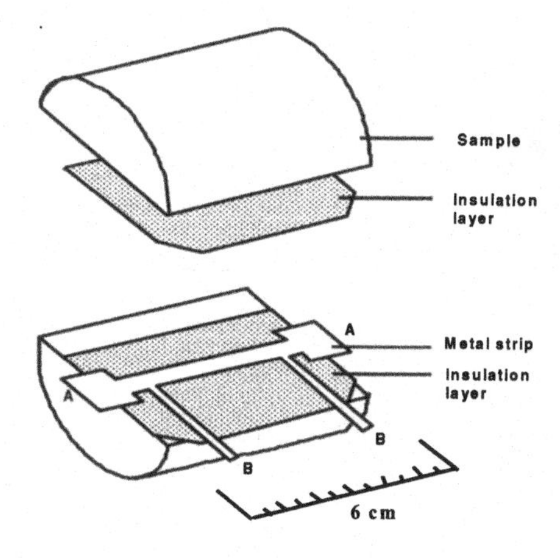


Figure 3. The THS sample arrangement.

The temperature rise of the strip is calculated from the temperature dependence of the strip resistance. Solving the heat equation for the given strip geometry gives the thermal conductivity of the sample. Dependent of the thermal diffusivity of the sample and the width of the strip, a typical measurement time is between 1-10 sec. The sample with diameter 50 mm and length 50 mm was cut in two parts, the surface polished and dried. Platinum was used as strip material and alumina used as insulation layers. At each temperature, four parallels were measured, and obviously erroneous results were rejected. The measurements were made with increasing and decreasing temperature in the range from 20 to 750 °C.

One example of measured thermal conductivity of a sample is shown in Figure 4, and the overall results are shown in Table 6.

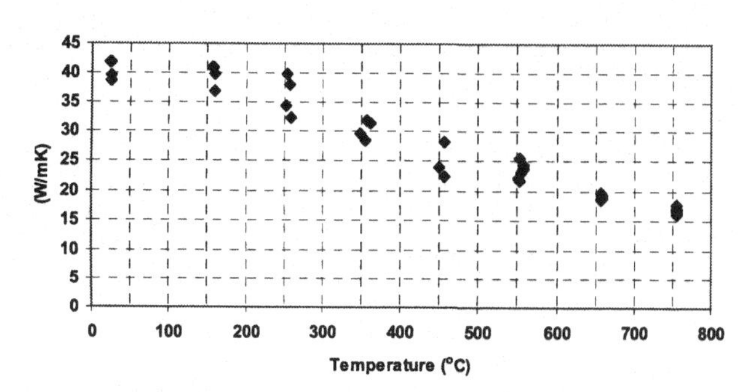


Figure 4. Thermal conductivity of a sample, measured by the THS-method.

Table 6. Thermal conductivity (THS) at 20 °C and 750 °C.

| Parameter | Range | Average | # samples | # types |
|---|---|---|---|---|
| Th.Cond. 20°C (W/mK) | 24 - 50 | 36.8 ± 7.4 | 14 | 13 |
| Th.Cond. 750°C (W/mK) | 10 - 23 | 18.9 ± 3.2 | 14 | 13 |

## Microstructure

Light microscopy studies of nitride-bonded SiC materials easily reveals the typically microstructure, where relatively large grains of SiC are surrounded by the microstructured $Si_3N_4$, as shown in Figure 5.

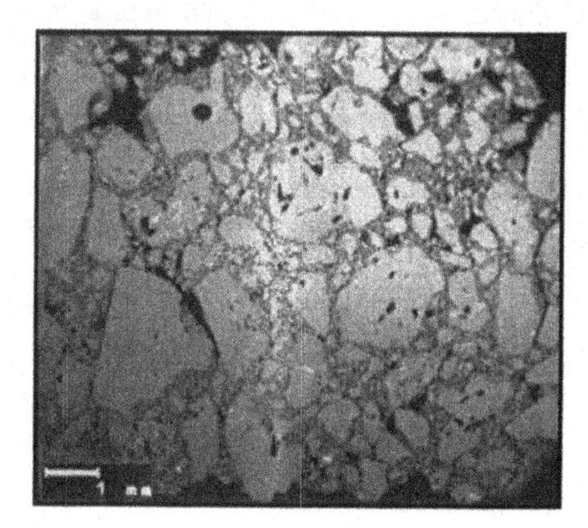

Figure 5. Microstructure of nitride-bonded SiC.

The apparent porosity is 16 ± 2 % and the pores are very small. Several pore size distribution measurements (Carlo Erba Macropores Unit 120, Carlo Erba Porosimeter 2000, and Micrometerics AccuPyc 1330 helium-pycnometer) have shown typical pore size radius to be from 0.1 – 10 μm, and most of the pores are below 1 μm.

A typical measurement of pore size distribution for samples taken from different positions is shown in Figure 6.

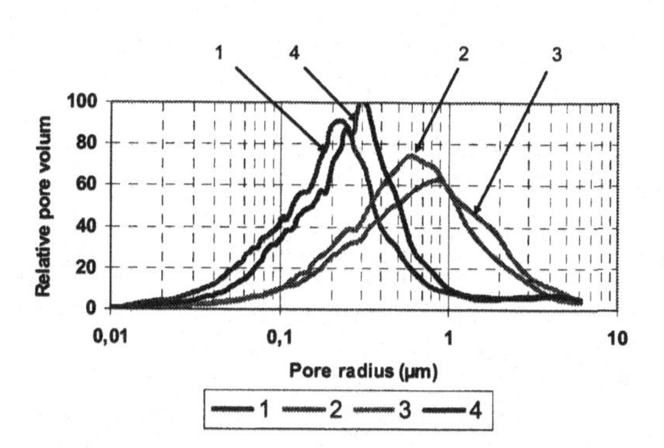


Figure 6. Pore size distribution from one block. Samples 1 and 4, are from the peripheral part of the block and samples 2 and 3 were taken from the central part of the block.

By assuming spherical pores the surface area of pores can be estimated (the assumption is not correct, but relative differences between materials will be correct). This parameter is important regarding the gas phase degradation/oxidation at the level above the electrolyte.

The total surface area investigated for 20 samples, and 11 different types, has been found to range from 0.12 – 1.24 $m^2/g$, with an average level of 0.52 ± 0.28 $m^2/g$. The results also showed the pore size surface area to be around 60 % larger at the peripheral part of the block than in the central part of the block, based on average values.

## Special test methods

*The oxidation resistance test* was performed by suspending the sample with dimension 50 x 30 x 15 mm from a balance in a resistance heated tube furnace. The furnace was heated at a rate of 300 °C/h to 950 °C and then kept at 950 °C for 100 hours. A constant flow through the furnace of synthetic air was maintained throughout the entire test. The course of oxidation in terms of weight change of the sample was recorded continuously during the test. The total weight change for the sample was recorded after 100 h exposure.

The weight change after 100 h exposure in flowing air at 950 °C of 24 samples and 15 different types have been found to range from 0.38 - 1.15 wt %, with an average level of 0.73 ± 0.24 wt %.

The weight change as a function of time in a typical measurement is illustrated in Figure 7.

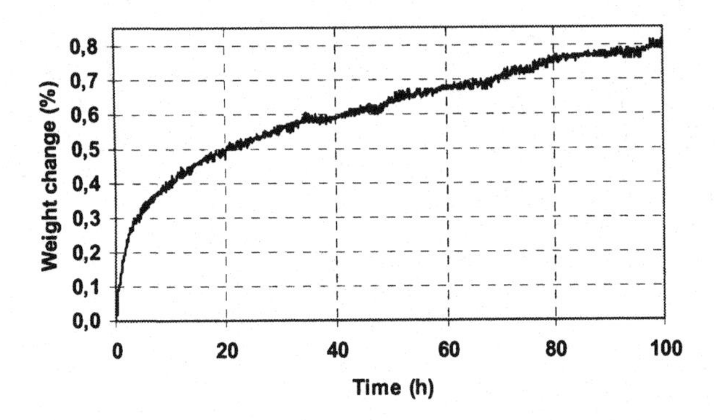


Figure 7. Typically weight increase during oxidation for 100 h in 950 °C in flowing air.

*The chemical resistance* was measured by using a special designed test cell set-up. The test cell, shown in Figure 8, consists of a graphite crucible lined with a sintered alumina tube (Alsint, Haldenwanger) and covered with a lid of graphite. A plate of $TiB_2$-graphite composite ($TiB_2$ - G, Great Lakes) covered the bottom of the crucible to ensure a stable level of aluminium in the crucible (Al wets $TiB_2$-graphite well), also acting as a current collector. A graphite anode (diameter = 15 mm) was placed in the middle of the cell and immersed 25 - 30 mm into the electrolyte. The test pieces (4 samples, dimension 10 x 10 x 110 mm) were placed symmetrically around the anode. The test cell was kept in a closed vertical tube furnace under argon atmosphere (outside the crucible). The temperature was kept constant at 955 °C measured with a thermocouple (Pt10Rh) placed in the crucible wall during the test. Before the test started, the temperature was also checked with a thermocouple placed in the middle of the electrolyte.

The initial composition of the electrolyte at the start of electrolysis was as follows (in wt %): 78 % $Na_3AlF_6$, 10 % $AlF_3$, 5 % $CaF_2$ and 7 % $Al_2O_3$. The duration of the test was 50 h and the current was 2 A. The anode was changed after 25 h. The carbon anode consumption was measured by weight. Normally, the anode consumption is 0.25 ± 0.02 g C/h during a test of 50 h.

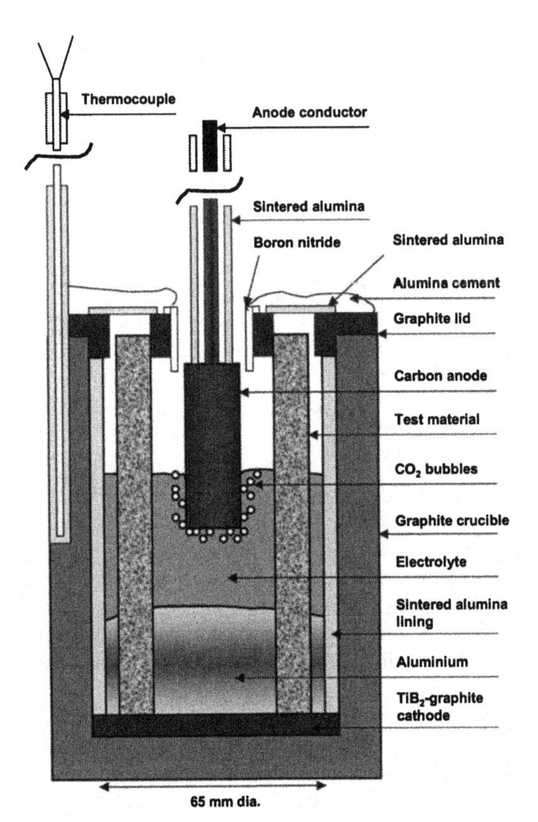


Figure 8. Test cell for SiC-based sidelining materials

After the test, the crucible with the test pieces was cooled for about 12 hours. The graphite lid was removed, and the rest of the test cell was heated up again in an open furnace. The test pieces were removed from the crucible when the bath melted. Adhering bath was removed mechanically and by washing in an aqueous solution of $AlCl_3$.

The degree of corrosion was related to the volume loss of the pieces during the test. The volume of the test materials was measured before and after the test by using ISO 5017 standard. The volume loss of each piece was estimated and a scale describing the degree of corrosion has been established (Table 7).

Table 7. The degree of corrosion based on volume loss.

| Volume loss (%) | Degree of corrosion |
|---|---|
| < 0.1 | 0 |
| 0.10 – 1.00 | 1 |
| 1.01 – 2.00 | 2 |
| 2.01 – 3.00 | 3 |
| 3.01 – 4.00 | 4 |
| 4.01 – 5.00 | 5 |
| 5.01 – 6.00 | 6 |
| 6.01 – 7.00 | 7 |
| 7.01 – 8.00 | 8 |
| 8.01 – 9.00 | 9 |
| > 9.01 | 10 |

Volume loss [%] =
[(volume before – volume after)/volume before] x 100%

Since the first test in 1997, we have performed chemical resistance tests of 400 samples from 37 types/suppliers. The overall results are as follows: The volume loss is measured from 0.07 – 20.65 % with an average value 5.03 ± 3.44 %. The average degree of corrosion is 5 on a scale from 0 – 10 (0 is best quality, 10 is worst quality). Good quality materials with respect to chemical resistance are defined as degree of corrosion from 0 – 3, middle range from 4 – 7, while poor quality materials are from 7 – 10.

The distribution related to the degree of corrosion for 400 samples tested is shown in figure 9.

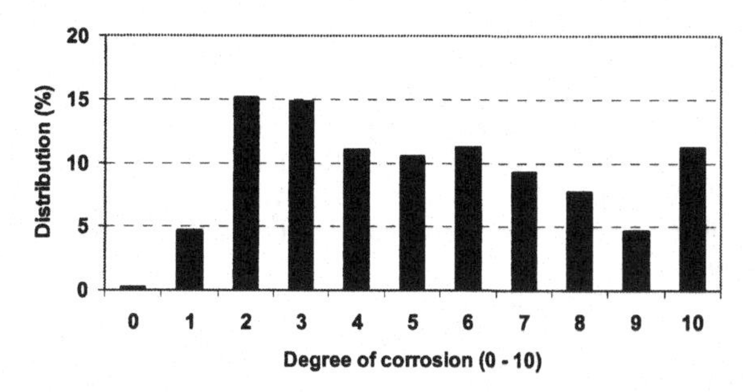


Figure 9. Distribution of degree of corrosion. 400 samples from 37 types/suppliers.

The distribution found showed 35 % to be good quality materials, 42 % in the middle range, while 24 % is poor quality materials.

### Variations across the blocks

The test work, and collecting of test results in a systematic manner, has revealed some typical fatures. The blocks are normally not homogeneous, and very often differences are observed between the peripheral part and the central part of the blocks. The relative differences in properties vary more or less from supplier to supplier, but some general remarks can be made, based on all materials tested. The properties that normally are not uniform across the block are shown in Table 6.

Table 6. Variations of properties as function of position within the blocks.

| Properties | Peripheral part | Central part |
|---|---|---|
| Apparent porosity | Lower | Higher |
| Density | Higher | Lower |
| Free Si | Lower | Higher |
| Alpha-$Si_3N_4$ | Higher | Lower |
| Beta-$Si_3N_4$ | Lower | Higher |
| Total oxygen | Lower | Higher |
| Pore surface area | Higher | Lower |
| Chemical resistance | Lower | Higher |

The parameter difference can be explained by the production method. The nitration of the block starts at the surface and moves to the central part of the block towards the end of the nitration process. The access of nitrogen to the central part of the block is limited by slow diffusion. Therefore, the apparent porosity is higher (and the density lower) in the central part of the block, and the content of free Si is also normally higher here. The variations are dependent on time, temperature and pressure of nitrogen gas during the nitration process, or on the production methods including raw materials and equipment the producers use.

The $Si_3N_4$ phases consist of $\alpha$-$Si_3N_4$ and $\beta$- $Si_3N_4$. The $\alpha$-phase is needle-like in structure, formed by a gas phase reaction between nitrogen gas and Si vapour and stabilized by small amounts of oxygen. The $\beta$-phase is shell-like in structure, formed by atomic nitrogen and silicon through a shrinking core mechanism [7]. The presence of liquid Si (temperatures higher than 1410 °C) provides faster $\beta$-phase formation. Therefore, this is also called the high temperature phase. The reaction between nitrogen and silicon is strongly exothermic. The temperature must be controlled by keeping the pressure of nitrogen within appropriate limits during nitration in industrial scale.

The reason for higher content of $Si_2ON_2$ in the central part of the block is the fact that oxygen on the Si and SiC surface is more likely to be trapped in the interior of the block, forming $Si_2ON_2$. This phase is then found when oxygen is present either by high oxygen level in the raw materials or oxygen getting access during nitration, for instance due to air leakage in the furnace.

### Examples of poor and good quality

Generally the bad quality materials are very inhomogeneous with respect to apparent porosity, density, content of free Si and showing large differences in the contents of the phases $\alpha$-$Si_3N_4$ and $\beta$-$Si_3N_4$, high levels of $Si_2ON_2$ and also large contents of oxides, as for instance $Al_2O_3$. These variations indicate poor production control during the process of making the materials. In the worst cases, high contents of free Si are found in the central part of the block, indicating poor nitration. Poor nitration may in worst cases lead to breakage during installation in the cell. Occasionally we also have observed large areas inside the block not to be nitrated.

When performing the chemical resistance test a large difference in chemical wear has been measured. It is especially important to detect the really poor quality materials. As an example, 4 samples from the same block were tested for chemical resistance, and as shown in Figure 10, one sample showed very high corrosion. This was due to high contamination of added oxides in one part of the block due to inadequate mixing.

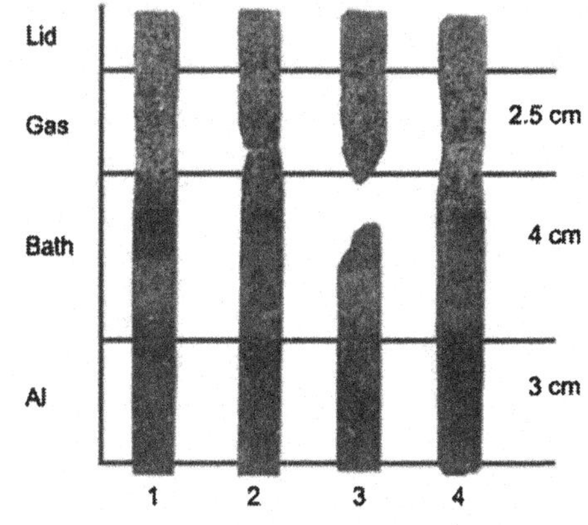


Figure 10. Example on a poor quality material.

In Figure 11 is shown a picture of four samples with average degree of corrosion 2, and four samples with average degree of corrosion 9 taken from two tests with different suppliers. There is obviously a large difference in rate of corrosion after 50 h exposure.

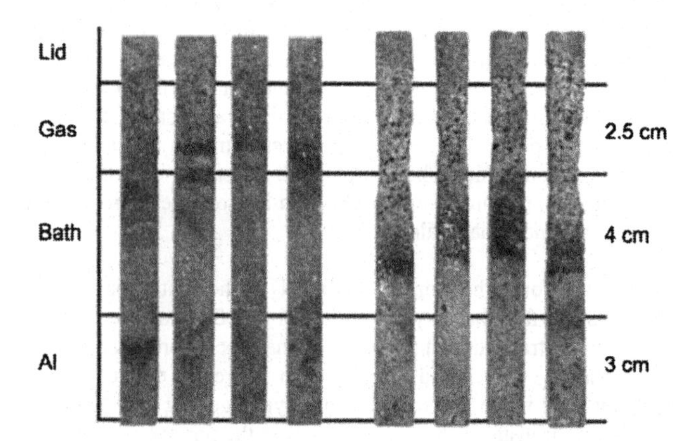


a) Type I – degree corrosion 2 b) Type II – degree corrosion 9.

Figure 11. a) Degree of corrosion 2 and b) degree of corrosion 9. See Table 7.

In some cases, we have observed a weight loss during heating up for the oxidation resistance method. This is probably due to burn-off of free carbon contained in the material, maybe originating from the organic binder used during mixing and pressing the blocks. This is illustrated in figure 12. It is not clear what effect this may have on the quality of the materials.

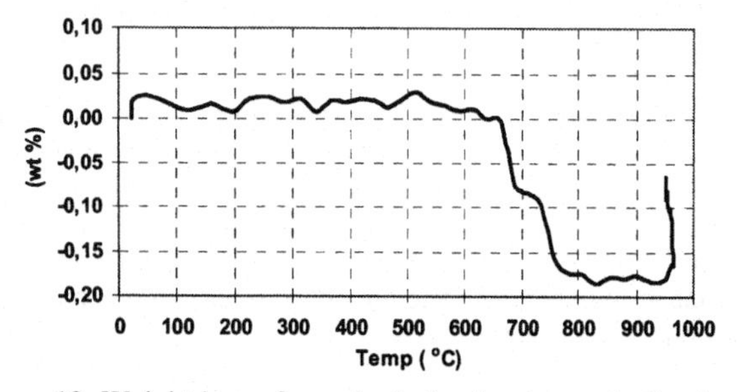


Figure 12. Weight loss of sample during heating up in flowing air as a function of temperature.

## Industrial relevance

First of all the test methods and the analyses performed are helpful for the producers to produce materials with good and stable quality as sidelining materials for use in aluminium reduction cells. There is no doubt that there exists a wide range of different qualities on the market. It is extremely important to produce a stable quality over time, one block with poor quality may be crucial for the stability, and in the worst cases for the lifetime of the cell.

The rate of oxidation/degradation in the upper part of the sidelining will certainly be different also in industrial cells with materials with degree of corrosion 10 and for instance degree of corrosion 2, especially for cells with unstable sideledge. The best protection for the SiC based sidelining is a stable frozen sideledge.

In high amperage cells, the importance of good quality sidelining materials will be more critical for meeting the higher demands in such cells. The tests have also shown that there is potential for improvements, and there is a need to develop even better materials in the future.

## Acknowledgement

The present work is a part of the current project "Thermodynamics Applied to High Temperature Materials Technology" (Thermotech), which is financed by the Research Council of Norway, the Norwegian aluminium industry, the Norwegian Ferroalloy Producers' Research Association, and Statoil. Also, thanks to all clients world-wide the latest 10 years including several SiC producers and primary aluminium companies.

## References

1. E. Skybakmoen, H. Gudbrandsen and L.I. Stoen, "Chemical resistance of sidelining materials based on SiC and carbon in cryolitic melts – a laboratory study", Light Metals 1999, pp 215-222.

2. E.Skybakmoen, "Evaluation of chemical resistance/oxidation of $Si_3N_4$-SiC sidelining materials used in Al electrolysis cells." Proceedings of Unified International Technical Conference on Refractories, Volume III, p 1330 – 1339, Cancun, Mexico, Nov 2001.

3. E. Skybakmoen, L. I. Stoen, J. H. Kvello and O. Darell "Quality evaluation of nitride bonded silicon carbide materials", Light Metals 2005, pp 773 - 778.

4. www.ceram.co.uk.

5. S.E. Gustafsson , E.Karawacki and M.N. Khan, "Transient Hot Strip Method for Determination of Thermal Conductivity and Thermal Diffusivity of Solids an Liquids", J. Phys. D; Appl. Phys. 12, (1979), 1411-1421

6. T. Log, "Thermal conductivity of carbon materials", Thesis 54, Institute of inorganic chemistry, The Norwegian Institute of Technology, Trondheim, 1989.

7. A.J.Moulsen, "Review reaction bonded silicon nitride: Its formation and properties," Journal of Materials Science, 14:1017-51, 1979.

**Light Metals 2008** *Edited by: David H. DeYoung*
*TMS (The Minerals, Metals & Materials Society), 2008*

# High Swelling Cold Ramming Paste for Aluminum Reduction Cell

G.Vergazova
Engineering and Technology Center , Krasnoyarsk Branch, Russian Engineering Company LLC., Russia

Keywords: swelling ramming paste; carbon cathode block; laboratory test;

## Abstract

The life and energy efficiency of a reduction cell largely depend on the sealing efficiency of the ramming paste that determines the extent and rate of electrolytic penetration into the cell bottom. The problem of the ramming paste quality and its densification in the joints is especially acute for more and more widely used graphitic and graphitized cathode blocks with their very low sodium swelling index and thermal expansion factor. High swelling capacity of ramming pastes can be an avenue of attack on the problem. The work presents a study to control the swelling capacity and binding of the ramming paste based on typical electrically and gas-calcined anthracite and pitch binder by selection of size distribution and liquid and solid additives, to improve the joints between graphitic or graphitized blocks.

## Introduction

Currently the marketplace offers new generations of cathode blocks with unique performance and prices however the industry has not invented yet a more efficient approach than use of cathode ramming pastes for joining these blocks in the electrolytic cell cathode bottom. Therefore the future and demand for new cathode blocks is closely tied to the technology for production and application of the cathode ramming pastes the significance of which is frequently underestimated.

As a rule the cathode blocks are made of carbon filler (electro- or gas-calcined anthracite or graphite) mixed with coal tar pitch. The last generations of blocks are made with petroleum coke, electro-calcined anthracite and graphite. Russian plants use standard ramming pastes with coal tar binder (tar-bonded) rammed at a room temperature (room temperature application). The pastes consist of carbon filler, additives and liquid coal tar binder additionally containing a solvent ensuring compactability of ramming pastes in a wide temperature range from 15 to 50°C. Presence of a solvent in the cold ramming paste recipe considerably reduces their strength and caking with the cathode blocks during preheat which significantly reduces the strength of the inner seams between the cathode blocks.

The use of graphitic and graphitized cathode blocks with low thermal expansion and low adhesion to ramming paste in the new high amperage cells once again raises the issue of the block-paste bond quality.

The key function of the inner seams is to prevent penetration (infiltration) of bath and metal under the cathode bottom. Besides the seams should withstand the exposure to alumina-cryolite melt and the attrition by metal particles. It is possible to improve quality of inner seams by use of adhesive pastes in the ramming paste/cathode block interface or by use of ramming pastes with high thermal expansion at preheat (high swelling pastes).

There are many known technologies for production of high swelling pastes many of which use various additives ranging from soot to titanium oxides and borides /1,2/.

The current paper offers findings of the studies of high swelling ramming pastes selected on the basis of the primary carbon raw material anthracite.

## Experimental & Results

The first series of tests was aimed at raw material selection for producing high swelling cold ramming paste.
The second series of tests was aimed at improving performance of high swelling cold ramming paste.

The most well known fillers for cold ramming paste are anthracites that have undergone heat treatment in gas furnaces or in electrocalciners. Since ramming paste is meant for joining cathode blocks of various grades from semi-graphitic to graphitized high attention was paid to electro-calcined anthracite.

The high swelling cold ramming paste recipe was designed using two different anthracite types:

- Type 1 – inertinitic anthracite with unique petrographical composition (Russian);
- Type 2 – vitrinitic anthracite (Chinese).

Type 1 is represented by semidull anthracite of inertinite lythotype of Russian occurrence with high content of fusinite and its components and low ash. Its key distinction from well known anthracites (including Chinese) is low real density and round shape of spongy particles.

Type 2 is represented by semibright anthracite of vitrinite lythotype with dense structure and flat-shaped particles.
Properties of raw and heat treated anthracites are shown in Table 1. Figure 1 shows macrostructure of Type 1 raw inertinite anthracite as compared to Type 2 raw vitrinite anthracite (Fig.2).

Table 1. Properties of raw anthracites

| Occurrence | Vitrinite micro-components group, Vt | Semivitrinite micro-components group, Sv | Fusinite micro-components group, F | Real density, g/cm$^3$ |
|---|---|---|---|---|
| Russian | 35 | 2 | 75 | 1,650 |
| Chinese | 65 | 11 | 22 | 1,760 |

Specific feature of Type 1 raw anthracite is its fibrous fusinite structure with closed porosity ensuring isotropic properties and high plasticity of the heat-treated material.

Type 2 raw anthracite is represented by vitrinite with specific flat shape of low porosity particles with high anisotropy and high elasticity of the material.

Study of Type 1 structure using X-ray diffraction technique and processing of diffraction curves has shown that the first two coordination spheres in Type 1 anthracite are alike whereas the third sphere substantially differs from the spheres of crystalline carbon (graphite). Thus heat treatment of Type 1 anthracite results in considerable changes of the structure (reduction of lattice spacing, growth of crystalline particles and stack thickness). Subjected to high-temperature treatment only 10-15% of Type 1 anthracite components are capable of graphitization. This factor ensures high heat resistance and strength of the inertinite calcined anthracite particles.

In Type 2 anthracites only the first coordination sphere is similar to that of graphite whereas the remaining two distinctly differ from the first one. Correspondingly an ordered structural change is achieved in Type 2 anthracites at higher temperatures (2500 °C). Meanwhile the key components of Type 2 anthracite, vitrites, are capable of graphitization. High temperature treatment of vitrinitic anthracites (Type 2) allows obtaining a strong material similar in properties to graphite.

For cold ramming paste composition design experiments anthracites with different petrographical composition, properties, and heat treatment were used, namely: Type 1 anthracite gas-calcined at 1320°C (GCA) and Type 2 anthracite electro-calcined at 2500°C (ECA).

Comparison of the two anthracites properties after heat treatment is shown in Table 2. A clear distinction of Type 1 anthracites (GCA) from other heat-treated anthracites can be seen in its low real density.

The impact of the filler nature is promimently manifested at the dry aggregate preparation for cold ramming paste. The aggregate of Type 1 gas-calcined anthracite has higher plasticity evidenced by relaxation coefficient value for Type 1 thermoanthracite (GCA). Type 2 electro-calcined anthracite (ECA) contains a considerable amount of graphitized material thanks to high calcination temperature nevertheless (unlike the elastic artificial graphite formed from petroleum coke) its plasticity balances out its elasticity though remaining inferior to that of GCA (I).

Thus compression properties of the two raw material types, Type 1 and Type 2, allow forecasting of high compactability without elastic expansion of the ramming pastes made from these raw materials.

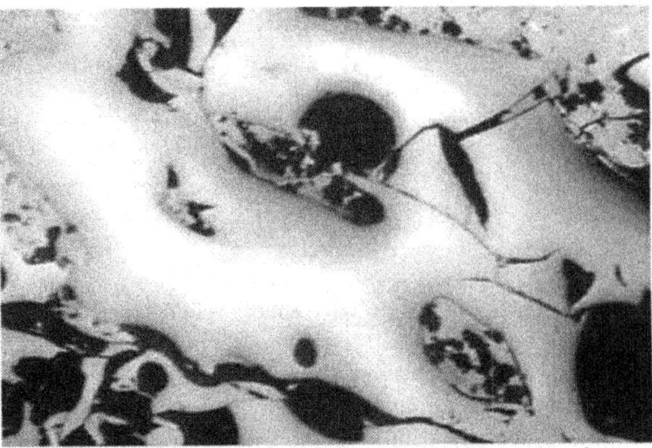

Fig.1. Macrostructure of Type 1 inertinitic anthracite

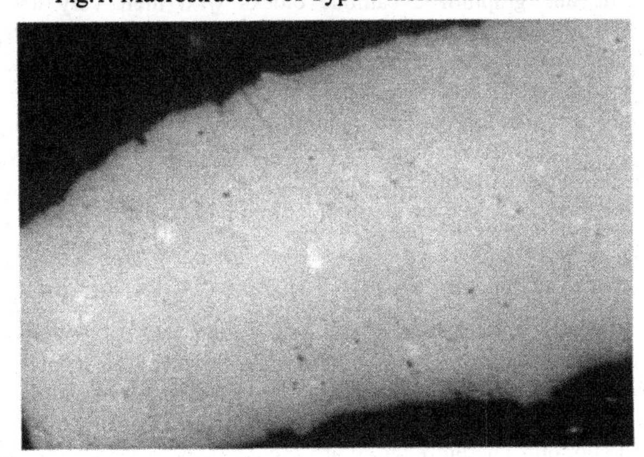

Fig.2. Macrostructure of Type 2 vitrinitic anthracite

**Test Results**

High ramming paste swelling factor was taken into account by the researchers for design of special granulometry filler aggregate. The key objective for developing granulometry is to ensure a dense compaction of the filler particles with minimal thickness of the binder interlayer, preserve this dense compaction during baking, and allow for paste expansion with minimal shrinkage as it bakes during the cathode preheat. Improvement of the caking ability and reactivity is gained by introduction of boron-containing additives.

Table 2. Performance, physical, and mechanical properties of the heat-treated anthracites

| Material | RD., g/cm$^3$ | A % | Pitch adsorbtion, % | $CO_2$ reactivity, mg/g.c | El/resistivity, μΩ.m | Relaxation coeff., % | Elastic expansion coeff., % | Particles microhardness, kg/mm$^2$ |
|---|---|---|---|---|---|---|---|---|
| **I** GCA | 1,745 | 3,5 | 39,0 | 0,150 | 960 | 9,9 | 2,6 | 240 |
| **II** ECA | 1,869 | 6,4 | 42,0 | 0,220 | 580 | 4,5 | 4,0 | 150 |

All the testing of the ramming paste samples has been carried out in accordance with Russian and ISO standards.

The impact of various filler types was studied by producing several cold ramming paste batches following the same recipe (same components and granulometry). The results of ramming pastes testing are shown in Table 3.

The test results for compaction of different filler pastes obtained using Fisher machine (Fischer Rammer RDC-194) as per ISO 17544:2004 (E) are shown on Figures 3 and 4.

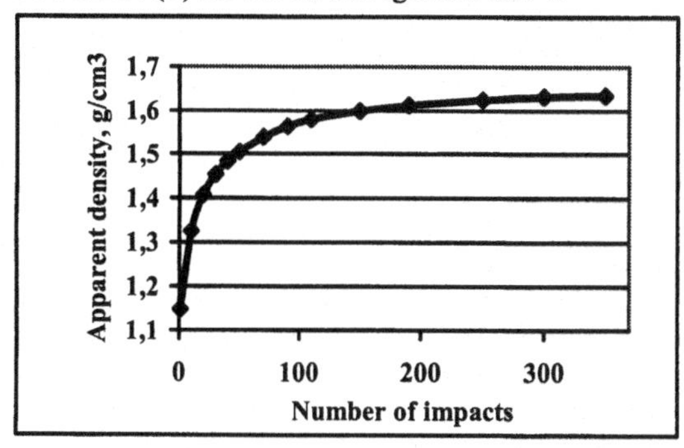


Fig.3 Compactibility of GCA (1) ramming paste at 25°C

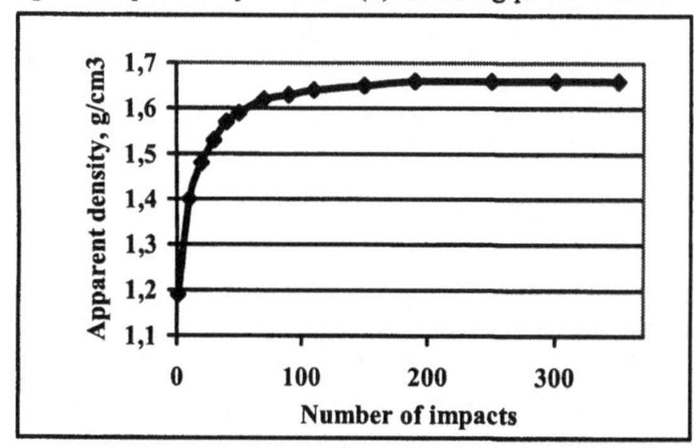


Fig.4 Compactibility of ECA (2) ramming paste at 25°C

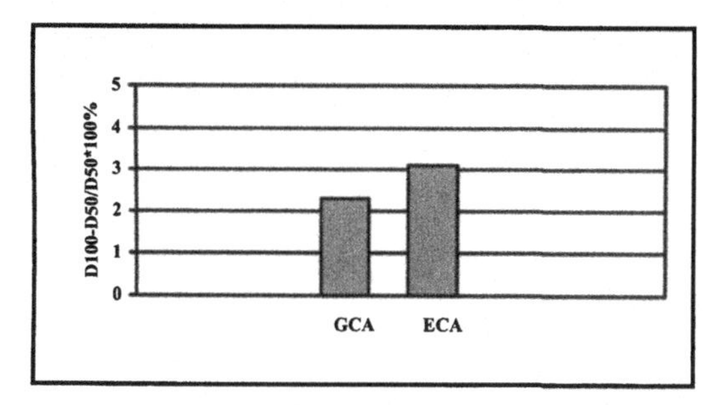


Fig.6 Fisher compactibility test results

Table 3 Physical and mechanical properties of cold ramming pastes made from different raw materials

| Properties | GCA (I) paste | ECA(II) paste | Typical paste |
|---|---|---|---|
| Green paste | | | |
| Apparent density, g/cm$^3$ | 1,590 | 1,600 | 1,600 |
| Baked paste | | | |
| Apparent density, g/cm$^3$ | 1,415 | 1,425 | 1,450 |
| Real density, g/cm$^3$ | 1,743 | 1,852 | 1,86 |
| Baking loss, % | 8,5 | 9,9 | 8,61 |
| Total porosity, % | 18,9 | 23,1 | |
| Open porosity, % | 18,0 | 17 | 18,0 |
| Residual volumetric expansion, % | 2,00 | 1,78 | 0,6 |
| Electrical resistivity, μΩ.m | 85 | 78 | 68 |
| Relative baking shrinkage, % | 0,08 | 0,17 | 0,14 |
| Compressive strength, MPa | 35,2 | 19,9 | 20,8 |
| Young's Modulus,GPa | 3,6 | 2,5 | 4,3 |
| Tensile strength, MPa | 4,5 | 2,6 | 1,8 |
| Coefficient of linear thermal expansion, $\alpha_L x10^{-6} 1/K$ | 4,2 | 3,3 | 4,0 |
| $CO_2$ reactivity: | | | |
| CRR,% | 35,0 | 40,0 | 32,4 |
| CRD,% | 48,5 | 40,5 | 35,0 |
| CRL,% | 16,5 | 19,5 | 32,6 |

Paste thermal expansion was tested with dilatometer in line with modified ISO 14427 method.

The dilatometer study findings for different paste types are shown on Figures 7, 8, 9.

Porosity of inner seams (as well as of cathode blocks) impacts not only mechanical properties of the cathode but also bath penetration rate and sodium infiltration with all the consequences to follow. That is why along designing of granulometry a special attention in this study was paid to the impact the raw materials have on the porosity of the baked sample. The pore size distribution in baked GCA (I) and ECA(II) paste samples are shown on Fig. 10 and 11.

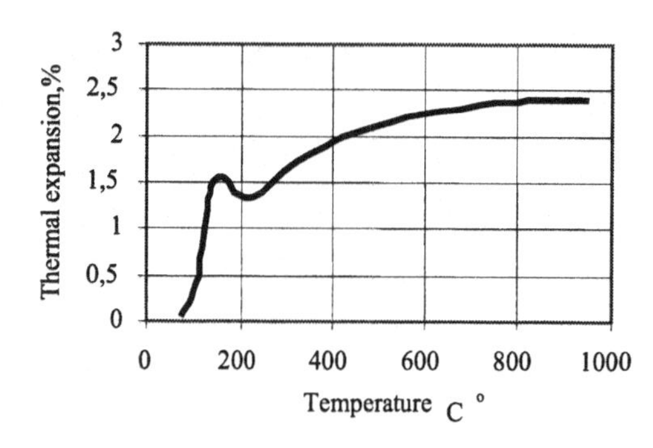


Fig. 7 GCA (1) cold ramming paste expansion curve

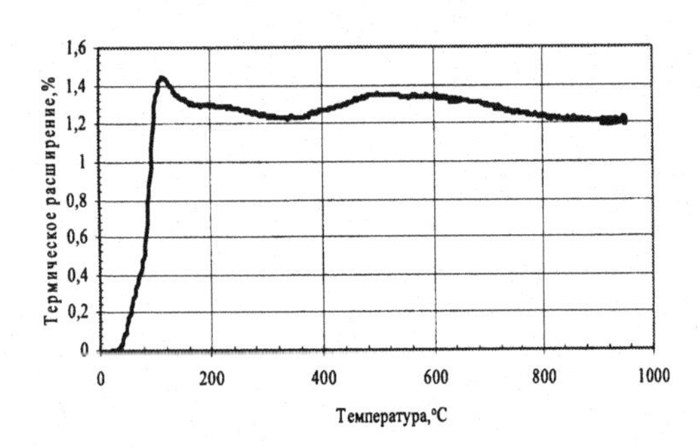


Fig. 8 ECA (2) cold ramming paste expansion curve

Taking into account that in the course of electrolytic cell operation oxygen can access inner seams through collector bar portholes some antioxidant additives are introduced in the paste. Table 3 also shows the $CO_2$ reactivity test results for pastes from various raw materials carried out in compliance with ISO 12988.

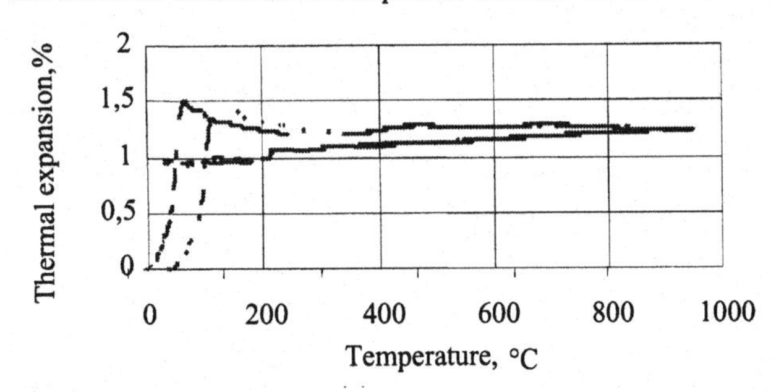


Fig.9 Dilatometric expansion curve for ECA (II) paste baking and then cooling to room temperature

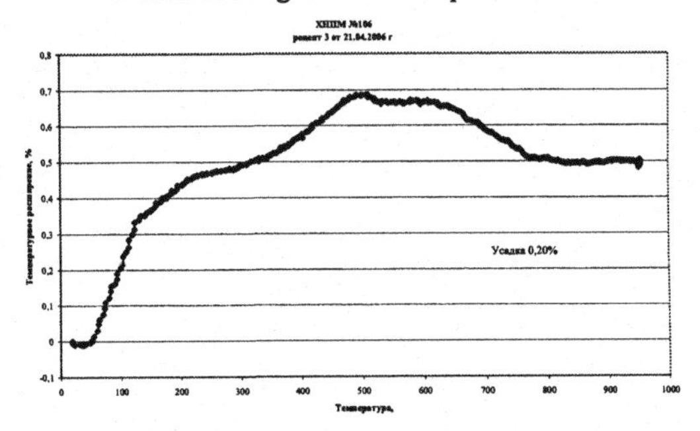

Fig.10 Dilatometric expansion curve for a typical paste baking

The results of the cathode bottom stress-strain state modeling with two types of ramming paste and three types of cathode blocks are shown in Table 4. The performance of cold ramming pastes from different anthracites in conjunction with three cathode block types (semi-graphitic, graphitic, and graphitized) has been studied on the subject of bath leak resistance using approach /4/. The test results are shown on Figures 13, 14, 15. Use of a typical paste with graphitic or graphitized cathode blocks leads to the bath leaks (Fig.16).

Table 4

| Paste | Cathode block stress level (30% graphite), MPa (E=2,9 GPa) | Graphitic cathode block stress level (100% graphite), MPa (E=4,5 GPa) | Graphitized cathode block stress level MPa (E=6,8 GPa) | Inner seam stress level, MPa | Paste ultimate strain, % |
|---|---|---|---|---|---|
| GCA(I) | 9,6 | 6,7 | 4,8 | 9,5 | 0,12 |
| ECA(II) | 9,6 | 6,7 | 4,8 | 10,4 | 0,10 |
| Typical | 9,6 | 6,7 | 4,8 | 12,4 | 0,05 |

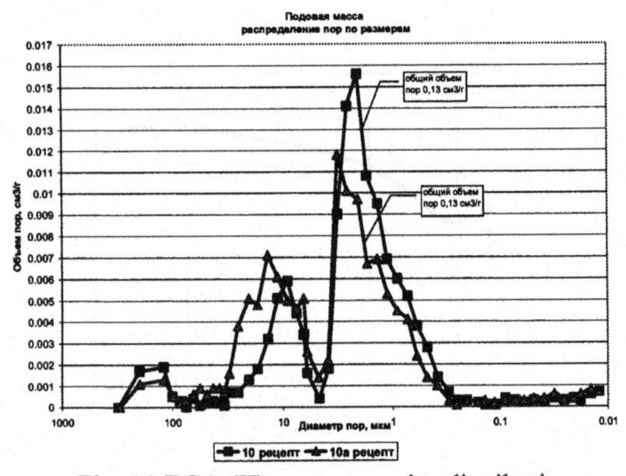


Fig. 11 ECA (II) paste pore size distribution

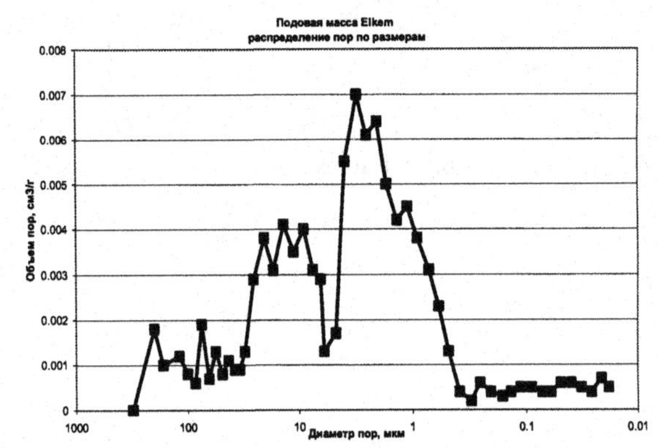


Fig. 12 Pore size distribution in GCA (1) paste

## Discussion of Results

The studies of ramming paste compactibility have shown that raw material properties manifest themselves right away starting with the compaction density. The electro-calcined anthracite based paste ECA (II) compacts faster, i.e. after 100 impacts using Fisher machine a high density is reached. To reach the same density GCA (II) paste requires a greater effort.

Pastes from gas-calcined anthracite (GCA Type I) have higher mechanical strength both compressive and flexural/tensile but at the same time have high electrical resistivity which is explained by the intraparticle porosity in raw anthracite.

It is considered that inner seam strength should not exceed the strength of the cathode block. From our viewpoint more important is carbon materials strength under a thermal load, i.e. deformation capability or the ultimate deformation determined as tensile strength over Young's modulus.

The cathode bottom stress-strain state modeling results with two types of ramming paste and three types of cathode blocks have shown that when cathode block stress is comparable with the ramming paste seam stress the failure probabilities of either are equal hence cathode reliability and operational durability considerably increase. The 10-year industrial use experience of GCA (I) paste only confirms this finding. In case inner seam stress is higher than that of the cathode block in conjunction with other adverse factors the cathode bottom failure will originate in the inner seams. Greater paste strength and higher thermal expansion compensate for the higher levels of inner seam stress.

The most interest has been generated by the dilatometric studies of various paste types. It should be noted that both pastes exhibit high expansion. However maximum expansion of a cold ramming paste was seen in GCA Type 1. (Fig.7) The structural features and closed porosity create a «squeezed pore» effect leading to higher paste plasticity even after baking, high flexural strength with strong caking to cathode blocks. The unique structure of inertinitic anthracites allows control over paste linear-volumetric expansion during baking without coking shrinkage through adjustments in granulometry. In turn particle porosity increases the filler aggregate specific surface area leading to reduced binder interlayer thickness. Correspondingly binder coking in such pastes occurs in a super-thin layer which does not result in paste shrinkage in general. High paste thermal expansion during baking is the reason for its stronger caking to cathode blocks which was noted in the course of experiment and industrial testing.

Cold ramming pastes of ECA Type 2 are characterized by a lesser volumetric expansion and insignificant shrinkage during coking in the range of 500-980°C (Fig.8). Higher expansion also improves caking to cathode blocks in the course of cell preheat and operation.

The raw material structural features reflect of porosity of the baked paste. The presented pore size distribution curves show that in baked ECA (II) paste pores under 10 micron are dominating with a higher total pore volume than in GCA (I) paste. This is explained by a higher volumetric expansion of GCA (I) paste and porosity of anthracite particles.

As can be gathered from the reactivity test results of the two studied paste types both pastes exhibit higher resistance to oxidation with GCA (I) paste having an edge. The difference in $CO_2$ reactivity of the baked pastes (Table 3) is caused by a high ash content in raw ECA(II) anthracite as well as by higher concentration of iron compounds in that ash.

For practical use of ramming pastes it is of significance to have cathode blocks compatibility with the ramming pastes. Considering high expansion capability of the two anthracite types ramming pastes both pastes have been tested with semi-graphitic, graphitic, and graphitized cathode blocks.

The GCA(I) pastes with strong caking to the cathode block surface have shown excellent results with all three cathode block types (semi-graphitic, graphitic, and graphitized). It is high mechanical strength of the baked Type (I) paste both tensile and flexural that leads to improved inner seam resistance to the stress levels higher than those for the cathode blocks. The key shortcoming of the GCA(I) paste, its higher electrical resistivity, can be reduced by corresponding additives.

The optimal combination of all properties from mechanical to thermo-physical can be seen in ECA(II) paste. Not very high mechanical (flexural and tensile) strength is compensated by greater expansion and caking to the cathode blocks of all three types (semi-graphitic, graphitic, and graphitized).

The bath resistance tests for cathode blocks and pastes presented here have shown great results (Fig.14, 15, 16). The pastes based on ECA (II) provide the best caking to all types of cathode blocks.

Fig.13 ECA (II) paste in contact with semi-graphitic block.

Fig.14 ECA (II) paste in contact with graphitic block.

Fig.15 ECA (II) paste in contact with graphitized block.

Fig.17 Typical paste in contact with graphitized block.

## Conclusion and Recommendations

Overall characteristics of the cold ramming pastes do not reflect their behavior in the course of cell preheat and operation.

A high swelling cold ramming paste from gas-calcined anthracite of inertinitic type with high strength comparable to that of the cathode block, low porosity with predominant pore volume formed by pores under 10 micron, low coking shrinkage and high residual expansion is suitable for bonding semi-graphitic and graphitic cathode blocks in the cell cathode bottom.

A high swelling cold ramming paste from electro-calcined anthracite with high residual volumetric expansion and minimal coking shrinkage, high strength on par with cathode block, low reactivity, low porosity with predominant pore volume formed by pores under 10 micron is suitable for bonding semi-graphitic, graphitic, and graphitized cathode blocks in the cell cathode bottom.

## References

1. Belitskus, D., 1978. *In Proc. TMS Light Metals,* pages 341-345.
2. Patent US № 2007/0138445 A 1,2007
3. Vergazova G.D., Sirazutdinov G.A. New carbon h pastes and for aluminum reduction cells. Moskow "Oil and gas», 1994, p. 85
4. Performance of cold ramming paste evaluated by resistance to bath effect, Aluminium of Siberia - 2007г., p.290

**Light Metals 2008** *Edited by: David H. DeYoung*
***TMS (The Minerals, Metals & Materials Society), 2008***

# Wear Mechanism Study of Silicon Nitride Bonded Silicon Carbide Refractory Materials

Ron Etzion, James B. Metson, Nick Depree
Light Metals Research Centre, University of Auckland, New Zealand

Keywords: Refractories, SNBSC, sidewalls, α and β $Si_3N_4$


## Abstract

Unused Silicon Nitride Bonded Silicon Carbide (SNBSC) bricks from different manufacturers were studied using x-ray powder diffraction (XRD), SEM-EDS, XPS and Solid state NMR. Powder XRD analysis reveals that the ratio of α to β $Si_3N_4$ varies significantly in different samples and in different zones of the brick. In samples which contain higher β $Si_3N_4$ its concentration decreases from the core to the exterior. The high β $Si_3N_4$ content in the core of some bricks could be attributed to the exothermic nature of the nitridation reaction and the consequent limited temperature control in the core of the brick. This would account for the higher internal temperatures conducive to β $Si_3N_4$ formation.
Core samples with high β $Si_3N_4$ content show higher corrosion rates in lab tests compared to samples with high α $Si_3N_4$ content. The corrosion test reveals that the degradation mechanism consist of a combination of bath penetration into the sample followed by gas attack.


## Introduction

In modern high amperage Aluminium reduction cells the service life and the cell efficiency are strongly influenced by the sidewall refractory lining. The use of $Si_3N_4$ bonded SiC refractories has becomes the state of the art for sidewalls and has largely replaced the traditional carbon materials. The advantage of these new sophisticated materials which have higher thermal conductivity than anthracitic materials is thinner cell lining giving increased cell capacity which can then accommodate larger anodes and hence increased productivity.

During cell operation a layer of frozen bath (frozen ledge) is formed protecting the sidewall from exposure to high temperature, corrosive liquid and vapour attack which can lead to chemical degradation. Anode effects and cell instability can lead to a situation were the frozen ledge will melt and the sidewall lining will be exposed to the molten bath. Thus there is considerable interest in the resistance of these materials to such corrosive attack.

Silicon Nitride bonded Silicon Carbide (SNBSC) refractories are formed by mixing SiC grains with Silicon powder and an organic binder, this mixture is sintered and shaped to bricks using vibration and pressing. The green bricks are fired in a furnace in an air atmosphere at low temperature (up to 300°C) and then in a Nitrogen atmosphere at temperatures up to 1400°C .During this stage a nitridation process converts the Si to $Si_3N_4$ which becomes the binder phase encapsulating the SiC grains. The final brick typically consists of 72-80 wt% SiC and 20-28 wt% $Si_3N_4$.
The aim of this work is to study the wear mechanism of commercial SNBSC materials using microstructural analysis of brick samples before and after exposure to laboratory and industrial scale reduction cell conditions.

### Material properties

The as manufactured SNBSC Blocks are typically observed to show color gradation which distinguishes an exterior zone and an interior zone [1]. Previous laboratory corrosion testing suggests the exterior zone is more reactive than the interior zone, particularly in the gas zone above the immersion level [1], although anecdotal evidence from smelters suggests the inverse is observed in reduction cell operation, i.e. blocks initially show good Cryolite resistance at the hot face and in the gas zone above the electrolyte, but then show accelerated failure once the exterior zone is breached. There are frequent comments in the literature on the "crumbly" nature of the refractories recovered after service in the cell. [2]

### Microstructure analysis

Eight SNBSC bricks from seven different manufactures have been analyzed using XRD to quantify the α and β $Si_3N_4$ content of the binder phase using the Gazzara & Messier calculation [3]. A cross section slice was cut from the brick and samples were taken both from the exterior and interior parts of the brick.

| maufacture | exterior | | interior | |
|---|---|---|---|---|
| | % α $Si_3N_4$ | % β $Si_3N_4$ | % α $Si_3N_4$ | % β $Si_3N_4$ |
| A | 87.28 | 12.72 | 88.24 | 11.76 |
| B | 72.08 | 27.92 | 35.10 | 64.90 |
| C | 74.69 | 25.31 | 60.20 | 39.80 |
| D | 89.55 | 10.45 | 92.45 | 7.55 |
| E | 91.70 | 8.30 | 90.30 | 9.70 |
| F | 80.02 | 19.98 | 75.66 | 24.34 |
| G | 85.70 | 14.30 | 86.40 | 13.60 |
| H | 48.51 | 51.49 | 24.61 | 75.39 |

Table 1: α and β $Si_3N_4$ content in samples from exterior and interior part of SNBSC brick cross section.

As it can be seen from the analysis results (table 1) there is uneven distribution of α and β $Si_3N_4$ along the cross section of the brick sample. High β $Si_3N_4$ content was usually observed in the interior part of the brick. In samples where high β $Si_3N_4$ content was observed usually $Si_2N_2O$ (about 5%) is also observed and even free Si was observed in a few cases. These last two phases were not observed in samples containing high α $Si_3N_4$
Further analysis of samples taken from an exterior, mid and core part of the brick cross section show reduction in β $Si_3N_4$ from core to exterior (Figure 1)

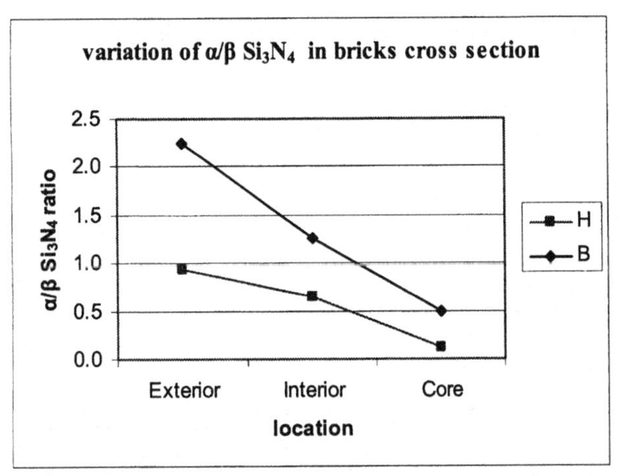


Figure 1: $\alpha/\beta$ $Si_3N_4$ ratio in samples taken from two SNBSC brick cross sections.

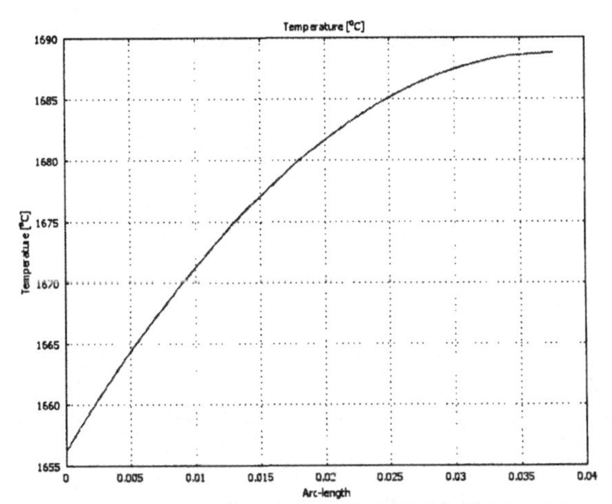


Figure 3: temperature profile of SNBSC brick cross section represent the distance from outer surface inward.

**Discussion of microstructure**

The high $\beta$ $Si_3N_4$ content in the internal part of some brick can be attributed to high temperature during the nitridation stage of the fabrication process. The nitridation reaction is an exothermic one as it can be seen in equation 1

$$3Si_{(s)} + 2N_{2(g)} = 3Si_3N_{4(s)}$$

$$\Delta H_{(1100°C)} = -824.7 \text{ kJ/Mole}$$

While the SNBSC materials are heated in the furnace at 1100°C the temperature in the core of the brick can rise above the melting point of Si due to the heat generated from the exothermic reaction and lack of heat dissipation from the brick core to the exterior surface. Since the $\alpha \rightarrow \beta$ $Si_3N_4$ conversion is observed at temperatures as low as 1450°C [4] this excess heat can result in formation of $\beta$ $Si_3N_4$ especially in the hot parts of the brick.

A computer model that represents the temperature profile of a brick cross section has been developed using the following assumptions: ambient temperature: 1100°C, Thermal conductivity at 1600°C= 6 W/(m*K), Si conversion rate: 4%/hour [5], Nitrogen speed at brick surface 2m/sec. As seen in Figure 2 and 3 the temperature on the outer surface of the brick could reach 1657°C while in the core it could reach 1688°C due to the heat generated from the exothermic reaction and the poor heat dissipation from the core of the brick to its exterior surface.

The formation of $\beta$ $Si_3N_4$ or conversion from $\alpha\, Si_3N_4$ depends on both on the temperature and the existence of liquid phase with in the SNBSC material during the nitridation reaction. Formation of liquid phase will enable formation/conversion to $\beta$ $Si_3N_4$ at a lower temperature.

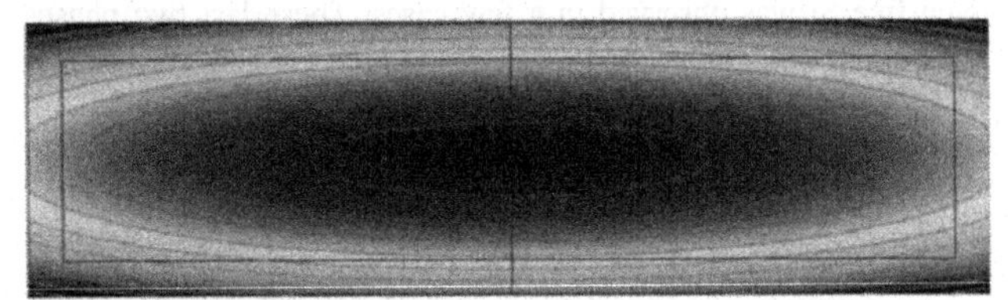

Figure 2: Temperature profile of cross section of SNBSC brick

**Experimental**

Industrial Corrosion test

Eight bricks samples (labeled A-H) were cut to a sample size (30x35x190mm) with grooves to allow mounting them on a stainless steel rack which was dipped into an industrial aluminium reduction cell between the anode block and the sidewall. The height of the rack was adjusted so the samples were half dipped in the bath and half exposed. The steel frame was insulated in order to prevent excessive heat loss through the frame. The frame was set in the crust and the experiment carried out for 77 hours. At the end of the experiment the samples were taken out and left to cool. The samples ware taken out of the metal frame cleaned from adherent frozen bath using Aluminium chloride solution. The clean samples (figure 4) were dried and the volume and density were measured using the ISO 5017 standard method, where the volume loss before and after the test was calculated.

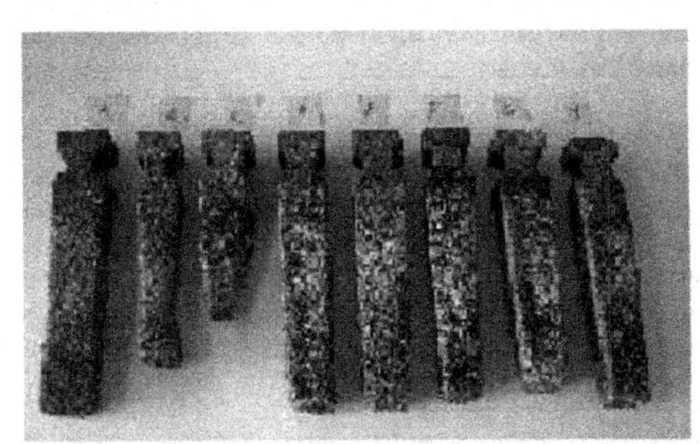

Figure 4: SNBSC samples after immersion of 77 hours in industrial Aluminium reduction cell

The volume loss of the samples comparing to their porosity and $\alpha/\beta$ $Si_3N_4$ ratio in the interior and exterior regions of the bricks from which the samples were taken are shown in Table 2.

| Sample | porosity | vol change (%) | interior α/β | exterior α/β |
|---|---|---|---|---|
| A | 14.13 | -17.61 | 2.53 | 6.86 |
| B | 15.10 | N/A | 0.54 | 2.58 |
| C | 14.31 | -56.08 | 1.51 | 2.95 |
| D | 11.69 | -15.82 | 12.25 | 8.57 |
| E | 12.88 | -17.24 | 9.31 | 11.05 |
| F | 11.85 | -16.21 | 3.11 | 4.00 |
| G | 12.16 | -9.42 | 6.35 | 5.99 |
| H | 11.42 | -14.44 | 0.33 | 0.94 |

Table 2: comparison between porosity, α/β $Si_3N_4$ ratio in the interior and exterior and corrosion of SNBSC samples (measured as volume loss) taken from brick from different manufacturers

Samples C and B showed relatively intense corrosion compared to the other samples (although Sample B broken during the trial and hence an accurate volume loss could not be obtained) both had a low α/β $Si_3N_4$ ratio in the interior and exterior compared to samples A, D-G. Sample H which had the lowest α/β $Si_3N_4$ ratio didn't show high corrosion in this test.

Laboratory corrosion measurements

Our setup (Figure 5) try to simulate the condition found in industrial Aluminium reduction cell. Four samples of dimensions 25x15x180mm were held symmetrically in a graphite crucible and supported by holes in the crucible lid.

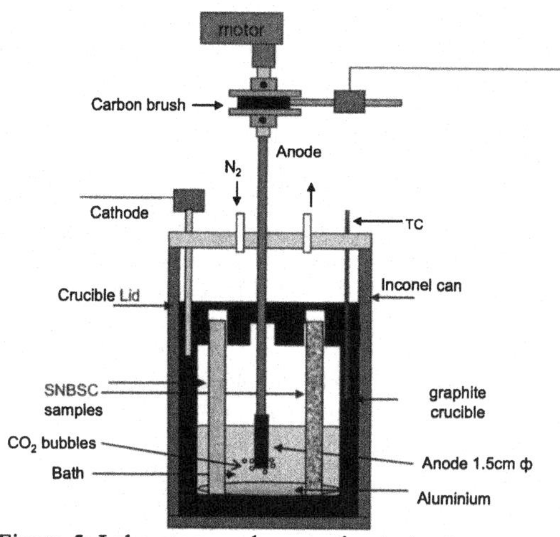


Figure 5: Laboratory scale corrosion test setup

The samples were exposed both to Aluminium metal, liquid bath and the gas phase above bath level. The bath composition at the beginning of the experiment was: 78% Cryolite, 10% $AlF_3$, 7% $Al_2O_3$, 5% $CaF_2$ the bath level was 9cm. A graphite anode (30mm diameter) was placed in the middle of the crucible and was immersed 4cm in the bath, this anode was connected to Inconel rod that was mounted to motor and carbon brash to allow rotation of the anode in order to give good distribution of the $CO_2$ bubbles, better stirring of the bath and to avoid sort circuit by carbon dust that builds up on the crust.

The samples were exposed to electrolysis environment for 48 hours at a bath temperature of 1000°C. Current of 15A was applied and the voltage range of 3.2-4.1 V. After the tests, the samples were taken out of the furnace and were allowed to cool, when cold they were soaked in $AlCl_3$ solution in ultrasonic bath to loosen the adhering bath which was removed using a knife. The degree of corrosion was assessed by measuring the volume change before and after electrolysis test using ISO5017 standard.

**Porosity measurements**

The porosity of samples from different parts of the brick cross section was measured using the ISO 5017 standard method. As reported by Skybakmoen et al [1] we also observed higher porosity in the interior part of the brick comparing to the peripheral part.

**Observation of corroded samples**

Samples that were placed in the bath for 48 hours without electrolysis show weight gain but no volume change and no visual signs of corrosion. The samples that were exposed both to bath and gas phase, by immersion half length in the bath, show high corrosion in the gas-bath interface and above bath level. Under bath level the samples show corrosion to a lesser degree.

To analyze the influence of porosity on corrosion the degree of corrosion (represented as volume loss in %) was plotted as a function of porosity. (Figure 6)

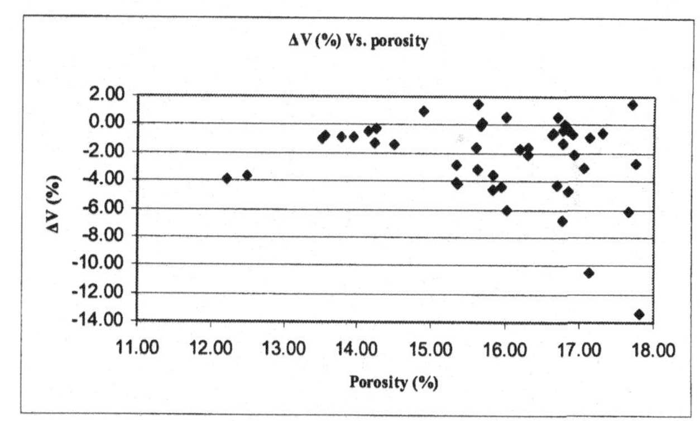


Figure 6: corrosion (measured by volume loss) as a function of porosity in SNBSC samples that were exposed to electrolysis.

As we can see from Figure 6, samples that show high corrosion typically had high porosity, but samples that had high porosity do not necessarily show high corrosion. This is in line with previous observations [6]. To examine the influence of β $Si_3N_4$ content on corrosion rate we plotted the volume change in % as a function of α/β $Si_3N_4$ ratio (Figure 7).

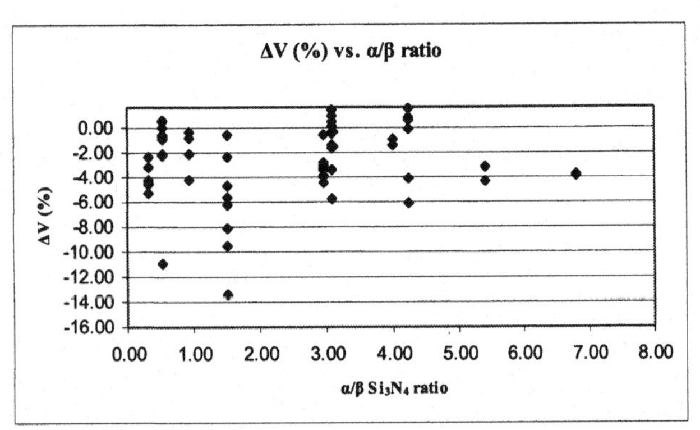


Figure 7: degree of corrosion (measured by % of volume loss) as a function of α/β $Si_3N_4$ ratio. The scattered results show no correlation between corrosion rate and specific phase composition of α or β $Si_3N_4$

The scattered results presented in figure 7 show no correlation between α/β $Si_3N_4$ ratio and corrosion rate, this might be since the α/β $Si_3N_4$ ratio varies along the brick cross section and the samples chosen for the corrosion test might be non homogeneous with α/β $Si_3N_4$ ratio varies within the sample. In order to test a homogeneous sample, smaller samples with more specific locations within the brick cross section were tested. Samples of dimensions 10x15x50mm were cut from the exterior, core, and mid section of brick (Figure 8)

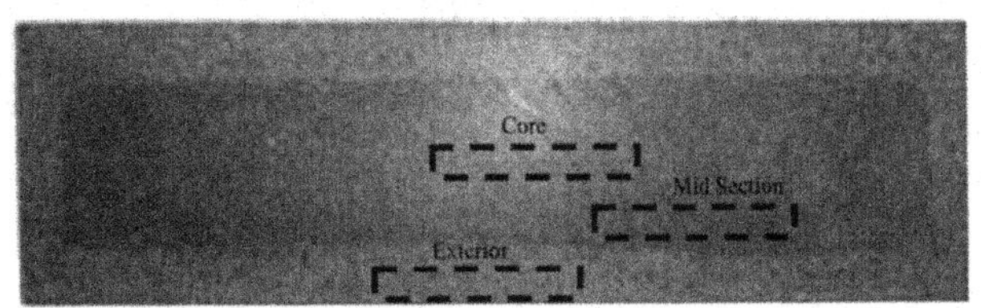


Figure 8: locations of samples taken from SNBSC cross section

Due to size of the samples, which is too small to test in the half immersed configuration, they were tested only in the gas phase. Four samples were placed symmetrically in graphite basket to allow the samples to be in contact with the corrosive gases; the basket with the samples was mounted on the anode rod and kept 5cm above bath level by a graphite spacer. Samples that were not exposed to molten bath showed little to no corrosion. In order to test the corrosion in this configuration the samples were presoaked in bath for 24 hours and clean of adhering bath prior to the electrolysis experiment. All the other parameters were kept as in the previous configuration.

When examining the degree of corrosion vs. α/β $Si_3N_4$ ratio (Figure 9) we can see higher correlation between high β $Si_3N_4$ content and high corrosion rate.

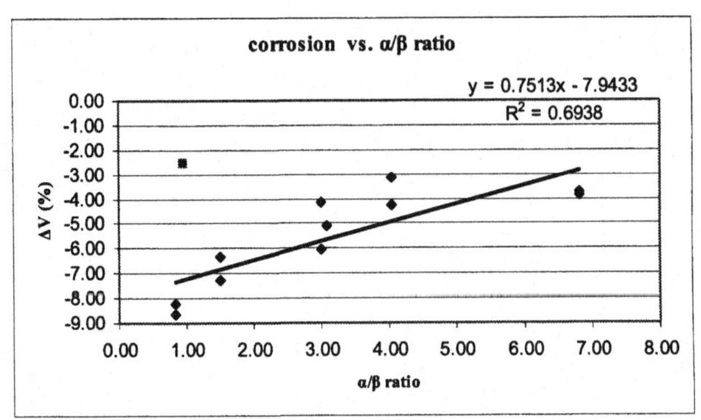


Figure 9: Correlation between corrosion rate and α/β $Si_3N_4$ ratio.

**Discussion on corrosion results**

Observation of the samples after electrolysis shows that the major attack occurs in the gas phase above bath level. The slow corrosion rate of samples which were exposed to the corrosive gases but not immersed in bath prior to exposure to electrolysis condition indicates that the corrosion is accelerated by oxidation of the binder phase by Na.

Open porosity has a contribution to the corrosion rate of the SNBSC material but this contribution is correlated with other factors such as the α/β $Si_3N_4$ ratio. Samples with higher β $Si_3N_4$ content show high corrosion rate compared to samples with high α $Si_3N_4$, this could be explained by the morphology of the β $Si_3N_4$ crystals which have elongated rod like shape [7] with high surface area compared to the round short α $Si_3N_4$ crystals. Hence these β $Si_3N_4$ crystals will oxidize faster.

The corrosion mechanism consists of a combination of attack mechanisms:

- Oxidation of the binder phase which is thermodynamically more prone to oxidation than the SiC grains [8] by Na ions which penetrates from the bath resulting in $SiO_2xNa_2O$ silicate species [1]. These silicate materials are more prone to oxidation compared to $Si_3N_4$ [9]
- Further oxidation of the binder phase by $CO_2$
- Attacked by corrosive gases like HF and $NaAlF_4$ on the oxidized glass to produce volatile $SiF_{4(g)}$
- Further oxidation of the remaining binder by $CO_2$ occurs and another cycle of attack by corrosive gases.

When comparing the corrosive gas attack on binder phase which was only oxidized by $CO_2$ without penetration of Na and formation of silicates (equation 2) to Attack on a sodium silicate (equation 3)

2) $SiO_2 + 4HF(g) = SiF_4(g) + 2H_2O$
$\Delta G_{(1000°C)} = -130.95$ kJ/Mol

3) $Na_2SiO_3 + 4HF(g) = 2NaOH + SiF_4(g) + H_2O$
$\Delta G_{(1000°C)} = -9344.4$ kJ/Mol

We can see that the silicate is less thermodynamically stable than the oxides, which explains the observation of higher corrosion rate on samples that were presoaked in bath prior to exposure to corrosive gases.

## Conclusions

During the fabrication of the SNBSC bricks the heat from the exothermic reaction should be taken into account of the process thermal control. Excess heating, especially in the core of the brick will results in formation of non uniform material in terms of α/β $Si_3N_4$ ratio, with usually high content of β $Si_3N_4$ in the core of the bricks. This phase due to its increase surface area is more prone to both oxidation and attack by corrosive gases which are found in the reduction cell atmosphere leading to failure of the brick. Temperature measurements in the core of the SNBSC brick during the fabrication process could give better understanding of the significance of the heat generated by the exothermic nitridation reaction and its contribution for formation of high β $Si_3N_4$ phase in the bricks.

These experiments suggest the importance of α/β $Si_3N_4$ ratio on the corrosion rate. However this ratio varies significantly between samples from different sources, and among samples from the same source. A wider range of α/β $Si_3N_4$ ratio bricks should be examined, and the cell gas atmosphere should be analyzed in order to understand better which species, found in the cell atmosphere, are those which contribute most to the corrosion rate of SNBSC materials.

## References


1. E. Skybakmoen, et al. *Quality evaluation of nitride bonded silicon carbide sidelining materials.* in *TMS-Light Metals.* 2005.
2. F.B. Andersen, et al. *Wear of Silicon Nitride Bonded SiC Bricks in Aluminium Electrolysis Cells.* in *TMS Light Metals.* 2004.
3. C. P. Gazzara and D.R. Messier, *Determination of Phase Content of $Si_3N_4$ by X-Ray Diffraction Analysis.* Bull. Amer. Ceram. Soc., 1977. **56**(9): p. 777-780.
4. E.T.Turkdogan, et al., *Silicon nitrides: Some physicochemical properties.* Journal of Applied Chemistry 1958. **8**: p. 296-302.
5. Molson, A.J., *Review Reaction-bonded silicon nitride: its formation and properties.* Journal of Material Science, 1979. **14**: p. 1017-1051.
6. Skybakmoen, E., H. Gudbrandsen, and L.I. Stoen. *Chemical resistance of sideline materials based on SiC and carbon in cryolitic melts -A laboratory study.* in *TMS-Light metals.* 1999
7. Messier, D.R., F.L. Riley, and R.J. Brook, *The α/β silicon nitride phase transformation* Journal of material science, 1978. **13**: p. 1199-1205.
8. Swain, M.V. and E.R. Segnit, *Reaction between cryolite and silicon carbide refractories.* Journal of the Australian Ceramic Society, 1984. **20**(1): p. 9-12.
9. Mayer, M.I. and F.L. Riley, *Sodium-assisted oxidation of reaction-bonded Silicon Nitride.* Journal of material science, 1978. **13**: p. 1319-1328.

**Light Metals 2008** *Edited by: David H. DeYoung*
***TMS (The Minerals, Metals & Materials Society), 2008***

# CHEMICAL RESISTANCE OF SIDELINING REFRACTORY BASED ON $Si_3N_4$ BONDED SiC

Richard LAUCOURNET[+], Véronique LAURENT[+], Didier LOMBARD[*]

[+] : ALCAN Centre de Recherches de Voreppe-FRANCE
[*] : ALCAN Laboratoire des Recherches et des Fabrications de St Jean de Maurienne-FRANCE

Keywords: sidelining, $Si_3N_4$ bonded SiC, chemical resistance, oxidation, air permeation

## Abstract

The state of art for the sidewall lining in modern electrolysis cells is $Si_3N_4$ bonded SiC refractories. The chemical resistance of sidewall materials is one the key factors of the pot lifetime. ALCAN intends to improve its electrolysis technology by identifying and selecting the most chemical resistant materials to the oxidation and chemical dissolution by the cryolite. The present study is aimed at improving the knowledge of corrosion mechanism of $Si_3N_4$ bonded SiC materials and determining the key parameters of materials structure which impact on the behaviour of these refractories in operation conditions. The chemical resistance in cryolite of $Si_3N_4$ bonded SiC samples was investigated through a study of the material behaviour in two oxidizing atmospheres: dry and steam. Moreover, the influence of the air permeability on the material oxidation was introduced.

## Introduction

In the modern technologies of aluminium production by electrolysis, the lifetime of sidelining materials become an important challenge. Nowadays, the refractory materials made of silicon carbide (SiC) are considered as a reference, because of their high chemical resistance to oxidation and to fluoride bath in comparison with carbon materials.

Some previous works [1, 2] have clearly shown that SiC bonded by silicon nitride have a better resistance than the materials bonded by oxides or nitride oxides.

$Si_3N_4$ bonded SiC refractories are made from a Reaction Bonded Silicon Nitride process [3, 4], which consists in the nitridation of silicon contained in a compact made of a mix of SiC grains and Si powder. Because of a nitridation step, an open porosity network is present and represents about 20 vol%.

Mostly, the main specifications are focused on the chemical composition, the density, the porosity, the mechanical properties and the thermal properties such as conductivity and expansion.

Today, there are a plenty of suppliers able to manufacture such material and it is needed for ALCAN to select the most competitive materials in lifetime but also in cost scope.

At first, this study addresses the corrosion mechanism of refractory located in the sidelining in order to determine the key parameters. In the last part, it is shown how the air permeation, as a new material property to consider, could influence the material behaviour in oxidizing conditions.

## Studied Material & Characteristics

All samples used in the experiments were taken in the core of a sidewall block. The characteristics of the material, given by the supplier datasheet are reported in the table 1

| | |
|---|---|
| SiC wt% | 75-80 |
| $Si_3N_4$ wt% | 22-25 |
| Oxide wt% | 0.5-2 |
| Si wt% | <1 |
| density | 2.55-2.80 |
| porosity | 15-20 |

Table 1: Typical material characteristics

An X-ray diffraction pattern of material illustrated in the figure 1, confirms the presence of the main phases: α-SiC, α-$Si_3N_4$ and β-$Si_3N_4$, and a good nitridation by the absence of Si.

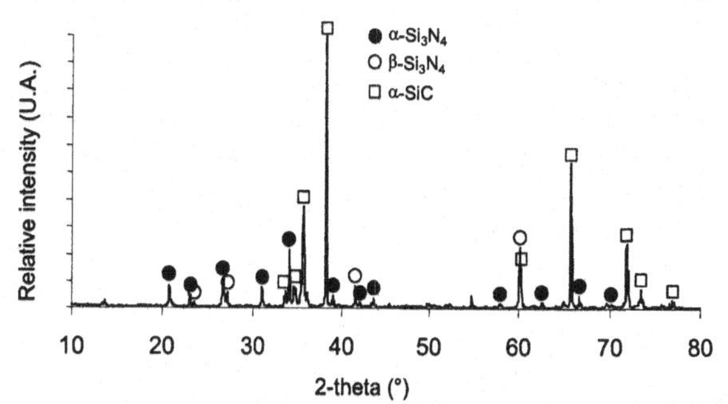


Figure 1: X-ray diffraction pattern of $Si_3N_4$ bonded SiC material

As it is shown by the SEM micrograph of the figure 2, the material microstructure is composed of coarse SiC grains (>0.5mm) bonded by a porous network of $Si_3N_4$. A magnification of binder (figure 3a & 3b) shows both types of $Si_3N_4$ crystalline structures with a very different specific area.

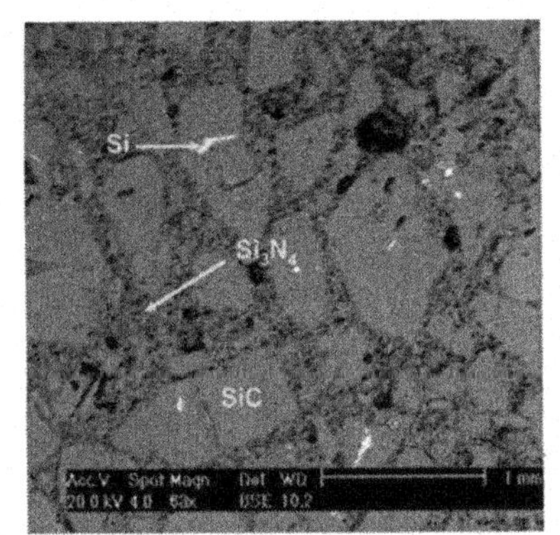

Figure 2: Material SEM micrograph

Figure 3a: β-$Si_3N_4$ Figure 3b: α-$Si_3N_4$

**Preliminary dissolution test**

A preliminary dissolution test was carried out in order to understand the role of the oxidation on the chemical dissolution of material by the fluoride bath.

Experimentation

Samples with size of 100×10×10 $mm^3$ were dipped into an alumina crucible filled with cryolite bath (CR=2.2, 5% CaF2) at 960°C. The test was performed during 24 hours under air or argon gas flow. The figure 4 details the experimental device.

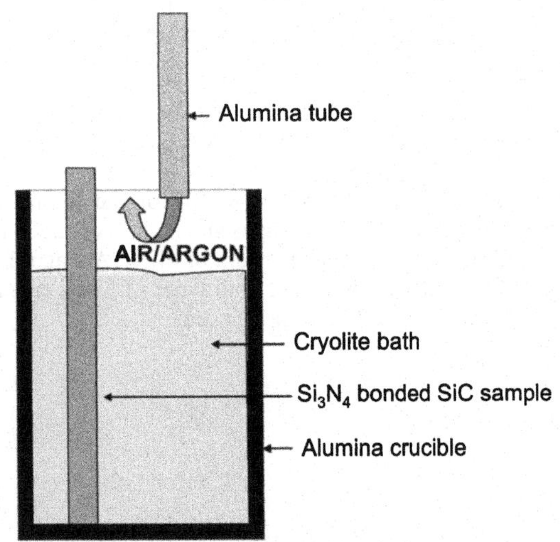

Figure 4: Experimental dissolution set up

Results and discussion

After exposition, the adhering bath was removed mechanically and the weight variations were measured. The results are reported in the table 2. The used samples are shown in the figure 5.

| Sample n° | Atmosphere | Weight Variation (%) |
|---|---|---|
| 1A | Air | -4.2 |
| 2A | | -3.8 |
| 3A | Argon | +11.2 |
| 4A | | +11.1 |

Table 2: Sample weight variation versus atmosphere after 24H in fluoride bath

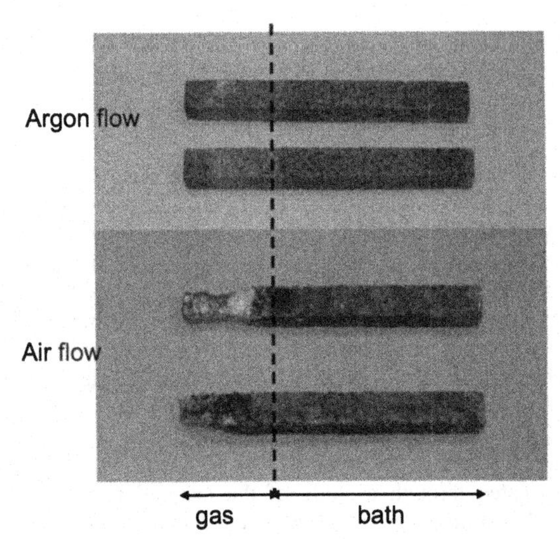

Figure 5: Samples after testing

Under argon, samples present a weight gain which corresponds to the bath infiltration into the porous material network. Under oxidizing atmosphere, a weight loss was measured in spite of bath infiltration, because of high corrosion in the gas phase above the bath level.

These results clearly establish that the corrosion by the bath is more severe in oxidizing conditions. They also confirm the previous Skykakmoen's observations [5] carried out from a much more complex test under polarization. Therefore, the oxidation may be considered as a key parameter in the $Si_3N_4$ bonded SiC material behavior in cryolite media and may increase the corrosiveness of evaluation.

SEM micrograph of the rupture surface of material after testing under air flow is shown in the figure 6. The porosity is filled by fluoride components and the SiC grains do not seem to be attacked after a short time test.

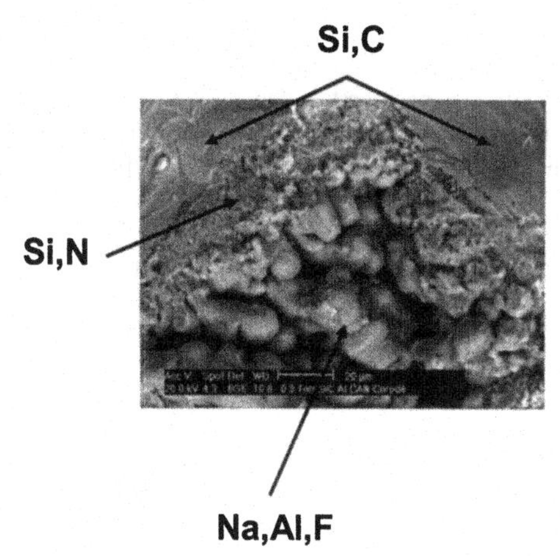


Figure 6: SEM micrograph of material rupture surface after dissolution testing under air flow.

**Material oxidation study**

Since the oxidation proved to be critical for the lifetime of the material, a focus on the oxidation mechanism was gone ahead. The studied parameters concerned atmosphere: dry air or steam, temperature and treatment duration.

Experimentation

Experiments were carried out in a tubular furnace. A steam generator was designed to distribute the steam from the inlet of the furnace chamber. A condenser was added at the outlet to recover the water. The samples (100×10×10 mm$^3$) were laid on alumina crank.

Results and discussion

At first, a holding time of 100 hours under dry air or steam was selected. The 800 to 1400°C temperature range was studied. The weight gain of samples was measured after treatment. The figure 7 shows the evolution of the average weight gain in both atmospheres versus the temperature.

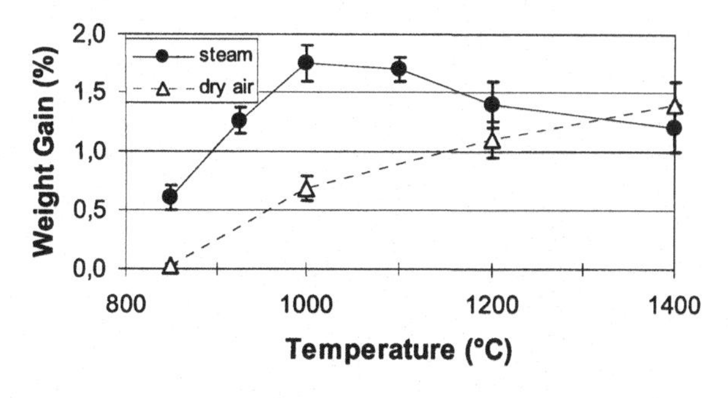


Figure 7: Evolution of the average weight gain under dry air or steam versus the temperature after 100 hours

*In dry air*

The oxidation may be considered as negligible below 900°C. In higher temperatures, the weight gain increases with the temperature. This evolution is in agreement with a previous study [6], which showed a similar trend with pure SiC or pure $Si_3N_4$ substrates. The well known oxidation reactions [7-8] are:

$$SiC(s) + 3/2O_2(g) = SiO_2(s) + CO(g) \quad (1)$$
$$Si_3N_4(s) + 3/4O_2(g) = 3/2\ Si_2N_2O(s) + 1/2N_2(g) \quad (2)$$
$$Si_2N_2O(s) + 3/2O_2(g) = 2\ SiO_2(s) + N_2(g) \quad (3)$$

In the case of silicon nitride, most of the studies [9-10] have revealed the formation of a thin layer of silicon oxinitride below silica layer.

*In steam*

The oxidation is more important in steam than in dry air in most of the temperature range studied. Maeda [11] and Moore [12] have already shown that water vapour promotes the oxidation. Moreover, Ogbuji [13] has suggested that the higher weight gain in wet atmosphere is mainly due to a more important permeability of silica layer to oxidizing species from water than from pure oxygen.

As shown in the figure 7, the maximal weight gain was obtained between 1000 and 1100°C. Indeed, below 1100°C, the weight gain increases with the temperature, whereas above this temperature, the weight gain decreases.

According to the literature, two approaches might explain this trend:

- The crystallisation of amorphous silica in cristobalite might give to the silica layer a lower permeability to oxidizing species. This mechanism was supported by Ogbuji [14].
- In the other hand, Fox et al. [7] explained that the water might lead to the volatilization of silica by the following reaction:
  $$SiO_2(s) + H_2O(g) \Rightarrow Si(OH)_4(g) \quad (4)$$
  That could explain why the weight gain measured at 1200°C was lower than at 1000°C after 100 hours.

In order to answer to this point, two isotherm treatments under steam at 1200°C and 1000°C were carried out. The figure 8 presents the experimental results.

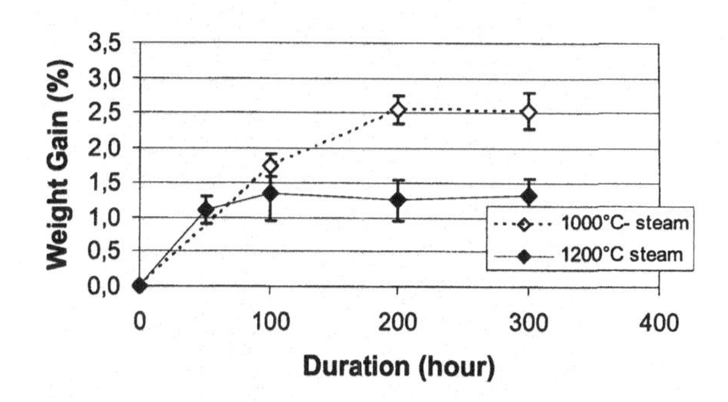


Figure 8: Evolution of the average weight under steam at 1000°C and 1200°C versus treatment duration

The oxidation evolutions confirm that at 1000°C the oxidation is higher than at 1200°C as the figure 7 has already shown. Note that the yield of oxidation is reached after 200 hours treatment at 1000°C whereas at 1200°C only 100 hours are needed, and no weight gain decreases after 100 hours. The most probable mechanism able to explain the evolution described in the figure 7 is the crystallization of amorphous silica into cristobalite, less permeable to oxidizing species.

We can conclude that the most severe experimental conditions able to promote the oxidation rate are:

- temperature range: 1000-1100°C,
- atmosphere: steam,
- treatment duration: 200 hours (valid for this specific sample format).

### Relation between material oxidation rate and dissolution rate in fluoride bath

Previously, it was demonstrated that the chemical corrosion is enhanced by the fluoride bath in oxidizing atmosphere. This second part is focused on the potential relation between weight gain obtained from oxidation and material volume loss resulting from a next step of dissolution in fluoride bath.

Experimentation
Samples oxidized under steam during 200 hours at 1000°C and 1200°C were submitted to the dissolution test under argon during 24 hours as it was described previously in the figure 4.

Results and discussion
The table 3 presents the weight gain of samples measured after oxidation and the weight loss measured after dissolution.

| Sample n° | Oxidation Temperature (°C) | Weight gain after oxidation (%) | Weight loss after dissolution (%) |
|---|---|---|---|
| 1B | 1000 | 2.55 | -20,5 |
| 2B | 1200 | 1.25 | -13,6 |

Table 3: Samples weight variation after wet oxidation and dissolution versus oxidation temperature

Both samples pictures are reported in the figure 9 after cleaning with an aqueous $AlCl_3$ solution.

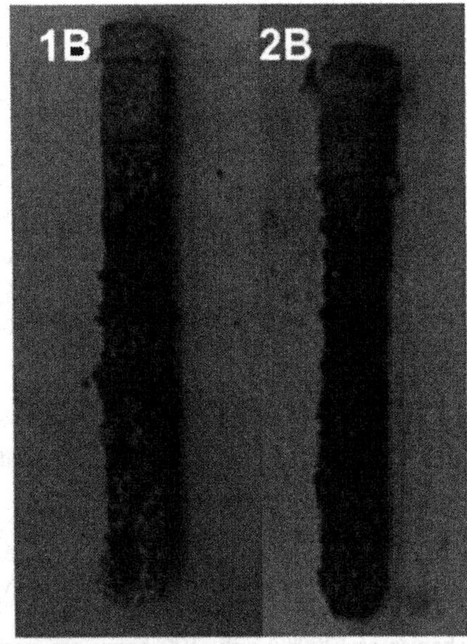


Figure 9: Samples picture after oxidation, dissolution and cleaning

According to the variation of weight after oxidation and dissolution, it is clear that there is a direct relation between oxidation rate and dissolution rate. This assessment is confirmed by the photograph of the figure 9: the attack of the sample 2B is more important than the attack of sample 1B.
The conclusion of this experiment is the higher gain weight is after oxidation, the stronger attack is observed in fluoride bath. That confirms the relation suggested by Schoennahl [2] based on another kind of cryolite resistance test.

### Air permeation impact on material oxidation

The best conditions to enhance the oxidation of $Si_3N_4$ bonded SiC material were previously described. It was moreover confirmed that the chemical resistance strongly depends on material oxidation.

In this part, the influence of the air permeation characteristic on material oxidation behavior is introduced.
The air permeation of refractory materials represents the porosity and the tortuous aspect of porous network as well. Therfore, this property is expected to influence the diffusion of oxidizing species inside the material. To clarify this point, an experimental device was designed to develop a steam gradient between the two faces of a refractory block.

Experimentation
The samples studied presented typical dimensions for air permeability measurement:

- diameter 50 mm
- thickness: 50 mm

The experimental device is described in the figure 10. The sample was held by a nickel alloy tube reamed at one of its end. Then, the sample was locked by a second tube by a compressive effect between both faces of the cylinder.
The device was introduced in a vertical furnace at 1000°C. Then, a water flow (1 droplet/10 seconds) was sent by a pump into the device. The water was immediately turned into steam and a pressure valve was added to keep 20 mbar as overpressure in order to push the steam through the sample.

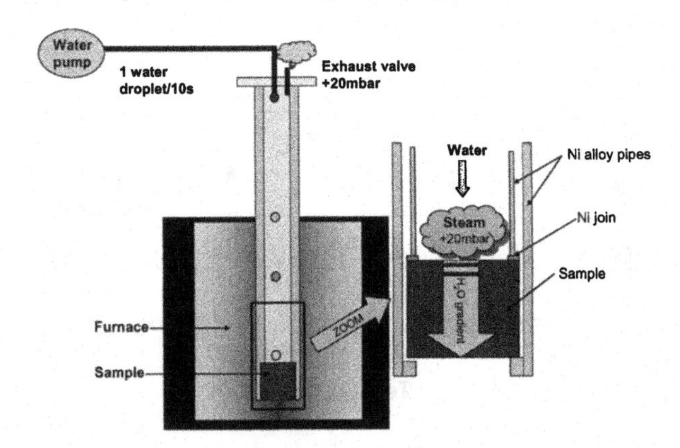


Figure 10: Experimental device developing a steam gradient in a refractory sample

At the end of the oxidation step, the device was cooled down and the sample was removed. The oxidation sample analysis is described in the figure 11. The central part of the sample was extracted by machining and turned into thin blades, which were finally milled into a fine powder. The oxygen and nitrogen contents were measured by an analyser of fusion gas (HORIBA EMGA-620W). The evolutions of oxygen and nitrogen contents versus the sample thickness were established.

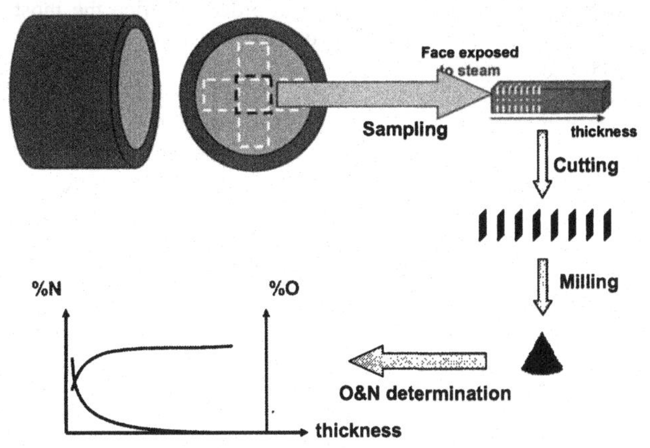

Figure 11: Sample analysis approach after oxidation

Results and discussion

The air permeation measures of five selected samples are reported in the Table 4. All of them were exposed to the test for 24 hours. Oxygen and nitrogen contents of a sample not oxidized were also determined as a reference: O=1.05%±0.25.and N=9.1%±0.3.

| Sample n° | Air permeability (nanoperm) | Density |
|---|---|---|
| 1C | 0.008 | 2,68 |
| 2C | 0.007 | 2,74 |
| 3C | 0.006 | 2,69 |
| 4C | 0.232 | 2,66 |
| 5C | 0.487 | 2,62 |

Table 4: Samples air permeation values

At first, the samples 1C, 2C and 3C which present very similar values of air permeation were studied to check the test reproducibility. The figure 12 shows nitrogen and oxygen contents evolution versus the thickness of samples, the zero abscise corresponding to the face directly exposed to the steam.

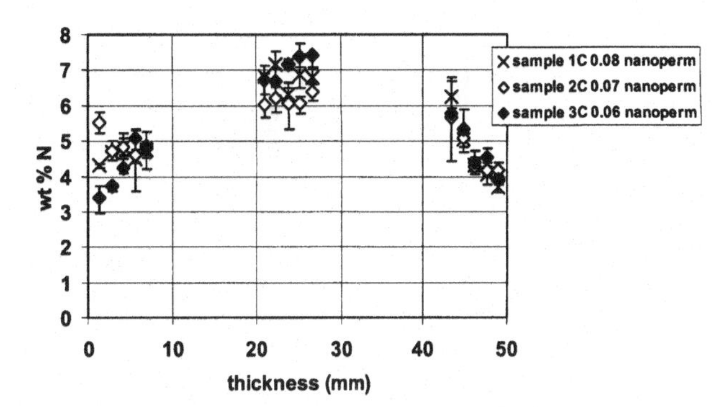


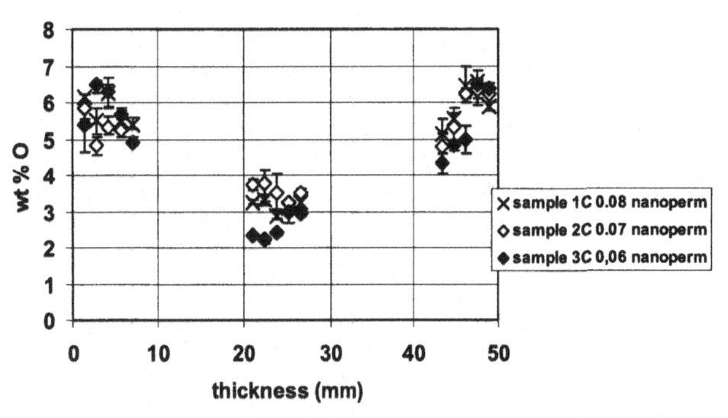


Figure 12: Oxygen (bottom) and Nitrogen (top) contents evolution versus thickness – samples 1C, 2C and 3C

Figure 12 shows :

- Evolutions of nitrogen and oxygen contents of the three samples are similar. The oxygen content is high in both zones close to the faces, whereas it is lower in the middle of the sample. Nitrogen content presents an opposite evolution. The middle of samples is less oxidized probably due to the more difficult access in the material for oxidizing species through the very narrow porosity network.
- Samples present some very close values of oxygen and nitrogen content versus the thickness which confirms that the experimental test seems to be reproducible.
- All the volume of samples is oxidized. Indeed, the reference values of oxygen and nitrogen are not found even in the middle of samples. In addition, the oxygen and nitrogen contents of both faces are very close, whereas a higher oxidation of the face exposed to the steam was expected.

The figure 13 shows nitrogen and oxygen contents evolution with the thickness of samples 3C, 4C and 5C.

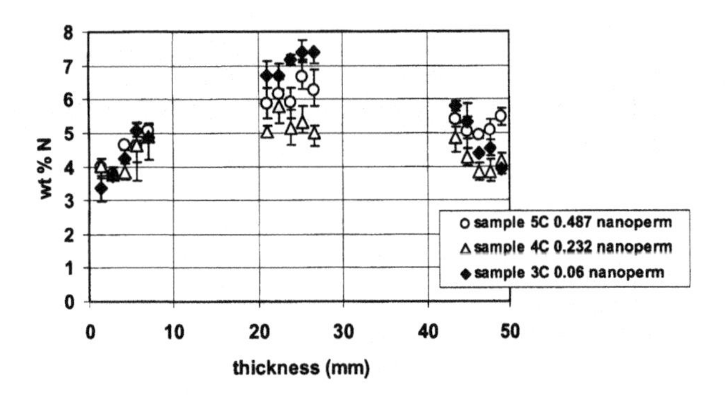


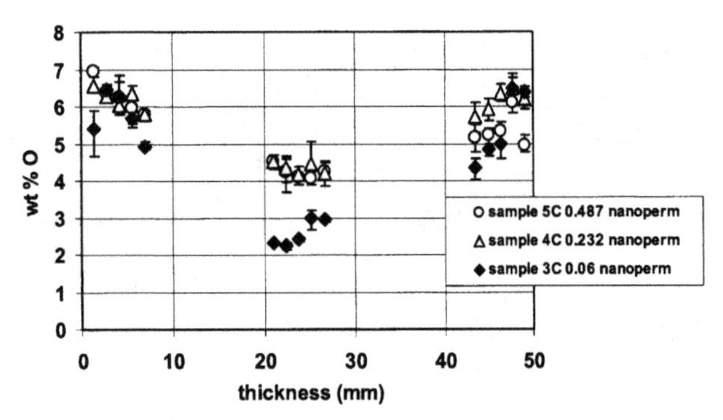


Figure 13: Oxygen (bottom) and Nitrogen (top) contents evolution versus thickness- samples 3C, 4C and 5C

Figure 13 shows:

- The air permeation level has no effect on the evolution of oxygen and nitrogen contents. Oxygen and nitrogen follow same trends: higher oxygen content near the external zone of samples and lower oxygen content in the middle. A similar conclusion may be made from nitrogen evolution that is logically reversed to the oxygen evolution.
- The air permeation value has no influence on the oxidation rate of zones near both sample faces. The oxygen and nitrogen contents are very close.
- Nevertheless, samples 4C and 5C are more oxidized than sample 3C. Indeed, the values of oxygen content were significantly higher (4.5%).

These results suggested that the air permeation might have an impact on the oxidation behaviour of $Si_3N_4$ bonded SiC. A low air permeation as 0.006 nanoperm does not lead to a same oxidation evolution inside the refractory than a high value as 0.481 nanoperm. However, with the current experimental conditions, it was not possible to find any difference between 0.232 and 0.481 nanoperm of air permeation. Moreover, the density of the studied samples was not exactly the same and might influence the oxidation as well.

## Conclusion

The present work was focused on the chemical resistance of $Si_3N_4$ bonded SiC material to oxidation and dissolution in cryolite bath. The chemical dissolution of material is shown to be enhanced in oxidizing media, even if the nature (liquid or gaseous) of the corrosive fluoride species was not identified.

There is a direct relation between the oxidation rate measured after preliminary oxidation step and the dissolution behaviour observed in cryolite bath. The setting of the material in steam environment at 1000°C was chosen in order to define the most severe experimental conditions and select the most competitive materials.

The air permeation of $Si_3N_4$ bonded SiC material was introduced as a factor of the oxidation behaviour. Additional works are needed to quantify the relation between this characteristic and material performance and lifetime.

## Acknowledgments

We thank A. El Bakkali from the Centre de Recherche sur les Matériaux à Haute Température CNRS UPR 4212 Orléans-France, for some micrographs and help.

## References


[1]: A.T.Tabereaux, A.Fickel, Evaluation of SiC Bricks, Light Metals, 1994

[2]: J.Schoennahl, E.Jorge, O.Marguin, S.Kubiak, P.Temme., Optimization of $Si_3N_4$ bonded SiC refractories for aluminium reduction cell, Light Metals, 2001

[3]: N.K.Reddy, J.Mukerji, $Si_3N_4$-SiC refractories produced by reaction bonding, J. Am. Ceram. Soc., 74[5] 1139-41, 1991

[4]: S.Y.Lee, Fabrication of $Si_3N_4$/SiC composite by reaction bonding and gas pressure sintering, J. Am. Ceram. Soc., 81[5] 1262-68, 1998

[5]: E.Skybakmoen, L.I.Stoen, J.H.Kvello, O.Darell, Quality evaluation of nitride bonded silicon carbide sidelining materials, Light Metals, 2005

[6]: D.S.Fox, Oxidation behaviour of CVD SiC and $Si_3N_4$ from 1200°C to 1600°C, J. Am. Ceram. Soc., 81[4] 945-50, 1998

[7]: D.S.Fox, E.J.Opila, Q.N.Nguyen, D.L.Humphrey, S.M.Lewton, Paralinear oxidation of $Si_3N_4$ in water-vapor/oxygen environnement, J. Am. Ceram. Soc., 86[8] 1256-61, 2003

[8]: N.S.Jacobsen, Corrosion of silicon based ceramics in combustion environments, J. Am. Ceram. Soc., 76[1] 3-28, 1993

[9]: R.E.Tressler, high temperature stability of non-oxide structural ceramics, Mat.Res.Sc.Bull., Sept. 1993

[10]: M.Backhaus-Rioult, V.Guerin, A.M.Huntz, V.S.Urbanovich, High temperature oxidation behaviour of high purity $\alpha$, $\beta$ and mixed silicon nitride ceramics, J. Am. Ceram. Soc., 85[2], 385-92, 2002

[11]: M.Maeda, K.Nakamura, T.Ohkubo, Oxidation of silicon carbide in a wet atmosphere, J.Mat.Sc., 23, 3933-3938, 1998

[12]: K.L.Moore, P.F.Tortorelli, M.K.Feber, J.R.Keiser, Observations of accelerated silicon carbide recession by oxidation at high water vapour pressures, , J. Am. Ceram. Soc., 83[1], 211-13, 2000

[13]: L.U.J.T. Ogbuji, The SiO2-Si3N4 interface, Part II: O2 permeation and oxidation reaction, , J. Am. Ceram. Soc., 78[5], 1279-84, 1995

[14]: L.U.J.T. Ogbuji, Effect of oxide devitrification on oxidation kinetics of SiC, J. Am. Ceram. Soc., 80[6], 1544-50, 1997

**Light Metals 2008** *Edited by: David H. DeYoung*
*TMS (The Minerals, Metals & Materials Society), 2008*

# SODIUM VAPOUR DEGRADATION OF REFRACTORIES USED IN ALUMINIUM CELLS

Asbjørn Solheim and Christian Schøning
SINTEF Materials and Chemistry, NO-7465 Trondheim, Norway

Keywords: Refractories, sodium, degradation

## Abstract

The bottom lining in aluminium cells can be degraded by several mechanisms. In the present work, reaction with sodium vapour that diffuses through the cathode was studied. The reaction paths depend on the oxygen level. At reducing conditions, metallic silicon is formed, whereas the presence of oxygen gives sodium oxide bound in the reaction product. Based on available thermodynamic data for the compounds present in the system $SiO_2$-$Al_2O_3$-$Na_2O$, each alkemade triangle in the phase diagram could be supplied with numbers for the equilibrium pressure of sodium (or sodium and oxygen), as well as lines showing the change in composition during attack. The calculations show that chamotte bricks normally end up mainly as nepheline. This is in accordance with practical observations. The reducing nature of sodium was verified through laboratory experiments. Autopsies of old linings have revealed the presence of silicon metal particles due to the reduction of $SiO_2$-containing compounds.

## Introduction

The bottom lining in aluminium reduction cells is made up of thermal insulation protected by refractories. The cathode carbon blocks may rest on an alumina bed, and the lining is sometimes supplied with steel plates as well as special barrier mixes designed to slow down the substances penetrating the cathode carbon. The materials used are mainly based on alumino-silicates, but substances rich in magnesium, such as olivine, are also commonly used. Although cathode failure is the main reason for shutting down cells, the service life partly depends on the quality of the lining materials. It is well known that both the dimensions and the physical properties of the materials change with time.

Several attempts have been made in understanding the properties and degradation mechanisms of refractories used in aluminium cells; see the review articles by Siljan *et al.* [1a-d, 2]. It seems clear that a number of mechanisms must be involved, possibly also reactions involving $SiF_4$ [3, 4]. It is clear that sodium is involved in the deterioration. Sodium, in the form of vapour or intercalation compounds, diffuses into the cathode and enhances the wetting between carbon and bath. Although the main role of sodium thus may be "opening up the path" for infiltration of bath into the cathode and lining, direct reaction between sodium vapour and the refractories also takes place.

The present paper focuses on the possible reactions between sodium vapour and the refractories used in the bottom lining of aluminium cells. The discussion is limited to materials in the system $SiO_2$-$Al_2O_3$. Upon reactions with sodium a number of compounds may be formed, as shown in Figure 1.

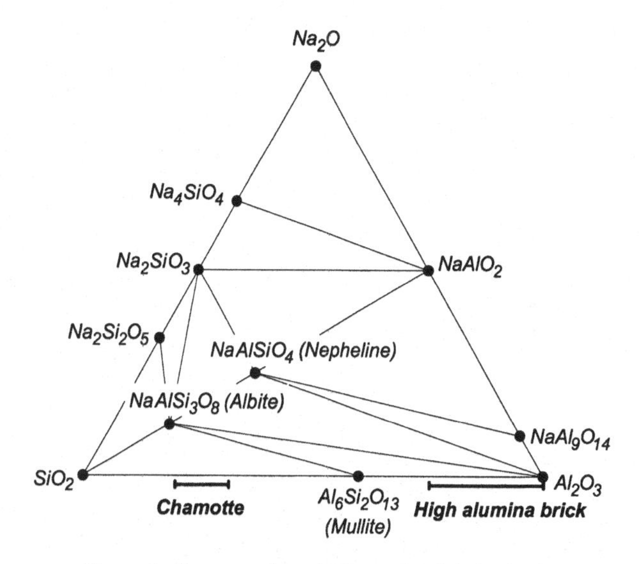


Figure 1. Compounds and alkemade triangles in the system $SiO_2$-$Al_2O_3$-$Na_2O$ (molar basis).

## Sodium Vapour

### Vapour Pressure

The sodium concentrations in metal and bath are determined by the following equilibrium,

$$\tfrac{1}{3}\mathrm{Al(l)} + \mathrm{NaF(l)} = \mathrm{Na(g)} + \tfrac{1}{3}\mathrm{AlF_3(s)} \tag{1}$$

The major part of the sodium formed in the cell dissolves into the bath and oxidizes at or near the anode, which is the main reason for loss in current efficiency in aluminium reduction cells [5]. The metal contains in the order of 100 ppm Na, depending on bath composition, type of cell, etc. [6]. The equilibrium vapour pressure of Na above the metal can be calculated by

$$p_{Na} = a_{NaF} \cdot a_{AlF_3}^{-1/3} \cdot \exp\left(\frac{-\Delta G_1^0}{RT}\right) \tag{2}$$

where p is the pressure [atm], a is the activity [-], $\Delta G^0$ is the change in standard Gibbs energy [$Jmol^{-1}$], R is the gas constant [$Jmol^{-1}K^{-1}$], and T is the absolute temperature [K]. Using thermodynamic data taken from JANAF [7] and activity data by Solheim and Sterten [8], the vapour pressure could be estimated as shown in Figure 2. The $NaF/AlF_3$ molar ratio in Hall-Heroult bath is commonly close to 2.2. The bath is somewhat less acid at the top of the metal due to the cathode reaction. From the data shown in Figure 2, it is reasonable to conclude that the vapour pressure of sodium at the cathode is normally about 0.02-0.03 atm, although the effect of $CaF_2$ on the activities of NaF and $AlF_3$ is not known.

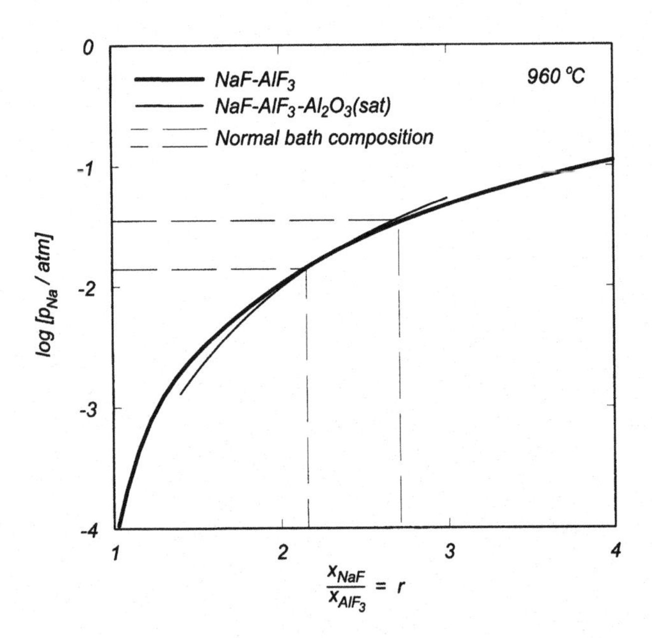


Figure 2. Vapour pressure of sodium above aluminium contacting bath as a function of the $NaF/AlF_3$ molar ratio at 960 °C, calculated with thermodynamic data from JANAF [7] and activity data from Solheim and Sterten [8] for the systems $NaF-AlF_3$ and $NaF-AlF_3-Al_2O_3(sat)$.

Transport of Sodium

Sodium formed at the metal-bath boundary (or at the metal-sludge boundary) is transported downwards through the cathode. The sodium transport probably takes place by gas diffusion in pores, as well as by diffusion of intercalation compounds $C_xNa$ formed in the cathode. Both processes are rapid, so sodium must be assumed to get through to the refractories almost immediately upon start-up of a new cell. The ordinary gas diffusion coefficients for the binary systems $Na-N_2$ and Na-Ar were calculated according to the Chapman-Enskog kinetic theory (see Bird, Stewart, and Lightfoot [9]) and the result is shown in Figure 3.

As will be clear from Figure 4, the sodium vapour formed close to the metal (vapour pressure 0.02-0.03 atm) does not condense until the temperature is below 550-600 °C. Some of the reactions proposed below proceed at considerably lower sodium pressure, so it can be assumed that the entire lining, including the insulation bricks, is vulnerable to attack by sodium. It should be emphasized, however, that the sodium pressure always decreases in the direction from the cathode towards the insulation bricks, since sodium is formed at the cathode and consumed in the lining.

**Reactions at Reducing Conditions**

Reactions between sodium and compounds in the refractory (or intermediate deterioration products) at reducing conditions can generally be written

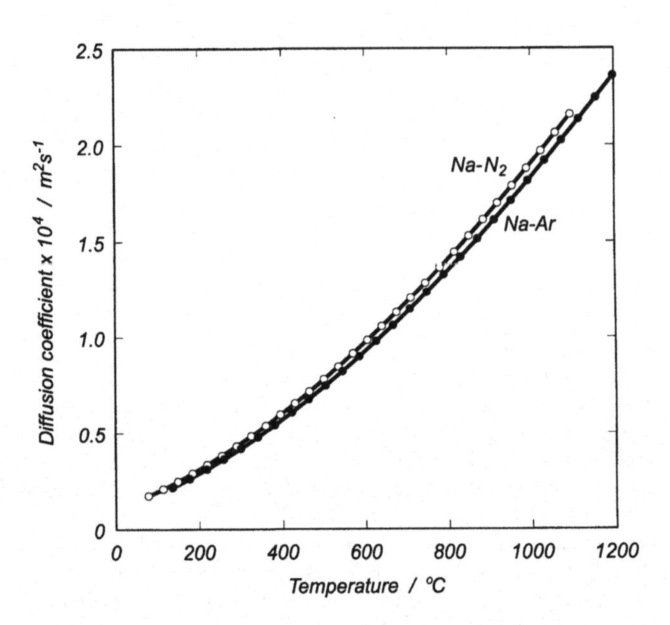


Figure 3. Ordinary gas diffusion coefficients at 1 atm total pressure for the binary systems $Na-N_2$ and Na-Ar, calculated by the Chapman-Enskog theory (see ref. [9]).

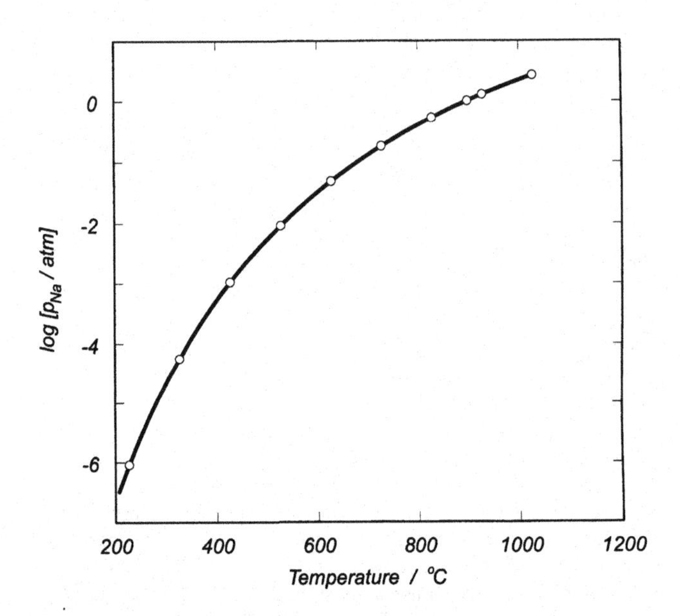


Figure 4. Vapour pressure of sodium above liquid sodium metal as a function of the temperature, calculated with thermodynamic data from JANAF [7].

$$\text{Na(g)} + \text{refractory} = \text{product} + \tfrac{1}{4}\text{Si(s)} \tag{3}$$

Assuming unit activity for the solid phases, the equilibrium sodium pressure becomes

$$p_{Na} = \exp\left(\frac{\Delta G_3^0}{RT}\right) \tag{4}$$

Similar reactions can be formulated with liquid aluminium as product,

$$Na(g) + refractory = product + \tfrac{1}{3}Al(l) \quad (5)$$

However, liquid aluminium is not stable in contact with the refractories, e.g.,

$$Al(l) + \tfrac{13}{12}SiO_2 = \tfrac{1}{6}Al_6Si_2O_{13} + \tfrac{3}{4}Si(s);\ \Delta G_6^0 = -128\ kJ \quad (6)$$

$$Al(l) + \tfrac{3}{8}NaAlSi_3O_8 = \tfrac{3}{8}NaAlSiO_4 + \tfrac{3}{4}Si(s) \quad \Delta G_7^0 = -112\ kJ \quad (7)$$

The change in Gibbs energy and the vapour pressure at 1200 K (927 °C) was calculated for all the alkemade triangles in Figure 1. The results are shown in Table I at the end of the paper. The computed vapour pressures were entered into the $SiO_2$-$Al_2O_3$-$Na_2O$ phase diagram, as shown in Figure 5. The arrows in Figure 5 indicate how the composition of the oxides changes during the sodium attack.

It was mentioned above that the maximum sodium pressure is 0.02-0.03 atm ($-\log p_{Na}$ = 1.5-1.7). From Figure 5 it can be found that chamotte ends up mainly as nepheline, mixed with smaller amounts of either $Na_2SiO_3$ or β-alumina (represented as $NaAl_9O_{14}$). If the chamotte contains less than 28 mol% $Al_2O_3$, and $\log p_{Na} < -2.22$, it will end up as a mixture of nepheline and albite. High alumina materials should normally end up as β-alumina with some nepheline, while it seems impossible to form $NaAlO_2$.

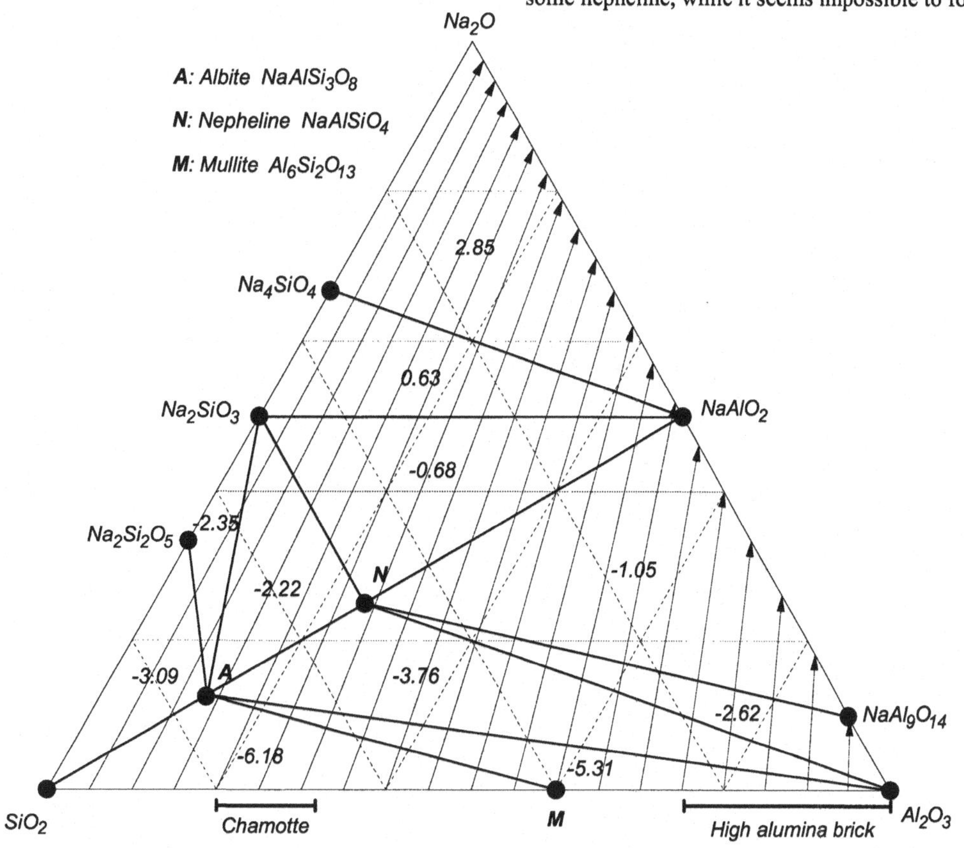


Figure 5. The system $SiO_2$-$Al_2O_3$-$Na_2O$ (molar basis). The numbers in the figure represent equilibrium values of $\log (p_{Na}$ / atm) at 1200 K during attack by sodium vapour at reducing conditions. The arrows indicate how the composition changes during deterioration.

The negative Gibbs energy in Eqs. (6) and (7) indicate that the reactions will be shifted towards the right hand side. By computing the sodium vapour pressure, assuming either silicone formation or aluminium formation within the same alkemade triangle (Figure 1), it was found that the calculated sodium pressure was higher in reactions forming aluminium. This means that formation of Si is thermodynamically favoured, but of course it can not be ruled out that Al is formed for kinetic reasons.

### Reactions at Oxidizing Conditions

Reactions between sodium and refractory compounds (or their deterioration products) at oxidizing conditions can generally be written

$$Na(g) + \tfrac{1}{4}O_2(g) + refractory = (Na_2O\ in\ product) \quad (8)$$

Again, assuming unit activity for the condensed phases, the equilibrium pressures of sodium and oxygen can be calculated by

$$p_{Na} \cdot p_{O_2}^{1/4} = \exp\left(\frac{\Delta G_7^0}{RT}\right) \tag{9}$$

These reactions were treated in exactly the same way as described for the reactions at reducing conditions described above. The calculated changes in Gibbs energy and the pressure product $\log p_{Na} + 1/4 \log p_{O_2}$ are given in Table II at the end of the paper, and the phase diagram showing the pressure products and the composition change patterns is shown in Figure 6 below.

sequentially. The reaction that requires the lower pressure proceeds until one of the reactant is consumed, followed by an increase in the sodium pressure until the next reaction is feasible, and so on. At a given extent of the attack, this pattern would produce a very sharp boundary between reacted and non-reacted material.

The amount of the different reaction products formed during sodium attack at reducing conditions was calculated, as shown in Figure 7. The starting material is 1 kg chamotte containing 36 wt% $Al_2O_3$. As can be observed in the figure, all $SiO_2$ in the original material disappears at $p_{Na} = 10^{-6.18}$ atm, while the remaining mullite reacts at $p_{Na} = 10^{-5.31}$ atm. It is interesting to note that the phases in the original chamotte vanish completely at a very low sodium pressure, but the reactions still proceed with increasing sodium pressure.

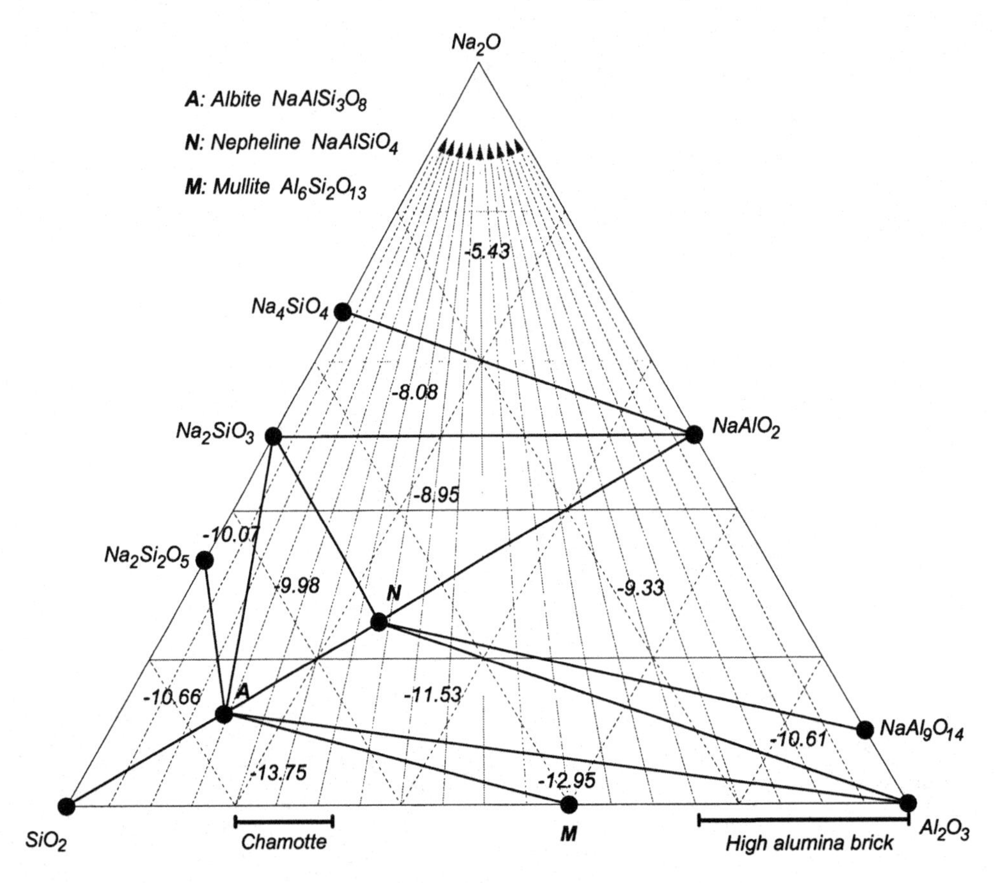


Figure 6. The system $SiO_2$-$Al_2O_3$-$Na_2O$ (molar basis). The numbers in the figure represent equilibrium values of log $p_{Na}$ + ¼ $\log p_{O_2}$ at 1200 K during attack by sodium vapour at oxidizing conditions. The arrows indicate how the composition changes during deterioration.

**Reaction Patterns**

The different reactions proceed at different equilibrium vapour pressures of sodium, and the vapour pressures are generally low. The temperature is relatively high (typically above 800 °C half-way between the cathode and the insulation). At these conditions, one would expect that the reaction rate is limited by sodium diffusion. If this is the case, the reactions will proceed

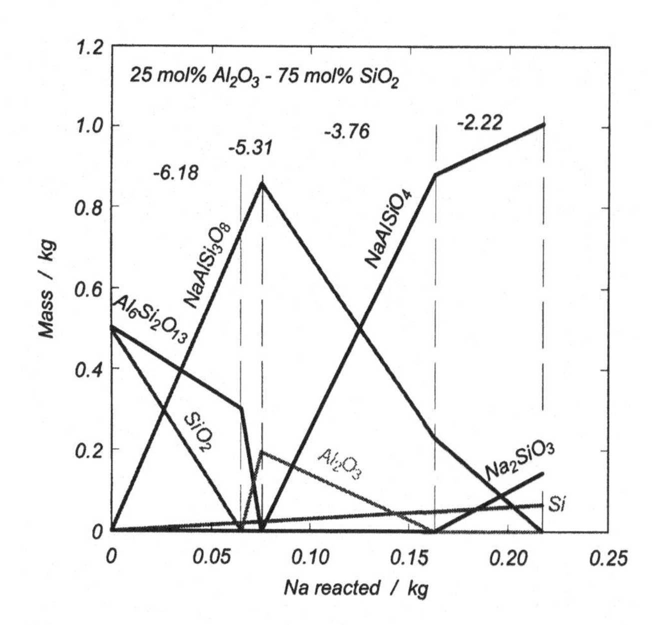


Figure 7. Mass of different phases during reaction with sodium at reducing conditions. The starting material is 1 kg chamotte with 64 wt% $SiO_2$ and 36 wt% $Al_2O_3$. The numbers in the figure are log ($p_{Na}$ / atm) at 1200 K.

### Experimental Evidence

Laboratory Data – Sodium Exposure

Cylindrical samples of different materials were exposed to sodium vapour for 4 h at 800 °C [11] (log ($p_{Na}$ /atm) = -0.40). The results showed that all commercial materials in the system $SiO_2$-$Al_2O_3$, except porous corundum, were attacked by the vapour. The reaction products were mainly $Na_2SiO_3$, $NaAlSiO_4$ (nepheline), $Na_2Al_2SiO_6$, glass, and silicon. This is in rough agreement with what can be expected from Figure 5. The compound $Na_2Al_2SiO_6$ is not shown in Figs. 5 and 6, due to lack of thermodynamic data. It should be kept in mind that the vapour pressure in these experiments was higher than what can be expected in practice, and $Na_2Al_2SiO_6$ will probably not be formed in an industrial cell.

In commercial materials, it was observed that the extent of deterioration decreased with increasing $SiO_2$-content. The attack looked similar as in tests using liquid fluorides. In some cases, there was a clearly visible reaction layer, and generally, the core of the sample was not attacked. The reason is probably that the volume increase due to the reactions, as well as the glass formed, leads to blocking of the pores in the original material, thus hindering diffusion of sodium vapour further into the material.

Autopsies of Spent Potlining

In autopsies of spent pot lining, Schøning and Grande [11] found that reactions with sodium had occurred in the refractory layer. Microscopy of a spent pot lining after 1800 days of operation gave evidence of the formation of silicon metal. A micrograph of the reacted refractory layer from that cell is shown in Figure 8. The light spots in the micrograph are silicon particles dispersed in the reacted refractory lining. Similar evidence for reaction with sodium were also found in cells after 300 and 3000 days in operation. [11] Sodium diffusion into the refractory layer resulted in a reduction of the $SiO_2$-content and the formation of sodium alumino-silicates and silicon, in line with the reactions given in Tables I and II.

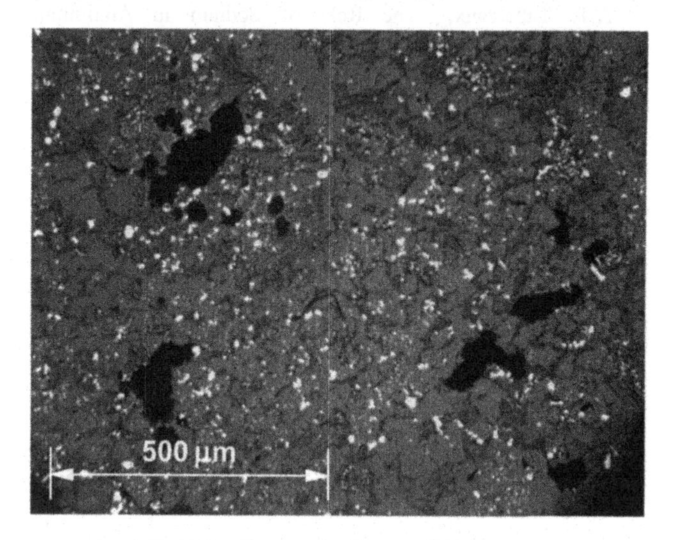


Figure 8. The presence of silicon particles (white spots) due to reduction of $SiO_2$-containing compounds in the refractory layer of a spent potlining.

### Acknowledgement

The present work is a part of the current project "Thermodynamics Applied to High Temperature Materials Technology" (Thermotech), which is financed by the Research Council of Norway, the Norwegian aluminium industry, the Norwegian Ferroalloy Producers' Research Association, and Statoil.

### References

1. O.-J. Siljan, T. Grande, and C. Schøning, "Refractories for Aluminium Electrolysis Cells", *ALUMINIUM*, **77** (2001).
   a) Part I: "Deterioration Mechanisms Based on Phase Equilibria", No. 4, pp 294/99.
   b) Part II: "Physical Properties of Penetrating Melt, Reduction by Metals and Volatile Fluorides", No. 5, pp. 385/90.
   c) Part III: "Laboratory Tests for Cryolite Resistance", No. 7/8, pp. 610/15.
   d) Part IV: "Comparison of Laboratory Investigations and Autopsies of Pot Linings", No. 10, pp. 809/14.
2. O.-J. Siljan, C. Schøning, and T. Grande, "State-of-the-Art Alumino-Silicate Refractories for Al Electrolysis Cells", *JOM*, May 2002, pp. 46/54.
3. A. Proshkin, A.M. Pogodaev, P.V. Polyakov, V.V. Pingin, and I.U. Patrachin, "Property Change of Dry Barrier Mixes Used in a Cathode of Aluminium Reduction Cells", Light Metals 2007, pp. 833/38.
4. A.V. Proshkin, A.M. Pogodaev, P.V. Polyakov, V.V. Pingin, and I.A. Yarosh, "Changes of Properties of Barrier Materials During Operation of Cells for Primary Aluminium Production", paper presented at Aluminium of Siberia 2007, Krasnoyarsk, Russia, September 11-13, 2007 (Proceedings, pp. 115/23).

5. P.A. Solli, T. Haarberg, T. Eggen, E. Skybakmoen, and Å. Sterten, "A Laboratory Study of Current Efficiency in Cryolitic Melts", Light Metals 1994, pp. 195/203.
6. A.T. Tabereaux, "The Role of Sodium in Aluminum Electrolysis: A Possible Indicator of Cell Performance", Light Metals 1996, pp. 319/26.
7. M.W. Chase (Ed.), NIST-JANAF Thermochemical Tables, 4th Ed. (*Journal of Physical and Chemical Reference Data*, Monograph No. 9).
8. A. Solheim and Å. Sterten, "Activity of Alumina in the System NaF-AlF3-Al2O3 at NaF/AlF3 Molar Ratios Ranging from 1.4 to 3", Light Metals 1999, pp. 445/52.
9. R.B. Bird, W.E. Stewart, and E.N. Lightfoot, Transport Phenomena, John Wiley & Sons, Inc., New York, 1960.
10. I. Barin, Thermochemical Data of Pure Substances, Third Edition, VCH Verlagsgesellschaft mbH, Weinheim, Germany, 1995.
11. C. Schøning and T. Grande, "The Stability of Refractory Oxides in Sodium-Rich Environments", *JOM*, February 2006, pp. 58/61.


Table I. Reactions with sodium vapour at 1200 K, at reducing conditions. The change in Gibbs energy was calculated from data in Barin [10].

| | $\Delta G^0$ kJ | $\log[p_{Na}]$ log [atm] |
|---|---|---|
| $Na(g)+\frac{35}{12}SiO_2+\frac{1}{6}Al_6Si_2O_{13} = NaAlSi_3O_8+\frac{1}{4}Si$ | -141.9 | -6.18 |
| $Na(g)+\frac{13}{8}Al_6Si_2O_{13} = \frac{35}{8}Al_2O_3+NaAlSi_3O_8+\frac{1}{4}Si$ | -121.9 | -5.31 |
| $Na(g)+\frac{1}{2}Al_2O_3+\frac{5}{8}NaAlSi_3O_8 = \frac{13}{8}NaAlSiO_4+\frac{1}{4}Si$ | -86.5 | -3.76 |
| $Na(g)+\frac{5}{4}SiO_2 = \frac{1}{2}Na_2Si_2O_5+\frac{1}{4}Si$ | -71.1 | -3.09 |
| $Na(g)+\frac{11}{2}Al_2O_3+\frac{1}{4}NaAlSiO_4 = \frac{5}{4}NaAl_9O_{14}+\frac{1}{4}Si$ | -60.2 | -2.62 |
| $Na(g)+\frac{3}{4}Na_2Si_2O_5 = \frac{5}{4}Na_2SiO_3+\frac{1}{4}Si$ | -54.1 | -2.35 |
| $Na(g)+\frac{3}{8}NaAlSi_3O_8 = \frac{1}{2}Na_2SiO_3+\frac{3}{8}NaAlSiO_4+\frac{1}{4}Si$ | -51.0 | -2.22 |
| $Na(g)+\frac{1}{8}NaAl_9O_{14}+\frac{1}{4}NaAlSiO_4 = \frac{11}{8}NaAlO_2+\frac{1}{4}Si$ | -24.2 | -1.05 |
| $Na(g)+\frac{3}{4}NaAlSiO_4 = \frac{3}{4}NaAlO_2+\frac{1}{2}Na_2SiO_3+\frac{1}{4}Si$ | -15.6 | -0.68 |
| $Na(g)+Na_2SiO_3 = \frac{3}{4}Na_4SiO_4+\frac{1}{4}Si$ | 14.5 | 0.63 |
| $Na(g)+\frac{1}{4}Na_4SiO_4 = Na_2O+\frac{1}{4}Si$ | 136.6 | 5.95 |

Table II. Reactions with sodium vapour at 1200 K, at oxidizing conditions. The change in Gibbs energy was calculated from data in Barin [10].

| | $\Delta G^0$ kJ | $\log\left[p_{Na}\cdot p_{O_2}^{1/4}\right]$ $\log\left[atm^{5/4}\right]$ |
|---|---|---|
| $Na(g)+\frac{1}{4}O_2(g)+\frac{8}{3}SiO_2+\frac{1}{6}Al_6Si_2O_{13} = NaAlSi_3O_8$ | -315.8 | -13.75 |
| $Na(g)+\frac{1}{4}O_2(g)+\frac{3}{2}Al_6Si_2O_{13} = 4Al_2O_3+NaAlSi_3O_8$ | -297.6 | -12.95 |
| $Na(g)+\frac{1}{4}O_2(g)+\frac{1}{2}Al_2O_3+\frac{1}{2}NaAlSi_3O_8 = \frac{3}{2}NaAlSiO_4$ | -264.8 | -11.53 |
| $Na(g)+\frac{1}{2}O_2(g)+SiO_2 = \frac{1}{2}Na_2Si_2O_5$ | -245.0 | -10.66 |
| $Na(g)+\frac{1}{4}O_2(g)+\frac{9}{2}Al_2O_3 = NaAl_9O_{14}$ | -243.8 | -10.61 |
| $Na(g)+\frac{1}{4}O_2(g)+\frac{1}{2}Na_2Si_2O_5 = Na_2SiO_3$ | -231.4 | -10.07 |
| $Na(g)+\frac{1}{4}O_2(g)+\frac{1}{4}NaAlSi_3O_8 = \frac{1}{2}Na_2SiO_3+\frac{1}{4}NaAlSiO_4$ | -229.3 | -9.98 |
| $Na(g)+\frac{1}{4}O_2(g)+\frac{1}{8}NaAl_9O_{14} = \frac{9}{8}NaAlO_2$ | -214.3 | -9.33 |
| $Na(g)+\frac{1}{4}O_2(g)+\frac{1}{2}NaAlSiO_4 = \frac{1}{2}Na_2SiO_3+\frac{1}{2}NaAlO_2$ | -205.7 | -8.95 |
| $Na(g)+\frac{1}{4}O_2(g)+\frac{1}{2}Na_2SiO_3 = \frac{1}{2}Na_4SiO_4$ | -185.7 | -8.08 |
| $Na(g)+\frac{1}{4}O_2(g) = \frac{1}{2}Na_2O$ | -124.6 | -5.43 |

**Light Metals 2008** *Edited by: David H. DeYoung*
***TMS (The Minerals, Metals & Materials Society), 2008***

# THE EFFECT OF CURRENT DENSITY ON CATHODE EXPANSION DURING START-UP

Arne Petter Ratvik[1], Anne Støre[1], Asbjørn Solheim[1], Trygve Foosnæs[2]
[1]SINTEF Materials and Chemistry, Energy Conversion and Materials, NO-7465 Trondheim, Norway
[2]Norwegian University of Science and Technology, Department of Materials Science and Engineering, NO-7491 Trondheim, Norway

Keywords: Cathode, Sodium Expansion, Cell Start-Up


## Abstract

During start-up of aluminium reduction cells, sodium penetration causes expansion in the carbon cathode, which may influence the lifetime of the cathode lining. Traditionally, the sodium expansion has been measured at cathode current densities up to 0.75 A/cm$^2$. However, it is well known that the current distribution in the cathode is non-uniform, and high local current densities may be experienced close to the sideledge, which commonly is associated with the W wear pattern. Hence, the sodium expansion may cause both local stresses in the cathode blocks, as well as in the total cell lining. The aim of this study is to determine the sodium expansion over a wider range of current densities. Typically, it is found that the sodium expansion starts to increase again above 0.7 A/cm$^2$, after the plateau reached at 0.2 A/cm$^2$. Apparently, this second increase continues outside the range of 1.5 A/cm$^2$ applied in this work.


## Introduction

Aluminium is produced by electrolytic reduction of $Al_2O_3$ dissolved in a cryolite-based electrolyte (bath). The electrolysis takes place in a cell containing the electrolyte and produced metal, see Figure 1. The cell is lined with carbon cathode blocks and carbon ramming paste (bottom), and SiC (sides).

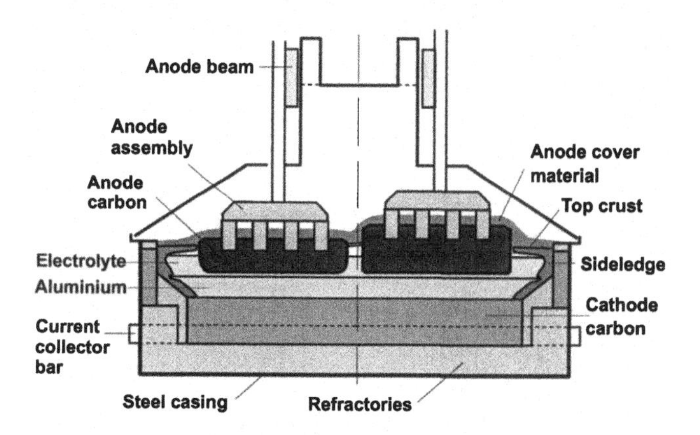


Figure 1. Schematic representation of an aluminium electrolysis cell with prebaked anodes.

The electrolyte, as well as the molten aluminium, contains dissolved sodium metal due to the following equilibrium [1],

$$3NaF\text{ (in bath)} + Al\text{ (l)} = AlF_3\text{ (in bath)} + 3\underline{Na} \quad (1)$$

Sodium penetrates into the cathode carbon structure. Besides "opening up the path" for the electrolyte by changing the wetting properties, this causes an expansion of the cathode block. Typically, the sodium penetration and expansion is most noticeable in the early start-up period of a new cell. A method for measuring this sodium expansion was first published in 1957 by Rapoport and Samoilenko [2]. Since that time, the method has been modified several times. One of the modifications is to apply pressure on the sample during the test [3,4].

An important issue for the aluminium industry today is the continuous effort towards increased cell productivity through prolonged cell lifetime and increased potline amperage. A consequence is a more rapid increase in cathodic current density during cell start-up. This will increase the rate of sodium uptake in the cathode blocks and, although it is an efficient way to increase the production capacity, it may have detrimental effects to cell lifetime. In addition, recent studies have indicated increased cathode wear at increasing current densities [5,6]. The aim of this study was to determine the sodium expansion over a wider range of current densities, using a laboratory scale apparatus and different commercial carbon cathode blocks.

## Experimental

The following cathode materials were tested:

1. One amorphous block, electrocalcined anthracite with 30 % graphite filler, baked to 1200 °C.
2. One graphitic block, all aggregates graphitized, baked to 1200 °C.
3. One graphitized block, petrol coke filler, whole block calcined above 2300 °C.
4. One graphite reference material.

The sodium expansion measurements were carried out in a RDC-193 apparatus (R&D Carbon Ltd., Switzerland), see the principal drawing in Figure 2. The cathode sample was placed in an anodically polarized graphite crucible containing bath with an initial cryolite ratio (CR) of 4.0. The bath consisted of 71.5 wt% $Na_3AlF_6$, 14.5 wt% NaF, 5.0 wt% $CaF_2$, and 9.0 wt% $Al_2O_3$. The total amount of bath was 765 g. The anode surface area exposed to the electrolyte was 65.5 cm$^2$. Since alumina feeding during the experiment was not possible, all the alumina was added at the start of the experiment. For the highest current densities, some excess alumina was added. A constant pressure of 5 MPa was applied to the sample by pressing the graphite extension placed on top of the cathode sample against a stop rod in the top of the furnace by a hydraulic power cylinder. The whole assembly was heated in a tubular furnace to 980 °C and the electrolysis was run until a maximum sodium expansion was reached or for two hours. The cathodic current density was varied between 0.05 and 1.5 A/cm$^2$. The expansion was measured by an extensometer, which was attached to the frame of the furnace and measured the position of the crucible support.

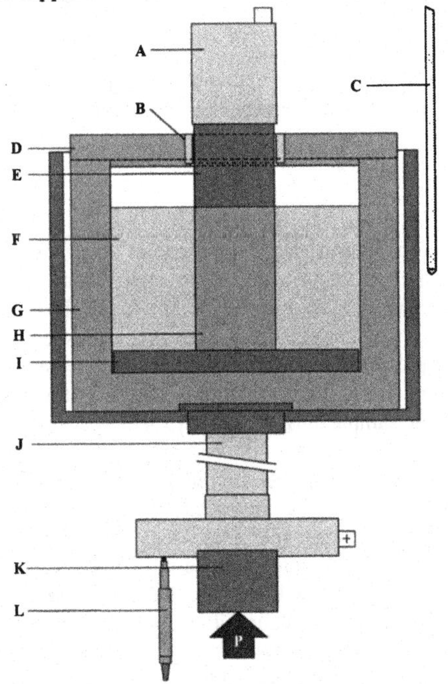


Figure 2. Principle drawing of the apparatus for measuring sodium expansion. A - heat resistant steel stop rod, B – insulating ring, C – thermocouple (Type K), D – graphite lid, E – graphite cylinder (sample extension), F – cryolitic melt, G – graphite crucible (anode), H – sample (30 mm dia. x 60 mm), I – alumina disk, J – heat resistant steel support, K – Hydraulic power cylinder, L – extensometer (LVD transducer).

## Results

### Physical Properties of the Materials

The measured physical properties of the cathode blocks are given in Table 1.

Table 1. Average physical properties.

| Material | Amorphous | Graphitic | Graphitized |
|---|---|---|---|
| Apparent density (g/cm$^3$) | 1.589 | 1.650 | 1.636 |
| Open porosity (%) | 14.3 | 17.8 | 19.6 |
| Electrical resistivity (μΩm) | 24.3 | 16.7 | 9.2 |
| Air permeability (nPm) | 1.1 | 6.5 | 2.3 |
| Compressive strength (MPa) | 31.3 | 23.3 | 27.0 |
| Young's modulus (GPa) | 12.2 | 7.8 | 7.3 |

### The Effect of Current Density on the Sodium Expansion

In Figures 3 to 5, the sodium expansion for the different cathode block materials are shown as a function of time for different current densities. Each curve is the average of two to four parallels of each material. The number of parallels can be read from Table 2, showing the sodium expansion as determined from the maximum of the expansion curves of all experiments. The maximum average sodium expansion for the tested materials as function of current density are shown in
Table 2 and plotted in Figure 6.

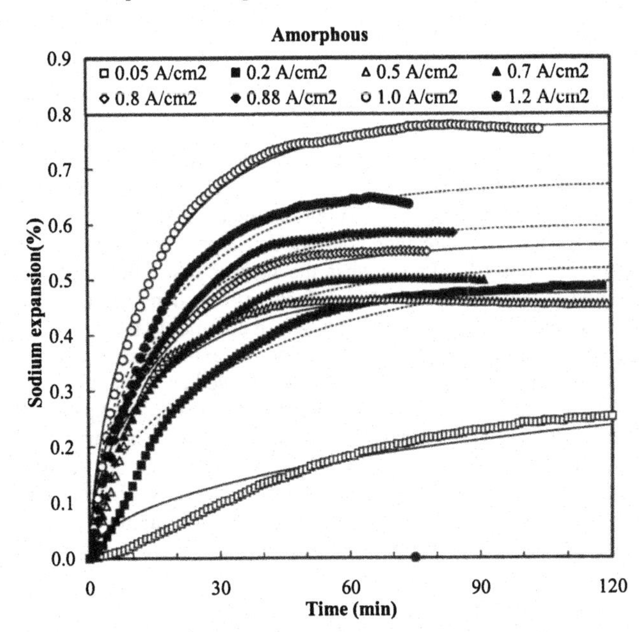


Figure 3. Sodium expansion versus time for an amorphous cathode block at different current densities. Symbols are experimental data based on 2-4 parallels, lines are modeled data. Applied pressure: 5 MPa.

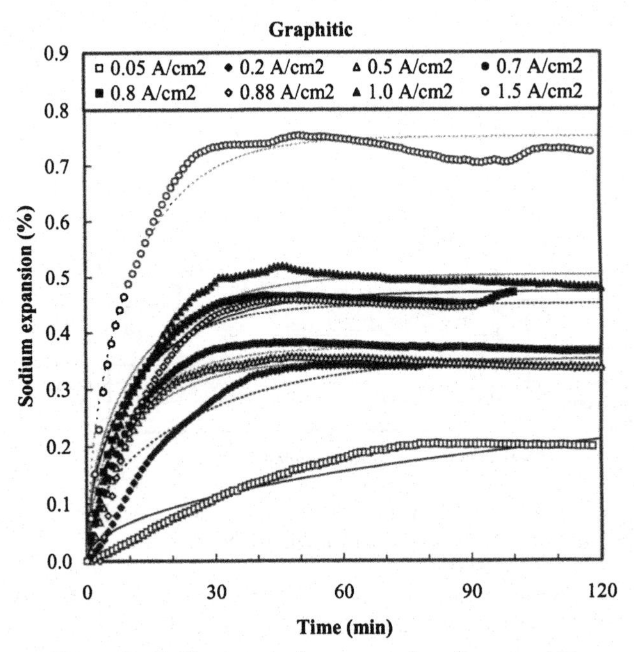


Figure 4. Sodium expansion versus time for a graphitic block material at different current densities. Symbols are experimental data based on 2-4 parallels, lines are modeled data. Applied pressure: 5 MPa.

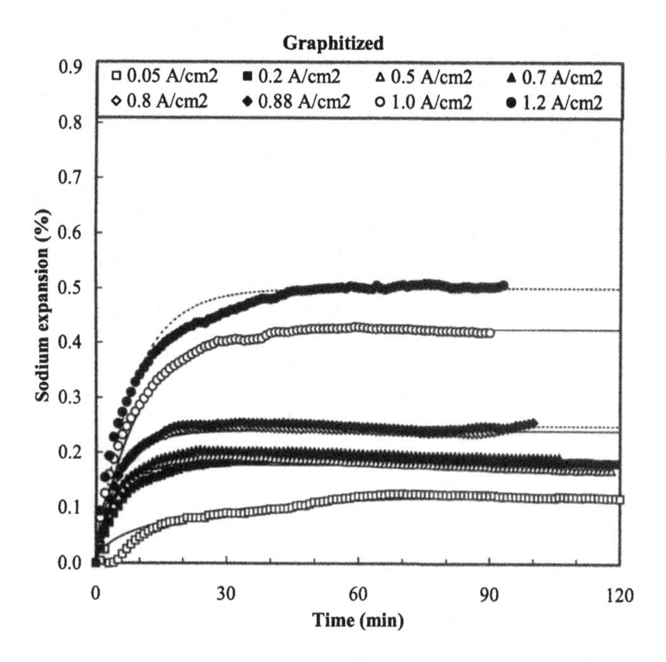


Figure 5. Sodium expansion versus time for a graphitized cathode block at different current densities. Symbols are experimental data based on 2-3 parallels, lines are modeled data. Applied pressure: 5 MPa.

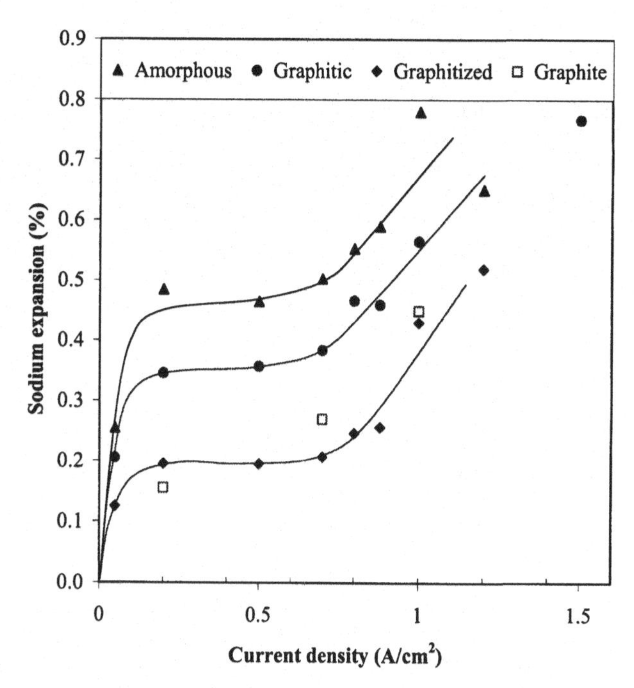


Figure 6. The maximum average sodium expansion of 2-4 parallels of tested cathodes versus current density. Applied pressure: 5 MPa.

Table 2. Maximum sodium expansion (percent) of cathode blocks at different current densities.

| Current density (A/cm²) | Maximum sodium expansion (%) | | | |
|---|---|---|---|---|
| | Amorphous | Graphitic | Graphitized | Graphite (ref) |
| 0.05 | 0.24 | 0.21 | 0.13 | |
| | 0.27 | 0.20 | 0.12 | |
| 0.2 | 0.50 | 0.31 | 0.20 | 0.17 |
| | 0.47 | 0.38 | 0.19 | 0.14 |
| 0.5 | 0.44 | 0.39 | 0.18 | |
| | 0.49 | 0.33 | 0.21 | |
| | | 0.35 | | |
| 0.7 | 0.44 | 0.38 | 0.24 | 0.21 |
| | 0.54 | 0.41 | 0.20 | 0.33 |
| | 0.53 | 0.36 | 0.18 | |
| 0.8 | 0.55 | 0.43 | 0.23 | |
| | 0.54 | 0.49 | 0.27 | |
| | 0.57 | 0.48 | 0.24 | |
| 0.88 | 0.51 | 0.44 | 0.23 | |
| | 0.66 | 0.46 | 0.31 | |
| | 0.54 | 0.47 | 0.23 | |
| | 0.65 | 0.47 | | |
| 1.0 | 0.79 | 0.52 | 0.46 | 0.45 |
| | 0.77 | 0.61 | 0.40 | 0.45 |
| 1.2 | 0.63 | | 0.50 | |
| | 0.67 | | 0.54 | |
| 1.5 | | 0.66 | | |
| | | 0.80 | | |
| | | 0.84 | | |

## Modeling of the Diffusion of Sodium into the Carbon

Zolochevsky *et al.* [7,8,9] have presented several models for the sodium expansion based on radial diffusion into a cylinder. The model that was found to give the best fit of the experimental data was:

$$\varepsilon_s = AC_0\left[1 - 4\sum_{n=1}^{\infty}\frac{L^2}{\beta_n^2(\beta_n^2 + L^2)}\exp(-D\beta_n^2 t/b^2)\right] \qquad (2)$$

where $\varepsilon_s$ is the sodium expansion, A is a material parameter describing the sodium expansion, $C_0$ is the bulk concentration of sodium in the melt, t is the time, b is the radius of the cylinder, D is the apparent diffusion coefficient, L=ba/D, where a is the surface exchange coefficient, $\beta_n$ ($n=1,2, \ldots, \infty$) are roots of $\beta J_1(\beta)-LJ_0(\beta)=0$. $J_0(x)$ is the Bessel function of the first kind of order zero and $J_1(y)$ is the Bessel function of the first kind of the first order.

The sodium expansion after infinite time is defined as $\varepsilon_s^{\infty}=AC_0$. The A coefficient is given as a material constant, but may also be regarded as the Na expansion coefficient for the given material. In this perspective, $C_0$ is in reality a representation of the sodium activity, $a_{Na}$. Since it is well known that the sodium expansion is reversible, considering $C_0$ as the Na concentration in the bulk melt is at best misleading. In reality, the factor that determines the measured maximum sodium expansion of a cathode material is $a_{Na}$ at the melt/electrolyte-cathode interface and the sodium expansion coefficient of the material.

To determine the sodium expansion coefficient or material constant A, we assumed that the activity of sodium at maximum sodium expansion (steady state) is independent of the tested material, *i.e.*, the sodium activity for a given current density is material independent. Using this approach, the material constant may be deduced from the observed differences between the maximum sodium expansion for the individual materials.

In the first approximation of the material constants, we assumed that the activity of sodium at current densities above 0.7 A/cm$^2$ is linearly proportional to the current density with intercept through origo. This is not an exact approach as current efficiency models show that there is a slight nonlinearity between current density and the activity. However, considering the scatter in the experimental data, this approximation was considered satisfactorily to deduce the material constants for the tested materials. When the sodium expansion coefficient, A, is determined, $C_0$ can be calculated from the maximum of the expansion curves. The deduced materials constants are included at the bottom of Table 3.

The model based apparent diffusion coefficients based on fitting Eq. 2 to the expansion curves are given in

Table 3. The apparent diffusion coefficient increases rapidly from low current densities and levels out for higher current densities (above 0.2 A/cm$^2$). As expected, the apparent diffusion coefficients increase inversely proportional to the Na content (or the sodium expansion) of the tested materials, Figure 7.

Table 3. Diffusion coefficient (cm$^2$/s) determined from sodium expansion measurements of cathode block materials at different current densities using Equation (2). Bulk CR = 4.0.

| Current density (A/cm$^2$) | D*10$^4$ Amorphous | D*10$^4$ Graphitic | D*10$^4$ Graphitized | D*10$^4$ Graphite ref |
|---|---|---|---|---|
| 0.05 | 0.4<br>0.2 | 0.8<br>1.0 | 2.0<br>19 | |
| 0.2 | 3.4<br>1.2 | 2.1<br>2.6 | 9<br>24 | 3.2<br>8.0 |
| 0.5 | 3.3<br>4.8 | 5.7<br>4.2 | 15<br>13 | |
| 0.7 | 2.5<br>3.5<br>2.1 | 4.6<br>5.3<br>8.4 | 24<br>47<br>16 | 14<br>22 |
| 0.8 | 3.3<br>2.5<br>2.7 | 4.1<br>4.2<br>4.1 | 15<br>31<br>21 | |
| 0.88 | 2.5<br>3.2<br>2.9<br>4.1 | 5.1<br>4.1<br>6.5<br>4.5 | 29<br>27<br>17 | |
| 1 | 4.2<br>4.6 | 5.7<br>5.5 | 12<br>50 | 5.9<br>14 |
| 1.2 | 4.1<br>4.0 | | 25<br>64 | |
| 1.5 | | 3.9<br>5.6<br>5.8 | | |
| | | | | |
| A (Na expansion coefficient) | 0.68 | 0.55 | 0.37 | |

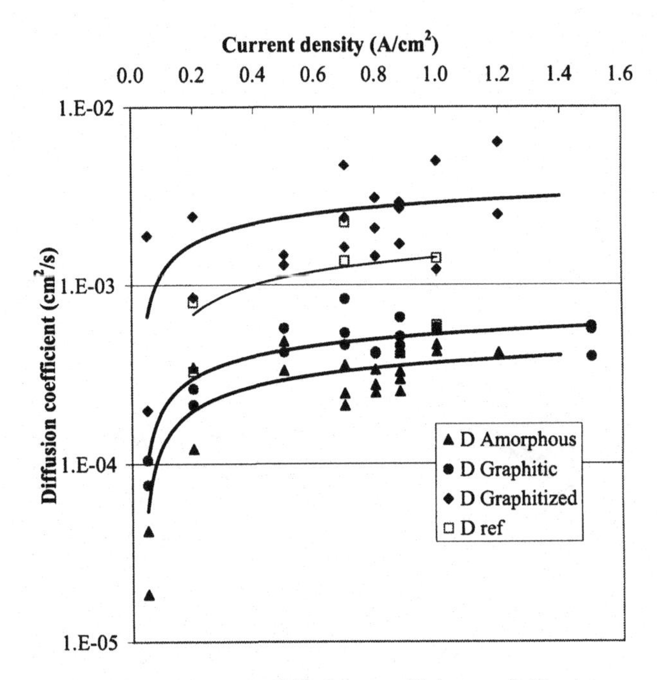


Figure 7. Apparent diffusion coefficients of Na into different cathode block materials as a function of the current density. Determined from model fitting of sodium expansion measurements.

## Discussion

### Tested Materials

In line with previous results, the amorphous material has a higher sodium expansion than the more graphitic materials. The graphite reference material has the same sodium expansion as the graphitized material. Normally, graphite materials show even lower sodium expansion, but the type of graphite and how it is heat treated will influence its sodium expansion.

The amorphous material contains more than 60 % anthracite. Even so, the difference between the amorphous material and the graphitic material is small compared to the difference between the graphitic material and the graphitized material. This indicates that the binder phase may be more important for the sodium expansion than the heat treatment of the aggregates used in the materials.

### Cathodic Current Density

The investigation of Na expansion at higher current densities is motivated both by the drive towards higher cell productivity through increased current density and the inherent variations in current density across the cathode. The higher current densities at the circumference of the cells, which is especially pronounced for graphitized blocks, may also cause unwanted effects during cell start-up; effects that may affect the long time stability of the cathode block.

From measurements and modeling, it is known that the current density at the cathode varies across the cell, see for instance [[10]], possibly by a factor of 100 or more, *i.e.*, from very low current densities in the middle of the cell to values exceeding 10 A/cm$^2$ along the sides. Two calculated current distributions, based on electrical conductivities typical for amorphous and graphite cathode blocks, are shown in Figure 8.

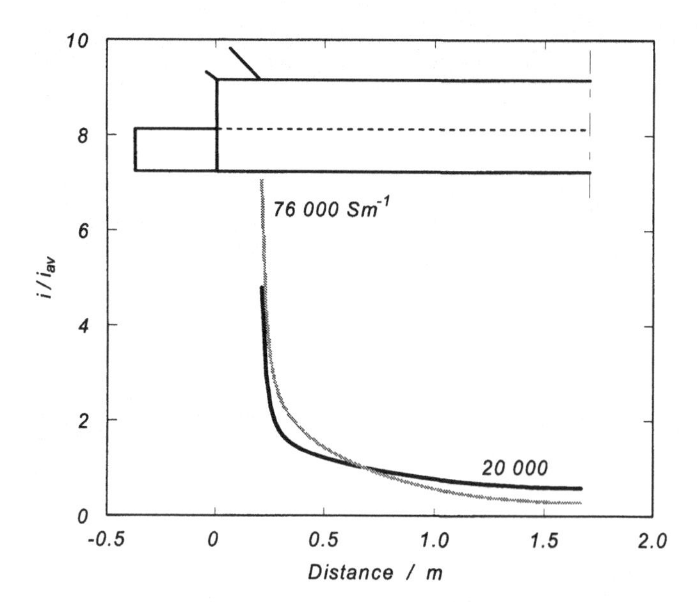


Figure 8. Example of calculated current distribution across the cathode assuming two different electrical conductivities of the carbon. The bottom ledge extends 0.2 m onto the carbon block.

Also, it is known that the sodium activity increases with increasing cryolite ratio, CR, which also is evident from Eq. 1. Hence, higher sodium expansion is experienced at higher CR. From basic electrochemistry, the NaF content in the diffusion layer towards the cathode is not independent of the current density. Higher current densities will lead to increased CR at the interface, resulting in higher $NaF/AlF_3$ ratio and higher Na activity. This is also in accordance with the observed increase in sodium expansion with increased current density.

All tested materials show a rapid increase in sodium expansion with increasing cathodic current density in the region 0-0.2 A/cm$^2$ (Figure 6). This is probably due competing reactions, e.g. the formation of aluminium carbide instead of aluminium at the cathode at low current densities [11]. Possibly, also the Na formation may be slower than the diffusion rate of Na into the electrolyte. The increase in the sodium expansion above this low current region is believed to be directly related to the increased cryolite ratio at the cathode surface due to diffusion phenomena related to the cathode reaction.

As a consequence, the cathode block will experience large differences in Na penetration and expansion during start-up. In addition to macroscopic stress over the cathode block, the block may also experience much higher microscopic stresses than previously anticipated. If this conveys to the wear rate experienced later in the service lifetime of the cathode, remains to be examined.

Apparent Diffusion Coefficients

The diffusion coefficient is not expected to vary with the current density (Figure 7). Also, the current model fitting of the expansion curves may not be representative for the true diffusion coefficients, as the model does not take into account a changing activity of Na at the interface between the electrolyte and the cathode. The low diffusion coefficients at low current densities are therefore partly related to the absence of sodium at the interface, as discussed above. The levelling out of the diffusion coefficients at higher current densities is in accordance with an almost linear relation between the activity of sodium, or more precisely, the cryolite ratio at the cathode-electrolyte interface, and the current density [12].

Also the fact that the rim of the cylindrical sample will experience equilibrium with the sodium activity at the interface before the mid of the sample may introduce errors to the model, *i.e.*, the observed expansion curves are only representative for the outer part of the sample cylinders, and as such, introduces inconsistencies to the model.

Implications for the Start-Up Strategy

Common practice today is start-up with bath only, without the presence of metal. The "only bath" strategy inherently means that the cryolite ratio becomes very high at the periphery of the cathode (Figure 8), which experiences a very high and rapid sodium penetration. Our experiments show that this may have a more detrimental effect than previously thought. Adding aluminium during start-up may counteract the fast Na penetration and rapid expansion since the formation of Na in this case predominantly takes place at the aluminium-bath interface. In theory, this should not be a problem, as carbon is not well wetted by Al. Na will then be distributed more evenly throughout the metal pad and the cathode carbon surface will experience a more balanced Na activity. Also, a slightly acidic bath may be beneficial, mainly because this gives lower concentrations of sodium, in addition to having a lower density than a neutral bath. A lower density may reduce or prevent the formation of a bath film between the aluminium pad and the carbon cathode.


**Acknowledgement**

Financial support from the Research Council of Norway and the Norwegian aluminium industry is gratefully acknowledged.


**References**


1. M. Sørlie and H.A. Øye, *Cathodes in Aluminium Electrolysis*, 2nd edition, Aluminium-Verlag, Düsseldorf 1994, 127.
2. M.B. Rapoport and V.N. Samoilenko, "Deformation of Cathode Blocks in Aluminium Baths during Process of Electrolysis", *Tsvetnye Metally,* 1957, 30 (2), 44-51.
3. J.M. Peyneau, J.R. Gaspard, D. Dumas, B. Samanos, "Laboratory Testing of the Expansion under Pressure due to Sodium Intercalation in Carbon Cathode Materials for Aluminum Smelters", *Light Metals,* 1992, 801-808.

4. H. Schreiner, H.A. Øye, "Sodium Expansion of Cathode Materials under Pressure", *Light Metals,* 1995, 463-470.
5. S. Wilkening, P. Reny, "Erosion Rate Testing of Graphite Cathode Materials", *Light Metals,* 2004, 597-602.
6. K. Vasshaug, T. Foosnæs, G.M. Haarberg, A.P. Ratvik, and E. Skybakmoen, "Wear of Carbon Cathodes in Cryolite Melts", *Light Metals*, 2007, 821-826.
7. A. Zolochevsky, J.G. Hop, T. Foosnæs, H.A. Øye, "Surface Exchange of Sodium, Anisotropy of Diffusion and Diffusional Creep in Carbon Cathode Materials", *Light Metals*, 2005, 745-750.
8 A. Zolochevsky, J.G. Hop, G. Servant, T. Foosnæs, H.A. Øye, "Rapoport-Samoilenko Test for Cathode Carbon Materials - I. Experimental Results and Constitutive Modelling", *Carbon* 41 (2003) 497-505.
9. A. Zolochevsky, J.G. Hop, T. Foosnæs, H.A. Øye, "Rapoport–Samoilenko Test for Cathode Carbon Materials - II. Swelling with External Pressure and Effect of Creep, *Carbon* 43 (2005) 1222-1230.
10 J.-M. Dreyfus and L. Joncourt, "Erosion Mechanisms in Smelters Equipped with Graphite Blocks – A Mathematical Modeling Approach", *Light Metals*, 1999, 199-206.
11 E. Skybakmoen, A.P. Ratvik, A. Solheim, S. Rolseth, and H. Gudbrandsen, "Laboratory Test Methods for Determining the Cathode Wear Mechanism in Aluminium Cells", *Light Metals*, 2007, 815-820.
12 A. Solheim and Å. Sterten, "Sodium in Aluminium and the Effect of Cathodic Overvoltage", *Proc. VIII. Al-Symposium*, Donovaly, Slovakia, 1995, 201-201.

**Light Metals 2008** *Edited by: David H. DeYoung*
***TMS (The Minerals, Metals & Materials Society), 2008***

# CHANGE OF ELECTRICAL RESISTIVITY OF GRAPHITIZED CATHODE BLOCK DURING AND AFTER ELECTROLYSIS IN ALUMINA MOLTEN SALTS

Yoshinori Sato[1], Hiroshi Imagawa[1], Manabu Hagiwara[2], and Noboru Akuzawa[2]

[1]SEC CARBON, LIMITED, 3-26 Osadano, Fukuchiyama, Kyoto 620-0853, Japan

[2]Dept. of Chemical Science and Engineering, Tokyo National College of Technology, 1220-2 Kunugida, Hachioji, Tokyo 193-0997, Japan

## Abstract

It can be considered that intercalation reaction between graphitized carbon and sodium in alumina molten salts gives rise to the erosion of graphitized cathode block used in the aluminium-reduction cell. The purpose of this investigation is to obtain knowledge about degradation of the cathode block by measuring the change of electrical resistivity of cathode samples during and after the electrolysis.

When the graphitized sample was applied as the cathode in the molten salts, the electrical resistivity of the sample showed sudden decrease immediately after start of electrolysis. The minimum value was seen at the progress point of 30 to 50 minutes in the electrolysis, which increased gradually afterwards.

Upon interruption of electrolysis, the electrical resistivity which had been lowered by the electrolysis showed sudden increase immediately after the interruption, and returned to the level of pre- electrolysis by the way of a plateau region. The pattern of the changes in electrical resistivity differed by the time of electrolysis, kind of raw material and final heat-treatment temperature (HTT) of the samples.

The longer the electrolysis time, the longer the time required to return to the original value. Samples prepared from needle like coke returned to a higher than original electrical resistivity while samples from lump coke returned to the original electrical resistivity. High HTT samples showed higher increase of electrical resistivity.

## Introduction

Graphitized blocks used as the cathode in aluminum-reduction are exposed to erosion during operation. One of the possible reasons of the problem may be the formation of intercalation compounds induced by the attack of sodium onto the cathode blocks.

It is well known that graphite reacts with alkali metals and the products are called alkali metal-graphite intercalation compounds (AM-GICs). Among alkali metals, heavy alkali metals (K, Rb and Cs) give stage 1 compounds with the composition of $MC_8$ (M=K, Rb, Cs)[1] and lithium also gives stage 1 compound with the composition of $LiC_6$[2]. Sodium, however, does not give rich compounds but very dilute compound of stage 8 with the composition of $NaC_{64}$[3]. In addition, the intercalation behavior of sodium into carbon materials is quite unique; less graphitized carbon materials are easily intercalated with sodium[4], while well graphitized ones are intercalated with only a very small amount of sodium.

The degradation behavior of carbon materials by intercalation has been reported by Inagaki and Tanaike[5],[6]. They observed that the particle size of needle-like petroleum coke during co-intercalation of alkali metals became small[7]. They concluded that this kind of degradation of coke was caused by the stress to the coke particles accompanied by enlargement of carbon-carbon interlayer distance with intercalation.

We have investigated the effect of heat-treatment temperature of carbon samples on the reactivity against metallic sodium by the measurement of XRD, ESR, and electrical resistivity. The results showed that carbon samples heat-treated from 1500 to 2400°C reacted with sodium and formed various types of GICs(stage 2+3 to stage 8), and in case of samples heat-treated below 1500°C, they degraded largely.[8]

This investigation aims to reach findings on how sodium in molten salts affects the degradation of cathode during electrolysis, by measuring the electrical resistivity of cathode during and after electrolysis.

## Experimental

### Materials

Cathode samples with two different kinds of coke (needle-like, or lump) and three different heat-treatment temperature (HTT) were prepared. Table 1 shows the sample code, material coke, HTT and electrical resistivity (ER) at room temperature (RT).

Molten salts consist of cryolite(Central Glass Co., Ltd.), NaF, $CaF_2$(Stella Chemifa Corporation), $Al_2O_3$(Sumitomo Chemical Co., Ltd.). Table 2 shows the composition of molten salts.

Table 1: Characteristics of the cathode samples.

| Sample code | Material Coke | HTT/°C | ER/mΩ·cm |
|---|---|---|---|
| A | Lump | 2800 | 0.93 |
| B | needle-like | 2800 | 0.72 |
| C-2000 | Lump | 2000 | 1.78 |
| C-2400 | Lump | 2400 | 1.33 |
| C-2800 | Lump | 2800 | 1.33 |

Table 2: Composition of molten salts

| component | ratio |
|---|---|
| Cryolite | 71.5% |
| NaF | 14.5% |
| $CaF_2$ | 5.0% |
| $Al_2O_3$ | 9.0% |

Experimental apparatus for electrolysis

Fig.1 shows the cathode sample and crucible used in experiment.

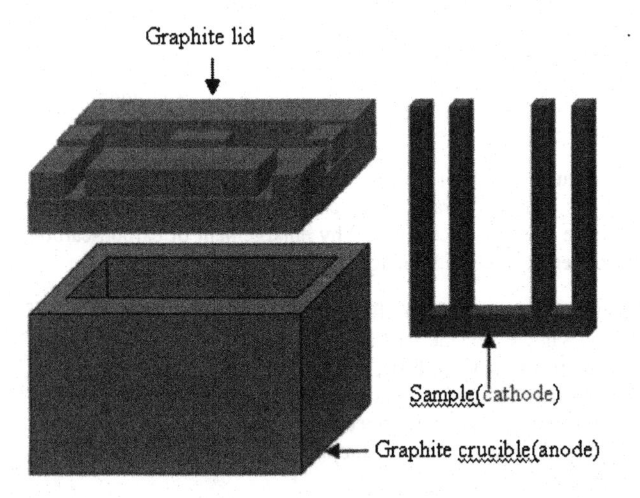


Fig.1 Schematic drawing of sample, graphite crucible and graphite lid.

Cathode sample had a characteristic shape, specially designed for the convenience of electrical resistivity measurement by the 4-terminal method. The steel bars (electric leads) were fixed to the terminals of graphite cathode by screws. Fig.2 shows the appearance of experimental apparatus for electrolysis.

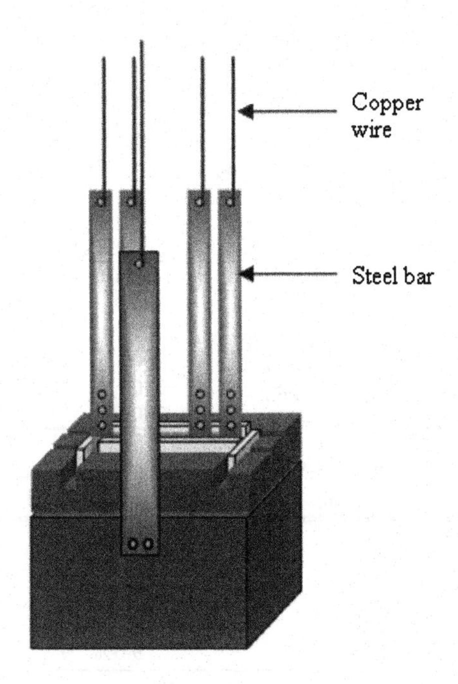


Fig.2 Appearance of experimental apparatus for electrolysis

Electrolysis

As Fig.3 shows, electrolysis was operated at 980°C.

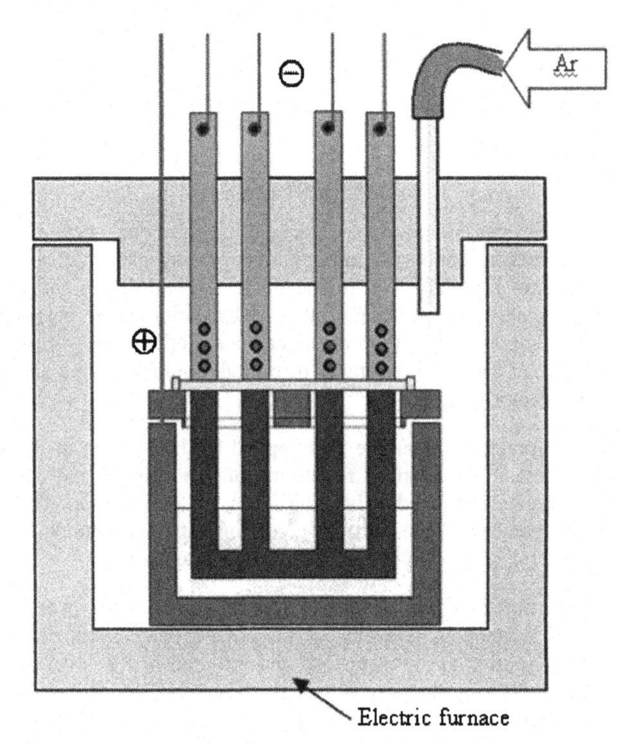


Fig.3 Construction of experimental apparatus

Measurement of electrical resistivity during electrolysis

Cathode sample was set in the graphite crucible, and then mixed salts, $Na_3AlF_6$-$Al_2O_3$-$CaF_2$-NaF were poured into the crucible (anode). After the measurement of electrical resistivity, the crucible was heated to 200°C and kept one night in the furnace. Then the crucible was kept at 400°C for a few hours in order to remove absorbed water and heated to 980°C under Ar atmosphere while measuring electrical resistivity. The current density of cathode was adjusted to be $1.25A/cm^2$. After confirming that the crucible temperature reached 980°C, electrolysis was carried out with several intervals.

Measurement of electrical resistivity after electrolysis

The basic pattern of the experiment was to electrolyze and to interrupt. The length of the series of electrolysis was 5 minutes, 10minutes, 15minutes, 20minutes and 30minutes. The electrical resistivity of cathode sample was observed after each interruption until no further change was observed.

## Results

Electrical resistivity during electrolysis

Fig. 4 shows the change of electrical resistivity during electrolysis in the case of Sample A and Sample B. Before electrolysis, the electrical resistivity of sample A was higher than that of sample B. It is reasonable because the orientation of needle-like coke (Sample B) is relatively high. The electrical resistivity decreased by the progress of electrolysis, and then turned to gradual increase after the minimum resistivity (at about 30 min of electrolysis).

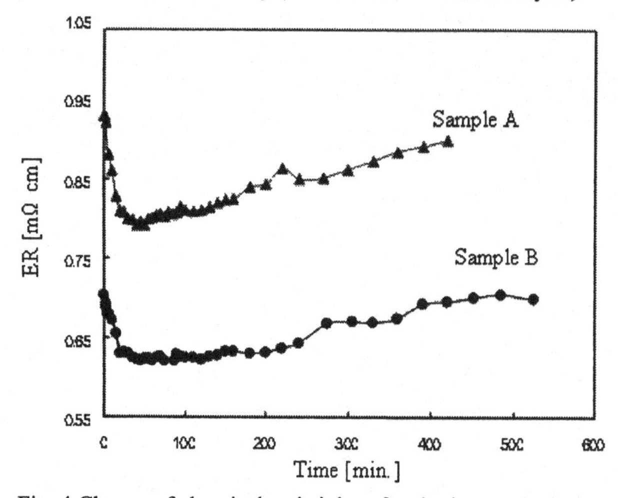


Fig. 4 Change of electrical resistivity of cathode sample during electrolysis in molten salts.

Change of electrical resistivity after interruption of electrolysis

Fig. 5 shows the change of electrical resistivity of sample A, after interruption of electrolysis. At first 5 minutes of electrolysis was carried out and then the electrical resistivity of the cathode sample was measured with time. After reaching a steady value of the electrical resistivity, 2nd experiment (10 min. of electrolysis and electrical resistivity measurement) was followed. And then 3rd (15min. of electrolysis), 4th (20 min. of electrolysis) and 5th (30 min. of electrolysis) were continuously carried out. The horizontal axis of Fig. 5 means the period after electrolysis. We have observed that, by the interruption of the electrolysis, the electrical resistivity increased rapidly with time and it finally returned to the original value, passing through a characteristic plateau region where the electrical resistivity increased very slowly with time. In addition, it was confirmed that the longer electrolysis period, the longer the time needed to recover the original value.

The result of sample B is shown in Fig. 6. The pattern of electrical resistivity change of sample B is essentially similar to that of sample A. However, the value of the electrical resistivity after a series of experiments exceeded the original value.

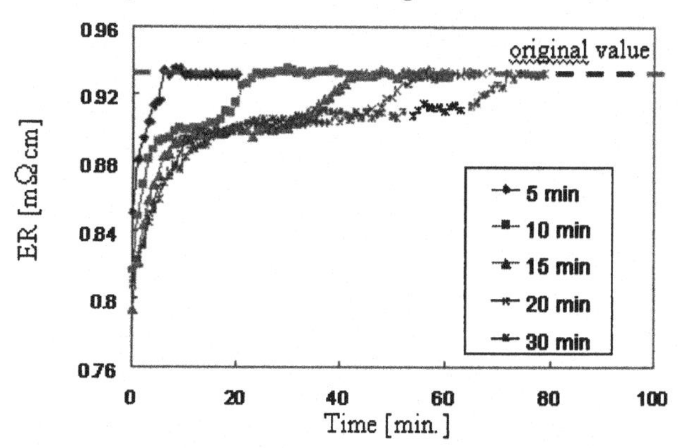


Fig.5 Change of electrical resistivity (ER) of sample A after interruption of electrolysis in molten salts by different electrolysis periods.

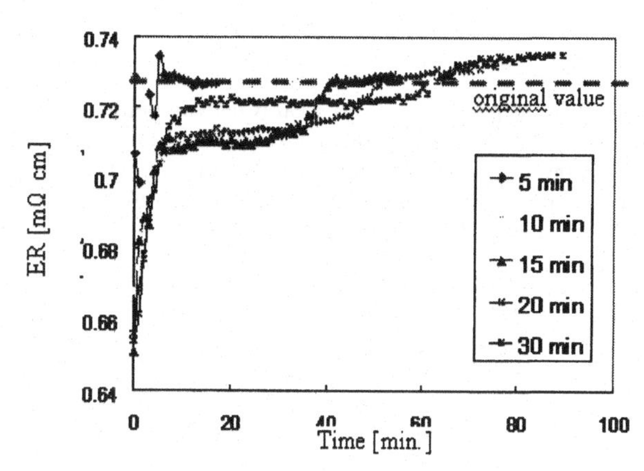


Fig.6 Change of electrical resistivity of sample B after interruption of electrolysis in molten salts by different electrolysis periods.

To study the effect of heat-treatment temperature, the samples C-2000, C-2400 and C-2800 were used, where numerals denote heat-treatment temperature, respectively. The results are shown in Figs.7-9. For all the samples, the behavior was qualitatively the same with those of Sample A and Sample B. It should be noted that sample C-2000, heat-treated relatively at low temperature, shows remarkable low electrical resistivity at the plateau (about 10 % of decrease compared to the original value), as can be seen in Fig.7.

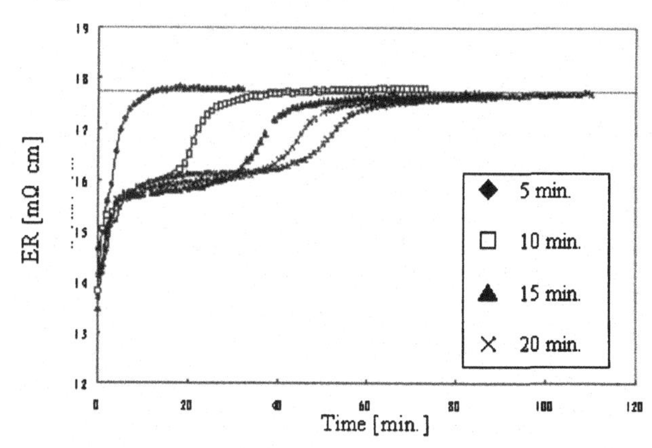


Fig.7 Change of electrical resistivity of C-2000 after interruption of electrolysis in molten salt by different electrolysis periods.

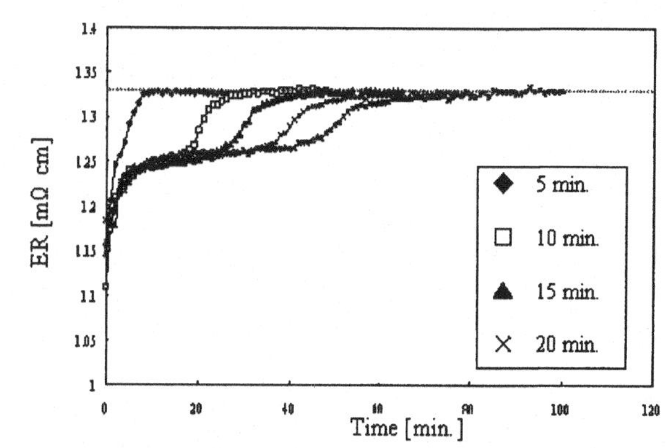


Fig.8 Change of electrical resistivity of C-2400 after interruption of electrolysis in molten salt by different electrolysis periods.

Temperature dependence of electrical resistivity before and after electrolysis.

Temperature dependence of electrical resistivity before and after electrolysis was measured. Only for sample A, a lot of data were taken in order to study temperature dependence in detail. For the other samples, data were taken at RT(20°C), 200°C, 500°C, 700°C and 980°C. The results are shown in Figs.10-14.

As seen in Fig.10, for sample A the electrical resistivity decreased, and then increased with temperature, the minimum electrical resistivity being at around 500°C. At 980°C, the electrical resistivity values of before and after electrolysis were the same. Below 980°C, however,the electrical resistivity after electrolysis is higher than that of virgin sample. For sample B shown in Fig.11, much larger increase of the resistivity by the electrolysis was observed.

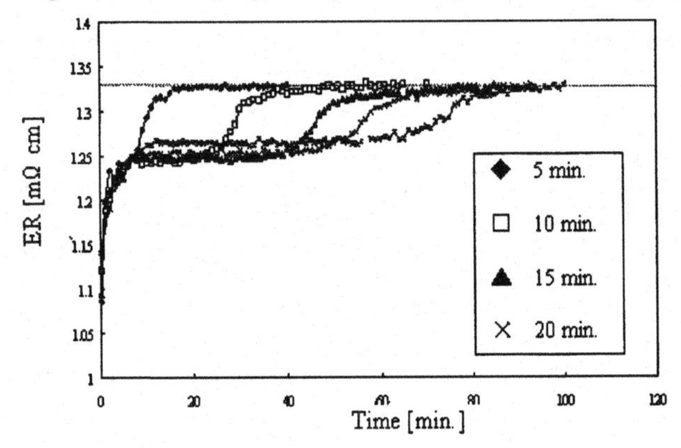


Fig.9 Change of electrical resistivity of C-2400 after interruption of electrolysis in molten salt by different electrolysis periods.

The results for the samples C-2000, C-2400, and C-2800 are shown in Figs.12-14. It can be seen that the degree of decrease of electrical resistivity due to electrolysis is larger for samples having higher HTT.

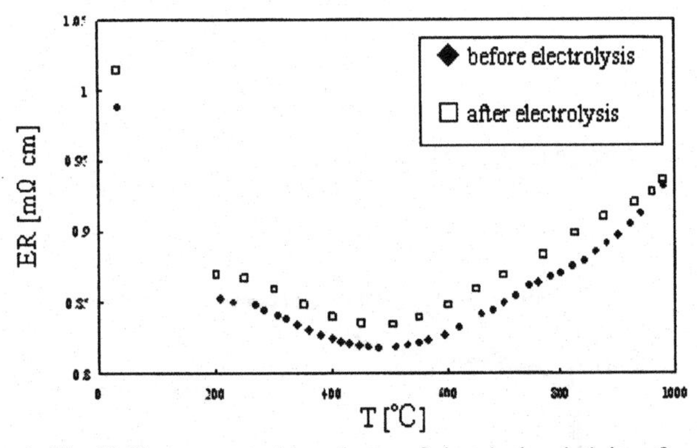


Fig.10 Temperature dependence of electrical resistivity of Sample A before and after electrolysis.

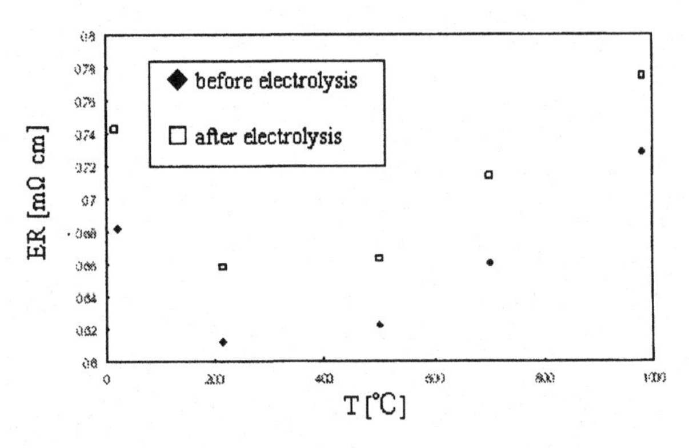


Fig.11 Temperature dependence of electrical resistivity of Sample B before and after electrolysis.

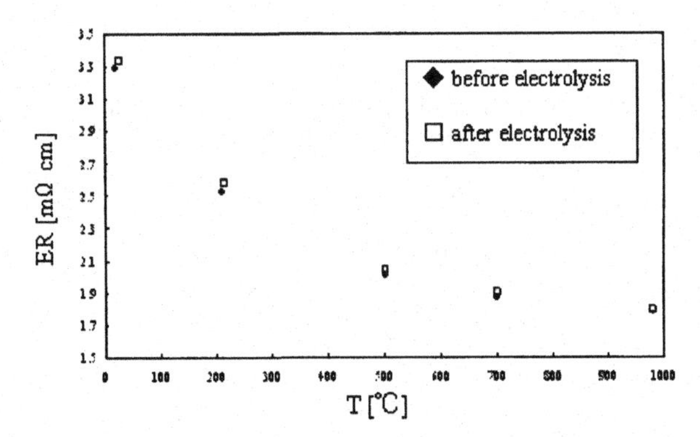


Fig. 12 Temperature dependence of electrical resistivity of C-2000 before and after electrolysis.

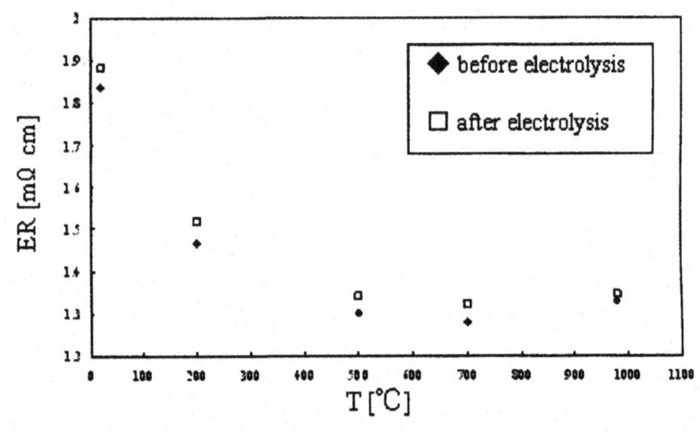


Fig. 13 Temperature dependence of electrical resistivity of C-2400 before and after electrolysis.

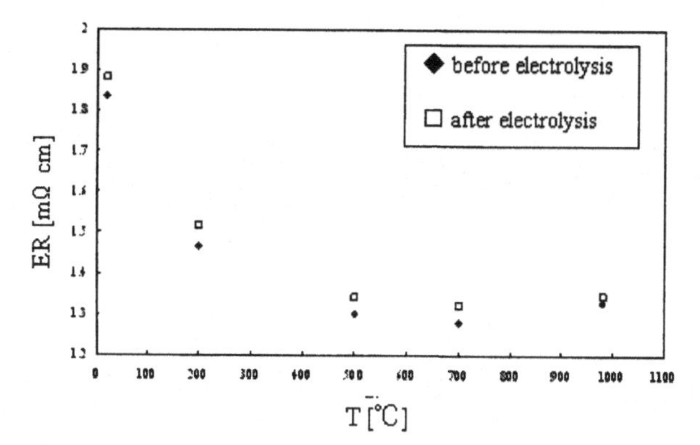


Fig. 14 Temperature dependence of electrical resistivity of C-2800 before and after electrolysis.

## Discussion

Behavior of electrical resistivity increase after electrolysis

In all samples, it was observed that the electrical resistivity decreased by electrolysis and then increased very rapidly by interruption of the electrolysis. Moreover, the longer the electrolysis time, the longer the time needed for electrical resistivity to return to the original value.

From these results, the mechanism can be considered to be as shown in Fig.15. Intercalation of sodium into cathode sample leads to decrease of electrical resistivity. By the interruption of the electrolysis, decomposition of intercalation compounds occured. This causes the increase of electrical resistivity. Finally, by this decomposition, it could be returned to original graphite crystal.

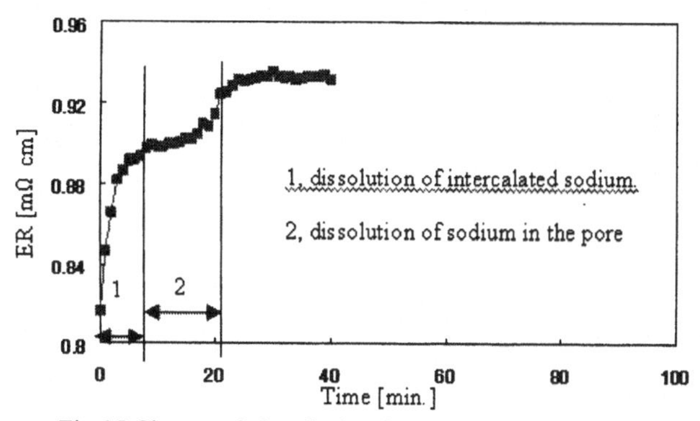


Fig.15 Change of electrical resistivity after electrolysis .

The reason that the electrical resistivity increases through the plateau region is not clear. But Brisson et al. [9] pointed out that there are two kinds of sodium (ionic sodium entered into the layer of graphite crystal and metallic sodium deposited in micro pores of the graphitized cathode) by XPS measurement. Accordingly, the present result may be explained by assuming two phases, i.e., rapidly decomposable and fairly stable ones. Just after the interruption of the electrolysis, the intercalated sodium rapidly dissolves into the molten salts accompanying sudden increase of the electrical resistivity, In contrast, the metallic sodium in pores dissolves relatively slowly in the molten salts.

On the other hand, there may be another explanation which does not take into account the role of sodium. Two types of phases exsist ;Phase I which consists of molten salts with relatively high concentration near the interface between cathode and molten salts, and phase II which consists of relatively low concentration inside the cathode. Phase I decompose rapidly by the interruption of electrolysis firstly and phase III gradually decomposes. A detailed study on characterization of the reactor after electrolysis, investigation applying chemical kinetics, etc. must be conducted to verify this theory.

Temperature dependence of electrical resistivity

The increasing ratio of electrical resistivity for samples A and B after electrolysis is shown in Table 3. Increasing ratio of sample B is higher than that of sample A. The results for the samples C-2000, C-2400 and C-2800 having different HTT are shown in Table 4. Increasing ratio is increasing with HTT. This result is consistent with that of samples A and B. These results suggest that the sample which has low degree of graphitization has high resistance against degradation in case of HTT over 2000°C.

This result can be explained from the point of view of graphite orientation of cathode sample. Samples prepared from needle-like coke or treated at high temperature have high orientation, but samples from lump coke or such treated at low HTT have low orientation of graphite. In electrolysis, if intercalation and de-intercalation of sodium in the molten salts gives rise to destroy the orientation of graphite, it can be explained that cathode sample from lump cokes or treated at low HTT has high resistance against degradation in the electrolysis. In case of the cathode from needle-like coke, degradation is caused rather easily by sodium intercalation due to the high orientation of graphite, while in cathode sample from lump coke, it is difficult to destroy the structure because of its low orientation of graphite.

Table.3 Increase of electrical resistivity after electrolysis (%) for Sample A and Sample B.

| | Sample A 1st | Sample A 2nd | Sample B |
|---|---|---|---|
| 20°C | 2.6 | 2.5 | 8.9 |
| 200°C | 2.0 | --- | 7.5 |
| 980°C | 0.6 | 0.7 | 6.4 |

## Conclusions

Electrical resistivity of the cathode samples decreased by the electrolysis in the molten salts at 980°C. Then it increased rapidlyat first till a characteristic plateau and finally returned to theoriginal value (or a slightly higher value). The results suggest that the intercalated sodium rapidly dissolves into the molten salts, and the metallic sodium in the pore dissolves relatively slowly in the molten salts.

Table.4 Increase of electrical resistivity after electrolysis (%) for Sample C-2000, C-2400, C-2800

| | C-2000 | C-2400 | C-2800 |
|---|---|---|---|
| 20°C | 1.5 | 2.6 | 108.7 |
| 200°C | 2.2 | 3.5 | 91.9 |
| 500°C | 1.8 | 3.2 | 17.7 |
| 700°C | 1.6 | 3.4 | 9.4 |
| 980°C | 0.5 | 1.2 | 12.2 |

Moreover, increasing ratio of electrical resistivity after electrolysis is high in case of cathode samples prepared from needle-like coke or higher HTT (>2000°C).

## Future Works

In order to confirm this conclusion, study on cathode samples heat-treated different HTT and prepared from needle-like coke shall be necessary.

It is also necessary to find more about sodium behavior during electrolysis by ascertaining characterization of cathode after electrolysis, by means of analysis methods such as XRD, Raman spectroscopy, magnetic resistance, EDX , and kinetic observation of the electrolysis reaction.

## References


1) K. Fredenhagen and G. Cadenbach, Z. anorg. Allgem. Chem., 158 (1926), 249-263.

2) D. Guerard and A. Herold, Carbon, 13 (1975), 337-345.

3) R. C. Asher, J. Inorg. Nucl. Chem., 10 (1959), 238-249.

4) A. J. Metrot and A. Herold, J. Chim. Phys. Physicochim. Biol., 1969, No. special: 71-79.

5) O. Tanaike and M. Inagaki, Carbon, 37 (1999), 1759-1769.

6) M. Inagaki and O. Tanaike, Carbon, 39 (2001), 1083-1090.

7) N. Akuzawa et al., Mol. Cryst. Liq. Cryst. to be published.

8) N. Akuzawa et al., Light Metals 2003 (2003)

9) Brisson, P.-Y., Darmstadt, H., Soucy, G., Fafard M. and Servant G. Carbon 2005 (2005)

ELECTRODE TECHNOLOGY

# Cathodes Manufacturing and Developments

*SESSION CHAIRS*

**Ketil A. Rye**
Elkem Aluminium
Mosjoen, Norway

**Morten Sørlie**
Elkem Aluminium ANS
Kristiansand, Norway

**Light Metals 2008** *Edited by: David H. DeYoung*
***TMS (The Minerals, Metals & Materials Society), 2008***

# CONSTITUTIVE LAWS OF CARBONEOUS MATERIALS OF ALUMINIUM ELECTROLYSIS CELL: CURRENT KNOWLEDGE AND FUTURE DEVELOPMENT

Donald Picard[1], Guillaume D'Amours[2], Mario Fafard[1]

[1]Aluminium Research Centre REGAL, Laval University; Quebec, Qc, G1V 0A6, Canada
[2]National Research Council of Canada, Aluminium Technology Centre; Saguenay, Qc, G7H 8C3, Canada

Keywords: Aluminium, Carbon, Constitutive laws, Hall-Héroult, Mechanical behaviour.

## Abstract

The Hall-Héroult cell behaviour at different stages of the electrolysis process is an important point to take into consideration in the design and the optimization the cell. Nowadays, numerical simulation has become a powerful and essential tool since *in situ* measurements are difficult to perform and cost expensive. For those numerical simulations, constitutive laws and their parameters' identification in laboratory are required for all the relevant physics of the cell materials. For the mechanical behaviour of the cell, many efforts have been done to characterize, to understand and to develop constitutive laws for the carboneous materials. Plasticity, viscosity, visco-elasticity, baking, etc., are examples of phenomena which have been addressed up to date, on both transient and steady state situations. This paper presents an overview on various level of modeling of the mechanical behaviour of the carbon cell lining material, from elastic to more complex like a thermo-(chemo)-visco-elasto-plactic one.

## Introduction

Aluminium producers invest time and money in research and development to reach the theoretical minimum energy requirement for the actual Hall-Héroult electrolysis cells [1], i.e. cells based on prebaked anodes. To reach this goal, all the aspects of the aluminium production must be optimized, including the cell materials. Therefore, knowledge of all the relevant physical behaviour of the cell materials is essential over the entire life of the reduction cell.

It is now well known that the start-up phase is crucial and could determine the performance and the life expectancy of the cell. Thus, in the last decade, efforts have been made to optimize it through the use of numerical modeling. Moreover, once the electrolysis process has began in the reduction cell, the prevailing high temperature, molten phases and the overall corrosive environment make *in situ* measurements very difficult to achieve. Therefore, numerical modeling has become an essential tool to evaluate all the phenomena taking place in the cell, from simple ones to the more complexes ones. To be relevant, actual models require the development of three-dimensional multi-physical constitutive laws.

### Overview of the numerical models

The problematic of the Hall-Héroult electrolysis cell being a multi-physical one, a wide range of numerical models have been proposed to assess those physics at different stages of the cell operation. However, most of the available models are developed to solve thermal, electrical, chemical and MHD problems, coupled or not and only few mechanical related models have been proposed. Briefly, Arkhipov *et al.* [2] have proposed a stationary thermo-mechanical and thermo-electrical (uncoupled) behaviour of a Soderberg cell. Sun *et al.* [3] have developed a thermo-chemo-mechanical three-dimensional model where all the material mechanical behaviour were considered linear elastic. Hiltmann *et al.* [4] have proposed a 3D thermo-mechanical model, where thermal load comes from a transient thermal model adjusted with their experimental data. None of those models have coupled all the important physics prevailing in the Hall-Héroult reduction cell at various stages of operation. Therefore, in previous studies of cell design (START-Cuve project), an integrated approach was adopted for the finite element simulation of cell preheating by developing the FE toolbox **FESh**$^{++}$ [5] where all of the important phenomena are coupled. The models developed in this FE toolbox required the integration of all the relevant materials constitutive laws, such as those related to the carboneous materials.

During the last decade, some works, mostly based on the finite element toolbox **FESh$^{++}$**, have been carried out to assess the start-up of the Hall-Héroult cell. While Richard *et al.* [6] have assessed the three-dimensional thermo-chemo-mechanical behaviour of the refractory concrete, Picard *et al.* [7], D'Amours *et al.* [8], and Zolochevsky *et al.* [9] (see also the related references) have oriented their work on the long term mechanical behaviour of the carboneous materials located in the cell lining, namely the carbon cathode blocks and the ramming paste, which ones are the main subjects of the review presented in this paper.

### Constitutive laws of carboneous materials

During the preheating phase, the confined carbon cathode blocks and the ramming paste are subjected to high temperatures inducing stresses, through the thermal expansion phenomenon of the materials. If the tensile or compressive stresses reach critical levels, a mechanical failure, mainly cracking, could occurs and ultimately provoking the end of the cell life prematurely. Therefore, the constitutive laws of the carbon materials must take into account all the relevant mechanical phenomena.

#### Ramming paste

The baked ramming paste has an elastoplastic behaviour similar to the carbon cathode blocks since both materials are made of carbon aggregates and a coal tar pitch binder. D'Amours *et al.* [8] have already detailed the elastoplastic model developed for the carbon cathode blocks. Basically, a failure envelope function of the shear stress and of the hydrostatic pressure was defined as :

$$F(\sigma_m, \rho, r(e,\theta)) = a_f \left[\frac{\rho r}{f_c}\right]^{\alpha_f} + \frac{m_f}{f_c}\left[\sigma_m + \frac{\rho r}{b_f}\right] - 1 = 0 \tag{1}$$

where $\sigma_m$ is the hydrostatic pressure, $\rho$ is deviatoric stress, $m_f$ is the friction parameter, $r(e,\theta)$ is the polar coordinate, $f_c$ is the absolute uniaxial compressive strength and $a_f$, $b_f$ and $\alpha_f$ are constants to be identified.

To compute both the elastic and the plastic strains during the hardening of the carbon cathode, a strength parameter $k$ controlling the evolution of the loading surface up to the failure envelope is introduced and expressed in function of the equivalent plastic strain $\epsilon_p$ and of a ductility parameter $d_h$. The following relation for the strength parameter $k$ is extracted from Etse and Willam [10] :

$$k(\epsilon_p, \sigma_m) = k_0 + (1 - k_0)\sqrt{\frac{\epsilon_p}{d_h}\left(2 - \frac{\epsilon_p}{d_h}\right)} \tag{2}$$

where $\epsilon_p$ is the accumulated equivalent plastic strain in the material and $d_h$ is the ductility parameter, which one is used to take into account the hydrostatic pressure effect on the ductility and the on plastic strain of the material. The ductility parameter is then itself function of the hydrostatic pressure; $d_h = f(\sigma_m)$.

Unlike the carbon cathode blocks, the ramming paste is green at the beginning of the start-up of an aluminium reduction cell. In its current state, the previous model developed for the cathode blocks can't be used for the ramming paste because it doesn't take into account the baking effect of this green material. During the manufacturing of the reduction cell, the ramming paste is compacted to form the peripheral seam of the lining. This compaction allows to the ramming paste to get some mechanical strength. During the preheating of the cell, the ramming paste is baked and several irreversible microstructural changes occur. Weight loss, swelling followed by a shrinkage and increase of the mechanical strength are some macrostructural changes that are observed during the ramming paste baking.

The irreversible baking effects on the evolution of the macrostructural mechanical properties of the ramming paste during the preheating phase have been taken into account through a baking index [11]. The concept of the baking index has already been summarized by Richard *et al.* [11]. Briefly, the baking index evolves proportionally with the uniaxial compressive strength, which one is function of the temperature, and is normalized to 1 at the maximum value of the compressive strength. The evolution of the baking index is shown in Fig. 1 for a specific ramming paste. In the context of cell preheating, some thermal and mechanical parameters are now function of the temperature, such as the thermal expansion of the ramming paste, while other ones are also function of the baking index $\xi$, such as the mechanical strength. Thus, the constitutive law is now a thermo-chemo-elasto-plastic one. In that case and by assuming small additive strains, the total stain rate can be decomposed as :

$$\dot{\epsilon} = \dot{\epsilon}_e + \dot{\epsilon}_p + \dot{\epsilon}_{th} + \dot{\epsilon}_{ch} \tag{3}$$

For the chemo-elasto-plastic component of the constitutive model, confined compression tests have been repeated for four different groups of ramming paste

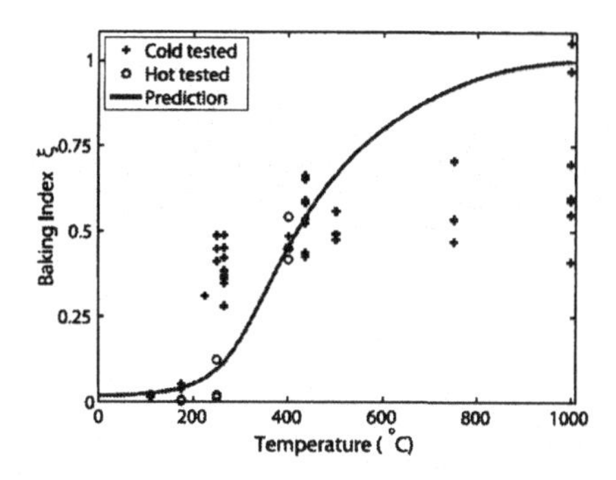


Figure 1: Evolution of the baking index.

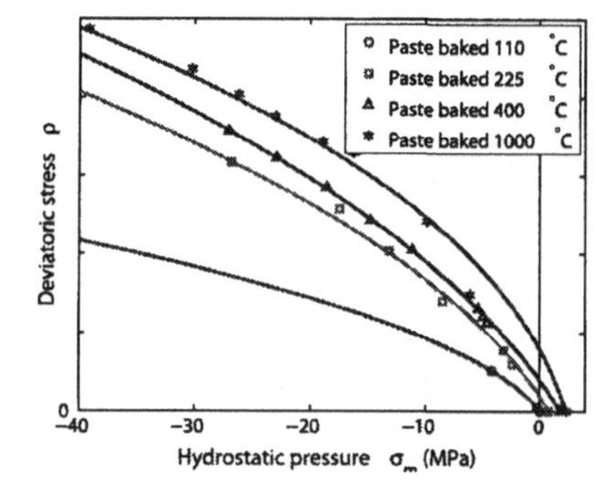


Figure 2: Evolution of the failure envelopes.

samples previously baked at different index levels corresponding to the temperatures of 110, 225, 400 and 1000°$C$. The measured and the approximated compressive meridians of the four different failure envelopes are shown in Fig. 2. To model the increase of the mechanical strength of the ramming paste in function of its baking index with the use of Eq. 1, it is necessary to express the majority of the parameters as a function of the baking index $\xi$. The new equation of the failure envelope thus becomes :

$$F(\sigma_m, \rho, r, k, c, \xi) =$$

$$\left\{(1-k)\left[\frac{\sigma_m}{f_c(\xi)} + \frac{\rho r(\theta)}{b_f(\xi)\, f_c(\xi)}\right]^2 + a_f(\xi)^{\frac{1}{\alpha_f(\xi)}} \frac{\rho r(\theta)}{f_c(\xi)}\right\}^{\alpha_f(\xi)}$$

$$+ \frac{k^{\beta_f(\xi)} m_f(\xi)}{f_c(\xi)}\left[\sigma_m + \frac{\rho r(\theta)}{b_f(\xi)}\right] - k^{\beta_f(\xi)} c = 0 \qquad (4)$$

Each one of the six different parameters in the previous equation, which are function of the baking index $\xi$, is

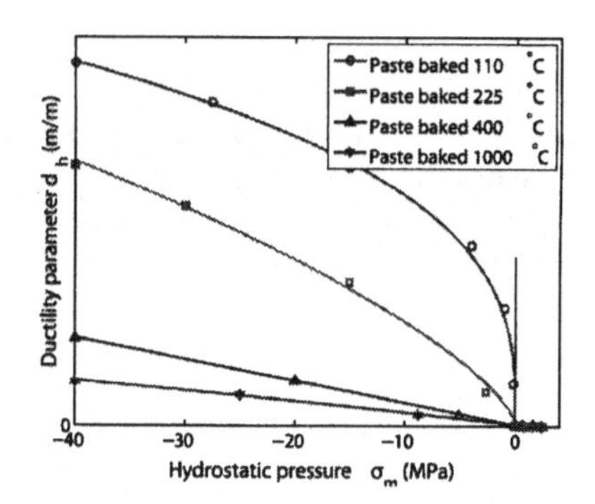


Figure 3: Evolution of the ductility curves.

expressed mathematically by a linear or quadratic function allowing the model to approximate the measured envelopes shown in Fig. 2.

A similar dependency of the ductility parameter $d_h$ towards both the hydrostatic pressure $\sigma_m$ and the baking index $\xi$ has to be defined to allow the elastoplastic model to reproduce adequately both the strains and the stresses of the ramming paste during a preheating. From the same uniaxial and confined compression tests performed on the ramming pastes previously baked at four different baking indexes, different curves representing the relationship between the ductility parameter $d_h$ and the hydrostatic pressure $\sigma_m$ have been measured and approximated. These four curves are shown in Fig 3. This figure firstly shows that the ramming paste is highly ductile when it is still green and when there is a compressive load applied. The figure also shows that the baking level has almost no influence when a tensile load is applied on a ramming paste sample; this one will always break in a brittle manner. Figure 3 finally shows that there is a decrease of the ductility when the ramming paste is baked.

Cathode

The elastoplastic constitutive law developed for the baked ramming paste by D'Amours *et al.* [8] can assess the hardening, softening and plastic strain of carbon materials. However, the long-term visco-elastic (creep/relaxation) behaviour was not taken into account. Therefore, Picard *et al.* [7] have proposed a consistent three-dimensional visco-elastic model, based on a thermodynamic framework to identify the model parameters previously determined with a phenomenogical approach, i.e. a rheological model that has been extended to the 3D case shown in Fig. 4. Instead of being defined by scalars, the parameters of the model are fourth order tensors and both the strains and the stress are second order tensors. Moreover, the number of Kelvin-Voigt elements constituting the proposed rheological model is, *a priori*, undetermined. The model has been developed with the

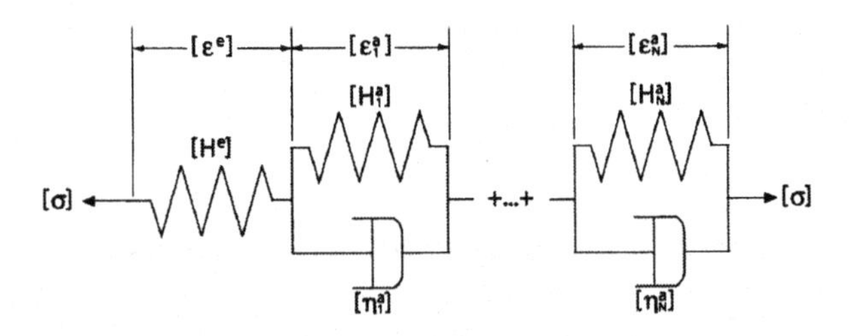


Figure 4: Three-dimensional Kelvin-Voigt rheological model. From [7].

hypothesis that the mechanical strains/stresses have no influence on the thermal, chemical, etc. variables. Thus, the mechanical problem can be uncoupled from the thermal and the chemical ones. The dependencies of external parameters (e.g. temperature) will then be taken into account in the finite element model throughout algebraic functions, as discussed later. Moreover, the carbon cathode has been assumed to be an isotropic material. Thus, the topology of all the positive semi-definite stiffness fourth order tensors of the rheological model can be assumed to be similar to the elastic one

$$\begin{aligned}[H^e] &= [H^e(E,\nu)] \\ [H^a_\alpha] &= [H^a_\alpha(E_{H^a_\alpha},\mu_\alpha)] \\ [\eta^a_\alpha] &= [\eta^a_\alpha(E_{\eta^a_\alpha},\varsigma_\alpha)]\end{aligned} \quad (5)$$

where $\alpha$ is the corresponding Kelvin-Voigt element, $E$ is the Young's modulus and $\nu$ is the Poisson's ratio. The parameters $E_{H^a_\alpha}$ and $E_{\eta^a_\alpha}$ are similar to a Young modulus and $\mu_\alpha$ and $\varsigma_\alpha$ to a Poisson's ratio. Considering a sustained uniaxial compressive load with free lateral stress, the stress-strain relations proposed by Picard *et al.* [7] are

$$\begin{aligned}\epsilon_1 &= -\frac{\sigma_1}{E_{H^e}} - \frac{\sigma_1}{9}\sum_{\alpha=1}^{N} K^{a^{-1}}_\alpha \frac{1-e^{-\lambda_{\alpha_1}t}}{\lambda_{\alpha_1}} \\ &- \frac{\sigma_1}{3}\sum_{\alpha=1}^{N} G^{a^{-1}}_\alpha \frac{1-e^{-\lambda_{\alpha_2}t}}{\lambda_{\alpha_2}}\end{aligned} \quad (6)$$

$$\begin{aligned}\epsilon_r &= \frac{\nu\sigma_1}{E_{H^e}} - \frac{\sigma_1}{9}\sum_{\alpha=1}^{N} K^{a^{-1}}_\alpha \frac{1-e^{-\lambda_{\alpha_1}t}}{\lambda_{\alpha_1}} \\ &+ \frac{\sigma_1}{6}\sum_{\alpha=1}^{N} G^{a^{-1}}_\alpha \frac{1-e^{-\lambda_{\alpha_2}t}}{\lambda_{\alpha_2}}\end{aligned} \quad (7)$$

where

$$K^{a^{-1}}_\alpha = \frac{3(1-2\varsigma_\alpha)}{E_{\eta^a_\alpha}} \quad ; \quad G^{a^{-1}}_\alpha = \frac{2(1+\varsigma_\alpha)}{E_{\eta^a_\alpha}} \quad (8)$$

and

$$\lambda_{\alpha_1} = \frac{E_{H^a_\alpha}(1-2\varsigma_\alpha)}{E_{\eta^a_\alpha}(1-2\mu_\alpha)} \quad ; \quad \lambda_{\alpha_2} = \frac{E_{H^a_\alpha}(1+\varsigma_\alpha)}{E_{\eta^a_\alpha}(1+\mu_\alpha)} \quad (9)$$

where $-\sigma_1$ is the constant compressive axial stress, $\epsilon_1$ the corresponding strain and $\epsilon_r$ the hoop strain. All other strain components are null. The positive coefficients $K^{a^{-1}}_\alpha$ and $G^{a^{-1}}_\alpha$, defined in Eq. 7, are related to the hydrostatic and deviatoric creep mechanisms respectively. The coefficients $\lambda_\alpha$, defined in Eq. 8, are also positive and are related to a relaxation time. Combined with experimental data [7], Eq. 6 and 7 are used to identify the parameters of all the fourth order tensors of Eq. 5 by the means of a least squares method through the use of a genetic algorithm available in Matlab. Both the axial and radial data must be used to identify the parameters [7]. Figure 5 shows an example of parameter identification results, based on a rheological model with 2 Kelvin-Voigt elements (Fig. 4). The experimental creep strains data are those of a virgin (non contaminated by the electrolysis process) semi-graphitic carbon cathode samples at room temperature. In the presented case, the model with two Kelvin-Voigt elements well represents the global creep behaviour of the material in both directions and the values of all the corresponding parameters are listed in Table I. The number of elements can be increased or lowered, if necessary, for other materials.

The constitutive laws proposed in this paper for the ramming paste and for the carbon cathode have been combined together in the finite element toolbox **FESh$^{++}$** to create a quasi-brittle thermo-visco-elasto-plastic constitutive law for the baked carbon cathode material. As mentioned earlier, the visco-elastic constitutive law [7] has been developed on the basis that the mechanical problem can be uncoupled from the thermal and the chemical

Table I: Semi-graphitic carbon cathode material parameters of a 3D rheological model with 2 Kelvin-Voigt elements

| | $E_{\eta_\alpha^a}$ | $\varsigma_\alpha$ | $E_{\eta_\alpha^a}$ | $\mu_\alpha$ |
|---|---|---|---|---|
| First element | 840306 | 0.07 | 712153 | 0.13 |
| Second element | 7562665 | 0.20 | 689502 | 0.39 |

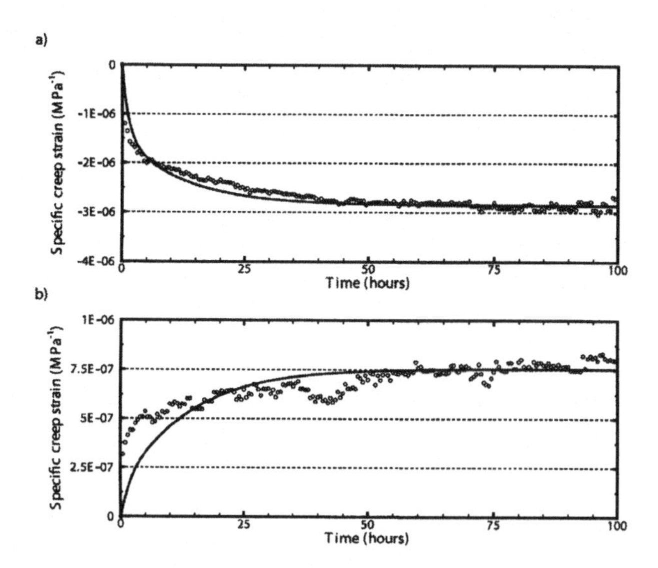


Figure 5: Parameter's identification results for the semi-graphitic carbon cathode material. a) Axial strain, b) Radial strain. o Experimental, — Model. From [7].

ones. Therefore, the proposed visco-elastic constitutive law can be used to identify the parameters of experimental creep tests at various temperatures. Thereafter, with the help of an interpolation method (polynomial, exponential, etc.), the visco-elastic parameters can be redefined with algebraic equations which are function of temperature. A similar procedure can be done to take into account the dependencies towards other parameters, such as the baking index [11], the chemical contamination level of carbon material, etc.

However, experimental data are still lacking to determine all the parameter dependencies for the proposed models. Also, most experimental data related to the mechanical behaviour of the carbon cathode blocks available so far under different conditions [9] are mostly uniaxial, which is insufficient for the three-dimensional visco-elastic parameters identification [7].

## Future development

The lining materials of the Hall-Héroult cell are at room temperature and then heated up before the bath pouring. Also, the carbon materials will be contaminated by different chemical species during the electrolysis process. The parameters of the constitutive laws proposed for the ramming paste and for the carbon cathode materials have been identified at room temperature. In fact, to characterize the ramming paste, the material was first baked at the desire baking index then cooled at room temperature before being tested. In both cases (paste and cathode), the lacking of high temperature experimental data is due to apparatus restrictions. Moreover, preliminary numerical results obtained within the START-Cuve project have shown the importance of taking into account the visco-elastic behaviour of the ramming paste during the preheating phase.

Therefore, in the upcoming months, new equipments developed in collaboration with manufacturers will be acquired. Those equipments will allow the characterization of the long term creep/relaxation behaviour of the carbon cathode materials and of the ramming paste (previously baked at various level), at room temperature up to 1000°C. The high temperature elastoplastic behaviour of baked carbon materials will also be studied. Moreover, the new equipments have been designed to test 101.4 mm diameter carbon material cylinder under a maximum compressive load of 35 MPa at 1000°C. Other equipments will also be developed to perform long term creep test under electrolysis on those relatively large samples.

The experimental data that will be gathered with the upcoming equipments will allow the establishment of a thermo-chemo-visco-elasto-plastic constitutive law for the carbon cell lining materials able to take into account all the relevant mechanical behaviours under the wide range of cell operation conditions. The laws are also planned to be adapted to the carbon anode.

## Conclusion

Understanding the mechanical behaviour of the carbon materials located in the cell lining is critical in order to avoid early mechanical failure, mainly during the pot start-up. Thus, experimental works were required to assess the behaviour of those materials, namely the ramming paste and the carbon cathode blocks, and for the development of three-dimensional constitutive laws. The 3D visco-elastic constitutive law recently developed by Picard *et al.* [7] has been combined, in the finite element toolbox **FESh**$^{++}$, with the thermo-chemo-elasto-plastic one previously proposed by D'Amours *et al.* [8] for the baked ramming paste. The new law is ready to use to run transient non linear thermo-chemo-mechanical numerical simulations.

Future experimental works are needed to identify the model parameters under various conditions such as at high temperature under electrolysis and at low baking index (ramming paste). Finally, upcoming works should also assess the carbon anode by using similar thermo-chemo-mechanical constitutive laws.

## Acknowledgments

The Natural Sciences and Engineering Research Council of Canada (NSERC), the Fonds québécois de la recherche sur la nature et les technologies (FQRNT), Alcan International Limited and the Aluminium Research Centre - REGAL are gratefully acknowledge for their financial support.

## References

[1] Choate, W. T., and Green, J. A. S., 2003. U.S. Energy requirements for aluminium production: historical perspective, theoretical limits and new opportunities. Tech. rep., U.S. Department of Energy, Energy Efficiency and Renewable Energy, Washington, D.C.

[2] Arkhipov, G. V., 2004. "Mathematical modeling of aluminium reduction cells in "russian aluminum" company". In Light Metals 2004: Proceedings of the 133rd TMS Annual Meetings, TMS Light Metals, The Minerals, Metals and Materials Society, pp. 473–478.

[3] Sun, Y., Forslund, K., Sørlie, M., and Øye, H., 2004. "3-D modelling of thermal and sodium expansion in soderberg aluminium reduction cells". In Light Metals 2004: Proceedings of the 133rd TMS Annual Meetings, TMS Light Metals, The Minerals, Metals and Materials Society, pp. 587–592.

[4] Hiltmann, F., and Meulemann, K.-H., 2000. "Ramming paste properties and cell performance". In Light Metals 2000: Proceedings of the 129th TMS Annual Meetings, TMS Light Metals, The Minerals, Metals and Materials Society, pp. 405–411.

[5] Désilets, M., Marceau, D., and Fafard, M., 2003. "Start-cuve: Thermo-electro-mechanical transient simulation applied to electrical preheating of a Hall-Héroult cell". In Light Metals 2003. Proceedings of the 132nd TMS Annual Meeting, TMS Light Metals, The Minerals, Metals and Materials Society, pp. 247–254.

[6] Richard, D., Fafard, M., and Désilets, M., 2003. "Thermo-chemo-mechanical aspects of refractory concrete used in a Hall-Héroult cell". In Light Metals 2003: Proceedings of the 132nd TMS Annual Meetings, TMS Light Metals, The Minerals, Metals and Materials Society, pp. 283–290.

[7] Picard, D., Fafard, M., Soucy, G., and Bilodeau, J.-F., 2007. "Three-dimensional constitutive creep/relaxation model of carbon cathode materials". *Transactions of the ASME, Journal of Applied Mechanics (In Press).*

[8] D'Amours, G., Fafard, M., Gakwaya, A., and Mirchi, A., 2003. "Mechanical behavior of carbon cathode : understanding, modeling and identification". In Light Metals 2003: Proceedings of the 132nd TMS Annual Meetings, TMS Light Metals, The Minerals, Metals and Materials Society, pp. 633–639.

[9] Zolochevsky, A., Hop, J., Foosnaes, T., and Øye, H., 2005. "Surface exchange of sodium, anisotropy of diffusion and diffusional creep in carbon cathode materials". In Light Metals 2005. Proceedings of the 134th TMS Annual Meeting, The Minerals, Metals and Materials Society, pp. 745–750.

[10] Etse, G., and Willam, K., 1994. "Fracture energy formulation for inelastic behavior of plain concrete". *Journal of engineering mechanics,* **120**(9), pp. 1983–2011.

[11] Richard, D., D'Amours, G., Fafard, M., Gakwaya, A., and Désilets, M., 2005. "Development and validation of a thermo-mechanical model of the baking of ramming paste". In Light Metals 2005. Proceedings of the 134th TMS Annual Meeting, TMS Light Metals, The Minerals, Metals and Materials Society, pp. 733–738.

**Light Metals 2008** *Edited by: David H. DeYoung*
***TMS (The Minerals, Metals & Materials Society), 2008***

# CATHODE HOT PATCHING TO PROLONG CELL LIFE AT DUBAL

Maryam Mohamed Khalifa Al-Jallaf, Ali. H. A. Mohammed, Joseph Antony,
Dubai Aluminium Company Limited, PO Box 3627, Dubai, U.A.E

Keywords: Cathode, Hot Patching, Corundum

## Abstract

The deterioration of cathode blocks may occur by general erosion, characterised by 'W' wear in more graphitised materials and by localised erosion, also known as potholing. Pothole failure is predominant in some cell technologies due to high amperage or use of certain cathode grades. The three main factors that contribute to pothole formation are weak areas in cathode, high current density and increased metal pad velocity.

Until 2003, Dubal focussed on using fused alumina, corundum, to hot patch the potholes in order to cease iron attack, thereby prolong cell life. The cost-benefit analysis was derived in terms of key performance indicators, labour and corundum cost to determine the break even point of extended cell life in a patched cell.

This paper summarises Dubal's experience with fused alumina hot patching in D18 cell technology with respect to stall location and describes the cost-benefit analysis.

## Introduction

The degradation of cathode blocks are attributed to general or localised erosion i.e. pothole, which often leads to cell failure. The three main contributors for pothole formation are weak areas in the cathode, high cathode current density and high metal pad velocity [1]. The weak areas could be created during the cathode production stages (manufacturing, handling, storage, etc) or during pot operation (pre-heat, start up and operation). The potential causes for high current density are anode spikes, alumina deposit on cathode surface and aluminium fingers. The third factor is due to localised metal velocity at the ledge toe; leading to an increase of physical abrasion of carbon and resulting in carbide dissolution [2]. The wear of cathode occurs due to three main mechanisms which are physical erosion, chemical corrosion and electrochemical corrosion [3]. Physical abrasion is attributed to metal pad movement caused by high magnetic field. The metal movement becomes crucial if the cell design is not magnetically compensated and if the cathode design is poor [4]. Chemical corrosion is the continuous formation and dissolution of aluminium carbide [4]. A combination of physical and chemical corrosion has a great impact on cathode degradation [5]. The formation of aluminium carbide is through electrochemical reaction rather than chemical reaction as a function of current density. The following equations represent the electrochemical reaction [3]:

$$4Al^{3+} + 3C + 12e^{-} \rightarrow Al_4C_3 \quad (1)$$

$$4Al \rightarrow 4Al^{3+} + 12e^{-} \quad (2)$$

In general, the degradation of blocks can not be prevented as the cell ages. Cathode failure mechanism in Dubal D18 cell technology is predominantly due to physical abrasion caused due to high metal pad velocities. However Dubal has experience in using fused alumina (corundum) to plug the potholes generated in the cathodes and extend cell life. This paper summarises Dubal experience with fused alumina hot patching.

## Fused Alumina (Corundum)

During the calcination stage of Bayer process, the gibbsite $Al(OH)_3$ gets converted to different crystallographic forms of transition alumina such as delta theta, gamma, etc [4]. A complete transformation of the intermediate phases at temperature > 1150°C results in formation of alpha alumina ($\alpha Al_2O_3$) [6] or corundum. Alpha alumina has a low rate of dissolution in the electrolyte, better thermal insulating characteristics and higher angle of repose. Smelter grade alumina usually contains 8-15 wt % ($\alpha Al_2O_3$) which helps to improve the quality of crust formation by forming an interlocking matrix [4, 6]. Heat treatment of alumina or bauxite at temperature of above 2000°C results in a product called fused alumina OR also called corundum in the industry. It has poor thermal conductivity and high density 3.8 g/cm$^3$ [7]. Refer to table 1 for the chemical composition and physical properties of corundum [8]. At DUBAL corundum is used in hot patching the cells experiencing cathode attack (generally detected by a sharp increase in iron content in pot metal).

**Table 1: Chemical and physical analysis [8]**

| Composition, Chemical Structure | | Limits – Unit |
|---|---|---|
| Total Alumina, | $Al_2O_3$ | 95.0 wt % Max. |
| Alpha Alumina, | $\alpha Al_2O_3$ | 85.0 wt % Min. |
| Silica Oxide, | $SiO_2$ | 1.0 wt % Max. |
| Titanium Oxide, | $TiO_2$ | 3.0 wt % Max. |
| Iron Oxide, | $Fe_2O_3$ | 0.20 wt% Max. |
| Magnesium Oxide, | MgO | 0.50 wt % Max. |
| Real Density | | 3.95 g/cm$^3$ Min |
| Moisture (0-300°C) | | 0.5 wt% Max. |

The particle size distribution required for the hot patching is represented in table 2:

**Table 2: Corundum Screen Analysis [8]**

| Screen Analysis (Cumulative) | Limits wt % | Comments |
|---|---|---|
| +4 mm | 0 | |
| +2 mm | 20-40 | Preferred |
| +1 mm | 99.0 | Minimum |
| - 1 mm | 1.0 | Maximum |

The most important quality criteria for accepting or rejecting corundum are the consistency of particle size distribution and alpha alumina content.

### Methodology of Cathode Hot Patching

Cells will be eventually cut out due to several possible reasons such as high iron, high cell voltage, upward cathode heaving or tap out [6]. The high iron cells can potentially be salvaged through the process of cathode hot patching. The process starts with identifying the cells with abnormal increase in the iron content due to direct contact of collector bars with metal or electrolyte. These cells would have an increase in collector bar temperature causing red bars and susceptible to tap out provided no action is taken.

The acknowledged procedure for preventing cell failure due to high iron is called hot patching and carried out as follows [9]:

1. Lift the anode stall to expose the cavity.
2. Use L shaped dip rod to scan the cathode surface for any cracks, as shown in Figure 1.
3. Repeat on other stalls until the attack location is identified.

**Figure 1: Scan cathode surface with L shaped Rod.**

4. Lower a bag of corundum weighing 20 kg above the attack location using pot skimmer tool or using newly developed assembly as shown is Figure 2.

**Figure 2: Discharge of corundum in the cavity.**

5. Guide the corundum particles into the cathode crack using the L rod.
6. Lower the anode block into the cavity to apply the pressure on the corundum for better cathode plugging.
7. Replace the anode.

In order to maintain a safe work practice, several safety precautions were ensured such as preheat of L Shaped Rod prior to cathode inspection, maintain a small opening in the pot hood to eliminate injuries, full usage of PPE to minimise direct exposure to bath, metal and radian heat and preheat the corundum before addition.

During the year 1997-1998, the rate of successful hot patching was recorded at 46 % out of a population of 160 pots, which were detected as on cathode iron attack. The cell life was extended by an average of 233 days thus saving 17000 cell days through hot patching, refer to Figure 3:

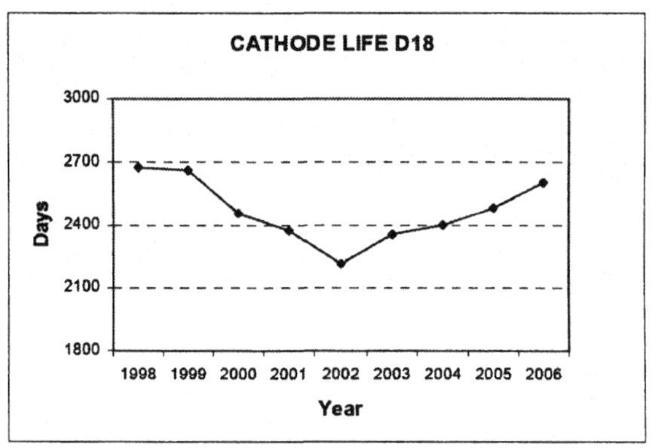


**Figure 3: Cathode average life in D18 cell technology [11].**

Moreover the gain in cell life due to hot patching has enhanced from about 233 days to 389 days by improving the effectiveness of hot patching. However the remaining pots were cut out due to unsuccessful patching [10]. The most common locations for heavy erosion and cathode attack in D18 cell technology were adjacent to anode; stall locations 1, 2, 3, 7, 8, 9, 10, 11 and 12; for age group pots of 1500-2100 days. Where as older pots with age > 2350 days were severely eroded at anode stall locations 1, 2, 8 and 17. Pots older than 2550 days were not considered economical for cathode hot patching [9]. The excessive erosion at the defined locations was due to high metal velocity of about 15-19 cm/s because these pots are not magnetically compensated. Moreover the erosion was rigorously escalated due to the chemical and the electro-chemical corrosion [3].

The number of cut out cells due to iron attack over the period of 1998-2006 is presented in Figure 4:

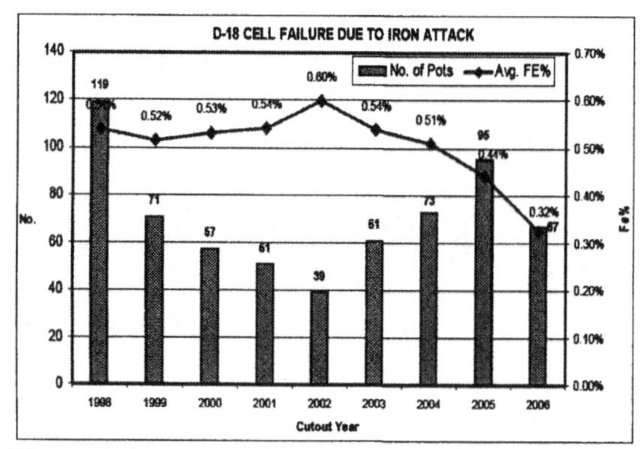


**Figure 4: D18 cell failure due to cathode Iron attack from the year 1998-2006 [12].**

## Techno-Economic Evaluation

The evaluation was carried out comparing the cost of hot patching, cell re-lining and average cell age at high iron between the 1998 and 2006. The results indicated significant offset in cost, refer to table 3 and 4.

**Table 3: Hot patching cost per D18 cell**

| Hot Patching Variable Cost | 1998 USD | 2006 USD | Difference |
|---|---|---|---|
| Energy Consumption | 90 | 81 | -9 |
| Current Efficiency loss | 3,351 | 6,567 | 3,216 |
| Corundum | 1,253 | 1,000 | -253 |
| Metal Analysis & Labour | 545 | 1,090 | 545 |
| **Total** | **5,239** | **8,738** | **3,499** |

**Table 4: Total pot relining and start-up cost**

| Pot relining cost (USD) | Year 1998 | Year 2006 |
|---|---|---|
| Lining materials | 55,041 | 68,096 |
| Shell Repair & Deck plates | 1,907 | 4,790 |
| Labour | 5,995 | 2,217 |
| Energy (pre-heat) | 435 | 435 |
| 72 hrs production loss | 1,907 | 5,552 |
| **Total** | **65,285** | **81,091** |

Hot patching cost has increased by 67% in 2006 compared to 1998 as shown in table 4, mainly contributed by increase in metal price. Total relining and start-up cost increased by 24% in 2006 compared to 1998, as shown in table 4.

Table 5 provides a cost-benefit analysis between 1998 and 2006.

**Table 5: Cost Benefit Analysis of Hot Patching**

| Cost | 1998 (USD) | 2006 (USD) |
|---|---|---|
| Pot relining | 65,285 | 81,091 |
| Average life | 2550 | 2508 |
| Cost per day | 25.6 | 32.3 |
| Hot Patching | 5,239 | 8,738 |
| **Breakeven** | **205** | **271** |

In 1998, the extended life through hot patching was 233 days compared to 389 days in 2006. The average gain in extended life was still higher than the breakeven. Life extension has a positive impact on spent pot lining generation as well as treatment cost.

## Discussion

The hot patching success rate and the gain in cell life are limited at cathode age above 2550 days. Because at higher cathode age, excessive erosion is covering large surface area which leads to increase hot patching cost i.e. material, effort, labour, etc along with higher risk of tap out.

The hot metal cost in 1998 was at $ 777 per tonne compared to an average cost of $1343 per tonne during the past three years. This was attributed directly to the increased cost of energy, alumina, etc over the years. This eventually leads to an increase to the attributed cost of hot patching.

Even though the cell life was increased but the overall involved cost was increased 67%. The process of patching was still profitable as the costs becoming explicitly high. However hot patching is also laborious and hazardous job most smelters do not apply this method mainly due to the economics and labour costs.

## Conclusion

The extension of cell life through cathode hot patching was economically justifiable during 1998. A recent evaluation confirms this finding with an advantage of 118 pot days beyond the break even point.

About 54% of high iron pots could not be successfully patched despite the effort and the material invested. However the techno-economical evaluation indicated feasibility of the hot patching process despite the increase in cost in terms of cathode lining, labour, energy and current efficiency deficiency cost. Therefore the cathode hot patching was still recommended to continue.

## Acknowledgement

The authors gratefully acknowledge the guidance, support and encouragement of Dubal management in carrying out this project and allowing sharing Dubal's experience globally. We also acknowledge contributions from associate professor Dr. Margaret Hyland and Dr. Venkatasubramaniam K. G. in reviewing the report.

## References

1. Cathodes in Aluminium Electrolysis-Morten Sorlie and herald A.Oye-2nd addition – pages 344-346

2. Light Metals 2005 – Eng Fui Siew et al – pages 757-762

3. Light Metals 2005 – Pretesh Patel et al – pages 763-769

4. Introduction to Aluminium Electrolysis Aluminium Smelter Technology – K. Grjotheim and H. Kvande – pages 179-184, 143, 63-68 and 75

5. Light Metals 1998 – Xian-an Liao and Harald A. Oye- pages 667-674

6. Theory Aluminium Smelter Technology – K. Grjotheim and B. Welch – pages 30-32, 35-36 and 142

7. http://www.kematechnik.hu/info/09_en.pdf

8. Dubal's Material Specification – MTSPE074

9. Report no. CIT 46/04/98-Cathode Hot Patching Method Improvement

10. Ref. Efficiency of cathode hot patching – Report CRS/Al/468)

11. Dubal Historical Data in Process Control potrooms archive system

12. Dubal Historical Data in Process Control Cell Lining archive system

# HIGH-PERFORMANCE PREPARATION PLANT FOR CATHODE PASTE

Berthold Hohl[1], Valery V. Burjak[2]

[1]Maschinenfabrik Gustav Eirich GmbH & Co KG, Walldürner Straße 50, 74736 Hardheim, Germany
[2]OAO Ukrgrafit, Severnoje Shosse 20, 69600 Zaporozhye, Ukraina

Keywords: Cathode Paste Preparation, Intensive Mixer

## Abstract

In the course of a modernization program, the Ukrainian company Ukrgrafit has converted part of their paste preparation line for cathodes and carbon blocks from conventional mixing technology to intensive mixing.

The new system was put into operation in the middle of 2006. Apart from the mixer, it also comprises heating of dry aggregates, feeding of pitch, addition of stearin, cooling of the mix in the mixer, intermediate storage of the finished mix, transport of the mix to the presses as well as the entire automation. The substantially improved homogeneity of the mix as well as the complete documentation of the preparation process clearly refined the quality of the final products.

The paper describes the criteria leading to the decision on the system as well as the realization of the task definition. Moreover, first operating results are presented.

## Introduction

The origins of today's Ukrgrafit works trace back to the thirties of last century. Initially, baked anode blocks for the aluminum electrolysis were produced. Since about 1970 Ukrgrafit have been able to produce a comprehensive range of carbon graphite products. In the year 1994 the company was converted to the open joint-stock company "Ukrgrafit".

With a workforce of approx. 3500 members mainly graphite electrodes, cathode blocks, carbon bricks for blast furnaces as well as metallurgical carbon bodies for the nearby aluminum smelter ZALK and for the production of ferroalloys are being produced today. The total production of carbon products is 150,000 t/year.

## Initial Situation

In production hall no. 1 various moldings of amorphous carbon, essentially cathode blocks, are produced by two hydraulic presses. For paste preparation originally mixing machines of the Soviet make "Anod-4" were used. A total of 10 of these machines provided the two presses (3550 t and 6300 t) with carbon paste via a complicated system of conveying systems and cooling drums.

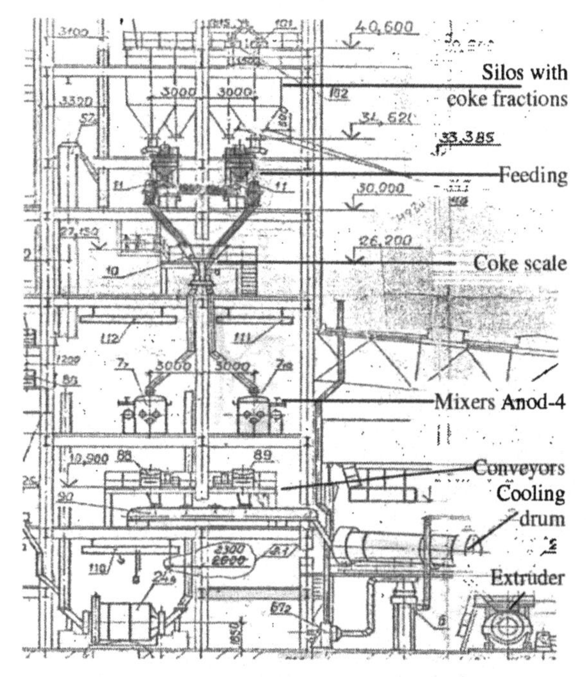


Figure 1. Existing preparation system in hall 1

The existing mixing and cooling system might have allowed regular production, however, at high process costs and varying product properties. In order to achieve a distinct improvement, Ukrgrafit decided to use the Eirich mixer for modernization.

The aims of modernization were in particular:

- Constantly high homogeneity
- Friable consistency of the mix without big agglomerates ("balls")
- Exact temperature control
- Documentation of all operating data
- Replacement of a large number of individual mixing machines by a single high-performance mixer

### Kind and Extent of Modernization

The most essential steps of modernization were:

- Dismantling a group of 4 existing mixers
- Adaptation of the building to the new equipment
- Installation of the new machines: coke heater EWK 09/41 + mixer-cooler DW29/4 + table feeder UE25
- Installation of new belt conveyors for transporting the paste to the presses
- Connection of the new system with the existing infrastructure
- Connection to the extracted air disposal system
- Installation and cabling of the new PLC control system and the power supply for the coke heater

Modification works started in January 2006. Since July 2006 the system has been fully in operation.

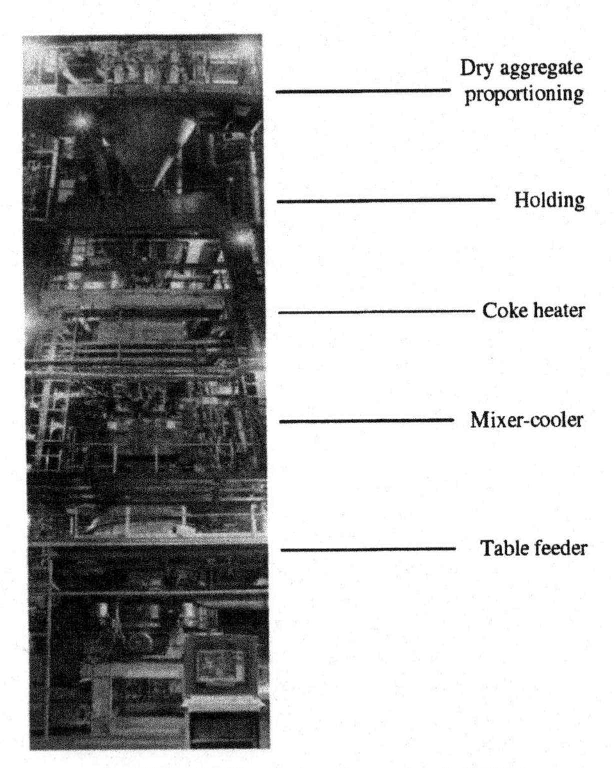


Figure 2a. New preparation system in hall 1

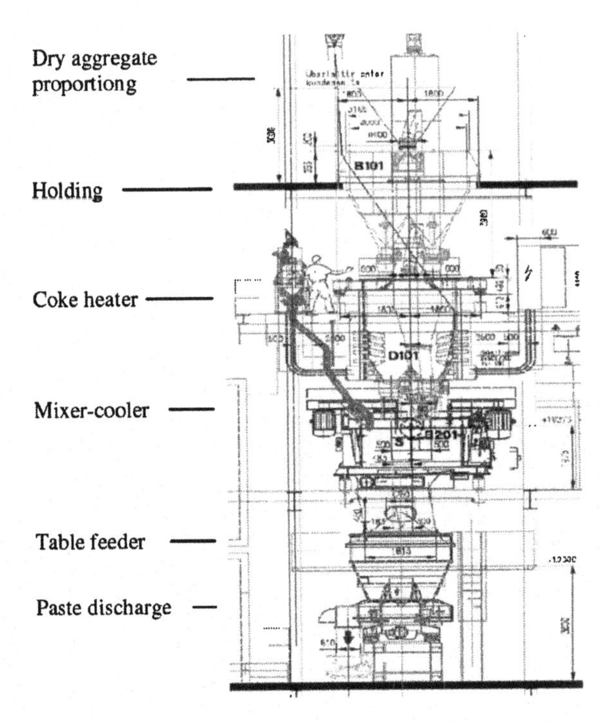


Figure 2b. New preparation system in hall 1

### Description of the High-Performance Mixing Process

The main elements of the new process are:

- electric resistance heater
- intensive mixer-cooler
- table feeder

Electric Resistance Heater EWK 09/41

This heating system uses the electric resistance of coke. The current flows through the coke particles so that the interior is heated up as well as the surface. These evenly heated particles stand for a good pitch impregnation. During a heating cycle a precalculated amount of energy is introduced. Therefore, temperature measurement is not necessary. There is no particle degradation in the heater thanks to the absence of any moving elements. The direct energy introduction enables the system to heat up the complete coke batch in less than 10 minutes!

Intensive Mixer-Cooler

Intensive mixers are famous for their rapid homogenization effect. The machine mainly consists of a rotating mixing pan and one or two eccentrically arranged high-performance rotors. By means of a frequency inverter the rotor speed can be varied depending on the actual process stage (dry mixing, wet mixing, cooling). The typical mixing and cooling cycle starts with the addition of the preheated coke fractions into the mixer. The addition of pitch and green scrap follows. As the energy content of all components is precalculated precisely, the mixing temperature is reached with high accuracy and repeatability. After a certain time, this "hot mixing period" is finished. An infrared camera reads the actual temperature of the material and the cooling phase starts by adding a precalculated amount of water. The total cycle time is only 15-17 minutes compared to 45-60 minutes with a conventional system without separate coke heater.

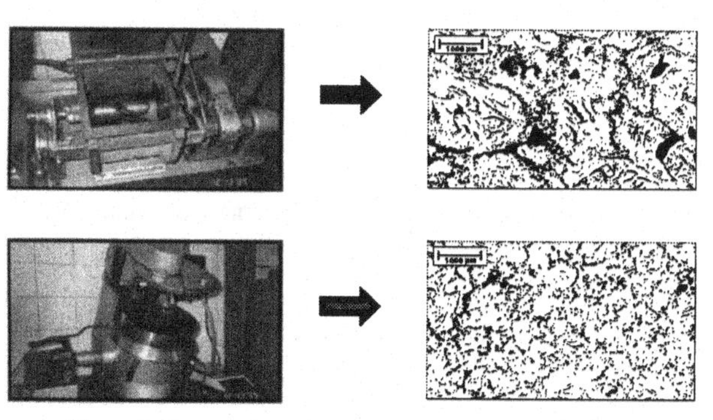

Figure 3. Comparison of the homogenization effect with Sigma kneader and intensive mixer

Figure 4. Mixing principle of the intensive mixer

Table Feeder

This machine was especially developed for these "critical" materials. The main function of the table feeder is to separate the preparation cycle from the press cycle. In the heated and insulated silo hopper up to 8 m³ of mix can be stored. The rotating silo floor serves as discharge unit. Installing the whole machine on load cells makes it possible to batch the carbon body "gravimetrically".

**Description of the New Preparation System**

Based on the existing building and the position of the presses, the Eirich mixing tower was erected approximately in the middle of the hall. For that purpose, a group of four existing mixers was dismounted. The remaining existing preparation line remained operational. A customer-provided feeding system provides a complete batch of dry material being fed via an intermediate hopper into the coke heater. Green scrap is fed via a separate scale directly into the mixer. The process steps "mixing" and "coke heating" are executed simultaneously, i.e. while a batch is being prepared in the mixer, heating of the dry materials for the next batch already takes place. The table feeder always stores as much material as it is necessary for the operation of the press(es). As soon as a certain minimum is undershot, the next mixer batch is being prepared. Decisive for the operation of the entire preparation line is always the amount of material required by the press(es) whereas the so-called press cycle program determines the details.

**Operating Results**

Paste Quality

As expected, the consistency of the carbon body has clearly changed. Instead of egg-shaped agglomerates, a friable, slightly plastic body is obtained.

Figure 5. Finished paste from old line

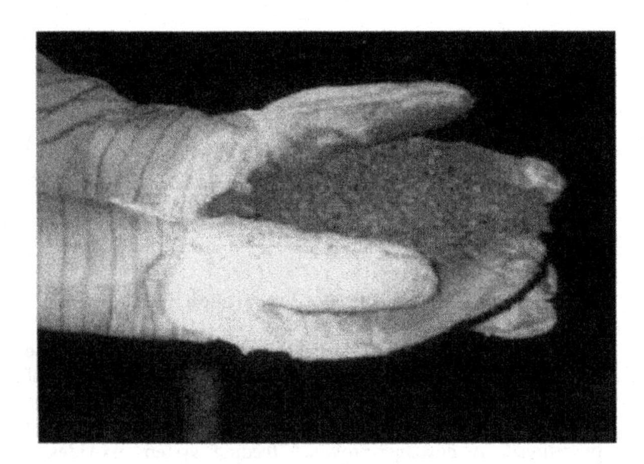

Figure 6. Finished paste from new line

Ukrgrafit were aware from the beginning that the forming process must be adapted to the modified paste characteristics, e.g. in view of compaction pressure, temperature behavior of the press mold etc. The process optimization of the smaller one of the two presses has been completed meanwhile and the reject rate (including baking and mechanical treatment) is just 0.3 % regarding wall side blocks and 4 % regarding cathode blocks. Gradually, also the big press is being adjusted. All in all, the obtained yield comes up to the expectations having been placed into the new technology.

| Table I. Current Operating Data of Preparation | |
|---|---|
| Output | max. 12 t/h |
| Temperature of preheated coke fractions | approx. 150 °C |
| Temperature of liquid pitch | 140 °C |
| Temperature of hot mix | 140 +/- 2 °C |
| Final product temperature | 110 +/- 2 °C |
| Cycle time of intensive mixer | 17 min |
| Electric heater energy input | 72 kWh/batch |
| Mixer filling | 2400-3000 kg/batch |
| Effective volume of table feeder | 8 m³ |
| Pitch addition / cathode blocks | 22 % compared to 24 % before modernization |
| Pitch addition / side wall blocks | 17 % compared to 18 % before modernization |

Table II. Comparison of Important Product Properties

| Product | | Before modernization | | After modernization | |
|---|---|---|---|---|---|
| (Cross section of brick) | Raw material | Density (g/dm³) | Standard deviation | Density (g/dm³) | Standard deviation |
| 575 x 425 | ECA | 1.63 | 0.02 | 1.69 | 0.01 |
| 575 x 360 | ECA* | 1.62 | 0.02 | 1.67 | 0.01 |
| 575 x 217 | ECA* | 1.63 | 0.02 | 1.67 | 0.01 |
| 580 x 222 | ECA* | 1.63 | 0.02 | 1.68 | 0.01 |
| 420 x 420 | GCA* | 1.62 | 0.02 | 1.66 | 0.02 |

* + 30 % of graphite added
ECA = electrically calcined anthracite
GCA = gas-calcined anthracite

## Advantages of the New System

- Higher green density
- Lower density variations in the forming process and the subsequent production steps
- Green production was optimized regarding environmental and safety issues
- Improvement of working conditions
- Reduced pitch consumption
- High machine availability and short downtime for maintenance
- Significant potential for further process optimization

## Summary and Conclusion

Ukrgrafit opted for the new technology to be in future competitive in view of quality and homogeneity of their products (cathode blocks) on the international market.
Thanks to close cooperation between supplier and customer, a smooth "transitional period" / conversion period could be guaranteed. After commissioning was completed, the paste quality quickly met the expectations.
The new mixing line leads to a significant improvement of the cathode production process thanks to better process control and product homogeneity.

The conversion of further production lines to high-performance intensive mixers is planned.

**Light Metals 2008** *Edited by: David H. DeYoung*
*TMS (The Minerals, Metals & Materials Society), 2008*

# TWELVE YEARS OF EXPERIENCE WITH A FULLY AUTOMATED GAS PREHEATING SYSTEM FOR SØDERBERG AND PREBAKE CELLS

Odd-Arne Lorentsen[1] and Ketil Å. Rye[2]
[1]Hydro Aluminium Metal, Technology and Operational Support, Norway
[2]Elkem Aluminium, Mosjøen, Norway

Keywords: Preheating, Start-Up, Early Operation

## Abstract

A preheating system with propane burners suitable for both Prebake and Søderberg cells was successfully developed more than 10 years ago at Elkem Aluminium. Two burners are controlled by a fully automated computer program with a predetermined heating rate, and they are pulsing according to the temperature on top of the cathode blocks.

The new preheating concept makes it possible to increase the temperature in the cathode lining in a smooth and uniform manner up to 970°C. By improving the start-up procedure, we now start the cells without an initial anode effect

The paper presents the experience with the preheating system and how to obtain a smooth start-up and early operation. Most of the operational results come from Elkem Aluminium Mosjøen, but some relevant results are also from Hydro Karmøy, which started with gas preheating some years later than Elkem.

## Introduction

Aluminium production cells are preheated in order to ensure baking of the ramming paste before liquid bath addition and to avoid/reduce thermal shock at bath addition, which may cause cracking of the cathode blocks. In recent years the increasing cost, environmental challenges and governmental regulations concerning handling and disposal of spent potlining (SPL) have forced many aluminum companies to work hard to obtain longer cathode lives and at the same time pushing the pots' capabilities by steadily increasing the line load.

Presently there are two main preheating methods being used, i.e. gas preheating and resistor bed heating (coke or graphite). The preheating method used by Elkem is gas, while Hydro Aluminium uses resistor (coke or graphite) and gas preheating. The gas preheating equipment used is the same as presently described.

Although gas fired systems in principle can offer a more uniform preheating of the cathode surface and lining, the relative simplicity of the electrical method is ensuring its continued use.

The gas preheating method in Mosjøen was developed from manually adjusted respiration burners to fully automated pulsing gas burners. Hydro in Stade (former VAW) used multi-flame gas burners from 1973 to 2003 and the Elkem's gas preheating system from 2003 until the closure of the plant in 2006. Elkem's gas preheating system was introduced at Hydro Karmøy in 2002 for Søderberg cells and Prebake cells in 2003 with good results. The equipment is also being tested at Hydro's plant in Sunndalsøra, because of environmental aspects of the start-up with coke preheating. Several other producers use gas preheating, but to our knowledge gas preheating has not been applied for cells larger than AP18. This equipment is also well suited for restarts, as presently being utilized by Søral Aluminium.

## Preheating

The fully automated gas preheating system has been described elsewhere [1, 2], being developed in collaboration between Elkem Aluminium Mosjøen and Industri-Teknik Bengt Fridh AB [3]. The equipment has later been sold to Hydro Aluminium, Søral, Alcan, Alcoa and Kubal [3]. In short it consists of two burners placed diagonally to each other (see Fig. 1) in the pot cavity and carefully adjusted with regard to angle and orientation to give the right heat distribution.

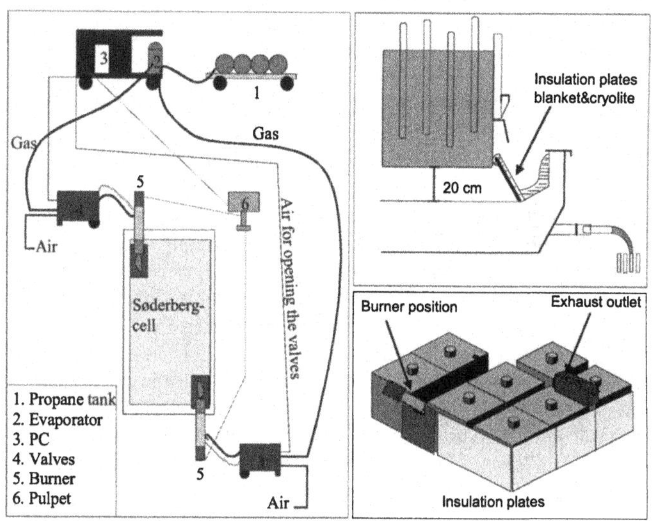


Figure 1: Sketch of gas preheating system and anode side covers.

The anode sides are protected by plates made of 2.5 cm Vermiculite on the inside glued to 2.5 cm of calcium silicate plate. These plates are prefabricated and give proper strength and insulation during the preheating. Fiber blankets (Firemaster 607® Insulfrax® or Saffil®) are placed on top of the anode channels and covered with ordinary anode cover material (crushed bath mixed with alumina), see Fig. 2. All fiber blankets are classified as non-hazardous.

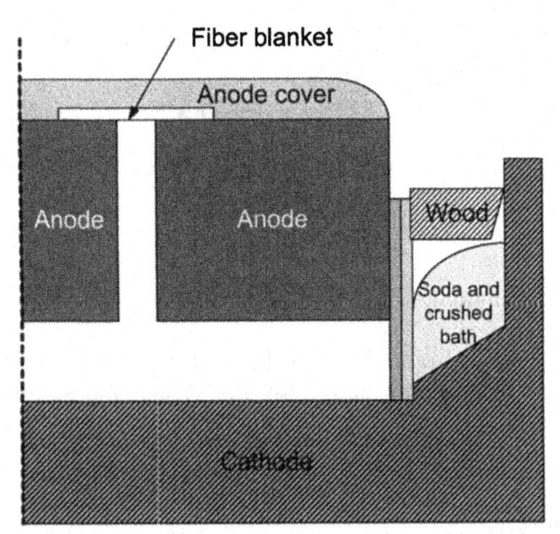


Figure 2: A prebake cell with insulation used during preheating.

Four K-type thermocouples (TC's) are embedded ~2 cm into the cathode blocks along the centre line. The thermocouples are protected by iron pipes or V-profiled sheaths and neither TC's nor sheaths/pipes are removed before bath addition. When the burners are started the computer program toggles the burners on/off between a small flame and a maximum output of 350 kW per burner ensuring even heating and circulation in the cell. The computer program monitors the TC's and uses the highest recorded temperature as basis for regulating the total output. The minimum flame is adjusted to be relatively small. Airburn damage to the cathode surface is eliminated by careful control of the air-to-fuel ratio to the burners. The cathode blocks in front of the burners are also protected from the flame by 2-3 mm steel plates (0.5*2 m) and the ramming paste by crushed bath/cryolite.

In addition to attaining a high final cathode surface temperature after the preheating is completed, one is aiming for an even cathode surface temperature *distribution* avoiding cold/hot spots. Adjusting the burner position and angle during preheating if lagging TC-readings are observed during preheating is therefore important. If a pot is started up while it's still cold in one end the cathode collector bars in the hot end takes most of the current and the relatively cold end freezes up. The pot becomes noisier and thermal gradients resulting from this uneven current distribution can cause cracks and corresponding electrolyte and metal penetration into the cathode blocks.

### Start-up

A cell with well-prepared coke/graphite resistor bed is considered by Hydro to give a uniform and good preheating, but it requires accurate leveling of the resistor bed and close anode follow-up during preheating. During start-up the coke floats to the top of the electrolyte causing fuming and unpleasant working conditions during removal of the preheating insulation material and coke skimming, while graphite sinks to the bottom and does not have to be removed. However, the start-up procedure after resistor bed preheating using coke is being re-evaluated because of environmental concerns.

One advantage of the gas preheating system described is that we remove nothing from the pot, except the burners, after the preheating period. The anodes are lowered from the high preheating position to about 6 cm from the cathode surface. Liquid bath is poured into the cathode and the pot is cut in as soon as the correct liquid level is attained. For Elkem pot sizes it is normally sufficient to add 2-3 tons of liquid bath, while Hydro Karmøy pots need 3-4 tons for cell start-up procedure. The start-up procedure can be performed in 15 minutes and a maximum pot voltage of 6-7 volts is normally obtained shortly after cut-in.

After cut-in the rest of the bath is added to the cell while the ACD is adjusted to a predetermined level. As soon as possible, a large quantity of sodium carbonate is added to the bath to reduce the $\%AlF_3$ to a target level of around 1-3%. Hydro Aluminium Karmøy also uses one of their feeder silos to add soda the first weeks of operation. After a few hours the pot can be started on alumina feed control. The feeders at the Prebake pots at Karmøy have to be installed after start-up, which is performed as soon as possible to avoid anode effect.

During start-up of aluminum electrolysis cell the cathodes are exposed to severe temperature and chemical stress caused by bath ingress, especially by the introduction of sodium to the cathode carbon [4]. There is a high probability that problems affecting cell life could be initiated during cell start-up and early operation.

Long anode effect duration (e.g. 60 minutes) during pot start-up can generate up to 20% of the total PFC emissions throughout cell life [5], assuming a average low AE rate of 0.05 AE/day. In addition, anode effects during cell start-up result in a high energy input, which may melt away the side insulation. External shell-cooling equipment [5] might be required to tackle with hot spots in the pot shell.

### Operational Results

The bath poured into the cell will dissolve the materials added to proctect the cathode during cell preheating (insulation material, steel plates, TC's etc). As a consequence, the metal purity will be deteriorated by these materials and it will take a couple of weeks to reduce the contaminants level and increase the ,metal purity (Fig.3).

The high Fe level is a disadvantage of the gas bake method and plants having Fe-contamination problems may find this a challenge, especially if several pots are started over a short time.

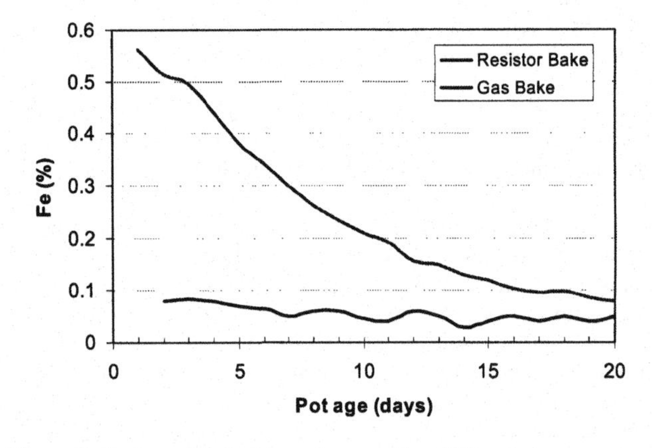


Figure 3: Fe in metal during early operation (Elkem).

The more gentle start of a gas preheated pot gives a better control of the bath temperature during early operation. Figure 4 shows that the average bath temperature during the first 10 hours after start was over 1000°C when using 24h resistor bake. Normal values presently with the gas bake is 980°C for the first 3 days of operation, also including the first 12 hours of operation after bath addition and before metal addition. Hydro Karmøy initially operates with higher temperature than Elkem because of a higher resistance set-point and a slightly lower acidity than Elkem in the first 2-3 days after start-up (see Fig.4)

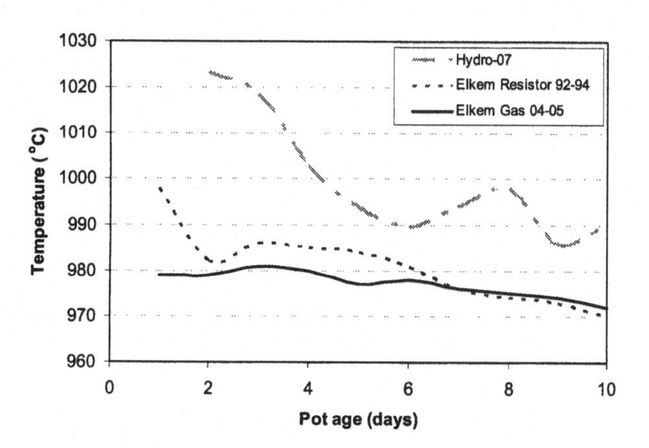


Figure 4: Average bath temperature in early operation.

Figure 5 shows the pot voltage at Elkem's plant for the same type of pots started at the same line load but with the two different types of preheating (i.e. 24h resistor vs. 72h gas bake). The pot voltage average of 10 pots at Hydro Karmøy is given as a comparison.

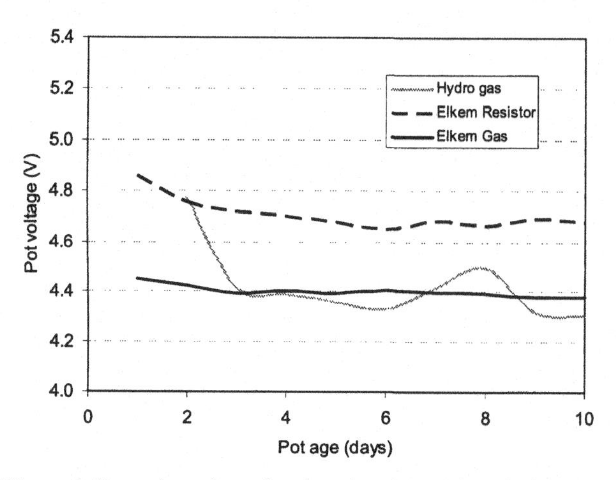


Figure 5: Pot voltage in early operation for Resistor bake vs. Gas bake. Comparable line load and pot type.

Figure 6 shows the results of a test run where two pots were closely monitored after start-up. Bath samples were taken with short time intervals and later analyzed with LECO for alumina content. While the feed rate for pot 481 was kept at 63% of nominal feed the alumina concentration dropped from 6% to less than 2% over a period of 12 hours until an anode effect occurred. It is evident that the trend resumes after the anode effect was quenched as the alumina concentration continues to drop over the next hours. For pot 340 the feed rate was adjusted to 75% and the alumina concentration remained at around 4-5%. These results indicate that metal is most likely produced as soon as bath is added to a pot after cut-in.

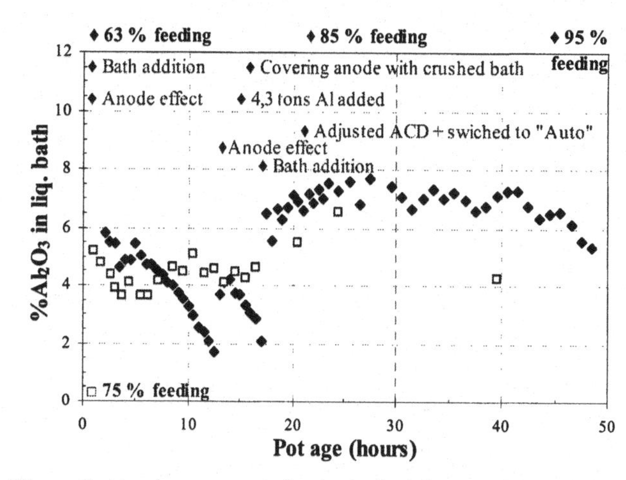


Figure 6: Alumina concentration in the bath in two of Elkem's PB pots after start-up. Work operations and alumina feed rate according to what used to be 1996 standard practice are indicated for pot 481(♦), while only feed rate is shown for pot 340 (□).

Historically an anode effect occurred shortly after start-up (often it was initiated intentionally). The purpose of provoking an anode effect was often to boost the bath temperature to avoid alumina muck formation and to avoid excessive side and bottom ledge formation in the first period of operation. Removal of coke from the preheating is also easier after an anode effect. Figure 7 shows how the AEF has changed over the last 13 years in Mosjøen. Average AEF for 10 pots being gas preheated and started at Hydro Karmøy in 2007 is also shown.

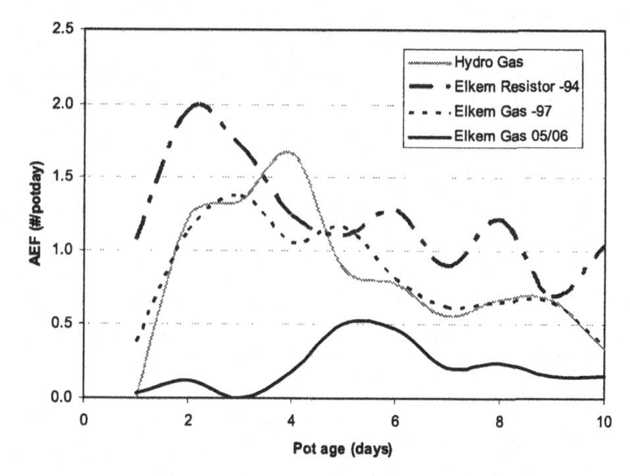


Figure 7: Anode Effect Frequency at Elkem's smelter in Mosjøen and 10 pots at Hydro Aluminium Karmøy the after start-up.

## Discussion

It had been argued [6] that high preheating temperatures may be harmful for the carbon refractory lining. Almost any crack or interfacial gap that may be present in the carbon lining and may open during preheating is a possible channel through which liquid bath and metal may rapidly reach the interior of the lining. There is however a growing understanding for increasing the preheating temperature closer to normal operating temperature [7]. This trend works contrary to the push for reducing preheating time to maximize production. A very high cathode surface temperature can easily be obtained even for short preheating times if the power output of the preheating equipment is high. It is however unlikely that the refractory and even the lower parts of the cathode blocks are sufficiently heated is such a case.

Zangiacomi *et al.* [8] preheated pots with gas burners for 36 and 72 hours, respectively, to 800 and 950°C and concluded that the subcathodic temperature is more dependent on preheating time than the final preheating temperature at the cathode surface. In addition they emphasized the importance of obtaining an even cathode temperature distribution before start-up, concluding that cells with low temperatures in the ends are usually more noisy and difficult to operate just after start-up than more uniformly preheated cells. Pots with longer preheating times also were more stable/less noisy.

Elkem data shows that the poorly preheated pots from their resistor bake require a higher operating voltage not only during early operation, but also later in normal operation. It has frequently been observed that the rule "if it starts high it stays high" is actually valid. Hence, the extra time spent on preheating may be a good investment which will not only help prevent cracks in the blocks that can reduce potlife, but also reduce the initial pot noise or voltage level and ease the process of ramping up to normal operating conditions. Hydro Aluminium has spent significant resources in improving and optimizing both coke and graphite based resistor preheating, but the voltage penalty just described is similar.

A resistor bed preheating operates at full line load towards the end of the preheating period and there is little chance of "speeding up" the last part of the preheating if temperatures for some reason are lagging behind. These situations may occur when the preheating started later than scheduled, while the start-up time is fixed. Graphitized cathodes are in addition more difficult to resistor preheat than semi-graphitised/anthracitic cathodes because of lower electrical resistivity and higher thermal conductivity. Therefore the graphite stripe solution was developed for a smooth start-up with low equipment cost. On the other hand, gas burners can be designed with enough redundancy to speed up the last part of the preheating, if necessary.

Not only do we aim for a high surface temperature, the sub-cathodic temperatures should also be as high as possible at the end of the preheating period. Low sub-cathodic temperatures (i.e. inadequate preheating of the deep lining) will act as a thermal drain on the bath added to the pot on start-up, and bath will therefore be cooled sufficiently to freeze on the cathode surface. An unnecessary cold cathode lining and bath freeze must be counteracted with higher pot voltage during early operation.

The sub-cathodic temperature is more dependent on the duration of the preheating than the final surface temperature. To attain stable pot control and enable lower pot voltage soon after bath addition the preheating curve should be driven to longer times.

Simulations [9] show that the large thermal inertia of the lining is such that even after 72 hours of preheating, the shell floor and most of the sidewall are still cold. The extent of baking of the ramming paste changes dramatically with the preheating time. Green paste deforms to accommodate the expansion of the cathode blocks, but it also shrinks when it starts to bake [9]. The timing of these processes must make sure that no gap will open where bath could penetrate into the lining during bath addition after the preheating is completed.

The longer preheating time permitted by the gas preheating equipment gives a better thermal soaking of the lining and a "start-up anode effect" is normally avoided when the final surface temperature of the cathodes is close to normal operating temperature (950°C). It is also beneficial to aim for reaching maximum cathode temperature a few hours before start up to obtain optimal thermal soaking. Kristensen *et al.* [10] (Nordural) is reportedly targeting the final surface temperature 12 hours before start-up.

A fast start-up gives less anode effects in the rest of the potroom caused by the current interruption many producers still use during start-up. This is of increasing importance now that anode effects are considered a serious environmental problem.

The start-up procedure after gas preheating allows the cathode temperature to stay high until the cell is started, and you do not need to handle hot equipment with the risk of getting burned or exposed to toxic gases removing hot insulation material and coke/graphite. Since the preheating insulation (see Figure 2) is not removed after start up, it gives a very gentle start-up with respect to work load, heat stress and working environment.

Initial cell voltage can be low due to the amount of heat available in both the cathode and the anode after preheating with our gas equipment. This must be beneficial taking into consideration that earlier initial anode effects often heated the bath to at least 1050°C, and the harmful effects this may give the cathode lining. Elkem has worked for many years to reduce the AEF, also just after start-up (see Figure 7). An improved start-up practice and a better alumina feeder [11] have made this possible. On the other side, Hydro at Karmøy had until recently no philosophy to reduce the number of anode effects the first days of operation, but they are now aiming to achieve reduced AEF and temperature just after start-up.

With full load resistor bake the first hours of operation required significant excess heat generation in order to heat the lining to steady-state temperatures and to prevent the bath from freezing. The need for extra voltage was significantly reduced after the introduction of gas preheating [1, 2]. This observation is recently supported by Kristensen *et al.* [10] who (using coke preheating) concluded that there is no need for excessive heat generation after start-up if the cathodes are adequately preheated. Also, eliminating the AE was found to reduce localized differential expansion.

A lot of bath is soaked into the cathode blocks and if excess electrolyte is not added the bath level will sink significantly after only a few hours of operation. The cell temperature then easily becomes very high because there is little surface area available in the side to transport away excess heat. This effect is also larger if the excess cell voltage is not reduced accordingly.

Aluminum producers use different approaches to adjust the bath acidity after start-up. Cathode expansion due to Na absorption after start-up can cause severe stresses in the cathode blocks Alba reported [12] to start their line 5 with basic bath (-4% excess $AlF_3$) for five days, causing severe cathode flaking only a few hours after soda briquettes addition.
Both Hydro and Elkem use less acidic bath than normal the first days of operation, and slowly increase the acidity during the first 30-50 days of operation by adding soda lost to the cathode lining, starting at about 1-3% excess $AlF_3$. Mosjøen data shows that no net aluminum fluoride addition is necessary in the first 50 days of operation. If normal $AlF_3$ additions are performed just after the acidity has been adjusted to normal there is a possibility that the cell will become more acidic than desired, so one should pay special attention to the cells for at least 100 days of operation. Lazzaro *et al.* [13] reported that the sodium diffusion rate from the top to the bottom of cathode blocks is low, giving a sharp sodium gradient in the first phase of early operation. Cathode expansion due to Na absorption after start-up can cause severe stress in the cathode blocks, especially if combined with a rapid cooling from very high temperatures.

Mosjøen and Karmøy both add aluminum metal to the pots 24 hours after bath addition. Special care should be taken to ensure than the temperature of the added metal is not too low. Temperature recordings in Mosjøen showed a temperature drop of 20°C after metal addition.

## Conclusions

Based on the results obtained after the gas preheating method was introduced, there is no doubt that it has improved the start-up of the pots in Elkem. The results at Hydro Aluminium at Karmøy look promising, and the pot operators are very pleased with the equipment. Of particular importance are reduced emissions during start-up and little heat exposure during removal of the preheating insulation. One should note that the development of the equipment at the plant in Mosjøen took two years and even then some months were necessary to adjust the equipment to the local conditions at the plant at Karmøy. The implementation of the gas preheating went well, and the equipment now runs smoothly.

All in all, the gas preheating equipment has given the following positive results:

1. Almost no need for follow-up during preheating
2. A more uniform cathode preheating (no hot-spots)
3. A quick start-up after the preheating has ended
4. No anode effect after start-up
5. No harmful emissions and exposure of PAH and fumes during start-up for the pot operators
6. No heat exposure for operators after the start-up
7. No wastes from the preheating insulation to dispose (left in the pot to dissolve)
8. No need for excess heat after the first hours of operation
9. Lower cell voltage after start-up
10. Aluminum production soon after start-up
11. Less noise after start-up (higher CE)
12. Presumably higher cathode life

## Acknowledgements

Elkem Aluminium and Hydro Aluminium Metal are acknowledged for the permission to publish. We are grateful to the information given by co-workers and foremen working with the gas preheating equipment. A special thanks is given to Kjetil Straumsheim at Hydro Aluminium in Årdal and Halvor E. Borgenvik at Hydro Aluminium Karmøy for their support.

## References

1. Lorentsen, O.-A., and Rye, K.Å., The New Elkem Aluminium Gas Preheating System for Prebake and Søderberg Cells, 9th International Symposium on Light Metals Production, ed. Jomar Thonstad, NTNU, Tromsø-Trondheim, Aug. 18.-21., 1997.
2. Lorentsen, O.-A., and Rye, K.Å., The New Elkem Aluminium Gas Preheating System for Prebake and Søderberg Cells, 6th Australasian Aluminium Smelting Technology Conf., Queenstown, New Zealand, 1998.
3. Modern Aluminium Cell Preheating System. A way to increase the Life Time of the Cathode Cell, pamphlet, Industri-Teknik Bengt Fridh AB, Företagsvägen 16 - 232 37 Arlöv – Sweden, info@industri-teknikbf.se, 2007-09-18.
4. Bearne, G.P. et. al., The CD200 Project - The Developement of a 200 kA Cell Design From Concept to Implementation, TMS Light Metals 1996, pp. 243-245.
5. Berndt, G. and Eick, I., Improvements in the electric heating of Hall-Heroult pots, TMS, Light Metals 2007, pp. 1001-1006.
6. Bentzen, H. Hvistendahl, J., Jensen, M, Melås, J. and Sørlie, M., Gas Preheating and Start of Søderberg Cells, TMS, Light Metals 1991, pp. 741-747.
7. Kvande, H., Cell start-up, The International Course on Process Metallurgy of Aluminium, Trondheim, June 3.-7., 1996.
8. Zangiacomi, C., Pandolfelli, V., Paulino, L., Lindsay, S. and Kvande, H., Preheating Study of Smelting Cells, TMS, Light Metals 2005, pp. 333-336.
9. Richard, D., Goulet, P., Dupuis, M. and Fafard, M., Thermo-Chemo-Mechanical Modeling of a Hall-Heroult Cell Thermal Bake Out, TMS, Light Metals 2006, pp. 669-674.
10. Kristensen, W.E., Hostkuldsson, G. and Welch, B., Potline startup with low anode effect frequency, TMS, Light Metals 2007, pp. 411-416.
11. Hvidsten, R. and Rye, K.Å., "Smart Feeders" for Alumina in a Hall-Heroult Prebake Cell,, TMS, Light Metals 2007, pp. 435-438.
12. Mahmood, M. and Mittal, A.C., Graphitized Cathode Flaking Phenomenon During Alba's Line-5 Start-Up, TMS, Light Metals 2007, pp. 577-581.
13. Lazzaro, G., Importance of Start-Up on Pot Life, Joint CIT Meeting, Rockdale, 13-15. Aug. 1996.

**Light Metals 2008** *Edited by: David H. DeYoung*
*TMS (The Minerals, Metals & Materials Society), 2008*

# CELL PREHEAT/START-UP AND EARLY OPERATION

Ketil Å. Rye
Elkem Aluminium, Mosjøen, Norway

Keywords: Preheating, Start-Up, Early Operation

## Abstract

The most common methods for preheating industrial aluminum pots are reviewed and advantages and disadvantages of each method are presented. Also, the present industrial trend of reducing pot turnaround times is commented and measurements of deep-lining temperatures at start-up are related to the pot behavior in early operation. In general it is seen that shorter preheating times is unfavorable for pot stability in early operation and this effect may limit the potential for further pot turnaround time reductions or lead to operational problems and/or reduced potlife.

## Introduction

Aluminium production pots are vessels built to contain liquid electrolyte at elevated temperature. The material which is in contact with liquid metal, and occasionally liquid electrolyte, is normally of carbonaceous nature. These lining materials have a high cost and it is of importance to avoid damage during start-up which may lead to premature pot failure or sub-optimal operational performance.

A good preheating of the lining materials will remove water in mortar and castables, blocks & refractories and will prevent flash pyrolysis & rapid gas evolution which may cause crack formation in blocks and past seams/joints. A good pot preheating prior to liquid electrolyte addition is therefore considered important at most smelters.

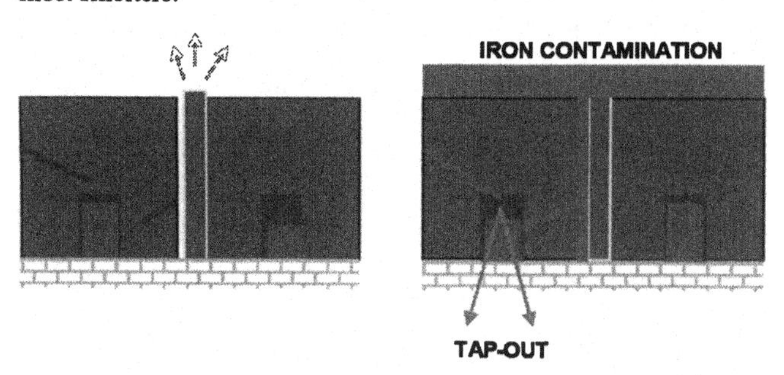


Figure 1. Schematic representation of crack formation due to thermal shock and/or flash pyrolysis at pot start-up.

Presently there are two main preheating methods being used, i.e. gas preheating/baking and resistor bed heating/baking. However, so-called "crash starts" where liquid electrolyte is added directly to a cold pot and the pot is cut in directly thereafter are also referenced in the literature [1], although this method is primarily used for restarting pots after patching/repairs or after periods of curtailment.

The gas fired systems will normally offer a more uniform temperature distribution, but the higher capital and operating cost of the gas bake will in many cases give the relatively simpler electrical method an advantage.

## Preheating

Resistor bake

The resistor bed method is based on using a layer of coke or graphite particles between the anodes and cathode block surface to provide ohmic voltage drop and act as a heating element [2,3,4]. Some plants use shunts to deflect a part of the electric current directly to the next pot without passing through the resistor bed. The shunts will enable a more gentle start of the preheating period and by gradually increasing the load passing through the resistor bed the preheating time is extended. A longer preheating time gives a better thermal soaking of the deep lining. A graphite bed or a thin bed of fine coke particles reduces the electrical resistance and may also be used to extend the preheating time.

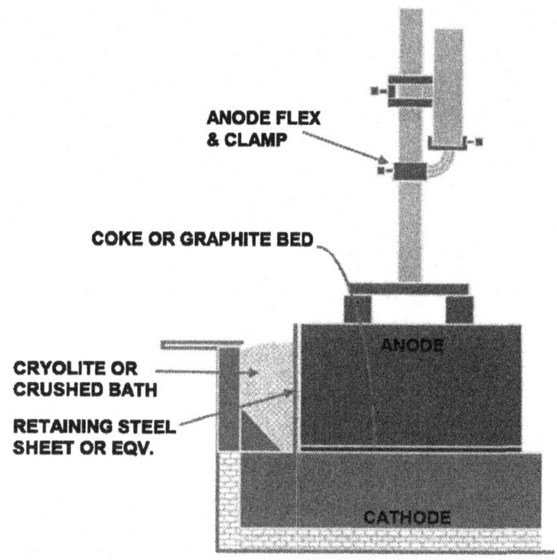


Figure 2. Resistor bake

As cathodes, anodes and anode rods are heated they will expand and move. Hence, to permit vertical movement, the anode rods are hooked up to the anode bridge/beam by means of temporary flexes. If flexes (or equivalent technical solutions) are not used the anode clamps must be loosened and re-tightened at regular intervals. A well-prepared coke/graphite resistor bed can give a uniform temperature distribution in the pot, but accurate leveling of the resistor bed and close follow-up during preheating is required to achieve this goal.

To control the preheating the cathode and anode currents (i.e. anode stem voltage drop) should be measured regularly and action taken immediately to correct deviations. Søderberg anodes should be milled/leveled to ensure good anodic and cathodic current distribution during resistor baking.

After pot start the coke floats to the bath surface and must be removed. The working conditions during coke skimming are unpleasant. Graphite particles tend to sink to the bottom, which precludes skimming.

The advantages of the resistor bake method are:

- Simple "low tech" equipment
- No need for external gas or fuel
- Shorter preheating time, i.e. reduced turnaround time (<= 1 day for full load preheat, 2 days with shunts)
- Limited cathode surface burn-off

The disadvantages of the resistor bake method are:

- Regular manual control of individual anode contact voltages and anode load is required
- Intense hot-spots can occur, especially when follow-up is inadequate
- Dust removal of resistor bed coke is normally necessary
- Short preheating time give inadequate thermal soaking of the lining; "start-up anode effect" commonly occurs

<u>Gas/Fuel bake</u>

A typical gas or fuel bake equipments consists of either two large propane, LNG or oil burners [5,6,7] or multiple small gas nozzles/burners [8]. Preferably, steel sheets are used to protect the cathode surface from direct flame exposure (see Fig 3).

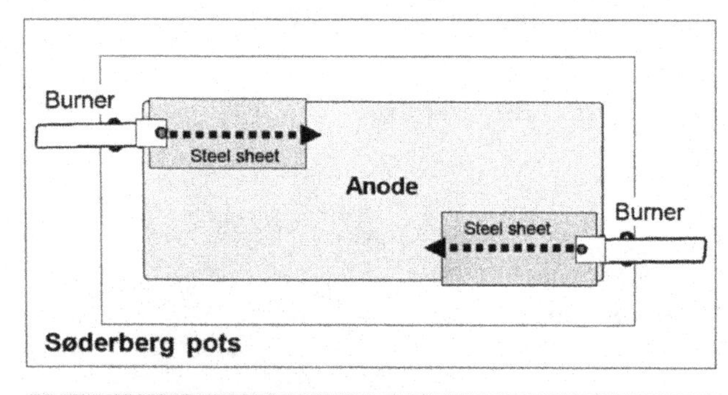


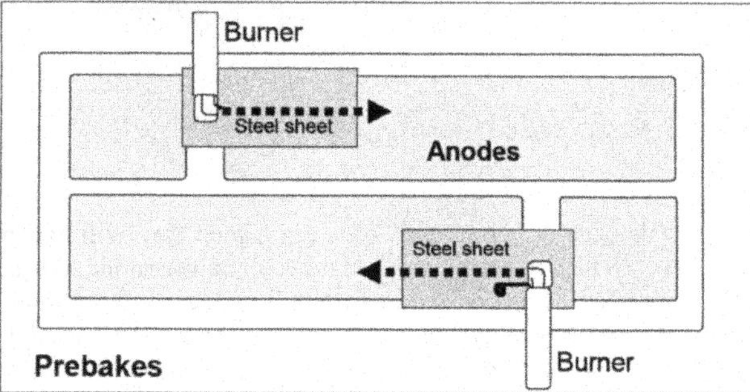


Figure 3. Two large propane burners for pot preheating

For large burners, computer control of the heat-up rate is typically achieved by embedding 4 to 6 thermocouples in the cathode surface. For multiple burners a manual control of the burner flame intensity, the number of burners operating and adjustments of nozzle angle/alignment can be made to control the heat-up rate and to correct for an uneven temperature distribution and/or hot-spots.

The pot needs to be properly wrapped and insulated to prevent excessive air-burn and heat-loss [2]. To contain the heat in the case of a multiple nozzle preheating, insulated steel side covers with slots for the burners are placed along the anode periphery.

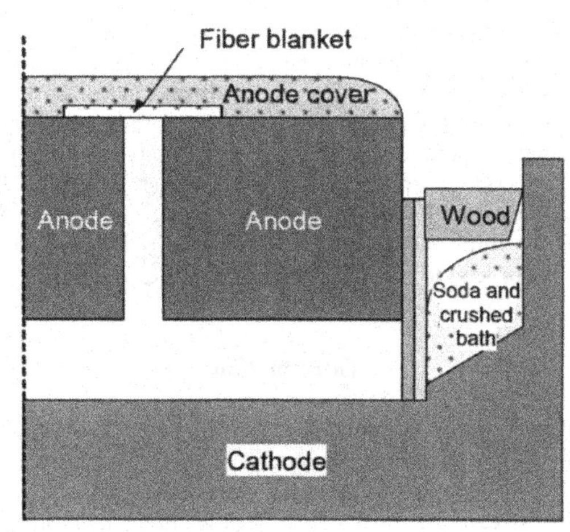


Figure 4. Pot wrapping/insulation of a PB-pot for preheating with two large propane burners.

In the case of the large propane burners used at Elkem Mosjøen [5,6,7] the anode sides are protected by composite plates made of vermiculite and calcium silicate (see Fig.4). These plates are prefabricated and give the proper combination of strength and insulation during the preheating. Fiber blankets, which are used for insulating the anode channels are covered with crushed bath mixed with alumina.

Advantages of the gas/fuel bake method are:

- Intense hot-pots are avoided and a more uniform surface temperature is achieved
- Good control of the heating rate
- No coke skimming after start-up
- Permits a longer preheating time, "start-up anode effect" is normally avoided
- Very little manual control needed for the computer controlled two-burner systems

Disadvantages of gas/fuel bake method are:

- More complex (and expensive) equipment needed
- External supply of fuel; extra logistics and cost
- More labor intensive pot wrapping/insulation procedure. Expensive wrapping materials may be needed
- Severe cathode burn-off may occur near the nozzles/burners. Steel sheets should be used to the limit the cathode burn-off. These steel sheets are normally non-removable, which gives high %Fe metal in early operation

Other preheating methods

With the metal bake method the pot is banked with cryolite/crushed bath and liquid metal is poured into the cold pot. Enough metal is added to give electrical contact with the anodes. Some plants let the metal solidify completely before load is applied and in some cases shunts are used to reduce the load.

Individual anode currents are monitored to prevent overload and burnoffs and liquid bath is added when sufficient metal temperature is achieved (950°C after approx. 3 days). One major disadvantage of the method is that metal will penetrate into open cracks and voids in the lining. This may harm the lining and reduce the potlife, but this method may be the best available method to restart a large number of pots.

A bath start ("crash start") consists of simply pouring liquid bath into the pot and applying load. Normally, the pot voltage is very high (30-40V) for the first hour after bath addition. Individual anode currents must be monitored to prevent burnoffs. Some hours after pull-up, the metal pad is added and the pot is in operation. This procedure is normally used for restarts only.

Temperature target - Surface temperatures

The ideal target for a perfect preheat is to get a surface temperature as close as possible to normal operating temperature (i.e. above 950°C) and with a stable thermal gradient through cathodes and refractory layers. There should be no hot-spots on the surface, i.e. all surface temperatures should be below 1000°C, and there should also be no cold areas on the cathode surface.

The duration and final temperature of the preheating is determined by the heat loss from the pot and the input power of the preheating equipment. According to G.Berndt et al [2] only 7 of the 30MWh's generated during preheating of the 175kA pots at Hydro Rheinwerk in Germany goes to increasing the temperature of the potlining materials, the rest is heat-loss. By improving the pot wrapping/insulation of pots for preheating they were able to reduce the preheating time by several hours, without detrimental results.

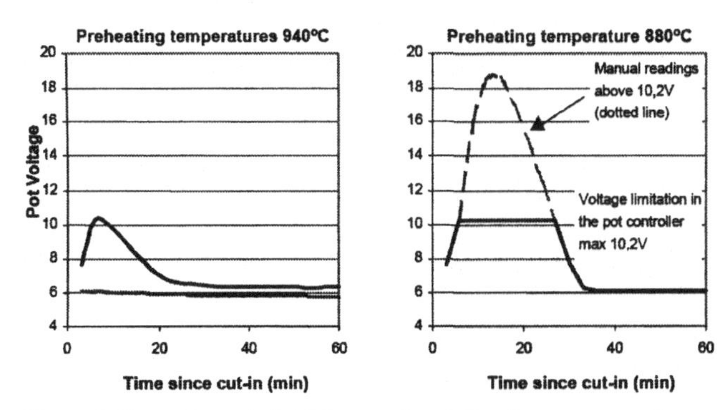


Figure 5. Pot voltage in initial operation following start-up of pots preheated to 940 and 880°C. Elkem Aluminum Mosjøen data.

Temperature target - Sub-cathodic temperatures

In our experience the sub-cathodic temperature should be as high as possible at the end of the preheating period, when the surface temperature reaches >900°C. It is the combination of a sufficient sub-cathodic temperatures (i.e. a good "thermal soaking") and a high final surface temperature that will prevent the "start-up anode effect" and reduce noise in early operation. A high cathode surface temperature in itself will not give this effect. The probable reason is that a low sub-cathodic temperature (i.e inadequate preheating of the deep lining) will act as a thermal drain on the bath added to the pot at start-up. Bath will freeze on the cathode surface and will result in a period of operation with significantly higher pot voltage. A higher graphite content of the cathodes will give higher sub-cathodic temperatures - provided the preheating equipment has the power to maintain the target surface temperature.

Figure 5 shows the initial pot voltage for three pots started at Mosjøen. One pot was started at 880°C after a period of operational problems with the preheating equipment and the pot voltage immediately increased to typical AE-levels. The two other pots were both started at 940°C and both gave acceptable initial pot voltages, but one pot behaved more ideally than the other, going in at 6V, while the other pot had a short period at 10V, before settling down.

It is not customary to place thermocouples inside the lining of ordinary pots for control purposes during preheating. Hence the optimal combination or target surface temperature and duration of the preheating, which gives the sub-cathodic temperature, must be found by experimentation with test pots or modeling work.

As an example, Figure 6 shows the temperature measured at the cathode surface and in the bedding layer beneath the cathodes during a three day preheating with the Elkem gas preheating system. The vertical temperature difference over the cathodes which is shown as "cathode thermal gradient" in Fig.6 is typically 100°C when the pot is started, i.e. only 50°C over the typical value after the pot has reached steady state. We do not see a period of very high pot voltage after bath addition when the sub-cathodic temperature is this high. However, if the preheating time is reduced to two days, the temperature gradient at start-up is typically 200°C and we do see a period of high pot voltage in the first minutes of operation.

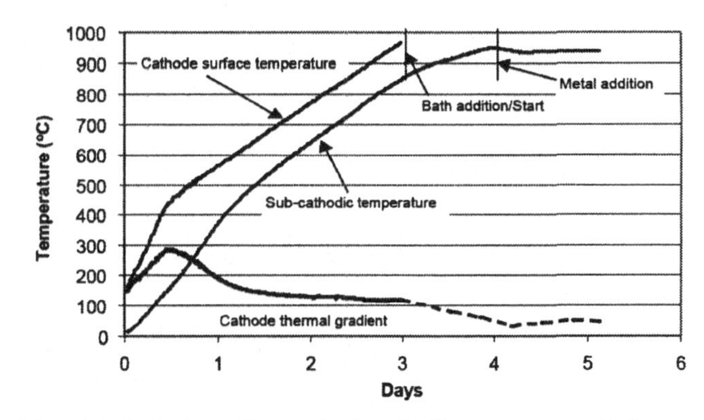


Figure 6. Cathode surface and sub-cathodic temperatures during a 3d preheating and the first 2 days of operation.

A similar response to a longer preheating time has also been reported by others. Zangiacomi et al.[9] preheated pots with gas burners for 36 and 72 hours, respectively, to 800 and 950°C and concluded that the sub-cathodic temperature is more dependent on preheating time than the final preheating temperature at the cathode surface, an observation which is generally supported in modeling work made by Richard et.al.[10]. In addition, the importance of obtaining an even cathode temperature distribution before start-up was emphasized. Pots with longer preheating times were more stable (less noisy) in early operation. Zangiacomi et al.[9] postulated that "In order to attain stable pot control soon after pull-up and minimize thermal gradients inside the blocks that may lead to cracking, the preheating curve should be driven to longer times and higher final temperature. Kristensen et.al.[4] also observed that an adequate preheating will enable a pot start-up with a low initial pot voltage.

Kvande[11] has summarized a set of quality criteria and ideal results for a good preheating, as shown in Table 1. The criterion for sub-cathodic temperature is based on experience acquired at Elkem Mosjøen and is added to Kvande's original table.

Table 1. Preheating quality indicators according to Kvande[11].

| Criteria | Ideal result |
|---|---|
| Final average cathode surface temperature | Close to normal operating bath temperature (950-970°C) |
| Final cathode surface temperature distribution | No hot spots. Relative Std.Dev less than 10% |
| Final vertical thermal gradients* | Sub-cathodic temperature at start-up >700°C for graphitic cathodes* |
| Relative Std.Dev for anodic current distribution | 10-15%. Very important quality indicator for resistor bed preheating |
| Relative Std.Dev for cathodic current distribution | 10-15%. Important quality indicator for resistor bed preheating |

*Elkem Al. Mosjøen data

## Bath soaking and early operation

Bath soaking period

After initial bath addition, cut-in, secondary bath addition and adjustment of the ACD to a predetermined level is done, the pot is in early operation. As soon as possible a large quantity of sodium carbonate should be added to reduce the $\%AlF_3$ to a target level of around 1 to 3%. The pot should then be allowed to operate without metal for some hours before the metal is added. This initial period of early operation is called "bath soaking".

The bath soaking period has several purposes:

- Clean-up; When the pot is running on bath only the dissolution of anode cover material, alumina, banking materials, etc. is more efficient than after metal is added. Some plants will add several shots of alumina to kill the "start-up anode effect". This alumina will settle on the cathode surface and will be more efficiently removed if the pot operates without a metal pad.
- Sealing cracks to prevent future metal infiltration; There will always be cracks/porosity in the lining. During soaking the bath will penetrate into these cracks/pores until it reaches an isotherm where it solidifies. This forms a seal against future metal infiltration.

It is believed that the bath soaking into the lining should have a high liquidus temperature (i.e. low $\%AlF_3$), for a couple of reasons;

- High ratio bath deposits more solid cryolite inside the cracks, thus forming a better seal. Low ratio bath may drain straight though a pot shortly after addition.
- High ratio bath gives a higher operational temperature, pushing the isotherms and the solid cryolite seal deeper inside the lining than in normal operation. This gives a seal more resistant to melting during later temperature excursions.

The bath soaking process takes time, but when metal is added the soaking is effectively cut off. The typical industrial range for soaking time is 4-36 hours. The lower sub-cathodic temperature given by an inadequate preheating implies that the soaking time should be longer in these cases.

Ramp-up period

After initial metal addition the pot is producing, but the temperature is higher than for a normal pot due to the lower $AlF_3$-content of the bath. The pot must now be brought into normal operation without undue delay, but without losing control and harming the lining. Practices for this period varies, but some rules of thumb are :

- The $\%AlF_3$ is ramped up from the soaking composition to normal operational values over a period of one to four weeks. Bath temperatures in the first days of operation are 25-35°C higher than normal.
- Na diffuses into the cathodes, leaving the bath enriched in $AlF_3$. Hence, $AlF_3$-additions are normally not needed during the first weeks. Rather, significant soda additions are generally needed to compensate for the Na soaking into the cathodes.
- If the preheating is close to optimal (i.e. no "start-up anode effect") a normal operation base resistance set-point can be used from shortly after metal addition. The bath temperature should be controlled with resistance modifiers, not by modifying the bath composition.
- If the preheating is non-optimal (with "start-up anode effect") a higher base resistance set-point may be needed for a short time until the isotherms in the lining are stable.
- Situations with simultaneous high $AlF_3$ (i.e above 8-10%) and high bath temperatures should be avoided. This combination will give a very high superheat, melt the protecting frozen crust in the sides completely and result in red sides.

Sometimes plants have "temperature set-points" in early operation that takes priority over ratio-control. This situation can lead to red sides if the soda addition in early operation is inadequate. The result is red sides and sometimes early failures.

## Effects on potlife and key operational parameters

The potlife of any given technology is normally not released and/or published by the Al-industry. Furthermore, there are many factors that can easily mask any effect a change in preheating equipment and/or preheating procedure may have on potlife. In increasing line load not only will the load increase in itself, but also the change in potlining design and materials choice influence potlife significantly. However, there are immediate effects that are easier to document, such as the effect on operating pot voltage.

Figure 7 shows the average pot voltage at the Elkem Mosjøen plant for the same type of pots started at the same line load but with two different types of preheating methods. The blue line is 24h resistor bake and the red line is a 72h gas bake.

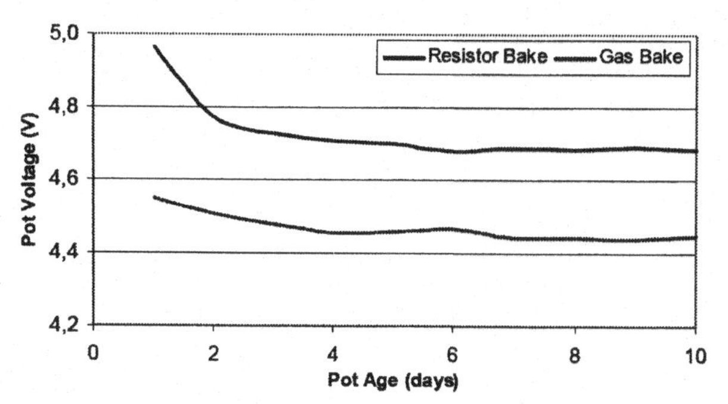


Figure 7. Pot voltage in early operation for pots preheated with 24h full load resistor bake (upper/blue) and 72h gas bake (lower/red) at Elkem Mosjøen.

The pot voltage was significantly lower in the case of the 72h gas preheating (these results are also reported in [12]). This advantage persisted for at least the first 2-3 years of operation, after which time several changes in overall potline parameters makes it hard to document the difference further.

A beneficial effect on pot voltage of a good preheating has also been reported by others. Kristensen et.al. [4] claimed that pots started with an improved preheating, start and early operation procedure operates 30mV lower than the reference pots. Improvements in %CE was also reported.

## Concluding remarks

A good preheating of Hall-Heroult pots can be obtained with both resistor bake and with gas/fuel burner systems. If the surface temperature at preheating completion is close to normal operating temperatures, and if the vertical thermal gradient through the cathodes at start-up is not too steep, the pot will not experience a period of very high pot voltage shortly after cut-in. A good preheating technology will also give long-lasting positive effects on pot voltage and other key operational parameters.

Technical personnel should regularly experience management attention and pressure to reduce the time that pots are inoperable. It is therefore important to be aware of the negative effects of cutting short the preheating time, bath soaking time and ramp-up period in order to increase pot production over the available service time of a given pot.

The author hope that some light has been shed on this topic by this paper.

## References


1. R.Vasanthakumar et al, "Reduction in pot turn around time", General Abstracts of the TMS Light Metals 2007, pp.9-11.

2. Berndt, G. and Eich, I., Improvements in the electric heating of Hall-Heroult pots, Light Metals 2007, pp. 1001-1006.

3 A.H. Mohammed, B.K. Kakkar, A. Kumar, "Improvement in cell preheat and start-up at Dubal, TMS Light Metals 2007, pp 997-1000.

4. Kristensen, E., Hostkuldsson, G. and Welch, B., Potline startup with low anode effect frequency, Ligth Metals 2007, pp. 411-416.

5. Lorentsen, O.A., and Rye, K.Å., The New Elkem Aluminium Gas Preheating System for Prebake and Søderberg Cells, 9th International Symposium on Light Metals Production, ed. Jomar Thonstad, NTNU, Tromsø-Trondheim, Aug. 18.-21., 1997.

6. Lorentsen, O.-A., and Rye, K.Å., The New Elkem Aluminium Gas Preheating System for Prebake and Søderberg Cells, 6th Australasian Aluminium Smelting Technology Conf., Queenstown, New Zealand, 1998.

7. Modern Aluminium Cell Preheating System. A way to increase the Life Time of the Cathode Cell, pamphlet, Industri-Teknik Bengt Fridh AB, Företagsvägen 16 - 232 37 Arlöv – Sweden, info@industri-teknikbf.se, 2007-09-18.

8. Bentzen, H. Hvistendahl, J., Jensen, M, Melås, J. and Sørlie, M., Gas Preheating and Start of Søderberg Cells, TMS, Light Metals 1991, pp. 741-747.

9. Zangiacomi, C. et al., Preheating Study of Smelting Cells, TMS, Light Metals 2005, pp. 333-336.

10. D. Richard, P. Goulet, M. Dupuis and M. Fafard, "Thermo-chemo-mechanical modeling of a Hall-Heroult cell thermal bake-out". TMS Light Metals 2006, pp. 669-674.

11. Kvande, H. "Cell Start-up", Int.Course on Process Metallurgy of Aluminum, Trondheim, 1996

12. O-A. Lorentsen and K. Rye. "Twelve years experience with a fully automated gas preheating system for Søderberg and prebake cells." TMS Light Metals 2008 – to be published.

**Light Metals 2008** *Edited by: David H. DeYoung*
***TMS (The Minerals, Metals & Materials Society), 2008***

# PROPERTIES OF PITCH AND FURAN-BASED $TiB_2$–C CATHODES

M. O. Ibrahiem*, T. Foosnæs and H. A. Øye
Department of Materials Science and Engineering,
Norwegian University of Science and Technology (NTNU)
7491 Trondheim, Norway

*) The Aluminium Company of Egypt (Egyptalum)
83642 Nag Hammadi, Egypt

Keywords: Pitch-based $TiB_2$–C, Furan-based $TiB_2$–C, Physical properties, Thermal behaviour.

## Abstract

Some characteristic properties of pitch and furan-based $TiB_2$-C bulk samples were studied. The open porosity of pitch and furan-based $TiB_2$-C bulk materials is 13.3 % and 34.6 %, respectively. The compressive strength of pitch-based samples is twice the value of furan samples (39 and 16.5 MPa, respectively).

The X-ray diffraction patterns of baked pitch and furan-based $TiB_2$-C samples showed the presence of TiC and $TiBO_3$ due to the oxidation of $TiB_2$ and/or the reaction between $TiB_2$ and carbon in the presence of diffusing oxygen. The TGA study in nitrogen atmosphere showed weight losses of 6.2 % and 2 % for green and baked pitch-based $TiB_2$-C samples, and 11.2 % and 7 % for green and baked furan-based $TiB_2$-C samples, respectively. There was a significantly different behaviour between pitch and furan-based $TiB_2$-C samples in thermal dilatometer experiments. Pitch-based $TiB_2$-C samples showed higher shrinkage than furan-based $TiB_2$-C samples after soaking.

## Introduction

One way to save energy in aluminium reduction cells is to reduce the cell voltage drop by narrowing the anode-cathode distance (ACD). Excessive reduction in ACD in conventional cells does not result in improved energy efficiency due to increased back reaction when the anode is moved closer to the mobile, uneven surface of the molten aluminium cathode. Hence, the poor wettability of carbon cathodes limits the reduction of ACD, because it is needed to maintain a metal pad of a certain height. Modifying the cathode surface to make it more wettable would allow lower ACD with a decreased thickness of the aluminium pad. One centimetre lower ACD lowers the energy consumption by 1.8 kWh/kg Al **[1]**.

Titanium diboride and composites based on $TiB_2$ have been investigated as wettable cathode materials for several decades **[2-11]**. These materials have the unique combination of beneficial properties such as high electrical conductivity, good wettability by liquid aluminium, and acceptable stability in the corrosive cryolite-aluminium environment.

## Experimental

Preparation of bulk samples

Bulk samples of pitch-based $TiB_2$-C composite were produced (50 mm diameter and 50 mm length) by mixing $TiB_2$ powder (70 %), coal tar pitch powder (20 %), anthracite particles (7.5 %) and carbon fibres (2.5 %) using the ball-milling dispersion method **[12]**. The mixed powder and a mould were heated at 155 °C for 1.5 hrs. The powder was hot pressed by applying a load of 30 kN for 30 seconds to get a uniform dense sample. After compaction, a number of samples were baked in a coke bed furnace. The baking cycle was up to 600 °C with slow heating rate (10 °C/hr), then 20 °C/hr until 1100 °C followed by a 5 hrs holding period. During cooling the rate was 35 °C/hr until room temperature was reached.

Bulk samples of furan-based $TiB_2$-C composite were produced by mixing $TiB_2$ powder (70 %), furfuryl alcohol (19.4 %), anthracite (7.5 %) and zinc chloride (0.6 %) using a conventional mixing method. The sample dimension was 50 mm diameter and 50 mm length. The starting materials were mixed together at room temperature using a glass rod. The paste was poured into an alsint alumina tube (50 mm inner diameter). The paste was cured at 80 °C for 24 hours. The cured sample was then calcined in a coke bed furnace. The baking cycle was slow heating (10 °C/hr) up to 600 °C, then 20 °C/hr until 1100 °C followed by a 5 hrs holding period. During cooling to room temperature the rate was 35 °C/hr.

Porosity and density

*Green density:*

The green apparent density was calculated from the dimensions and weight of the sample using the formula:

$$\rho_a = m / v \qquad (1)$$

Where: $\rho_a$ = apparent density, g/cm$^3$
m = dry mass, g
v = calculated sample volume, cm$^3$

*Baked density and open porosity:*

The baked hydrostatic density and the open porosity of $TiB_2$-carbon composites were determined according to ISO 12985-2. The apparent baked density of a material is defined as the ratio of its dry mass to volume. The volume is determined by measurement of the Archimedes force (the mass of the displaced liquid) applied to the sample saturated with water after boiling. The sample dimensions were 20 mm in diameter and 20 mm in length. The baked density and the open porosity are given as the following:

$$\rho_{baked} = \rho_w[m_1/(m_3-m_2)] \quad (2)$$

$$\varphi_p = 100\cdot(m_3-m_1)/(m_3-m_2) \quad (3)$$

Where: $m_1$ = dry mass, g
$m_2$ = immersed mass, g
$m_3$ = mass of sample after saturation by boiling, g
$\rho_w$ = water density, g/cm$^3$
$\varphi_p$ = open porosity, %

Electrical resistance

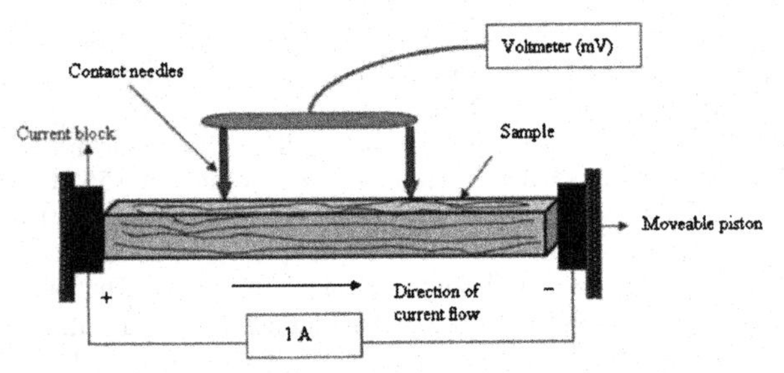


**Figure 1.** Schematic arrangement for the measurement of the specific electrical resistance by the voltage drop method.

The electrical resistance of bulk samples was measured using the voltage drop method. The sample dimension was 50·10·10 mm$^3$. The measurement was done on the four sides of the sample as shown in Figure 1. The samples were turned to give a total of 4 readings for each side. The total electrical resistance of the sample was calculated using the following equation:

$$R = (V\cdot A) / (L\cdot I) \quad (4)$$

Where: R = specific electrical resistance, μΩm
V = voltage drop between two contact points, volt
A = area, mm$^2$ = W·H
I = current, constant 1A
L = distance between two contact points, 10 mm

Cold compression strength

Compression strength of the samples was determined by using a Digimax C-20 press, Form + Test Prufsysteme, Germany. The sample (50 mm in diameter x 50 mm in height) was placed between two compression plates. The upper load cell was lowered with a velocity of 2 mm/min until the sample broke. The compressive strength of the material is defined as:

$$\sigma_c = F_{max} / A \quad (5)$$

Where: $\sigma_c$ = compressive strength, N/mm$^2$
$F_{max}$ = total load on specimen in failure, N
A = area, mm$^2$

Thermal stability and oxidic attack

*TGA:*

The TGA analysis was performed for green and baked pitch and furan-based $TiB_2$-C samples. The samples were powdered using a mortar and pestle. The samples (0.5 g) were put in a pre-weighed Pt crucible. The reference material was $Al_2O_3$. Thermal gravimetric analysis was performed between 20 °C and 1000 °C, with a heating rate of 5 °C per minute in nitrogen atmospheres.

*Thermal dilation:*

The thermal dilation of pitch and furan-based $TiB_2$-C samples were performed using Dilatometer DIL 402C. The samples dimensions were 5 mm in diameter and 5 mm in length. The dilatometer program was started, and the sample was heated from 20 °C to 1000 °C at a rate of 1 °C /min followed by 5 hrs holding time. The sample was cooled to room temperature using the same rate. The temperature and the dimensional change were logged every 10$^{th}$ second. During heating and cooling the atmosphere was argon (99.99 %) to prevent air burn. The coefficient of thermal expansion (CTE) α is an expression for the dilation of the material as a function of temperature:

$$\alpha = \frac{1}{L}\frac{dL}{dT} \quad (6)$$

Where: L = Original length of sample, mm
dL/dT = Slope of dilation curve between 300 °C and 700 °C.

**Results and discussions**

X-ray diffraction and TGA

The X-ray diffraction patterns for the as-received $TiB_2$ powder and pitch and furan-based $TiB_2$-C samples are given in Figure 2. X-ray patterns were studied for pitch and furan-based $TiB_2$ composites before and after baking in a coke bed furnace at 1000 °C. The samples of pitch and furan-based $TiB_2$ before baking prove the existence of an additional phase, which mainly consists of carbon. The presence of carbon is due to the presence of anthracite particles and the carbonaceous binder. After baking, two new phases were identified corresponding to titanium carbide (TiC) and titanium borate ($TiBO_3$). Two other expected phases, $TiO_2$ and $B_2O_3$, were not detected.

A thermodynamic study of the system $TiB_2$, C and air was performed by Dionne et al. **[13]**. They showed that for a carbon activity ($a_c$) equal to 1, a direct reaction between $TiB_2$ and carbon was not favoured:

$$TiB_2\,(s) + 3/2\,C\,(s) = TiC + ½\,B_4C\,(s) \quad (7)$$
$$\Delta G_{1300\,K} = 99\ kJ/mol$$

However the reaction between $TiB_2$ and C is possible in the presence of oxygen. They studied the effect of varying partial pressures of oxygen on the equilibrium. A low partial pressure (around $5x10^{-22}$ atm) of oxygen is needed to initiate the reaction to form TiC from $TiB_2$ and C. Therefore, it is reasonable to suppose that there was enough oxygen left during baking to favour the reaction between $TiB_2$ and C to form TiC and $B_2O_3$ (Equation 8).

$$2\ TiB_2\ (s) + 3\ O_2\ (g) + 2\ C = 2\ TiC\ (s) + B_2O_3\ (l) \quad (8)$$
$$\Delta G_{1250\ K} = -1752\ kJ/mol$$

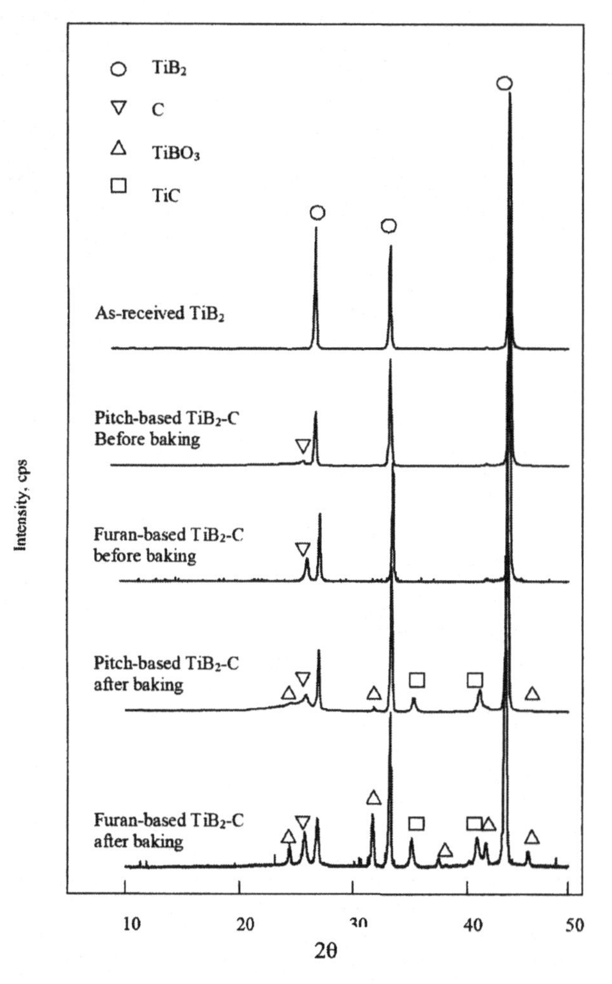


**Figure 2.** X-ray diffraction patterns for as-received $TiB_2$ powder and $TiB_2$–C samples.

The oxidation behaviour of $TiB_2$ is an important factor for different $TiB_2$ commercial applications. It is well known that $TiB_2$ oxidizes according to Equation 9.

$$TiB_2\ (s) + 5/2\ O_2\ (g) = TiO_2\ (s) + B_2O_3\ (s) \quad (9)$$
$$\Delta G_{1300\ K} = -1406\ kJ/mol$$

Titanium diboride may be oxidized to $TiBO_3$ but no thermodynamic data are available according to Equation 10.

$$4\ TiB_2\ (s) + 9\ O_2\ (g) = 4\ TiBO_3\ (s) + 2\ B_2O_3\ (l) \quad (10)$$

Where, TiC will be the stable reaction product in the absence of oxygen (Equation 8). It is however found that $TiB_2$ is oxidized to $TiBO_3$ with low partial pressure of oxygen (10 – 0.05 ppm $O_2$ in Ar) [14]. Titanium borate was detected by XRD analysis but titanium oxide and boron oxide were not detected. During oxidation above 450 °C, $B_2O_3$ is liquid, but it solidifies during cooling **[15]**. From the XRD results taken on $TiB_2$, which was oxidized at 700 °C and 800 °C, Kulpa and Troczynski **[14]** and Koh, Lee and Kim **[16]** claimed that $B_2O_3$ which formed was crystalline. They further suggested that $B_2O_3$ evaporated extensively above 1000 °C, but its vapour pressure is $4.79 \times 10^{-8}$ atm at 1300 K. However, other workers **[17-18]** suggested that $B_2O_3$ formed after oxidation was amorphous, so that it could not be detected by XRD.

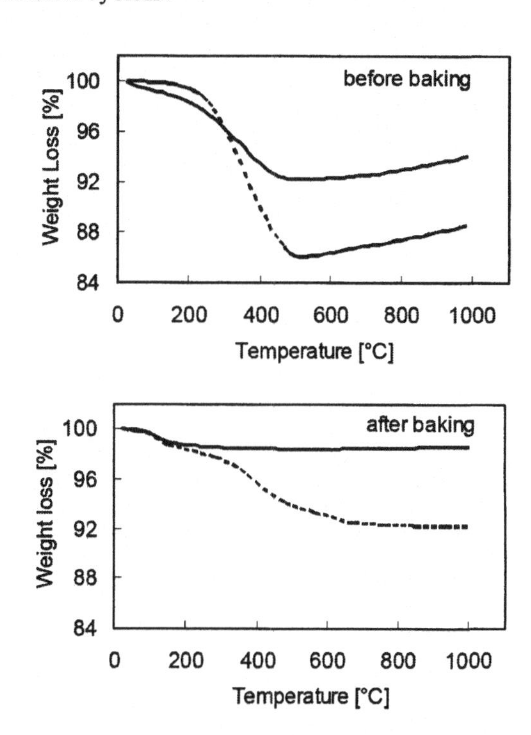


**Figure 3.** TGA plots of green and baked pitch and furan-based $TiB_2$-C samples in nitrogen; fully drawn line: pitch-based $TiB_2$-C sample; broken line: furan-based $TiB_2$-C sample.

Figure 3 shows the TGA curves of green and baked pitch and furan based $TiB_2$-C samples in nitrogen atmosphere. For green samples, weight loss occurred in two steps between 20 °C and 500 °C, the total weight loss being about 8 and 14 % for pitch and furan based $TiB_2$ composites, respectively. A small weight loss occurred at 20 °C - 190 °C due to the liberation of volatile substances, probably water vapour from pitch, furan and $TiB_2$. The second dramatic weight loss was at 190 °C - 500 °C caused by further liberation of volatiles from pitch and furan resin. The volatiles are mainly water, ammonia, methane, hydrogen and light hydrocarbons from the pitch and carbon dioxide, carbon monoxide and water from the furan resin. This weight loss is followed by a weight gain starting after 500 °C. The total weight gain is about 1.8 and 2.8 % for pitch and furan based $TiB_2$-C, respectively. The increase in weight can be attributed to the formation of new phases as a result of Equations 8, 9 and 10. Titanium carbide has high density (4.93 g/cm$^3$) **[19]**. The overall weight loss became 6.2 % and 11.2 % for pitch and furan based $TiB_2$-C composites, respectively.

The TGA curve of a baked pitch-based $TiB_2$-C sample in nitrogen atmosphere shows that there is a weight loss of about 2 % between 40 °C - 170 °C due to the liberation of volatile substances. This is followed by constant weight until 1000 °C.

Baked furan-based $TiB_2$-C samples showed a two step weight loss between 20 °C and 600 °C followed by constant weight; the total weight loss being about 7 %.

A TGA study of $TiB_2$ at different temperatures and XRD patterns have been reported **[20]**. Table 1 summarizes the results from these experiments. A significant weight increase of $TiB_2$ powders at each temperature was observed. The formation of $B_2O_3$ in the liquid state was reported. In XRD analyses only $TiB_2$ was identified below 500 °C. With increasing temperature in the range 500 °C - 1000 °C new oxides appear, such as $TiO_2$ (from 600 °C) and $TiBO_3$ (from 800 °C).

**Table 1.** Weight increase and identified phases in TGA experiments **[20]**.

| No | Atmosphere | Weight increase, % | Temp., °C | Phases, XRD |
|---|---|---|---|---|
| 1 | Air | 6 | 500 | $TiB_2$, |
| 2 | Air | 11 | 600 | $TiB_2$, $TiO_2$ |
| 3 | Air | 25 | 700 | $TiB_2$, $TiO_2$ |
| 4 | Air | 20 | 800 | $TiB_2$, $TiO_2$, $TiBO_3$ |
| 5 | Air | 20 | 900 | $TiB_2$, $TiO_2$, $TiBO_3$ |
| 6 | Air | 10 | 1000 | $TiB_2$, $TiO_2$, $TiBO_3$ |

Physical properties of $TiB_2$-C composites

Some observations were done before and after baking of pitch and furan-based $TiB_2$-C bulk samples in a coke bed at 1000 °C. Before baking, cured furan-based $TiB_2$-C samples were much harder than pitch-based $TiB_2$-C samples. Pitch-based $TiB_2$-C samples remained dimensionally stable after baking but small shrinkage cracks developed along the outer periphery of the samples. For both materials a yellow/brown colour was observed on the outer surface of the samples associated with the oxidation of $TiB_2$ to $TiO_2$ and $B_2O_3$.

Some characteristic properties of the pitch and furan-based $TiB_2$-C bulk samples are given in Table 2. Before baking, the apparent density of pitch based $TiB_2$-C is higher than that of furan-based $TiB_2$-C since the density of coal tar pitch (1.32 g/cm$^3$) is greater than that of furfuryl alcohol (1.133 g/cm$^3$). The baked samples became denser and more ordered and increased packing in the binder phase leads to shrinkage of specimen. Furthermore, the weight of the samples increased due to oxidation of $TiB_2$ particles.

**Table 2.** Physical properties of $TiB_2$-C composites at room temperature.

| Property | Unit | Pitch-based $TiB_2$-C | Furan-based $TiB_2$-C |
|---|---|---|---|
| Green apparent density | g/cm$^3$ | 2.48 | 2.34 |
| Baked hydrostatic density | g/cm$^3$ | 2.62 | 2.44 |
| Open porosity (baked samples) | % | 13.3 | 34.6 |
| Compressive strength (baked samples) | MPa | 39 | 16.5 |
| Sp. el. resistance (baked samples) | μΩm | 36.7 | 39.7 |

The open porosity of pitch and furan-based $TiB_2$-C materials is 13.3 % and 34.6 %, respectively. The large difference in the values of open porosity for the two materials is due to the material processing and the precursor carbonization. Shrinkage of the resin during curing and carbonization produces slit-shaped pores which need to be filled by a liquid matrix precursor for subsequent carbonization if the porosity, density and mechanical properties of the composite are to be improved. The lower porosity of pitch-based $TiB_2$-C samples is the main reason that the compressive strength is twice the value of furan samples. The compressive strength of pitch and furan-based $TiB_2$-C materials were 39 and 16.5 MPa, respectively. The large pores reduce the material strength.

The electrical resistivity of pitch and furan-based $TiB_2$-C bulk samples were 36.7 and 39.7 μΩm at room temperature. The values are of the same order as anthracitic carbon bottom blocks with 30 % graphite. At operating temperatures of the aluminium reduction cell the effective electrical resistivity will become lower than that measured at room temperature **[21]**. Furthermore, the pores of the material will be filled with aluminium carbide and electrolyte or aluminium resulting in further reduction in the electrical resistivity. Aluminium has a resistivity at 950 °C of 0.285 μΩm **[22]**.

Thermal expansion

A thermal dilation curve of the pitch-based $TiB_2$-C sample is given in Figure 4. The sample expands from room temperature until 75 °C due to the release of residual stresses from the forming process. As the temperature is increased, pitch in the sample tends to soften and it becomes completely plastic resulting in slumping of the sample up to 340 °C. The sample expanded suddenly at 350 °C due to trapped light binder volatiles. The sample height was then reduced after the release of volatiles and at 490 °C the transition from a plastic to solid started. The sample maintained its height until 560 °C. The release of volatiles continued during the coking process until 1000 °C. The changes occurring before the coking process are not considered, so the maximum thermal shrinkage of the sample after the soaking period was approximately 1.9 %. During cooling, the sample showed linear shrinkage.

A thermal dilation curve for the furan-based $TiB_2$-C sample is given in Figure 5. The sample started with expansion from room temperature up to 170 °C due to the release of residual stresses from the forming process as well as light volatiles. As the temperature increases further strong shrinkage is observed until 400 °C due to polymerization and liberation of water formed during the polymerization-condensation reactions. After 400 °C, carbonization begins accompanied with the release of volatiles. No change in the specimen dimensions is observed between 715 °C and 1000 °C. The specimen shows expansion by soaking at 1000 °C for 5 hrs due to the oxidation of $TiB_2$. The maximum thermal shrinkage of the sample after the soaking period was approximately 0.4 %.

There is a difference between pitch and furan-based $TiB_2$-C samples behaviour in thermal dilatometer experiments. Pitch-based $TiB_2$-C samples showed higher shrinkage than furan-based $TiB_2$-C samples before and after the soaking period. The thermal expansion coefficient of pitch-based $TiB_2$-C is larger than that of furan-based $TiB_2$-C material. During soaking, furan-based $TiB_2$-C showed an expansion of 0.09 %. Pitch-based $TiB_2$-C showed a shrinkage of 0.04 %. This was possibly due to the increased $TiB_2$ oxidation at 1000 °C.

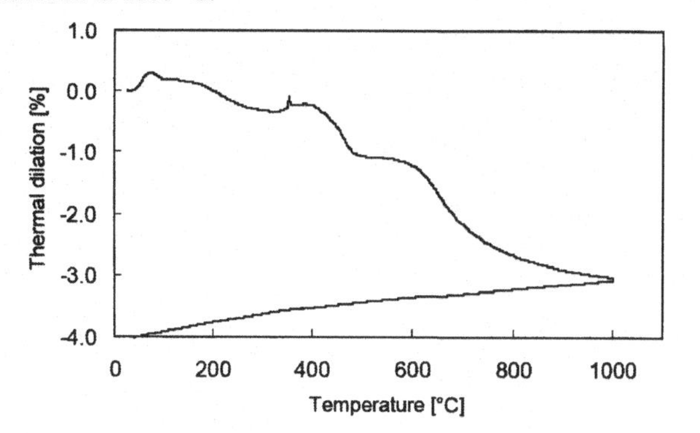


**Figure 4.** Linear dimensional changes of pitch-based $TiB_2$- C during baking.

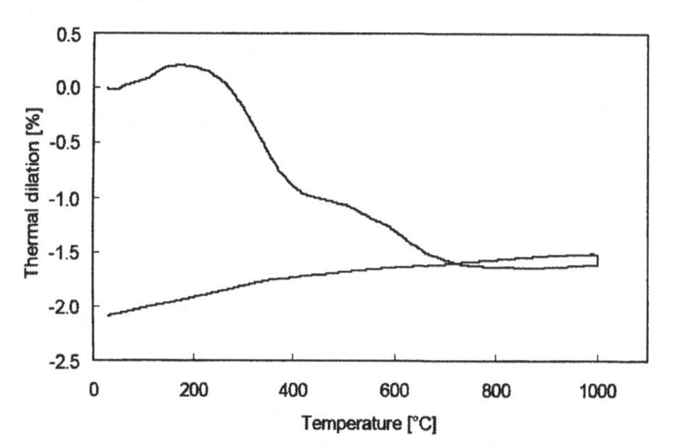


**Figure 5.** Linear dimensional changes of furan-based $TiB_2$- C during baking.

During cooling to room temperature, the two materials showed linear shrinkage. The linear thermal expansion coefficient was calculated from the slope of the cooling curve between 300 °C and 700 °C. The linear thermal expansion coefficients of pitch and furan-based $TiB_2$-C composites are $4.9 \cdot 10^{-6}$ and $3 \cdot 10^{-6}$ $K^{-1}$, respectively. The thermal expansion coefficient of the graphitized carbon used in this work is given by the manufacturer as $3 \cdot 10^{-6}$ to $3.5 \cdot 10^{-6}$ $K^{-1}$ in the range 20-1020 °C. The linear thermal expansion coefficient of furan-based $TiB_2$-C sample is the same as for graphitized carbon. The difference in the linear thermal expansion coefficient for the pitch-based $TiB_2$-C and the carbon material is slightly higher.

## Conclusions

The X-ray diffraction patterns of baked pitch and furan-based $TiB_2$-C samples showed the presence of TiC and $TiBO_3$ in the coating due to the oxidation of $TiB_2$ and/or the reaction between $TiB_2$ and carbon in the presence of diffusing oxygen. Some characteristic properties of pitch and furan-based $TiB_2$-C bulk samples were studied. The carbonization process strongly affects the physical properties of the bulk samples. Before baking, the apparent density of pitch and furan-based $TiB_2$-C samples were 2.48 and 2.34 $g/cm^3$, respectively. After baking, the densities of the two materials are higher than the green densities (2.62 and 2.44 $g/cm^3$) as a result of densification and shrinkage during baking. The open porosity of pitch and furan-based $TiB_2$-C bulk materials is 13.3 % and 34.6 %, respectively. The massive difference in open porosity between the two materials refers to material processing and precursor carbonization. The compressive strength of pitch-based samples is twice the value of furan samples (39 and 16.5 MPa, respectively). The specific electrical resistance is of the same order as for anthracitic carbon bottom blocks with 30 % graphite (36.7 and 39.7 μΩm for pitch and furan-based $TiB_2$-C samples, respectively). The TGA study in nitrogen atmosphere showed weight losses of 6.2 % and 2 % for green and baked pitch-based $TiB_2$-C samples, and 11.2 % and 7 % for green and baked furan-based $TiB_2$-C samples, respectively.

There was a significantly different behaviour between pitch and furan-based $TiB_2$-C samples in thermal dilatometer experiments. Pitch-based $TiB_2$-C samples showed higher shrinkage than furan-based $TiB_2$-C samples after soaking. The maximum thermal shrinkage of pitch and furan-based $TiB_2$-C samples was 1.9 % and 0.4 %, respectively. During soaking, furan-based $TiB_2$-C samples showed expansion of 0.09 %, in contrast, pitch-based $TiB_2$-C samples shrunk 0.04 %. This was possibly due to $TiB_2$ oxidation at 1000 °C. The linear thermal expansion coefficient of the furan-based $TiB_2$-C material is the same as for graphitized carbon ($3 \cdot 10^{-6}$ $K^{-1}$). The difference in the linear thermal expansion coefficient of the pitch-based $TiB_2$-C material ($4.9 \cdot 10^{-6}$ $K^{-1}$) and the carbon material ($3.5 \cdot 10^{-6}$ $K^{-1}$) is slightly high.

## Acknowledgement

Financial support has been provided by the Norwegian Research Council and the Norwegian Aluminium Industry through the CarboMat program, for which the authors are grateful.

## References


[1] H. Kvande " Energy Balance", Fundamentals of Aluminium Production 2004, Volume 1, Trondheim, Norway, May 10-21, 2004, p. 197-216.

[2] S. K. Das, P. A. Foster and G. J. Hildeman: U. S. Patent 4 308 114, 1981.

[3] A. Y. Sane: U. S. patent 4 595 545, 1986.

[4] B. Mazza, A. Bonfiglioli, F. Gregu and G. Serravalle, "Process Aspects in Aluminium Reduction Cells with Wettable Cathodes", Aluminium, 60 (1984), p. 760-763.

[5] C.J. McMinn, "A Review of RHM Cathode Development", Light Metals 1992, p. 419-425.

[6] R. C. Dorward, "Energy Consumption of Aluminium Smelting Cells Containing Solid Wetted Cathodes", J. Appl. Electrochem, 1(1983), p. 569-575.

[7] L. G. Boxall, A. V. Cooke, and H. W. Hayden, "$TiB_2$ Cathode Material: Application in Convential VSS Cells", J. Metals, 36 (1984), p. 35-40.

[8] A. J. Gesing and D. J. Wheeler, "Screening and Evaluation Methods of Cathode Materials for Use in Aluminium Reduction Cells in Presence of Molten Aluminium and Cryolite up to 1000 °C", Light Metals 1987, p. 327-333.

[9] L. G. Boxall, W. M. Buchta, A. V. Cooke, D. C. Nagle and D. W. Townsend: U. S. Patent 4 466 996, 1984.

[10] A. Y. Sane, D. J. Wheeler and C. S. Kuivila: U. S. Patent 4 560 448, 1985.

[11] J. J. Duruz: U. S. Patent 5 004 524, 1991.

[12] M. O. Ibrahiem, T. Foosnæs and H. A. Øye, "Stability of

$TiB_2$ –C Composite Coatings", Light Metals, 2006, p. 691-696.

[13] M. Dionne, L. Gilies and A. Mirchi, "Microscopic Characterization of a $TiB_2$-Carbon Material Composite: Raw materials and Composite Characterization", Metallurgical and Materials Transactions A, (23A), October 2001, p. 2649-2656.

[14] A. Kulpa and T. Troczynski, "Oxidation of $TiB_2$ Powders Below 900 °C", J. Am. Ceram. Soc., 1996, 79 (2), p. 518-520.

[15] D. B. Lee, Y. C. Lee and D. J. Kim, "The Oxidation of $TiB_2$ Ceramics Containing Cr and Fe", Oxidation of Metals, 2001, Vol. 56, Nos. 1/2, p. 177-189.

[16] Y. Koh, S. Lee and H. Kim, "Oxidation Behaviour of Titanium Boride at Elevated Temperatures", J. Am. Ceram. Soc., 2001, 48 (1), p. 239-241.

[17] M. G. Barandika, J. J. Echeberria and F. Castro, "Oxidation Resistance of Two $TiB_2$-Based Cermets", Mater. Res. Bull., 1999, 34 (7), p. 1001-1011.

[18] P. O. By, Diploma Thesis, Met. Inst. NTH, 1991.

[19] "Handbook of Chemistry and Physics", 73rd edition, 1992, Lide DR (ed.), CRC Press.

[20] G. Pettersen, "Development of Microstructure During Sintering and Aluminium Exposure of Titanium Diboride Ceramics", PhD Thesis, NTNU, 1997.

[21] M. Sørlie, H. Gran and H. A. Øye, "Property Changes of Cathode Lining Materials During Cell Operation", Light Metals 1995, p. 497-506.

[22] Metals Reference Book, Ed. E. J. Smithells, 5th Ed., Butterworths, London, 1978, p. 947.

**Light Metals 2008** *Edited by: David H. DeYoung*
***TMS (The Minerals, Metals & Materials Society), 2008***

# ELECTRODEPOSITION OF $TiB_2$ FROM CRYOLITE-ALUMINA MELTS

Dmitry Simakov[1], Sergei Vassiliev[2], Parviz Tursunov[2], Nelli Khasanova[2], Viktor Ivanov[1], Artem Abakumov[2], Anastasiya.Alekseeva[2], Evgenii Antipov[2], Galina Tsirlina[2]
[1] Engineering and Technological Centre, Ltd., Krasnoyarsk, Russia
[2] Laboratory for Basic Research in Aluminium Production, M.V.Lomonosov Moscow State University, Moscow, Russia

Keywords: Titanium diboride, Electrodeposition, Wettable cathode, Cryolite melt

## Abstract

Formation of $TiB_2$ and various by-products under cathodic polarization is studied in relation to in situ deposition of coatings on graphite directly from cryolite-alumina bath. This procedure can provide a cheap cathode coating having very good wettability. In order to evaluate the nature of products, the voltammetry and multistep potentiostatic technique are applied in combination with XRD, SEM and local analysis. For CR between 2 and 3 and alumina content of 2-8 wt.%, current efficiency of $TiB_2$ formation is found to pass a maximum at current densities of ca. 0.4-0.5 A/cm$^2$. Parallel deposition of $Ti_2O_3$ and silicides in certain potential regions is confirmed. Wetting test demonstrated the formation of homogeneous Al film on as-deposited coatings even under open circuit conditions. Wettability was much better as compared to samples with brushed $TiB_2$-based coating. A disadvantage of the deposited coatings is their dendrite microstructure which may be improved in future.

## Introduction

Titanium diboride attracts stable attention because of its ability to behave as a wettable cathode for aluminum electrolytic production [1]. $TiB_2$ fabrication technologies are rather complex and energy consuming, and the resulting material is still too expensive for large-scale operation of bulk diboride cathodes or even mechanically deposited coatings. Cathodic electrodeposition from melts is very promising cheap strategy to cover carbon or metallic cathodes with refractory materials like $TiB_2$. Electrodeposition of titanium diboride from fluoride [2,3] and mixed chloride-fluoride [4-7] oxygen-free melts was reported, as well as from borate halide-free melts [8,9]. The majority of researchers studied $TiB_2$ plating on metallic substrates. The proposed procedures are basically suitable for plating on the bottom and walls of aluminum production bath (see Ref. [9] for example), but require the subsequent technological steps (removal of electrolyte from the bath, and its infill with cryolite melt, with unavoidable interruption of electrolysis). In addition to the evident complications of technology as a whole these steps can result in oxidation of $TiB_2$ coating and is mechanical cracking induced by the difference in support and coating thermal-expansion coefficients. This makes very attractive the idea to perform the deposition of $TiB_2$ directly from cryolite-alumina melt immediately after bath starting. Devyatkin et al. [10] reported a principal possibility of titanium diboride electrodeposition on various metals from cryolite melt with addition of titanium and boron oxygen-containing compounds. This bath composition is already similar to alumina-cryolite melt, as some oxofluoride complexes surely appear. The aim of the present work is to study systematically the effect of alumina content, CR and current density on $TiB_2$ plating processes in the context of possible technological decisions.

## Experimental

Electrochemical measurements were performed in melts of cryolite ratios CR = 2.0-3.0 and different alumina contents (2 – 8 wt. %) containing also 5 wt. % $CaF_2$. Melts were prepared using technical grade cryolite (phase composition as determined by quantitative X-ray analysis: 19,2 % $Na_3AlF_6$ + 73.6 % $Na_5Al_3F_{14}$ + 7,2 % $AlF_3$) and analytical grade NaF, $CaF_2$, $Al_2O_3$. Titanium and boron were introduced as oxides ($B_2O_3$ and $TiO_2$) or complex fluorides ($NaBF_4$ and $Na_2TiF_6$), total concentration of additives being 1-3 wt. %. The stoichiometric molar ratio of titanium and boron was fixed (Ti:B = 1:2). All experiments were performed at 970-980 °C. Electrochemical cells consisted of graphite crucibles serving sometimes as auxiliary electrodes. Most usually vertical graphite working and auxiliary electrodes were used. The reference electrode was Pt wire, but all potentials below are recalculated versus an aluminum reference electrode accounting for the difference value obtained in test experiment with aluminum reference electrode (porous alumina test-tube immersed in the same melt, with liquid aluminum and tungsten current collector). Cyclic voltammograms of graphite electrode in background melt were always measured before adding titanium and boron compounds into the melt and starting each series of measurements. This procedure gave a possibility to check the purity of the melt and to recalibrate the platinum reference electrode using aluminum deposition/dissolution potentials.

All electrochemical measurements were performed using Autolab PGSTAT30 equipped with Booster 20A for high-current measurements.

For cyclic voltammetry, the anodic and cathodic limits were varied in order to cover the potential regions of certain set of electrode processes. Scan rates were 1-10 mV/s. To obtain steady-state polarization curves, the potential of working electrode was fixed, and the current-time curves were registered. The next potential step was started when current started to change slower than for 2% per minute. For coulometric analysis after potentiostatic or galvanostatic cathodic deposition products, the stepwise potentiostatic mode was applied. In the course of the subsequent steps (shifted towards more and more positive potential values) the potentiostatic dissolution of film components took place. Step potentials were chosen on the basis of cyclic voltammetry data, in order to provide the complete oxidation of each possible individual phase contained in the coating. The ratio of charges spent for partial dissolution of titanium diboride and for film deposition was considered as the current efficiency of plating process.

To tests the wettability of coating with aluminum, the samples were immersed into the mixture of Al (100 g) with low-melting potassium containing electrolyte (CR = 1.3, 47.35 g of calcinated KF, 52.65 g of $AlF_3$) in alumina crucible at 750-800 °C. Melt acted as a fluxing agent preventing the formation of oxide layer on the external aluminum surface and promoting the removal of the residual solid cryolite-alumina melt from $TiB_2$ coating. Immersion time was 30-40 s (into the melt) and then 15 – 30 s (into aluminum metal).

For phase analysis of products formed in the course of preparative galvanostatic or potentiostatic deposition as well as after cyclic voltammetry, X-ray powder diffraction with a focusing Guinier-camera FR-552 ($CuK_{\alpha 1}$-radiation, Ge was used as an internal standard) and Huber G670 Guinier diffractometer ($CuK_{\alpha 1}$-radiation, image plate detector) were used. The microstructure characterization and elemental analysis were performed with Leo Supra 50 VP Scanning Electron Microscope (SEM), Electrothermal Atomization Atomic Absorption Spectrometer with Zeeman background correction (Kvant ZETA, Cortec, Russia) and the Eagle-II μProbe x-ray microfluorescence spectrometer (EDAX, USA).

## Results and discussion

Our challenge is to deposit titanium diboride coating directly from industrial electrolyte with the additives of titanium and boron compounds. As it follows from the literature [10,11], complications can go from high volatility of boron fluoride compounds and from formation of passive insulating TiO film when titanium compounds in the melt are partly reduced. The preliminary gravimetric tests demonstrated no essential loss of melt weight for 4-8 wt. % $Al_2O_3$ content and up to 2 wt. % $B_2O_3$ content, i.e. for these melt compositions the problem of volatile boron fluorides was found to be minor.

Another series of preliminary experiments (Fig.1) was related to electrochemistry of boron and titanium individually. For B(III)-containing melt, cyclic voltammetry (Fig. 1a) demonstrates the appearance of redox responses preceding aluminum reduction. XRD conforms that $CaB_6$ or $SiB_6$ are formed (note that silica impurities are unavoidable in industrial bath). For Ti(IV) compounds, two characteristic redox processes are observed (Fig.1b). The process at more positive potentials corresponds to a single electron reduction with formation of $Ti_2O_3$. Deeper reduction (still at potentials more positive than the onset of aluminum evolution) results in formation of titanium aluminides or silicides, as one can judge from XRD. These observations agree with the literature data [12].

Two redox processes are observed also when both B(III) and Ti(IV) additives are introduced simultaneously (Fig.2). As we can conclude from characteristic potential regions, the process at more positive potentials is the formation of Ti(III) compounds, when another one is the target $TiB_2$ deposition process. These conclusions are unambiguously confirmed by XRD data. At variance with to conclusion given in Ref. [10], we found that Ti(III) formation preceded diboride deposition in the overall range of melt compositions under study. Just the impossibility to avoid this process induces the most sharp limitations of the lower current density for galvanostatic deposition of diboride. In addition, just Ti(III) formation is the main reason of the decrease of $TiB_2$ current yield.

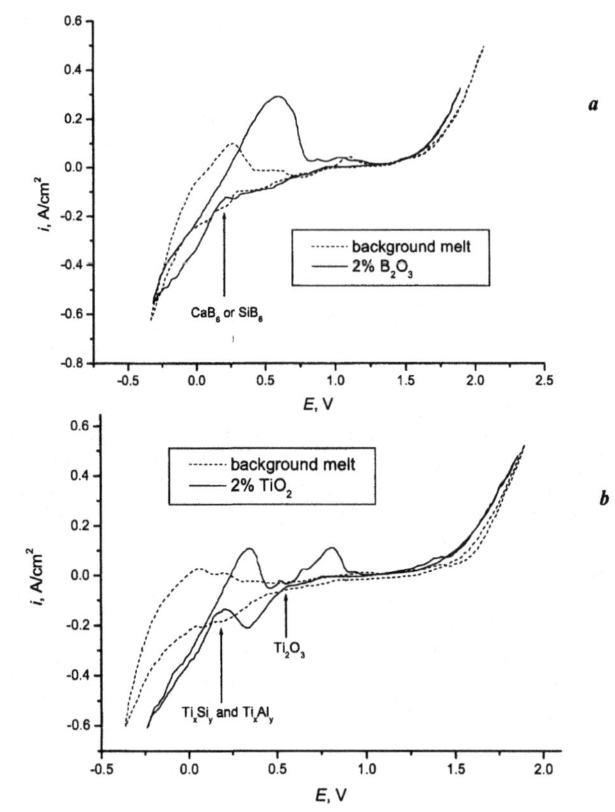


Fig. 1. Cyclic voltammograms of graphite electrode in cryolite-alumina melt (CR = 2.7, 8 wt. % $Al_2O_3$, 5% $CaF_2$) containing 2 wt. % $B_2O_3$ (a) and 2 wt % $TiO_2$ (b). Scan rate 2 mV/s.

Solubility of $Ti_2O_3$ in the melt is much lower as compared to $TiO_2$. When the concentration of Ti(IV) is high enough, and a lot of Ti(III) is formed in the vicinity of electrode surface, the formation of needle-like $Ti_2O_3$ crystals starts immediately, and the formation of $TiB_2$ phase appears to be inhibited (Fig.3). $Ti_2O_3$ solubility increases with CR, and this is the reason why at high concentration of additives like 2 wt. % ($B_2O_3$ + $TiO_2$) $TiB_2$ coating can be formed exclusively at CR 2.9-3.0.

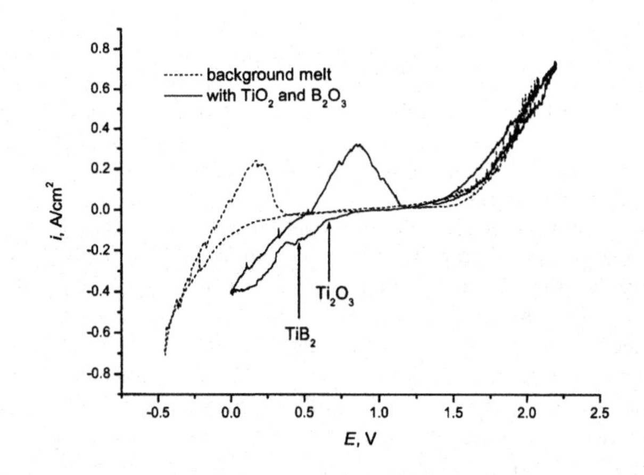


Fig. 2. Cyclic voltammograms of graphite electrode in cryolite-alumina melt (CR = 2.9, 8 wt. % $Al_2O_3$, 5 wt. % $CaF_2$) containing 2 wt. % of ($B_2O_3$ + $TiO_2$) stoichiometric mixture. Scan rate 10 mV/s.

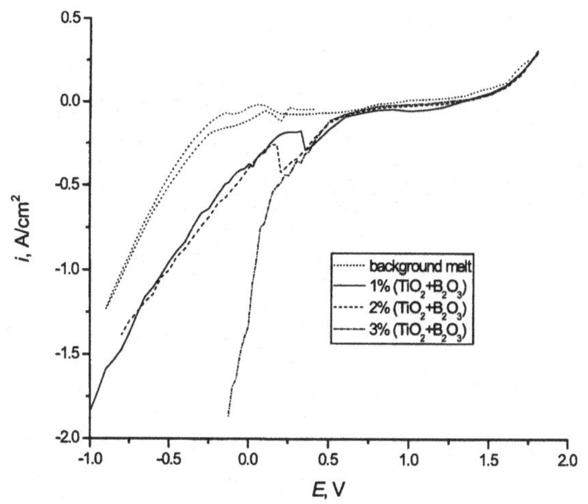

Fig. 3. Steady-state polarization curves of graphite electrode in cryolite-alumina melt (CR = 2.9, 8 wt. % $Al_2O_3$, 5 wt. % $CaF_2$) at various concentrations of stoichiometric mixture ($B_2O_3$ + $TiO_2$): 1 (solid), 2 (dashed), 3 (dot-dashed) wt. %.

According to coulometric data, 0.35-0.55 A/cm$^2$ range of current density is optimal for deposition of $TiB_2$ coating in mass transport configuration of our laboratory cell (Fig.4). At more low current densities, Ti(III) compounds are formed (the limiting diffusion current of this single electron process in this configuration is 0.2-0.3 A/cm$^2$). At higher current densities, diboride formation is suppressed as well, and the predominating products of electrolysis are aluminum and titanium aluminides.

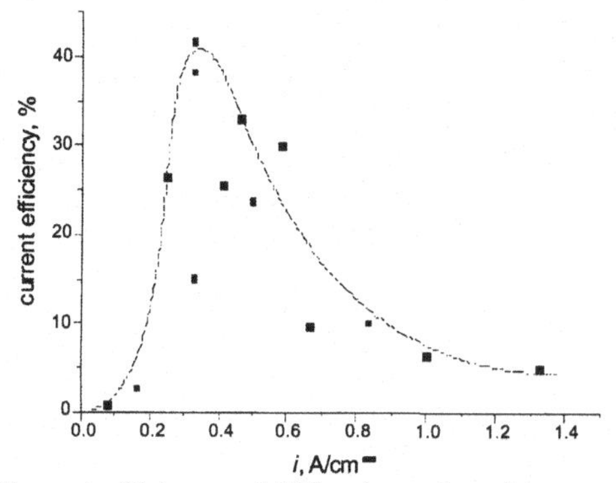

Fig. 4. Current efficiency of $TiB_2$ electrodeposition on graphite from cryolite-alumina melt (CR = 2.9, 8 wt. % $Al_2O_3$, 5 wt. % $CaF_2$, 1 wt.% of ($B_2O_3$ + $TiO_2$) stoichiometric mixture).

Typical coatings formed from cryolite-alumina melt are thick, but porous (Fig.5). Thickness of these black coatings are about an order higher as compared to the thickness estimated from electric charge spent for deposition (even if one assumes 100% current efficiency). Quantitative XRD analysis confirms that $TiB_2$ content in the coating never exceeds 15%. A loose dendritic morphology of coatings formed at graphite surface is confirmed by SEM data (Fig.5). XRD demonstrates that the major fraction of coating consists of solidified melt ($Na_3AlF_6$ and $Na_5Al_3F_{14}$ phases). The samples formed in the melts with 8 wt. % $Al_2O_3$ contain also some fractions of codeposited oxides: alumina (corundum) and $Ti_2O_3$. To suppress codeposition of these insulating components, one should use the melts with alumina content never exceeding 2-4%.

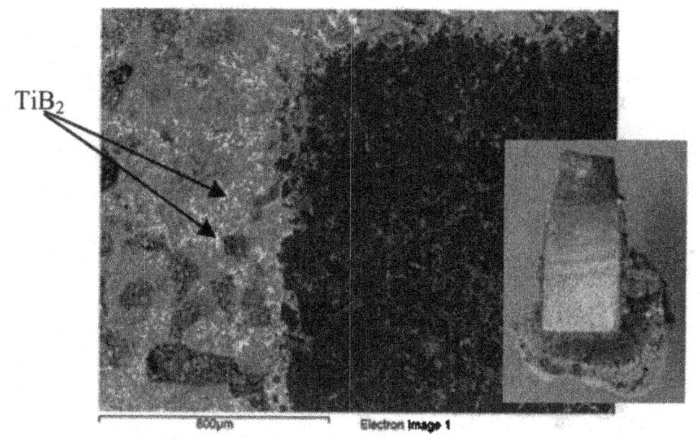

Fig. 5. SEM image of coating cross-section. $TiB_2$ was deposited from cryolite-alumina melt (CR = 2.9, 8 wt. % $Al_2O_3$, 5% wt. $CaF_2$, 1 wt. % of ($B_2O_3$ + $TiO_2$) stoichiometric mixture) at current density 0.45 A/cm$^2$, deposition time was 60 min. Bright fragments correspond to $TiB_2$ dendrites. Insert presents the photo of electrode cross-section.

Wettability test (Fig.6) demonstrated that when the loose coating contacts liquid aluminum the major portion of dendritic layer is destroyed., but despite of this fact the wettability is excellent. In contrast to mechanically grafted cathodes prepared with the use of diboride powder, the electrodeposited coating is wettable under open circuit and do not require cathodic polarization or any other types of pretreatment. SEM data (Fig.7) confirm the existence of a thin (about 100 μm) but dense diboride layer with good adhesion, it forms during the initial steps of deposition and later gives a start to dendritic growth. From industrial point of view the thickness of this dense layer should be increased to provide the stable operation of the coating for a long period. At this stage the attempts to vary the basic melt composition (CR and alumina content), current density, the nature of boron- and titanium-containing reagents ($B_2O_3$ + $TiO_2$ или $NaBF_4$ + $Na_3TiF_6$) were helpless for the improvement of the coating morphology. However there are still some compositional details to be varied. Our further experiments are concentrated on the problem of dendrite suppression, in order to open the door for the use of so nicely wetting coating in practice.

Fig. 6. Graphite electrode coated with electrodeposited $TiB_2$ film (deposited from the melt with CR = 2.9, 8 wt. % $Al_2O_3$, 5 wt.% $CaF_2$, 1 wt. % of ($B_2O_3$ + $TiO_2$) stoichiometric mixture, at current density 0.45 A/cm$^2$ for 60 min). General view (left photo) and cross-section (right photo) after wetting test.

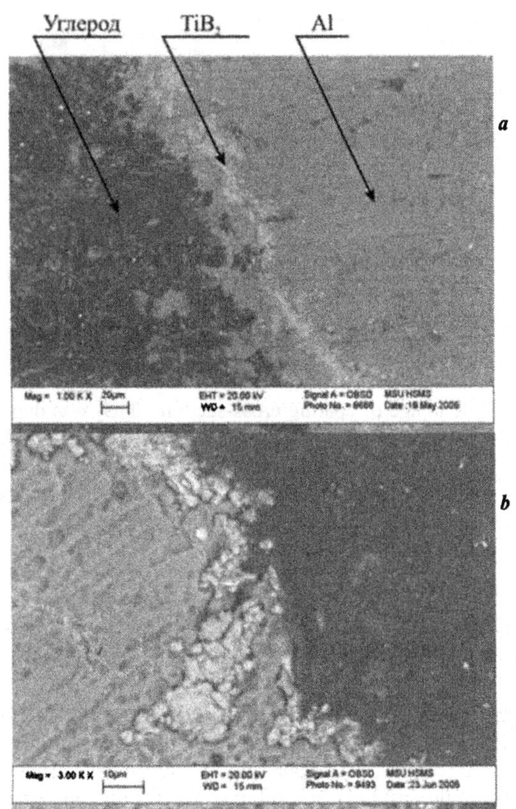


Fig.7. SEM images of the interfacial graphite/diboride/aluminum region before (a) and after (b) partial etching of aluminum.

## Conclusion

We demonstrated for the first time a possibility to deposit the wettable titanium diboride-containing coatings in situ from industrial bath, with the use of conventional melt composition. Deposition can be initiated immediately after starting the bath under its normal mode, by means of adding the oxides of boron and titanium. When the deposition is completed, one should simply stop to add the reagents. After some short period the concentration of both additives will decrease up to typical impurity level, and starting from this moment the bath will operate as usual.

Diboride coatings can be deposited in a wide CR range (2.3-3.0) and in the presence of the pronounced quantities of alumina. The quality of coating is better at higher CR and lower alumina contents. We found that the total concentration of boron and titanium oxides should not exceed 1-2 wt. %. The current efficiency never exceeds 40%, and this value is achieved when the current density is 0.35-0.55 $A/cm^2$. All these conclusions are summarized schematically in Fig.8.

The electrosynthetic protocol we reported above provides the coatings with excellent wettability but low mechanical stability resulting from dendritic growth of the target phase. The work on suppression of dendrite formation is in progress.

This study is carried out within the framework of ETC "RUS Enginering" – MSU agreement.

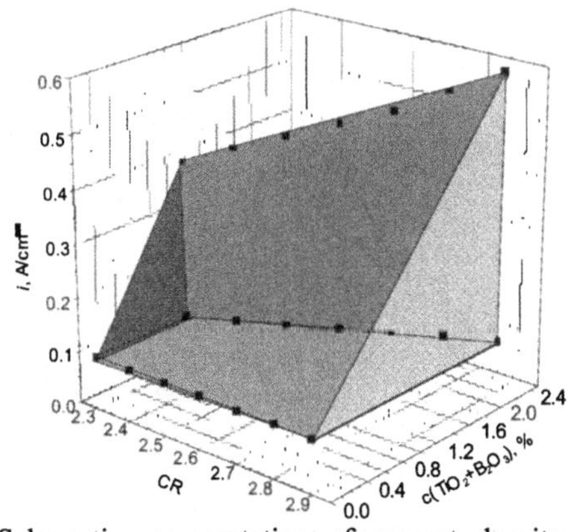


Fig. 8. Schematic representation of current density and bath composition intervals favoring the formation of $TiB_2$ coating. Diagram is constructed on the basis of experimental data for cryolite-alumina melts containing 8% wt. $Al_2O_3$ and 5% wt. $CaF_2$.

## References

1. R P. Pawlek, *"Aluminium Wettable cathodes: an update"*, Light Metals, 2000, 449-453.
2. G. Ett, E. J. Pessine, *"Pulse current plating of $TiB_2$ in molten fluoride"*, Electrochim. Acta 44 (1999) 2859-2870
3. J. Li, B. Li, L. Jiang, Z. Dong, Y. Ye, *"Preparation of highly preferred orientation TiB2 coatings"*, Rare Metals, 25 (2006) 111-117
4. Y. Ban, Z. Wang, Z. Shi, H. Kan, S. Yang, X. Cao, Z. Qiu, *"Preparation of $TiB_2$ inert cathode on graphite by electrodeposition process for aluminum electrolysis"*, Light Metals, 2007, 1055-1059
5. M. Makyta, V. Danek, G. M. Haarberg, J. Thonstad *"Electrodeposition of titanium diboride from fused salts"*, J. Appl. Electrochem. 26 (1996) 319-324
6. F. Lantelme, A. Barhoun, E. M. Zahidi, J. H. von Barner, "Titanium, boron and titanium diboride deposition in alkali fluorochloride melts", Plasmas & Ions 2 (1999) 133-143
7. J. Li, B. Li, Z. Dong, *"Electrodeposition of [001] oriented TiB2 coatings"*, Materials Letters 59 (2005) 3234-3237
8. D. Schlain, F. M. McCawley, Ch. Wyche, *"Electrodeposition of titanium diboride paintings"*, J. Electrochem. Soc. 116 (1969) 1227-1228
9. Biddulph R.H., Wickens A.J., Creffield G.K., European Patent Application EP O 021 850, 7.01.1981
10. Devyatkin S.V., Kaptay G., *"Chemical and Electrochemical Behaviour of Titanium Diboride in Cryolite-Alumina Melt and in Molten Aluminum"*, Journal of Solid State Chemistry, 154 (2000) 107-109
11. S. V. Devyatkin, G. Kaptay, *"Physical, Chemical, and Electrochemical Behavior of Boron Oxide in Cryolite-Alumina Melts"*, Russian Journal of Applied Chemistry, 75(2002) 565-568.
12. S.V.Devyatkin, G.Kaptay, J-C.Poignet, J.Bouteillon, *"Chemical and Electrochemical Behaviour of Titanium Oxide and Complexes in Cryolite-Alumina Melts"*, High Temp.Material Processes 2 (1998) 497-506

**Light Metals 2008** *Edited by: David H. DeYoung*
***TMS (The Minerals, Metals & Materials Society), 2008***

# CREEP DEFORMATION IN $TiB_2$/C COMPOSITE CATHODE MATERIALS FOR ALUMINUM ELECTROLYSIS

Jilai Xue, Qingsheng Liu, Baisong Li
Department of Nonferrous Metallurgy, University of Science and Technology Beijing
Xueyuan Road 30, 100083 Beijing, CHINA

Keywords: Aluminum electrolysis, $TiB_2$/C cathodes, Creep deformation

## Abstract

Creep deformation of $TiB_2$ / carbon composite cathode materials was investigated in a modified Rapoport apparatus. Experimental data for the carbon based cathode materials with varying amount of $TiB_2$ were obtained against varying external pressure. In general, the creep strain increases with increased temperature and prolonged time of aluminum electrolysis. The addition of $TiB_2$ in the carbons can, however, reduce the creep deformation. This effect can be further explained by fitting the experimental data into Burger's viscoelastic creep model.

## Introduction

In the first half of this century, Hall-Heroult process will likely remain the major process for primary aluminum production. Carbon and graphite is still the dominant materials for cathode used in aluminum reduction cells. However, research efforts have been made toward improving the cathode properties for energy saving and lang service life [1-10]. For the carbon cathodes, sodium penetration is widely considered as the main cause for cathode deterioration [1,6]. The poor wettability by aluminum also limits the possibility to reduce the cell voltage drop between the anode and cathode. As for today the most favored wettable material is $TiB_2$.

The intention of this work is to find out a medium-term, economically acceptable solution for the wettable cathode technology. The main attracting point of the $TiB_2$/C composites material is its possibility to realize the concept of a wettable cathode by an industrial process similar to the currently used one to fabricate the carbon cathodes, so that such a cathode could retain most of advantages of the pure carbons but have a better aluminum-wettability [4, 5] and sufficient resistance to sodium attack. Using this strategy, the initial investment for cathode fabricating facilities could be reduced, and the technical obstacles in implementation of the new cathode in a exsisting smelter using conventional Hall-Heroult process would be minimized.

This paper represents a continued effort to make better cathodes with appropriate addition $TiB_2$ into the carbon or graphite materials. The experimental investigation will reveal information on the creep behaviors in the $TiB_2$/C composite cathodes during lab aluminum electrolysisl, and the effect of $TiB_2$ addition on the creep deformation.

## Experimental

Preparation of cathode samples
Table I gives the compositions of the samples of $TiB_2$/C composite cathodes used for aluminum electrolysis tests. All samples were made of electrocalcined anthracites + graphite (industrial products), $TiB_2$ powder (chemical grade) and carbonaceous binder (industrial product). The content of $TiB_2$ varied from 10 to 20 wt%, while the binder was kept at a constant amount in the composite mixtures.

Table I. Compositions of $TiB_2$/C Cathode Samples (wt%)

| Sample No. | $TiB_2$ | Anthracite+graphite | binder |
|---|---|---|---|
| CT0 | 0 | 80 | 20 |
| CT10 | 10 | 70 | 20 |
| CT20 | 20 | 60 | 20 |

The samples were formed in laboratory using a process similar to the industrial one. The carbons, $TiB_2$, and binder materials were mixed and the pastes were pressed to a cylindrical sample with a pressure up to 35 MPa. The green samples were 50 mm in length and 25 mm in diameter and were dried at about 150 °C. They were then heat treated up to 1200 °C in a vessel packed with fresh carbonaceous powders.

Set-up for creep measurement
Figure 1 shows the modified Rapoport [11-15] system used for the measurements of creep deformation during aluminum electrolysis. The external load (A) was applied through a loading fram (B) and a loading extension rod (E) to the testing sample (I). The measuring extension pins (D) was fixed into the cathode sample while it can move freely independent of the load through a hole in the middle of B.

The expansion or strain (displacement) of the sample during tests was measured by a LVD transducer (C) located on the top of the furnace, and the measured signals were logged once a minute into a computer system. The external load was provided by a constant pressure system, which can maintain the pressure at a given value for a period of time. During the testing, the applied pressure was increased each time in steps of 2 MPa.

The experiments were conducted in a vertical tube furnace, where a graphite crucible (G) containing the electrolyte melt (H) was placed in the heating tube (K). The temperature was controlled with a thermocouple in the crucible, and the furnace was flushed with the argon gas (99.99%) though the gas inlet (M) and the gas outlet (F). The testing sample rested on a machined alsint support (J) and acted as a cathode during electrolysis, while the graphite crucible as an anode. The constant current for electrolysis was provided by a DC power supply.

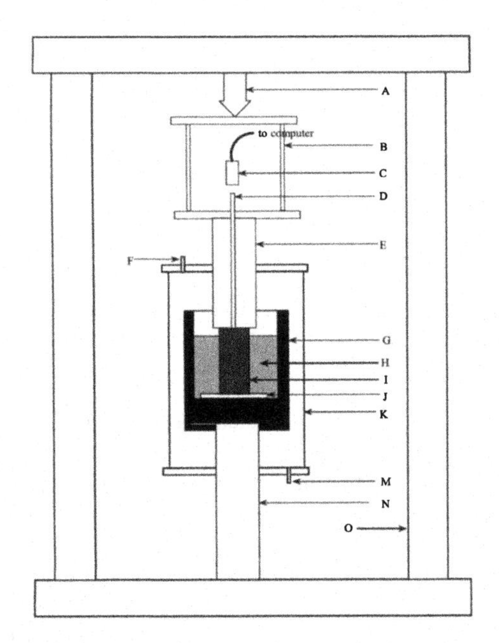


Figure 2. A modified Rapoport apparatus used for creep strain measurement

A - Load, B - Loading fram, C - LVD transducer ((MICRO-EPSILON, range 10 mm, resolution 1 μm), D - Measuring extension pin, E - Loading extension rod, F - Gas outlet, G - Graphite crusible, H – Melt, I - Testing sample, J - Sample support, K - Heating tube, M – Gas inlet, N - Crucible support, O - Steel fram

The aluminum electrolysis was carried out at 0.45 $A/cm^2$ in a cryolitic melt with 8% $Al_2O_3$, 5% $CaF_2$ and a cryolite ratio of 4. The cathode sample was 40 mm immersed in the melt and was subject to creep measurement 2 h after starting the electrolysis.

## Results and Discussion

Effects of temperature and external pressure

Creep strain curves against testing time at 30 □ are shown in Figure 2 – Figure 4 with varying content of $TiB_2$. In the first stage or in the period of 10 min to 15 min from the starting time, these curves exhibit the feature of a typical transient creep: the creep rate (the curve slop) is high at first but soon decreases. This is followed by the secondary stage of a steady-state creep (flat section of the curves), where the creep rate is small and the strain increases slowly with increase testing time.

Figure 5 – Figure 7 show the creep strain curves vs. testing time obtained with the loading pressures of 2MPa, 4MPa, 6MPa, respective, at the temperature of 960□, in which the two stages of creep deformation process appear again. No third stage i.e., accelerating creep was recorded, in which the strain could become so large that it results in a fracture in the testing samples.

The creep strain is highly dependent on stress, temperature and testing time. In an aluminum reduction cell, the stress can form from thermal expansion and sodium expansion of cathode blocks, and the constrain of the cell steel shell. The operating temperature and the external pressure are, therefor, the most significant factors in determining the creep deformaton of the cathode materials.

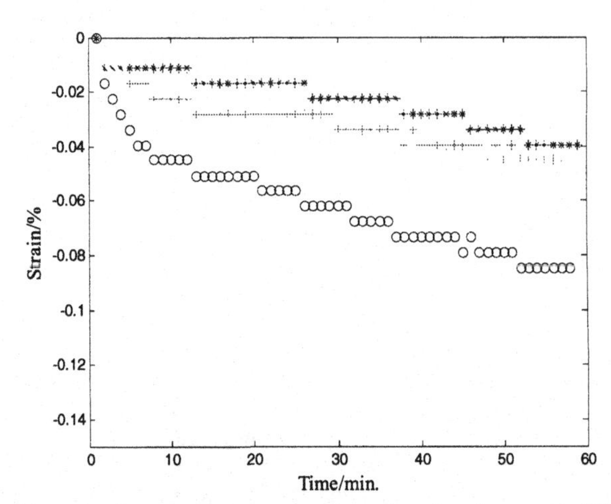


Figure 2. Creep strain curves with testing time for sample CT0 at 30°C (loading pressure: 2MPa, 4MPa, 6MPa, respectively, from top to bottom)

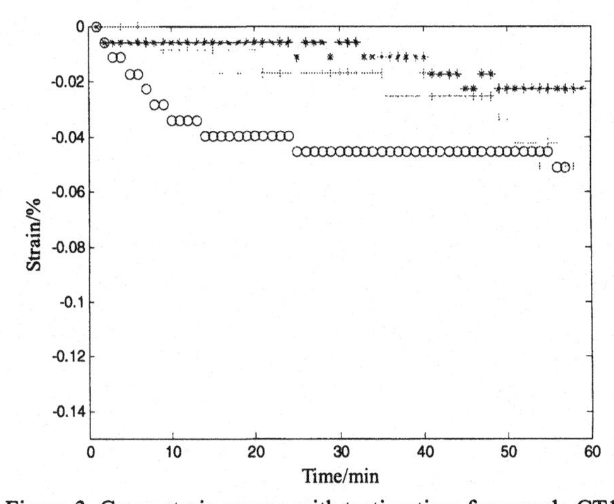


Figure 3. Creep strain curves with testing time for sample CT10 at 30°C (loading pressure: 2MPa, 4MPa, 6MPa, respective, from top to bottom)

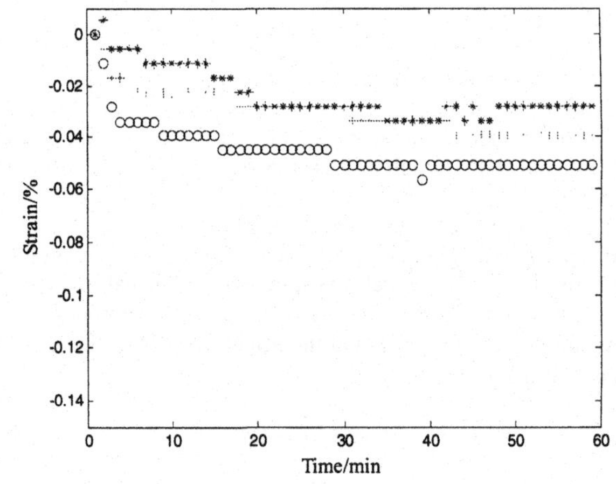


Figure 4. Creep strain curves with testing time for sample CT20 at 30°C (loading pressure: 2MPa, 4MPa, 6MPa, respective, from top to bottom)

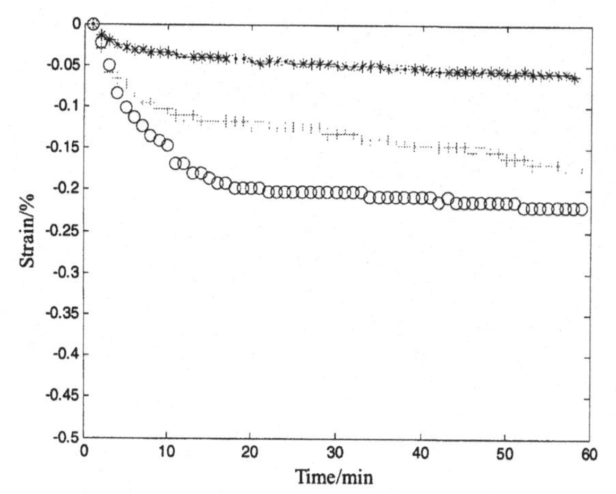


Figure 5. Creep strain curves with testing time for sample CT0 at 960°C (loading pressure: 2MPa, 4MPa, 6MPa, respective, from top to bottom)

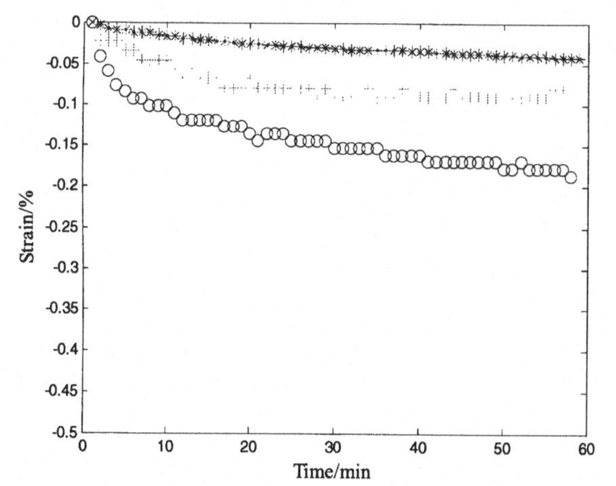


Figure 6. Creep strain curves with testing time for sample CT10 at 960°C (loading pressure: 2MPa, 4MPa, 6MPa, respective, from top to bottom)

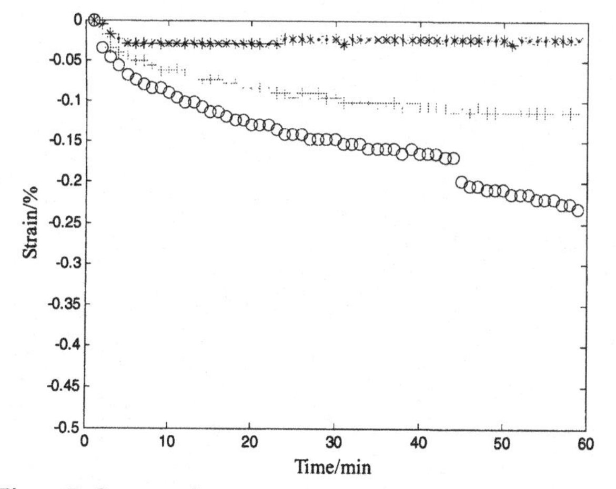


Figure 7. Creep strain curves with testing time for sample CT20 at 960°C (loading pressure: 2MPa, 4MPa, 6MPa, respective, from top to bottom)

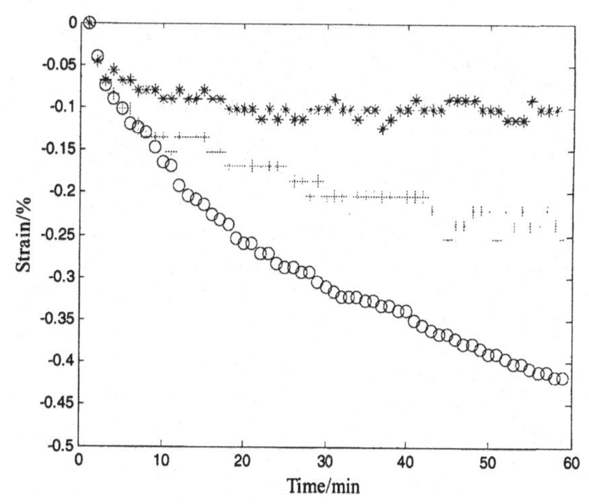


Figure 8. Creep strain curves with testing time for sample CT0 during aluminum electrolsis at 960°C (loading pressure: 2MPa, 4MPa, 6MPa, respective, from top to bottom)

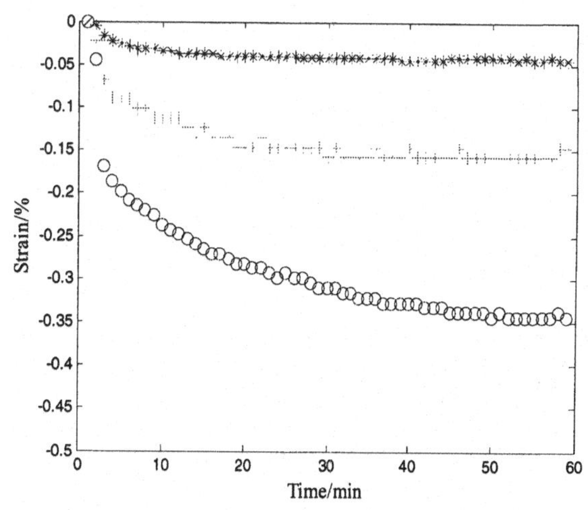


Figure 9. Creep strain curves with testing time for sample CT10 during aluminum electrolysis at 960°C (loading pressure: 2MPa, 4MPa, 6MPa, respective, from top to bottom)

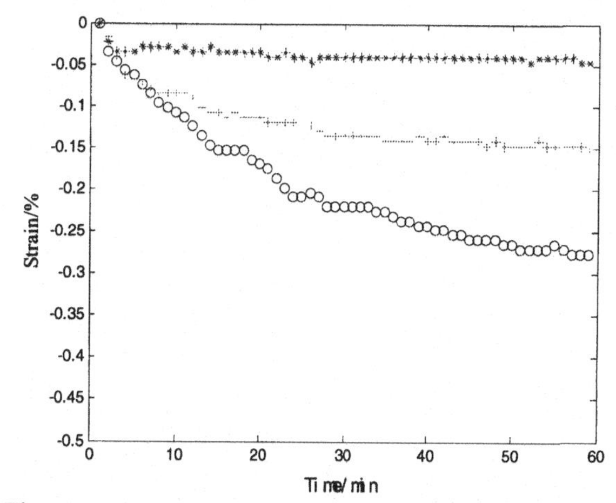


Figure 10. Creep strain curves with testing time for sample CT20 during aluminum electrolysis at 960°C (loading pressure: 2MPa, 4MPa, 6MPa, respective, from top to bottom)

For the same cathode materials at elevated temperatures, the values of the creep strain with varying loading pressure (Figure 5, 6 and 7) are all much higher than those at the room temperature (Figure 2, 3 and 4).

In general, the creep strain increases as the external pressure rises, including the level of the stady-state creep. However, it is noticed that the transcient creep, i.e., the creep rate at the first stage, shows the most significant change. A sharp slop of the curves at the first stages was found when the pressure of 6 MPa was applied. And also the transcient creep became the major part of the total creep strain.

Effect of the electrolysis process

Figure 8, 9 and 10 show the values of the creep strain measured during aluminum electrolysis are higher than those in Figure 5, 6 and 7, where the same materials have been used. The main reason for this difference could be the metalic sodium generated by the reactions

$$3\ NaF(l) + Al(l) = AlF_3(l) + 3\ Na(in\ C) \quad (1)$$

$$Na(in\ Al) \rightarrow Na(in\ C) \quad (2)$$

The sodium penetrates into the carbon hexagonal carbon layers by intercalation or absorption/capillary condensation in the pores of the materias, which results in carbon stucture deterioration. Other fluorides from the cryolitic melts can also penetrate into the carbon cathodes. These changes in material microstructures and property may lead to the creep deformation and possibly material damage.

Compared with the ones with no electrolysis, the transcient creep at the first stage of the strain curves during aluminum electrolysis increases the most with the pressure of 6 MPa (Figure 8 – 10). While the strain curves during electrolysis with lower external pressure, in constrast, still show the creep pattern similar to the ones without electrolysis, in which the strain curve often flat out after the transcient creep stage. This suggests that the sodium induced creep deformation may increase so much with a higher stress that may develop within the cell cathode linings during cell operation, that may eventually approach the fracture failure strength of the cathode material.

Combined with the data of sodium expansion that have been investigated extensively [7, 8], the creep behavior can provide key information in cathode material selection and cathode structure design for prolanged cell service life [13-16].

Effect of the $TiB_2$ addition

The purpose of addition of $TiB_2$ is to improve the wettability by liquid aluminum on the cathodes. However, the $TiB_2$ addition should have no harmful to or even have benifit to the sodium resistance of the cathodes.

The strain curves in Figure 5 to Figure 10 show that the cathode samples with $TiB_2$ additions have smaller creep strain than those without $TiB_2$, for either in elelctrolysis or not. The ones with 10% (Figure 9) and 20 % $TiB_2$ additions (Figure 10) have significant lower values in the creep strain than the pure carbon (Figure 8), and in addition, their transciet creep rates are smaller than the pure carbon. Previous investigation [7, 8] has revealed that the $TiB_2$ has good chemical stability and resistance to sodium attack at elevated temperatures. When it is well dispersed in the binder phase that is usually the weakest part of the cathode, the structure strength and the resistanc to sodium attack of the composite cathode can be improved, so does the resistance to the creep deformation.

Modelling the creep strian in $TiB_2$ /C material

According to the characteristic behaviors of the creep strain, a four-elemental Burger model [17] is selected to fit the experimental data measured for the $TiB_2$/C composite materials. It is assumed that the total strain is composed of three components: elastic strain, viscoelastic strain and viscoplastic strain. The total strain $\varepsilon$ can be written as

$$\varepsilon(t) = \sigma_0 \left[ \frac{1}{E_1} + \frac{t}{\eta_1} + \frac{1}{E_2}\left(1 - e^{-t/\tau}\right) \right] \quad (3)$$

where $\tau = \eta_2 / E_2$ , σ is the applied stress or load, $E_1$ the instant elastic module, $E_2$ and $\eta_1$ the elastic module constant and the viscosity of, respectively, of the Kelvin element. $\eta_2$, the viscosity associated with unrecoverable strain.

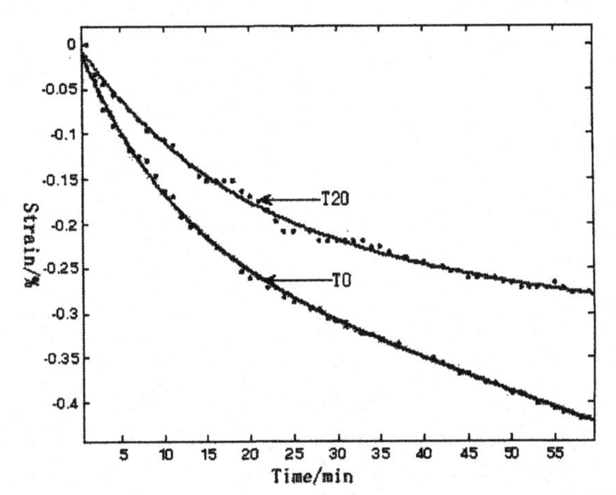


Figure 11. Comparison between the experimental data from creep measurements and the curves from Burger's creep strain model

Table II. Material parameters from Burger's model

| No. | $E_1$(MPa) | $E_2$(MPa) | $\eta_1$(PMa•S) | $\eta_2$(PMa•S) | $R$ |
|---|---|---|---|---|---|
| CT0 | 15950 | 4.682 | 4.758 | 0.8109 | 0.9972 |
| CT20 | 2392 | 4.285 | 18.67 | 1.263 | 0.995 |

The material parameters obtained from the curves of the model in Figure 11 are listed in Table II. It is obvious that the elastic modules $E_1$ and $E_2$ with $TiB_2$ addition decrease, while the $\eta_1$ and $\eta_2$ increase. The model can discribe the creep rate at the first stage and the stady-state stage on the creep strain curves.

## Conclusions

- The creep strain of $TiB_2$/C composite cathodes increases with increased loading pressure at constant temperature, in which the transcient creep at elevated temperature is significant larger than the one at room temperature.
- Aluminum electrolysis generates sodium and causes larger creep deformation than the one with no electrolysis.
- The $TiB_2$ addition can contribute to a reduction in creep deformation due to the better chemical stability and stronger mechanical strength for both thermal and sodium-induced creep strain. This effect can be further explained by applying Burger's model to discribe the creep behaviors for the $TiB_2$ /C composite cathode materials.

## Acknowledgement

Financial support from National Natural Science Foundation of China (NSFC), and Fund for Doctor Degree Study in Ministry of Education of China is gratefully acknowledged.

## References

1. M. Sorie and H. A. Oye, Cathodes in Aliminium Electrolysis, 2nd edi., Aluminium-Veriag, Dusseldorf, (1994).
2. J. Xue and H. A. Oye, "Sodium and Bath Penetration into $TiB_2$-Carbon Cathodes during Laboratory Aluminum Electrolysis," *Light Metals 1992*, Warrendale PA, USA: TMS, 1992, 773-778
3. J. Xue and H A. Oye, "Investigating Carbon/ $TiB_2$ Materials for Aluminum Reduction Cathodes," *JOM,* 44(11)(1992), 28-34.
4. J. Xue and H.A. Oye , "Wetting of Graphite and Carbon/$TiB_2$ Composites by Liquid Aluminum," *Light Metals 1994*, Warrendale AP, USA: TMS, 1993, 631-637
5. J. Xue and H. A. Oye, "$Al_4C_3$ Formation at the Interface of Al–Graphite and Al-Carbon/$TiB_2$ Composites," *Light Metals 1993*, Warrendale PA, USA: TMS, 1994. 211-217.
6. H. A. Oye, "Materials Used in Aluminum Smelting," *Light Metals 2000*, Warrendale PA, USA: TMS, 2000, 3-15.
7. J. Xue, Q.-H. Liu, J. Zhu and W.-L. Ou, "Sodium Penetration into Carbon-Based Cathodes during Aluminum Electrolysis," *Light Metals 2006,* Warrendale PA, USA: TMS, 2006, 651-654.
8. J. Xue, Q.-H. Liu, and W.-L. Ou, "Sodium Expansion in Carbon/$TiB_2$ Cathodes during Aluminum Electrolysis," *Light Metals 2007,* Warrendale PA, USA: TMS, 2007, 1061-1066.
9. M. Dionne, G. Lesperance and A Mirtchi, "Wetting of $TiB_2$-Carbon Material Composite," *Light Metals 1999*, Warrendale PA, USA: TMS, 1999, 389-394.
10. R. P. Pawlek, "Aluminum Wettable Cathode: An Update," *Light Metals 2000*, Warrendale PA, USA: TMS, 2000, 449-454.
11. M.B. Rapoport, V.N. Samoilenko, "Deformation of Cathode Blocks in Aluminum Baths during Process of Electrolysis", *Tsvet Met 1957*, 30(2), 44-51.
12. D. S. Newman, O. Hahl, , H. Justnes, S. Kopperstad and H. A. Oye, "A Technique for Measuring In Situ Cathode Expansion (Rapoport Test) during Aluminum Electrolysis," *Light Metals 1986*, 685-688.
13. A. Zolochevsky, J. G. Hop, G. Servant, T. Foosnas and H. A. Oye, "Rapoport-Samoilenko Test for Cathode Carbon Materials I. Experimental Results and Constitutive Modeling," *Carbon*, 41(2003), 497-505.
14. A. Zolochevsky, J. G. Hop, G. Servant, T. Foosnas and H. A. Oye, "Creep and Sodium Expansion in a Semi-graphitic Cathode Carbon," *Light Metals 2003*, Warrendale PA, USA: TMS, 595-602.
15. A. Zolochevsky, J. G. Hop, G. Servant, T. Foosnas and H. A. Oye, "Surface Exchange of Sodium, Anisotropy, of Diffusion and Diffusional Creep in Carbon Cathode Materials," *Light Metals 2005*, Warrendale PA, USA: TMS, 2005, 745-750.
16. J. Crank, The Mathematics of Diffusion, Second Edi., Oxford: Oxford University Press, 1999.
17. H. J. Frost and M. F. Ashby, Deformation-Mechanism Maps: The Plasticity and Creep of Metals and Ceramics, Pergaman Press, 1982

# PENETRATION OF SODIUM AND ELECTROLYTE TO VIBRATORY COMPACTION $TiB_2$ CATHODE

Zhaowen Wang[1], Yungang Ban[1], Zhongning Shi[1], Bingliang Gao[1], Dingxiong Lv[2], Chenggui Ma[2], Hongmin Kan[1], Xianwei Hu[1]
1. Northeastern University, Mail box 117, Shenyang City, Liaoning Province, 110004, CHINA
2. Design & Research Institute of Northeastern University, Shenyang City, Liaoning Province, 110006, CHINA
Keywords: Penetration; Vibratory compaction $TiB_2$; Sodium and electrolyte

## Abstract

Vibratory compaction $TiB_2$ and 30 mass% graphite contained block were used as cathodes in aluminum electrolysis. In the same condition, After 5 h electrolysis, The depth of electrolyte penetrated into the 30 mass% graphite cathode was about 13 mm, it was deeper than that into $TiB_2$ cathode which was less than 1 mm. SEM and EDS analysis showed that Na, F, Al penetrated slightly into the vibratory compaction $TiB_2$ cathode much less than that into the 30 mass% graphite cathode. Vibratory compaction $TiB_2$ cathode can decrease sodium penetration effectively, but it can not hinder sodium penetration thoroughly. The penetration resistance mechanism of vibratory compaction $TiB_2$ cathode to sodium and electrolyte was also discussed in this paper.

## Introduction

The structure of the industrial aluminum electrolysis cell was basically without significant change since the development of Hall-Héroult process. The anode and cathode are still made of carbon, for the active carbon cathode, there are some problems, and they include the side reaction of formation of $Al_4C_3$, which will increase the voltage drop across the carbon bottom. Therefore, in industry it needs to maintain a molten aluminum pad on the cathode surface. But liquid aluminum does not wet the carbon bottom well, a certain height of molten aluminum must always be present to assure a good physical and electrical contact with cell bottom, this layer of liquid aluminum is subject to electromagnetic forces under the influence of the high cell current. The anode-cathode distance can therefore be limited not less than 4~5 cm, so too much energy will be required to produce aluminum [1]. In recent years, much innovative research has been done about inert cathode. The research showed that $TiB_2$ maybe have more super properties than any other materials to replace carbon as inert cathode, such as high hardness(3350Hv)0.5N, high melting point(2980℃), low electric resistance(9μΩ), good absorption cross section for thermal neutrons(1524 barns), high wettability and good thermal shock resistance. More importantly, it is wetted by molten aluminum and it is resistant towards chemical attack by both aluminum and cryolite-alumina melts, which should create a lower anode-cathode distance, a decrease in electrolyte penetration, a reduction in energy consumption and potentially an increase in pot life. Besides, the thermal expansion coefficient (4.6μm/m/℃) of titanium diboride is close to that of graphite (4.3μm/m/℃) [2-8]. However, the high price of pure $TiB_2$ limits its large-scale application. A layer of $TiB_2$ (about 10mm) was pressed on the surface of the carbon for preparation of vibratory compaction $TiB_2$ cathode merely, so the cost of production reduces largely. Some charcoal ink factories developed production technology of vibratory compaction $TiB_2$ cathode in china in recent years. In order to investigate penetration resistance ability of $TiB_2$ cathode to sodium and electrolyte, vibratory compaction $TiB_2$ cathode and 30 mass% graphite contained block were elected as cathode materials in penetration experiments, compared the penetration resistance ability to sodium and electrolyte after 5 h electrolysis. The penetration resistance mechanism of vibratory compaction $TiB_2$ cathode to sodium and electrolyte was also discussed in this paper.

## Experiment

Vibratory compaction $TiB_2$ cathode (Φ50 mm) and 30% graphite cathode (Φ50 mm) were used as cathode materials in the experiments. The $TiB_2$ layer of vibratory compaction $TiB_2$ was 10 mm. The dimension graphite crucible used in the experiments was as following: outside diameter 110 mm, inside diameter 90 mm, highness 120 mm, thickness 10 mm. The composition of the electrolyte used was 87% $Na_3AlF_6$, 3% $CaF_2$, 5% LiF, 5% $Al_2O_3$ as in mass fraction, all the reagents used in the experiments were analytically pure. Vertical mode electrodes were adopted in the experiments, namely graphite crucible used as anode and containing electrolyte in the meantime, cathode material at the top of the cell and its bottom immersed in electrolyte 10 mm. In order to obtain high cathode current density, corundum tube was used at around of the cathode materials. The temperature of electrolysis was held at 980℃ by DWT-720 operating instrument (made in Shanghai sixth automated instrument plant). The cathode current density was held at 0.8 A/cm$^2$ by MPS-702 regulated power supply (Beijing tradex electronic technology corp.). The whole experiment was carried out under the protection of argon. The experimental facility was showed in figure 1. After 5 h electrolysis, phenolphthalein ($C_{20}H_{14}O_4$) reagent solution (1 g $C_{20}H_{14}O_4$, 100 mL $C_2H_5OH$, 100 mL $H_2O$) was used to detect the penetration depth of sodium. Phase analysis was proceeded on X' Pert Pro polycystic X-ray diffraction apparatus (made in panalytical B.V. Netherlands). Morphology and element distribution analysis were proceeded on SSX-550 scanning electronic microscope (made in Shimadzu Corporation, Japan).

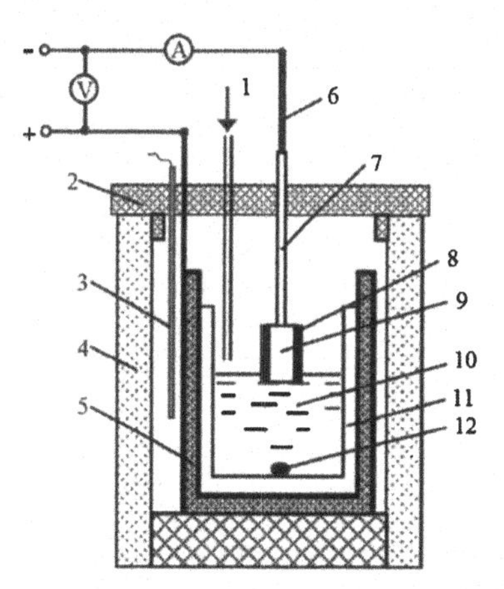


Fig.1 Experimental setup for penetration of sodium and electrolyte into cathode materials
1—Argon; 2—Heat retaining setup; 3—Electric thermo-couple; 4—Heat device; 5—Rustless steel crucible; 6—Cathode guard bar; 7—Corundum protector tube; 8—Corundum tube; 9—Cathode material; 10—Electrolyte

molten salt; 11—Graphite crucible; 12—Reduction aluminum

## Results and discussions

Penetration of sodium to vibratory compaction $TiB_2$ cathode

After electrolysis, cut cathode material from lengthwise direction to measure sodium penetration depth, namely put filter paper on the axle wire of cross-section of the cathode materials, the filter paper immerged in the phenolphthalein reagent solution beforehand. Because NaF or Na becomes magenta when they fall across phenolphthalein, From the off-color depth of phenolphthalein determined the penetration depth of NaF or Na. Figure 2 showed the penetration depth of NaF or Na. In the same condition, after 5 h electrolysis, the depth of electrolyte penetrated into the 30 mass% graphite cathode was about 13 mm, it was deeper than that into $TiB_2$ cathode which was less than 1 mm, illustrated that $TiB_2$ cathode have more superior penetration resistance ability to sodium than 30 mass% graphite cathode.

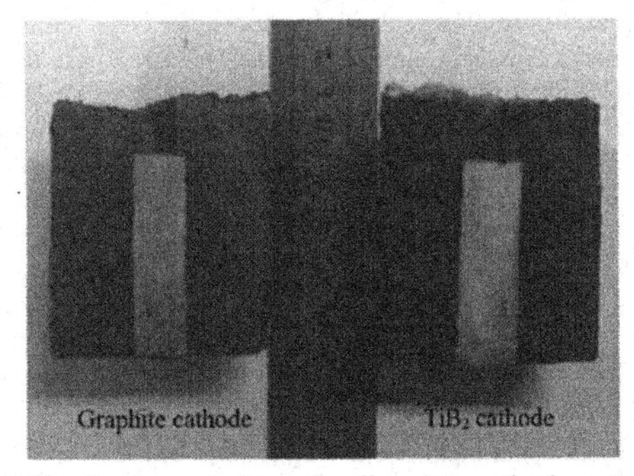


Fig.2 Depth of penetration of sodium into cathode materials

The penetration mechanism of sodium in vibratory compaction $TiB_2$ cathode and 30 mass% graphite cathode can be described as unsteady diffusion, which can be determined with Fick's law of diffusion, square of the penetration depth was in proportion to time. There are two dominating viewpoint about penetration mechanism of sodium and electrolyte to cathode materials. One is pure cryolite or aluminum didn't wet the surface of the cathode, so they can not penetrate into the micropore of carbon cathode, when aluminum dissolved in the cryolite which promoted the wettability wholelly, so sodium and electrolyte can penetrate into cathode materials. The other is the reaction of aluminum substitute sodium in the cryolite melt was proceeding uninterruptedly, at certain temperature, sodium penetrated in carbon materials and react with carbon, generated carbon-sodium chemical compound, the compound can be wetted by cryolite-alumina melt, so electrolyte can penetrate into carbon cathode materials[1].

In the electrolysis process, the aluminum generated at the cathode dissolved in the electrolyte partly, the wettability of the electrolyte and carbon become well, infiltration capacity increased in carbon. Furthermore, $Na^+$ may discharge at the surface of the cathode and form sodium, or aluminum substitute sodium in the electrolyte melt.

$$Na^+(l)+e^-=Na(l) \quad (1)$$

$$Al(l)+3NaF(cryolite)=3Na(l)+AlF_3(cryolite) \quad (2)$$

So sodium and aluminum accelerate the penetration of sodium and electrolyte to cathode materials.

At the same time, the penetration of sodium and electrolyte to cathode was also effect by electrocapillarity. Polarity of electrode was a pacing factor, the same carbon material, when it was used as anode, gas generate on its surface which exclude electrolyte, when carbon material used as cathode, the wettability become well when electrolysis.

Vibratory compaction $TiB_2$ and 30% graphite cathode are both porous materials. Electrolyte flow direction negative polarity through pore space when direct current pass cathode materials. This dynamics phenomenon is electrocapillarity, the mechanism can be described as in figure 3.

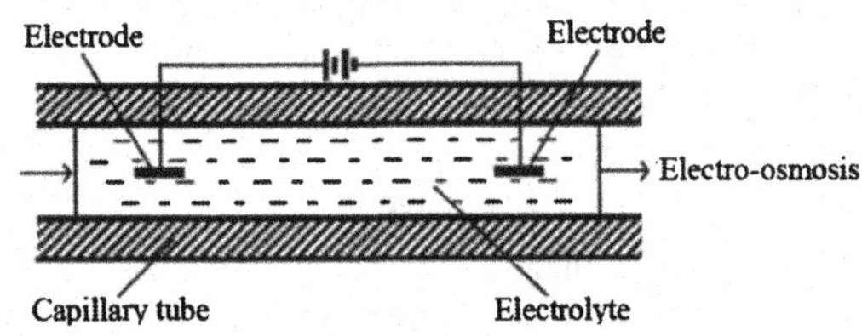


Fig.3 Mechanism of penetration of sodium and electrolyte into cathode

According to the standpoint of Flumkin, electrocapillarity can be interpreted by the theory of electric double layer. Electrocapillarity decrease boundary tension between the melt and carbon, the wettability become well, and electrolyte can penetrate into carbon, infiltration capacity increasing with the enlargement of the current density.

XRD analysis of cross-section of cathode materials

After 5 h electrolysis, the XRD analysis showed that NaF and $Na_3AlF_6$ existed in the graphite cathode, but NaF and $Na_3AlF_6$ were not found in the graphite of vibratory compaction $TiB_2$ cathode. Vibratory compaction $TiB_2$ cathode can decrease sodium penetration effectively.

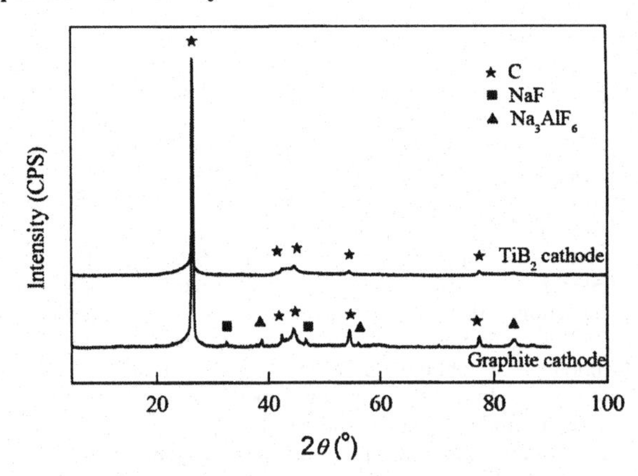


Fig.4 XRD patterns of cross-section of vibratory compaction $TiB_2$ cathode and graphite cathode after electrolysis

SEM and EDS analysis of cross-section of cathode materials

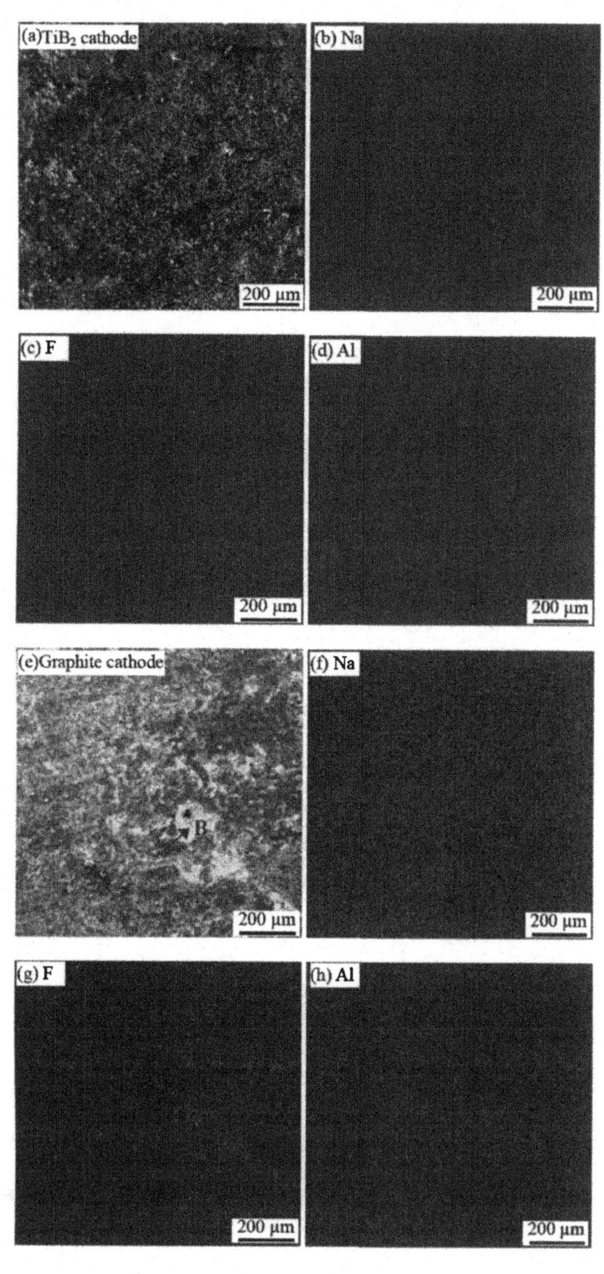


Fig.5 Morphology of cross-section of cathode materials and element distributions after electrolysis

Figure 5 showed the morphology of cross-section and element distributions of cathode materials 10 mm away from bottom after 5 h electrolysis. Figure 6 showed EDS analysis of A and B point in figure 5. There was only little sodium element in the base stock of vibratory compaction $TiB_2$ cathode. Vibratory compaction $TiB_2$ cathode can not hinder sodium penetration thoroughly; only slow up the penetration rate in a certain extent. $TiB_2$ cathode can be wetted by liquid aluminum, there was a layer of liquid aluminum at the surface of cathode, sodium generated by chemical or electrochemical reaction must get through liquid aluminum and $TiB_2$ layer before reaching the graphite base of vibratory compaction $TiB_2$ cathode. Diffusion rate of sodium in liquid aluminum and $TiB_2$ layer was very slow, so absorption rate of sodium by graphite base was slow. Actually, $TiB_2$ layer can not effect the equilibrium concentration of sodium in base cathode, only delay sodium absorption and decrease concentration gradient of sodium in graphite base.

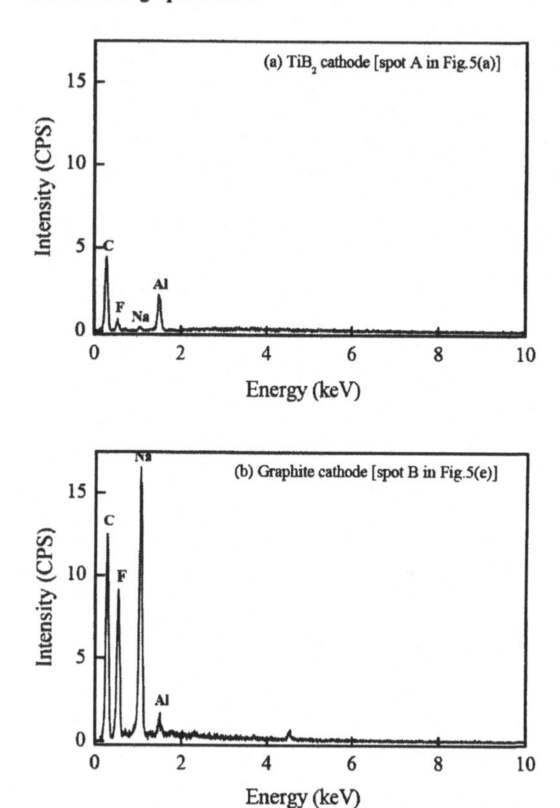


Fig.6 EDS spot analysis of cross-section of cathode

## Conclusions

1. Compared penetration resistance of vibratory compaction $TiB_2$ cathode and 30% graphite cathode to sodium and electrolyte. SEM and XRD analysis indicated that vibratory compaction $TiB_2$ cathode can resist penetration of sodium and electrolyte effectively than 30% graphite cathode.
2. $TiB_2$ cathode decreased sodium penetration effectively, but it can not hinder sodium penetration thoroughly.


## Acknowledgements

The authors wound like to acknowledge the financial supported by "National Natural Science Foundation of China (50674031)"


## References


1. Qiu Z.X. *Prebaked Aluminum Electrolyte*. Bei-jing: Metallurgy Industry Press, 2005, 465-590.

2. Ibrahiem M O, Foosnæs T, Øye H A. Stability of $TiB_2$–C Composite Coatings [A]. Travis J G Light Metals [C]. Warrendale: Minerals, Metals & Materials Society, 2006. 691–696.

3. Feng N X. Penetration of cryolite melt and sodium into cathode carbon blocks during electrolysis [J]. Acta Metallurgica Sinica, (6)1999: 611–617.

4. Marilou M C. Plant Experience with an Experimental Titanium Dioxide [A]. Alton T T. Light Metals [C]. Warrendale: Minerals, Metals & Materials Society, 2004. 399–404.

5. Qiu Z X, Kai G. $TiB_2$-coating on Cathode Carbon Blocks in Aluminum Cells [A]. Euel R C. Light Metals [C]. Warrendale: Minerals, Metals & Materials Society, 1992. 431–437.

6. Welch B. Future Materials Requirements for the High-energy Intensity Production of Aluminum [J]. J. Met., 2001. 53(2): 13–17.

7. Keller R. Wetting of Carbonaceous Cathode Material in the Presence of Boron Oxide [A]. Joseph L A. Light Metals [C]. Warrendale: Minerals, Metals & Materials Society, 2001. 753–756.

8. Sorlie M, Øye H A. Cathode in Aluminum Electrolysis [M]. Dusseldorf: Aluminum-Verlag Gmbh, 1994. 282–361.

**Light Metals 2008** *Edited by: David H. DeYoung*
***TMS (The Minerals, Metals & Materials Society), 2008***

# EFFECTS OF GRAIN GRADUATION ON TAPPED PACKING EFFICIENCY IN PREPARING $TiB_2$-C COMPOSITE MATERIAL FOR ALUMINUM ELECTROLYSIS

LÜ Xiao-jun[1], LI Jie[1], LAI Yan-qing[1], LI Qing-yu[1,2], TIAN Zhong-liang[1], FANG Zhao[1]

(1.School of Metallurgical Science and Engineering, Central South University, Changsha 410083, China;
2. School of Chemistry and Chemical Engineering, Guangxi Normal University, Guilin 541004, China)

Keywords: aluminum electrolysis, $TiB_2$-C composite material, Furnas model, linear packing, packing efficiency, grain graduation

## Abstract

The effects of the size distribution of coarse-fine binary particles and the volume fraction of fine particles on the tapped packing efficiency and porosity of $TiB_2$ ceramic powder were investigated according to the Furnas model and linear packing density theory. The results show that the tapped packing efficiency is higher whereas the porosity is lower when the size ratio ($R$) of coarse particles to fine particles increases. In addition, the tapped packing efficiency is also governed by the volume fraction of coarse or fine particles, and reaches its peak value when the volume fraction of fine particles is 30%~40%. Comparing the experimental results to the values calculated by Furnas model, it was confirmed that Furnas model can suit $TiB_2$ ceramic powder system, and the selection of the parameter $C_2$ in Furnas model was discussed according to experimental results.

## Introduction

In the present design of the Hall-Heroult cell, carbon material has been used as its cathode. However, a serious drawback of the carbon cathode is its non-wetting by molten aluminum, necessitating a substantial pool of metal. To avoid shorting between the metal and anode, the anode-cathode distance must be kept at a safe range, and a voltage drop between anode and cathode increases. It results in high energy consumption and low energy efficiency[1]. Therefore, there have been maded efforts to improve the $TiB_2$ inert wettable cathode in recent years. As pure $TiB_2$ ceramic cathode has many disadvantages, such as difficult in adhesion to the matrix and sintering, poor resistance to thermal shock and high cost and so on, these drawbacks are the main obstacle for the application of pure $TiB_2$ ceramic material. While, $TiB_2$-C composite cathode material, which are easy to join to the matrix and capability of being produced and fabricated economically into required shapes and so on, is regarded as the most potential one of inert wettable cathode materials for aluminum electrolysis[2]. In previous researches, studies about $TiB_2$ powder particle size distribution and the volume fraction of coarse-fine particles in $TiB_2$-C composite cathode has not been seen. The importance of particle packing in ceramic processing is generally appreciated because of its influence on the sintered mechanical properties, the shrinkage and density on sintering and the microstructure and properties of the final ceramic product. A high maximum packing density is directly dependent upon the particle size distribution[3-6]. In order to improve mechanical properties of materials, when a certain amount of small spheres is added to pores among large spheres, packing of different size spheres would bring into low porosity of packing.

To resolve the cracking resulted from excessive fine $TiB_2$ powder or unreasonable particle size ratio in the $TiB_2$-C material, and improve the density and mechanical properties of sinered material, the effect of the particle size distribution and the volume fraction of coarse-fine particles on the packing efficiency and porosity of $TiB_2$ powder system was studied in this paper. Theoretical particle size distribution of the most dense green compact was obtained. At the same time, on the basis of Furnas model[7] and the linear packing theory[8,9], $C_2$ parameter value in Furnas model was decided to suit $TiB_2$ powder system. Empirical equation about the most reasonable particle size ratio and forecasting the packing efficiency or porosity of $TiB_2$ particles packing was provided.

## Particle Packing Model

For the same diameter of the spheres, no matter what packing, the minimum porosity of packing is no less than 25.95%, and has nothing to do with the spheres size. The packing mechanism of the Furnas model is such that smaller particles are introduced and distributed to the interstices of larger packed particles so that the porosity is reduced. According to Furnas model, packing efficiency($E$) is the function of the volume fraction of coarse-fine particles and the size ratio in a binary mixture, and may be calculated through the following formular:

$$E = E_c + (1 - E_c)E_f F_1(\varphi_f)F_2(R) \quad (1)$$

Where, $E_c$ and $E_f$ are the packing efficiency of the coarse and fine particle fractions, respectively, $F_1$ is a function of the volume fraction of fine particles, $\varphi_f$; $F_2$ is a function of the size ratio of coarse particles to fine particles, $R$. The function $F_1$ is:

$$F_1(\varphi_f) = \left|e\varphi_f \ln \varphi_f\right|^{C_1} \quad (2)$$

Where $e$ is the base of a natural logarithm and $C_1$ is a constant. After comparing $C_1$ values for different binary systems, it was found that by taking $C_1$ as a function of $E_c$. The function is:

$$C_1 = \frac{5}{4E_c} \quad (3)$$

The fitting function $F_2$ is:

$$F_2 = \exp\left(\frac{-C_2}{R}\right) \quad (4)$$

Where $C_2$ is a constant, a smaller $C_2$ leads to a more rapid increase of the $F_2$ value when the particle size ratio increases. In Furnas model, when $E_c = E_f$=0.5~0.6, the $C_2$ value may equal 4.

After obtaining the functions of $F_1$ and $F_2$, eqn (1) is combined with eqn (2), (3) and (4), to get the full descriptive equation for calculating the packing density of a binary mixture:

$$E = E_c + (1 - E_c)E_f \left|e\varphi_f \ln \varphi_f\right|^{\frac{5}{4E_c}} \exp\left(-\frac{4}{R}\right) \quad (5)$$

Furnas model describes the packing behavior of the ideal binary particles, and reflects that packing efficiency decreases with the reduction of the size ratio of coarse to fine particles, while packing efficiency also depends on the volume fraction of coarse and fine particles. But, due to different systems or different particle size, initial particles packing efficiency $E_c$, $E_f$ are also different. For a binary powder system having $E_c \neq E_f$, eqn (5) can still be used when the volume of the coarse powders is dominant in the mixture. For a system in which the volume of fine particles is dominant, $E_f$, may be used as the first term in eqn (5), instead of $E_c$.

Linear packing density model is based on the assumption that the particles are not affected by particles agglomerates, electrostatic and van der waals forces in the process of particles packing. However, the actual particle packing process is not the ideal state of model assumptions, so it exists the deviations between the model and experimental results. In order to match closely the actual situation, A.B.Yu[8] investigated that the effect of absolute particle size on particle packing be modelled by use of an initial porosity to take into account the packing of mono-sized particles and the concept of packing size ratio as a measure of the particle-particle interaction in forming a packing of mixed powders. When the packing is dominated by gravity and the size of mono-sized particles is measured by the average sieve diameter (d), the porosity of mono-sized particles is expressed by the following eqn (6),

$$\varepsilon_d = \varepsilon_0 + (1 - \varepsilon_0)\exp\left(- a d^{\,b}\right) \qquad (6)$$

Where $\varepsilon_0$ is the limiting porosity when gravity is dominant. The parameters in eqn (6) is listed in table 1.

Table 1 Parameters in eqn.(6) and (7) for the packings

| parameter | Poured packing | Tapped packing |
|---|---|---|
| $\varepsilon_0$ | 0.567 | 0.433 |
| $a$ | 0.247 | 0.446 |
| $b$ | 0.749 | 0.579 |
| $p$ | 0.293 | 0.593 |

The equivalent size ratio of fine to coarse particles is introduced to measure interaction between two components, and relation between the equivalent size ratio of fine to coarse particles and the actual size ratio of fine to coarse particles is represented by a powder law function.

$$r_{ij} = R_{ij}^{\;p} \qquad (7)$$

Where $r_{ij}$ is the equivalent size ratio of fine to coarse particles, $R_{ij}$ is the actual size ratio of fine to coarse particles, parameter $p$ is given in table 1.

Because the density of $TiB_2$ powder is big, the packing is dominated by gravity. Eqn(5) is combined with eqn(6), and (7), to calculate the packing efficiency of $TiB_2$ powder system.

### Experimental

The sintered pure $TiB_2$ ceramic material (density ≥ 96%) is broken and screened into the different size powder. To select a size as the main powder particles, fine particles distribute the voids among the main particles. Mixing uniformly, measuring cylinder is filled with powder in the action of gravity. The quality of powder is weighed, the tapped density is obtained by volume weight method. In the experimental, FZS4-4 vibration density analyzer is used to test the tapped density under the same vibration frequency and amplitude. The effect of particles size ratio and volume fraction of binary particles powders on tapped packing efficiency was investigated.

Tapped packing efficiency = tapped packing density / powder density

Comparing the values calculated with the experimental results, the suitability of applying Furnas model in $TiB_2$ powder system is tested, and parameter $C_2$ in model is adjusted to fit the experimental data, so the proposed equation makes predicting the packing density of the mixture system more accurate.

### Experimental Results and Discussion

According to eqn.(6) and table 1, the tapped porosity and the packing efficiency for different size of mono-sized particles may be calculated, the results calculated are listed in table 2.

Table 2 the tapped packing efficiency ($E$) calculated of different particle size of $TiB_2$ powder

| Particle size/μm | 2030 | 265 | 195 | 77 | 50 |
|---|---|---|---|---|---|
| $E$ | 0.567 | 0.567 | 0.567 | 0.565 | 0.559 |

The diameter of coarse particles is 2030 μm, mixed with different volume fraction of fine particles, 265μm and 77μm, respectively. The experimental results are listed in table 3.

Table 3 the experimental tapped packing efficiency for the system of coarse particles (2030μm)

| $R$ | $\varphi_f$ | | | | | | | | |
|---|---|---|---|---|---|---|---|---|---|
| | 0.1 | 0.2 | 0.3 | 0.4 | 0.5 | 0.6 | 0.7 | 0.8 | 0.9 |
| 7.7 | 0.592 | 0.622 | 0.637 | 0.640 | 0.630 | 0.614 | 0.596 | 0.580 | 0.570 |
| 26.2 | 0.61 | 0.665 | 0.688 | 0.689 | 0.684 | 0.67 | 0.651 | 0.633 | 0.615 |

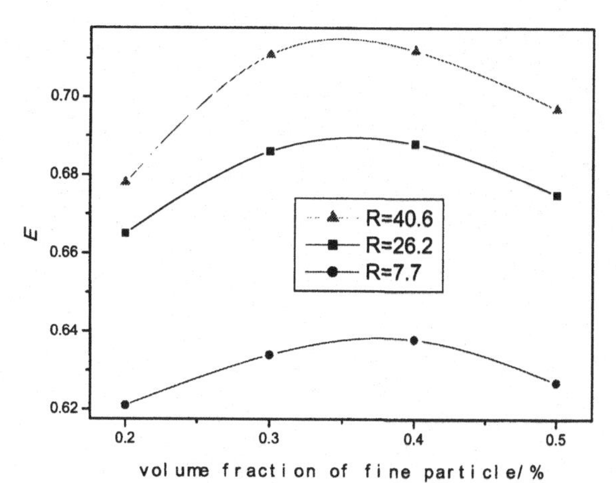


Fig. 2 The effect of volume fraction of different fine particle size on tapped packing efficiency

As can be seen from Fig.1, when the size ratio of coarse to fine particles reduces, the tapped packing efficiency of the system correspondingly decrease. When the volume fraction of fine particles in the system is small, the tapped packing efficiency is small. this is because the part of void among coarse particles is not still filled by fine particles, the porosity of the packing is large. Fine particles gradually distribute the interparticle gap with the increase of volume fraction of fine particles, so the vibration packing efficiency increases. When the volume fraction of fine particles was 30-40%, the vibration packing efficiency is maximum, increasing continuely fine particles, the vibration packing efficiency decreases. The main reason is that the specific surface area of the particles increases with the increase of fine particles, thus, the fine particles are easy to form agglomeration and interparticle friction increase since interparticle forces such as electrostatic and van der waals forces become more important. The vibration force can not effectively overcome interparticle friction in the process of packing, so that only a handful of fine particles can enter the gap between coarse particles, while, the majority of fine particle adhesion on the surface of coarse particles, or agglomerate, or become a bridge, so the packing porosity increases, and the tapped packing efficiency decreases.

As can be seen from Fig.2, with the decrease of the size ratio of coarse to fine particles, the tapped packing efficiency of the system reduces significantly, and it is trend that the volume fraction of fine particles with the maximum tapped packing efficiency increases. This phenomenon is well in according with observations reported elsewhere[10].

To confirm the validity that Furnas model is applied to $TiB_2$ powder system. The parameter $C_2$ value of the equation suited $TiB_2$ powder system is obtained. When $C_2$ equals 4, according to eqn.(5) and (7), the experimental results and the values calculated of different size ratio are listed in table 4.

Table 4 the experimental results and the values ($E$) calculated of different size ratio ( $C_2$=4 )

| $R$ | $\varphi_f$ | | | | | | | |
|---|---|---|---|---|---|---|---|---|
| | 0.2 | | 0.3 | | 0.4 | | 0.5 | |
| | calculated | measured | calculated | measured | calculated | measured | calculated | measured |
| 4.4 | 0.602 | 0.601 | 0.612 | 0.610 | 0.613 | 0.612 | 0.608 | 0.606 |
| 7.7 | 0.622 | 0.622 | 0.638 | 0.637 | 0.641 | 0.640 | 0.632 | 0.630 |
| 10.4 | 0.634 | 0.621 | 0.654 | 0.640 | 0.657 | 0.642 | 0.646 | 0.632 |
| 26.2 | 0.670 | 0.665 | 0.699 | 0.688 | 0.704 | 0.689 | 0.686 | 0.675 |
| 40.6 | 0.683 | 0.674 | 0.716 | 0.705 | 0.721 | 0.710 | 0.703 | 0.694 |

As can be seen from table 4, for $TiB_2$ powder system, when the size ratio of coarse particles to fine particles $R$ equals 4.4 and 7.7, the value calculated can closely fit the experimental data. It is shown that the empirical equation (5), linear packing equation (6) and equation (7) are suitable for this system. However, when $R$ >10.4, the experimental data is smaller than the values calculated. it indicates that the size of fine particles is smaller with the increase of the size ratio $R$ (coarse/fine), when the size of the main particles is unvariable. The function of a gravity-dominated packing become worse with the reduce of the particle size, the contribution of interparticle forces such as electrostatic and van der waals forces in packing become more significant. the particles are easy to agglomerate, or become a bridge, so the capacity of fine particles distributing the void among the coarse particles in packing become poor. Thus it exists the deviations between the values calculated and experimental results.

If $C_2$ equals 3, 5 and 6, comparing the values calculated to the experimental data in Fig.3, 4 and 5, it is found that the curve of $C_2$=5 closely fits the experimental data through adjusting contant $C_2$ value.

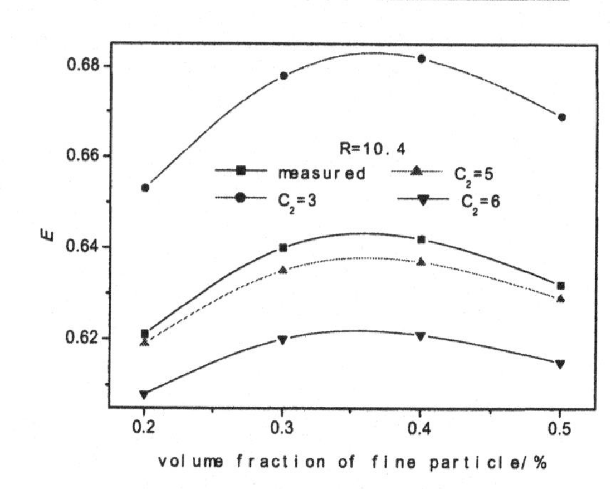


Fig.3 $C_2$ versus the tapped $E$of system (R=10.4)

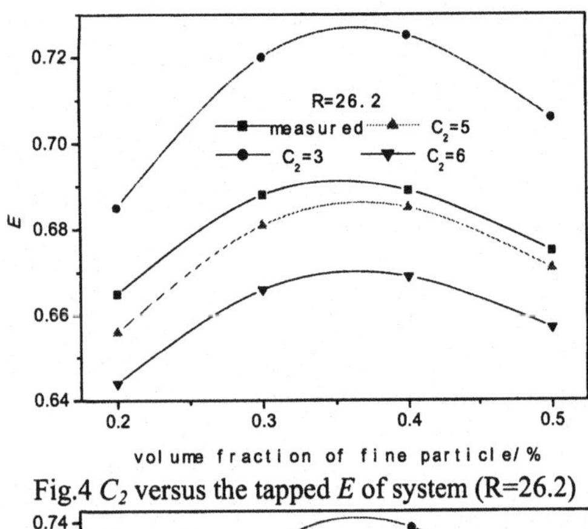


Fig.4 $C_2$ versus the tapped $E$ of system (R=26.2)

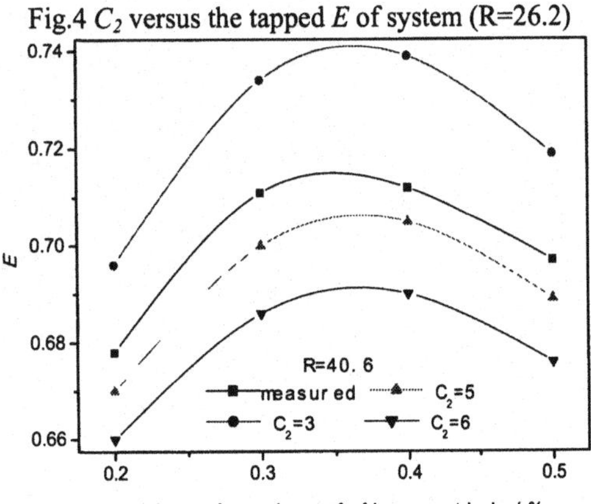


Fig.5 $C_2$ versus the tapped $E$ of system (R=40.6)

## Conclusions

1) in the $TiB_2$ powder system, with the increase of the size ratio of coarse particles to fine particles ($R$), the tapped packing efficiency increases, and the tapped porosity decreases. The tapped packing efficiency reaches the maximum when the volume fraction of fine particles is 30% to 40%.

2) when the size ratio of coarse particles to fine particles ($R$) is smaller ($R \leq 7.7$), empirical equation (5) is suitable for the system.

3) when the size ratio of coarse particles to fine particles ($R$) is larger than 10.4, empirical equation suited for the system may be expressed the following formular:

$$E = E_C + (1 - E_C)E_f \left| e\varphi_f \ln \varphi_f \right|^{\frac{5}{4E_C}} \exp\left(-\frac{5}{R}\right).$$


## Acknowledgement

Project Supported by the National Basic Research Program of China (2005CB623703) and Applied Basic Research in Guangxi Province (GUIKEJI0639032).


## References


[1] LIU Ye-xiang. Research Progress of Inert Anode and Wettable Cathode For Aluminum Electrolysis[J](in Chinese). Light Metals, 2001(5): 26-29.

[2] Curtis J McMinn. A Review of Rhm Cathode Development[A]. In: Euel R Cutshall, Eds. Light Metals1992. Warrendale PA, USA: TMS, 1991: 419 ~ 425.

[3] Guo A, Beddow J K, Vetter A F. A Simple Relationship Between Particle Shape Effects and Density, Flow Rate and Hausner Ratio[J]. Powder Technology, 1985,43(3): 279-284.

[4] Bierbrauer J, Edel Y. Dense Sphere Packing from New Codes[J]. Journal of Algebraic Combinatories. 2000,11(2): 95-100.

[5] Dinger D. R. One-Dimensional Packing of Spheres (Part □): Monodisperse Distributions[J]. Am Ceram Soc Bull, 2000, 79(2):71-76.

[6] Dinger D. R. One-Dimensional Packing of Spheres (Part □): Monodisperse Distributions[J]. Am Ceram Soc Bull, 2000, 79(4):83-91.

[7] Zheng J M, Carlson W B, Reed J S. The Packing Density of Binary Powder Mixtures[J].Joural of the European Ceramic Society, 1995,15(5):479-483.

[8] Yu A. B., Bridgwater J, Burbidge A. On the Modeling of the Packing of Fine Particles[J]. Powder Technology, 1997,92(3):185-194.

[9] Stovall T, Delarrard F, Bull M. Linear Packing Density Model of Grain Mixture[J]. Powder Technology,1986,48(1):1-12.

[10] Westman, A.E.R.& Hugill, H.R., The Packing of Particles[J], J.Am.Ceram.Soc.,1930, 13(10):767-7.

## Inert Anode

*SESSION CHAIRS*

**Odd-Arne Lorentsen**
Hydro Aluminium
Porsgrunn, Norway

**Jomar Thonstad**
Norwegian University of Science and Technology
Trondheim, Norway

**Light Metals 2008** *Edited by: David H. DeYoung*
***TMS (The Minerals, Metals & Materials Society), 2008***

# INERT ANODES: AN UPDATE

Rudolf P. Pawlek
TS+C, Avenue du Rothorn 14, CH-3960 Sierre, Switzerland

Keywords: primary aluminium, electrowinning, inert anodes, cermets, metal anodes, low temperature electrolyte

## Abstract

This overview covers the development of inert anodes for the primary aluminium industry in the period 2003 -2007. It continues the review of cermets, including their mechanical and physical properties, their behaviour and their manufacture, especially Cu ($NiO$-$NiFe_2O_4$) cermets in cryolite melts. However, the overview focuses particularly on the manufacture and behaviour metal anodes, including steels and Ni-Fe alloys. These alloys must be passivated before use in electrolysis, otherwise they will dissolve in the cryolite electrolyte. Low temperature electrolytes are used to avoid aggravated corrosion; KF-based electrolytes have proved suitable for low temperature electrolysis. Multi-polar electrolysis in a slurry electrolyte using metal electrodes could be a solution in the future. Inert anodes were tested on laboratory and batch scales, but no information about industrial scale tests is yet available.

## Introduction

The development of inert anodes for the Hall-Heroult process to produce aluminium has been reviewed several times in the past [1-11]. The use of an inert anode instead of carbon eliminates the formation of greenhouse gas $CO_2$, when oxygen liberated from alumina reacts with the standard carbon anode. It also eliminates the perfluorocarbon (PFC) by-products and other polluting emissions such as PAHs. This update is a continuation of [3].

## Cermets

While oxides as inert anodes for the aluminium industry have received only limited attention during the last years [12, 13], more detailed studies concern the fabrication of Al-Ni-Cu-O cermets [14], Al-Ti-Fe-O cermets [15] and Sn-Sb-Fe-O rutile type materials [12]. Samples of the composition as reported in [14] were fabricated close to theoretic density and were tested in lab scale experiments. The authors observed that significant deterioration of the anode was due to an increasing NiO content in the anode and to the generation of a new NiO-phase in the anode itself which resulted in swelling and delamination. However, delimanation took place only in the outer part of the anode due to electrolyte composition. The authors of [15] fabricated a new inert anode based on $Al_2O_3$, $TiO_2$ and Fe as raw materials. During electrolysis they found a new phase $Fe_2TiO_5$ in the anode. The dispersed metal phase increases the electrical conductivity in the anode, and the composition Al-Ti-Fe-O showed good corrosion resistance and stable cell voltage during the electrolysis test. During the examination of the ceramic degradation in cryolite alumina melts, Abakumov et al [12] found the following tendencies for the Sn-Sb-Fe-O rutile materials: selective iron dissolution; possibility to decrease the quantity of dissolved iron by decreasing its content in the ceramic sample; and downtrend for tin content in the melt after initial growth. The slow decrease of the tin concentration in the melt is probably due to the formation of volatile tin compounds.

The composition Fe-Ni-$Al_2O_3$ as inert anode material was tested by Shi et al. [16] and Cao et al. [17], who examined the manufacture of such anodes. They added Fe and Ni powders to $Al_2O_3$, then pressed this at 160 MPa and sintered at 1500°C. The addition of Fe and Ni increases the electrical conductivity, which changes from semi-conducting to metallic behaviour with increasing temperature. Such inert anodes were tested by Shi et al [16] in a 300 A bench scale electrolysis cell for 10 hours at 960°C. During electrolysis the metallic phase Fe-Ni was oxidised by the nascent $O_2$ to form $Fe_2O_3$ and NiO. The dissolved iron and nickel were then deposited into the aluminium metal. In addition, $Fe_2O_3$ and NiO reacted with alumina to form $FeAl_2O_4$ and $NiAl_2O_4$. This change of the anode surface composition may result from the consumption of iron and nickel. In the anode layer surface $Al_2O_3$ was left alone as a protection. After this, loss of Fe and Ni continued by diffusion towards the anode surface, where they oxidised and then dissolved in the cryolite-based electrolyte. The Fe-Ni-$Al_2O_3$ composite anode presented good thermal shock resistance, good electrical conductivity, as well as low corrosion rates. Qiu, Shi and Xu [18] tested Fe-Ni-$Al_2O_3$ inert anodes in a low temperature electrolyte comprising 8wt.% NaCl and 4 wt.% $CaF_2$, besides cryolite and alumina. Analysis of the anode layers showed a shell of $Na_3AlF_6$ and $FeAl_2O_4$, while the core consisted of complicated phases including $Al_2O_3$, NiO and $FeAl_2O_4$ ceramic and of $Ni_3Fe$ and $Ni_3Al$ metallic phases. These phases in the core were dispersed from each other. The wear rate of such anodes was reported to be 19 mm per year at an electrolysis temperature of 850°C.

In an attempt to develop an inert anode, Qiu, Shi and Xu [19] proposed a Fe-Ni-Co-$Al_2O_3$ cermet, in which Fe, Ni and Co were metal powders and $Al_2O_3$ was the ceramic material. While Fe, Ni and Co serve only for the electrical conductivity, $Al_2O_3$ was the sinter-aid to form aluminate spinels. These spinels are good electrical conductors and can be bonded together to form a solid anode. Nascent oxygen on the anode surface enhances the formation of spinels. Nascent oxygen forms spinels such as $NiFe_2O_4$, $FeAl_2O_4$, $NiCo_2O_4$. In addition Fe and Ni will form intermetallic compounds such as $Ni_3Fe$. Unreacted alumina on the anode surface may dissolve in the molten bath, but it does not contaminate the metal purity. The corrosion rate of such anode was calculated from the mass loss to be 24 mm per year.

Xie [20] prepared $Cu_2O$-$10CuAlO_2$- xCu cermets as candidate inert anodes for aluminium production via hot pressing, and studied their electrical conductivity. Such materials can act as electrical conductors when the copper content of the mixture is more than 15 wt.%. The electrical conductivity increases with decreasing particle size of the metal phase and the porosity. The particle size of the metal phase decreases with increasing pressure, but increases with increasing temperature.

However, more attention was paid to $NiFe_2O_4$ as an inert anode material. Galasiu et al. [21] prepared such inert anodes either by mechanical mixing of the components on one side or by coprecipitation as hydroxides of Ni and Fe and then calcining the mixture at 1350°C on the other side. The densities of the mixes prepared by coprecipitation were higher than with mechanical mixing. Their apparent porosity was lower, their electrical conductivity was better, and the quality of inert anodes manufactured by coprecipitation was better than that of those manufactured through mechanical mixing of the components. Hot-pressing sintering for the manufacture of $NiFe_2O_4$ was preferred by Wang [22], while Liu [23] simulated the grain graduation composition of $NiFe_2O_4$ spinel biscuit on the ball mill arrangement theory by means of packed density experiments. Liu calculated beforehand four main grains to be used in the manufacture to obtain the closest packing of the anode material. He calculated that the inert anode should have the following composition: 42% coarse grain, 18% medium grain and 40% fine powder. Yan et al. [24] used NiO, $Fe_2O_3$ and nitrides to produce $NiFe_2O_4$-based ceramic bars. They used solid-state reaction and compacting-sintering technique for the manufacture. Abakumov et al [12] discovered selective dissolution of iron from spinel oxides, especially from nickel ferrite, and Lai [25] reported that the solubility in melts of Ni and Fe from $NiFe_2O_4$ is 0.0085% Ni respectively and 0.070% Fe. Yan et al. [24] reported that $NiFe_2O_4$ dissolves into the electrolysis baths non-stoichiometrically with dissolved Fe/Ni mass ratios always greater than stoichiometric ratio of $NiFe_2O_4$. A bath ratio of 1.38 was used in the tests. Bath penetration occurred along the grain boundaries where the Fe and Ni from the $NiFe_2O_4$ could diffuse and dissolve into the bath, causing formation of intra-pores between the grains within the reaction zone until the bath within the pores was saturated both with Fe and with Ni. Therefore, the use of $NiFe_2O_4$ bars with large grain size and an optimized particle size distribution would be beneficial for minimizing bath penetration. However, depletion of Fe in the $NiFe_2O_4$ matrix would lead to porous microstructure within the reaction zone, thus facilitating bath penetration. Luo et al. [26] also observed that nickel ferrite is a kind of N-type semiconductor and its mechanism of electrical conduction makes $Fe^{3+}$ much more easily corroded by the electrolyte during electrolysis. The corrosion mechanism is proposed to be: $NiFe_2O_4 + 2AlF_3 = NiO + 2FeF_3 + Al_2O_3$.

$NiFe_2O_4$ with Cu was the next step for research. Tao et al.[27] manufactured test anodes by hot pressing. The relative density of their anodes was better than 99% and the electrical conductivity was about 75 $\Omega cm^{-1}$ at 900°C bath temperature. They found that copper enriches in the outer surface of the anode bar due to high copper mobility. This, however, results in an increase in copper contamination in the produced aluminium. Laboratory electrolysis tests were also reported by Wu [28], who determined the corrosion rate of such anodes to be 3 mm per year. Lai and co-workers [29] also measured the electrical conductivity of Cu-$NiFe_2O_4$ cermets and they found in experiments at 960°C bath temperature, that the electrical conductivity increases from 9.09 $Scm^{-1}$ to 9.45 $Scm^{-1}$ with copper content in the cermet increasing from 5% to 10%.

Then Zhang et al. [30] examined the effect of the copper content on the microstructure and mechanical properties of Cu/(10NiO-$NiFe_2O_4$) cermets. Such cermets were produced by cold pressing and 1500°C sintering temperature. The relative density of the samples increases with a Cu-content of up to 5 wt.%; the bending strength increases with increasing Cu-content of up to 15 wt.%, and the fracture toughness increases with increasing Cu-content up to 20 wt.%. Then Wang et al. [31] tested 5Cu-(10NiO-$NiFe_2O_4$) inert anodes in three different electrolysis bath compositions: a traditional $Na_3AlF_6$-$AlF_3$-$Al_2O_3$ high temperature bath; KF-$AlF_3$-$Al_2O_3$ low temperature bath; and $K_3AlF_6$-$Na_3AlF_6$-$AlF_3$-$Al_2O_3$ multiple low temperature bath. The cermet inert anode was corroded at a stoichiometric ratio and spalled off layer by layer. From the anodic corrosion the total weight of the three anode components (Fe, Ni and Cu) was higher in the high temperature bath electrolyte than in the two low temperature electrolytes, and it is mooted that the multi cryolite system has potential interest for further electrolysis tests. Tian et al. [32] examined the corrosion effect with increasing copper content from 0 to 20 wt.%. Their results indicate that the Cu content in the anodes had little effect on the steady state concentrations of Ni in the electrolyte; the concentration of Fe decreases from 304 ppm to 206 ppm and Cu increases from 21 ppm to 70 ppm with increasing Cu-content from 0 to 20 wt. %. This result was also confirmed by Lai et al. [33]. The cermet composition 5Cu/10NiO-90$NiFe_2O_4$ showed the best corrosion resistance.

Also Liu et al. [34] examined nickel ferrite spinel as an inert cermet anode for aluminium electrolysis and they added $TiO_2$ and $MnO_2$ to $NiFe_2O_4$ in order to improve its properties. The total metal amount (75% Ag and 25% Cu) was 15% of the anode weight. Chemical corrosion was found to be: 0.18% solubility of NiO in $Na_3AlF_6$ melt containing 5 wt.% $Al_2O_3$ and 0.003% $Fe_2O_3$. Electrochemical corrosion was explained as follows: $2Al + NiFe_2O_4 = NiAl_2O_4 + 2Fe$. The addition of $TiO_2$ or $MnO_2$ may affect the corrosion resistance of the spinel in cryolite- alumina mix molten salts.

Laurent and Gabriel [35] proposed the use of an inert anode consisting of NiO-$NiFe_2O_4$-Cu-Ni, and they prepared this from mixed oxide powders and metal powders, isostatically pressed in cold state and then sintered at 1350°C under a controlled atmosphere containing at least one inert gas. Lai et al. [33] determined the total impurity concentration of such cermets to be 378 ppm in the bath. The corrosion behaviour was examined by Luo et al. [36], who observed that Cu and Ni were present as $Cu_xNi_y$ alloy in the anode. The nickel of $Cu_xNi_y$ was primarily oxidized in oxygen at 900°C. In the polarization experiment the Ni content of the anode decreased significantly and the residual metal was Cu. Nickel ferrite spinel only corroded in the presence of oxygen, and $Fe_2O_3$ of the nickel ferrite dissolved in the electrolyte.

Another alternative as inert anode could be the use of Ni-$NiFe_2O_4$ composition. Lai [37] examined such cermets with respect to their densification affected by ball milling time, particle size of raw powders, contents of metallic phase, sintering atmosphere and temperature. Weak reductive atmosphere is favourable to densification as well as sintering temperature up to 1300°C. Li [38] examined the thermal stress under complex boundary conditions using a finite-element software. The corrosion behaviour under electrolysis conditions was examined by Lai [39] and Qin [40]. While Lai [39] used an artificial neural network-based solution methodology for modelling the anode corrosion, Qin [40] used electrolysis parameters. The results show that the corrosion rates increase rather slowly with $Al_2O_3$ concentration near saturation in the electrolyte but that depletion of alumina may lead to catastrophic corrosion. The anodes perform well in cryolite ratio range from 2.2 to 2.4 at about 960°C. Dissolved aluminium and high current density have a detrimental effect on anode performance. Principally, the corrosion mechanism appears to be aluminothermic reduction with dissolved aluminium, fluoridation of the anode matrix and subsequent dissolution in the bath. Lai et al. [29] measured the electrical conductivity, which increases from 24.39 $Scm^{-1}$ to 69.41 $Scm^{-1}$ with increasing Ni-content from 5 wt% to 20 wt.%.

Then Lai et al. [41] prepared Ni-NiO-$NiFe_2O_4$ cermets with various NiO and Ni contents. They investigated the corrosion behaviour in $Na_3AlF_6$-$Al_2O_3$ melts. Post electrolysis micro examination of anodes showed that nickel may be oxidized first and then transferred to the melt by chemical dissolution. Solubility tests demonstrated that with increasing NiO concentration in the ceramics the iron content in electrolyte melts decreased and the nickel concentration in the melt increased. However, if nickel in the cermet was oxidized first by oxygen to form NiO then the corrosion would be a chemical dissolution of newly formed NiO-$NiFe_2O_4$ and such process would slow the cermet corrosion significantly. The cermet composition: 5 Ni-10NiO-$NiFe_2O_4$ behaved best in electrolysis tests. The average value of total impurity concentrations for cermets with the composition 17Ni/(NiO-$NiFe_2O_4$) was found by Lai et al. [33] to be 304 ppm in the electrolyte.

In order to improve the electrical conductivity of $NiFe_2O_4$ cermets, Liu et al. [42] added $TiO_2$, which proved to significantly improve the conductivity. Zhang et al. [43] added of SiC whiskers to $NiFe_2O_4$ and concluded that the appropriate sinter technique for such a composite was 1180°C and 6 hours. Then Xi et al. [44-46] added $V_2O_5$ to $NiFe_2O_4$ cermets. This formed $Ni_2FeVO_6$, which has a low melting point, and allows liquid phase sintering. The formation of $Ni_2FeVO_6$ increased corrosion resistance only when adding up to 1 wt.% of $V_2O_5$; any more $V_2O_5$ proved to be detrimental. Samples with grain size graduation have better properties than those without. With appropriate grain graduation, silver additions may form a network structure in the ceramic structure. The inert anodes showed best properties with 0.50-0.355 mm grain size. Furthermore $V_2O_5$ additions to $NiFe_2O_4$ cermets improve the corrosion resistance of inert anodes. Among other inert anode materials, Lai [25] also proposed the use of $ZnFe_2O_4$ and $ZnAl_2O_4$. However, the solubilities of zinc and iron, which form zinc ferrite, are 0.0313% and 0.070% respectively, while that of zinc aluminate is 0.0265%.

Also, some more hitherto unreported Alcoa patents have been found. Inert anodes with complicated structure were found by Ray, Liu and Weirauch [47] with addition of CoO to $NiFe_2O_4$ and by Ray et al. [48] with the addition of ZnO to $NiFe_2O_4$. In another invention, Dawless et al. [49] propose to conduct the nascent oxygen bubbles upwards in such a way that the dissolution of alumina into the bath will be optimized. Sulphur impurities can be removed by for example submerging a purifying electrode in the bath [50]. Of much more practical use could be patent [51]. Here a solid material comprising from about 40 to 80 wt.% cryolite, 2 to 25 wt% alumina and about 5 to 25 wt.% cementitious binder circumscribes an anode system in an electrolysis cell. These are inert anodes which are fixed at a top metal plate. When suspending such a protected inert anode block into an electrolysis bath it is necessary that a remaining solid structure thickness of between 40 and 70% is maintained to provide a sufficient anode surface for cell electrolysis.

Hydro Aluminium was also active, proposing the manufacture of a stable inert anode [52] consisting of especially nickel, iron and chromium oxides and a metal such as copper and/or silver being the binder. These electrodes are arranged in a vertical or inclined position in the electrolysis cell [53] where the cell design facilitates the separation of aluminium and the evolved oxygen. The anodes can also be shaped in teeth form [54] to increase the active anode surface.

Aluminium Pechiney patents show that. besides the described electrode composition in [35] Tailhades et al. [55] also propose an inert anode comprising a spinel mixed oxide and a metal which is entirely or partly reducible by the electrolysis process. Lamaze [56] developed an assembly which comprises an inert anode in the form of a pocket, a connecting conductor, a mechanical connection, and a soldered metal joint which is arranged between one surface of the open extremity of the anode and an extremity of the connection of the conductor.

**Metallic anodes**

During recent years, non-ferrous and ferrous metal alloys were tested as inert anodes in laboratory scale electrolysis cells. Of special interest were studies of anode bubble gas behaviour and of anodic overvoltage, and the development of low temperature electrolytes.

Cassyre et al [57] observed in lab tests that the form of gas evolution is much the same for tin oxide, copper and/or copper-nickel anodes: tiny bubbles escape continuously without much coalescence, forming a foam around the electrode. The bubble diameter was estimated to be about 0.1 mm. which is 10 to 30 times smaller than on graphite anodes. The depth of the gas layer under the anode grows regularly with increasing current density until reaching 2 – 2.5 mm at 1 $A/cm^2$. Increasing current density over 1 $A/cm^2$ does not increase the penetration of the gas under the anode. Gao and co-workers [58] confirmed this finding. The average bubble diameter decreases with increasing current density. The bubble generation process on a metal anode can be divided into three steps: anode oxidation, oxygen generation and bubble departure. The bubble diameter was reported to be about 5.5 mm at a current density of 0.05 $A/cm^2$ and about 2.5 mm in the case of a current density of 0.18 $A/cm^2$. Then Kiss et al. [59] simulated the convection of the bubble laden electrolyte layer around inert anodes. The results have shown that good wetting of the inert anode results in the generation of smaller bubbles. Smaller bubbles create weaker fluctuation of the anode coverage. The average value of the covering of inert anodes does not exceed 22-30%.

In experiments Thonstad et al. [60] tested the anodic overvoltage of inert anodes made of 40Cu-30Ni-30Fe (wt%). Stirred melts yielded straight Tafel lines and an overvoltage at 1 $A/cm^2$ in the order of 0.4V.

As described in [3], the corrosion of inert anodes is very much dependent on the bath conditions. The lower the bath temperature the slower is the anode corrosion. This spurs the development of optimum low bath temperature compositions. Operating temperatures below 950°C and $AlF_3$-NaF-KF-$CaF_2$-$Al_2O_3$ bath compositions have been reported by [61, 62]; operating temperature below 940°C and $AlF_3$-NaF-KF-$CaF_2$-$Al_2O_3$ bath composition has been reported by [63-65] ; operating temperature at 870°C and $Na_3AlF_6$-$AlF_{3(in\ excess)}$ -$Al_2O_3$ bath composition has been reported by [66]; operating temperature at 870°C and NaF-$AlF_3$-$BaF_2$-$CaF_2$ bath composition was reported by [67]; operating temperature at 850°C, and NaF-$AlF_3$-NaCl-$CaF_2$-$Al_2O_3$ bath composition was reported by [18, 68]; operating temperature between 660-860°C, and bath composition NaF-$AlF_3$-KF-LiF was reported by [69]; operating temperature 750°C and NaF-$AlF_3$ bath composition was reported by [60]; and operating temperatures below 700°C and $AlF_3$-KF-$Al_2O_3$ bath composition was reported by [31, 70-73]. However, electrolytes including potassium exclude the use of carbon lining.

Non-ferrous metal alloys were tested with only minor importance. While Antipov et al. [74] examined the electrochemistry of Ni, Fe

and Cu as well as their binary alloys NiFe, $Ni_3Fe$, $Ni_3Al$, NiCu and $Ni_9Cu$, Glucina and Hyland [75] used aluminium bronze.
Due to availability and price, steels and other ferrous metals and alloys were most tested as inert anodes in aluminium electrolysis cells. While Cao et al. [76] used steel anodes for the manufacture of aluminium-iron master alloys, Galasiu et al. [77] used chrome-nickel steels as inert anodes. The latter found that with increasing chrome content (up to 24 wt.%) and increasing nickel content (up to 19.4 wt.%) corrosion decreases, and at the highest mentioned concentrations corrosion was "zero". However, before electrolysis the anodes were exposed to a thermal treatment in air. Lainer [78], Lyakishev et al. [79] and Kim et al. [80] used steels as inert anodes. They noted that these materials cannot be used without preliminary preparation, otherwise massive oxide layers form on the anode surface causing subsequent exfoliation. Probably, under electrolysis conditions, the oxygen evolving on the anode surface forms an oxide film first which prevents the anode body from dissolution, followed by an increase of cell voltage, after which the film on the anode surface increases the overvoltage of oxygen liberation.
Nickel-based alloys were often used as inert anode materials for aluminium electrolysis. The following compositions were reported: Ni-Al [81], Ni-Cu [81], Ni-Fe [66, 68, 81-88], Ni-Cu-Al [89, 90], Ni-Cu-Cr [89], Ni-Fe-Cu [60, 63, 64, 89, 91-93], Ni-Fe-Co [61], Ni-Fe-Cu-Al-Nb [63], Ni-Fe-Co-Al-Nb [63].
Degradation of metals as inert anodes was observed by Simakov et al. [81] according to either low or high resistance of solid products. Low resistance products tend to promote internal corrosion, resulting in formation of a porous metal layer. High-resistance products cause pronounced suppression of oxygen evolution and a rather high ohmic drop. Thus for successful anode operation it seems essential to minimize the content of low conductive phases and films. But the metal anode surface also needs protection against dissolution. For Ni-Fe-based anodes, it was proposed to create a barrier layer inhibiting the diffusion of fluoride and oxygen ion species to the anode core [94], the outer layer may also be an integral nickel-iron oxide [66]. The protection may also be created by applying a colloidal slurry [95, 96]. This may also be a cobalt oxide layer [91, 97] or a nickel-cobalt mixed oxide coating [88].
It was also observed that metal corrosion can be slowed by using low bath temperature and dissolved high alumina concentration in the bath. Better alumina dissolution results from stronger bath circulation [98], which can be generated by the escape of evolved oxygen through flow-through openings in the electrodes [99] or by sloped inter-electrode gaps in such a way that the flow of the electrolyte is assisted by anodically produced oxygen bubbles [100]. In this respect the anode form plays an important part. This may consist of a number of active anode members spaced apart and parallel to one another [101] or a grid-like open structure with V-shaped configuration in cross section. [102].
Inert metal anodes have their biggest advantage when used in conjunction with aluminium wettable cathodes in electrolysis cells with oxygen-resistant cell side walls [103]. The cathodes may have a V-shaped section [104] and may be equipped with an aluminium collector reservoir [105]. Antille et al. [106] modelled a semi-industrial scale electrolysis cell for a current load of 25 kA, while von Kaenel and de Nora [107] made a technical and economical evaluation for this electrolysis cell.
The US Department of Energy [108] sponsored another project: this was to produce aluminium using low temperature electrolyte baths (temperatures less than 700°C) continuously fed with alumina, together with metal anodes and wettable cathodes. According to Barnett, Mezner and Bradford [109, 110], several anodes and cathodes, e.g. permanent electrodes, sit in the "mushy" (partly solid phase low temperature) electrolyte. While the anodes consisted of a cast Cu-Fe-Ni plate, the cathodes were titanium diboride slabs [92], and a 6% slurry of alumina was maintained in the molten electrolyte. The electrolytic cell was preferentially equipped with a reservoir for colleting molten aluminium [111] and the anodes were equipped with slots to fix the cathode slabs [112]. In general, multicellular, bipolar electrodes have the advantage of power saving and absence of anode effects [113]. The cast anodes have the advantage that they produce a very hard protective coating during use in the cell. Bergsma et al. [69] also proposed a homogenisation phase for about 8 hours at 1100°C in air. By homogenisation, the normally two metallurgical phases in the cast state change into a single phase. This has the big advantage that it offers a more uniform microstructure for the anode surface, with less structures subject to oxidation. This mushy electrolyte electrolysis was successfully tested in lab scale tests.

**Conclusions**

Many groups are researching a wide range of cermets, as well as steels and nickel alloys, but steels and Ni-Fe-based alloys must be passivated before use in electrolysis, otherwise they will dissolve in the cryolite electrolyte. However, a radical change from conventional pot line design may be necessary.
Low temperature KF-based electrolytes reduce solubility and rate of chemical attack, so multi-polar electrolysis in a slurry electrolyte using metal electrodes could be a solution in the future.
Inert anodes have their biggest advantage when used in conjunction with aluminium wettable cathodes in electrolysis cells with oxygen-resistant cell walls.
Inert anodes were tested on laboratory and batch scales, but no information about industrial scale tests is yet available.

**Acknowledgement**

The author acknowledges with thanks the help of Mrs. G. Brichler and Mr. A. Bushnell during the preparation of this manuscript.

**References**

[1] H. Kvande, "Inert anodes – latest developments", Industrial Aluminium Electrolysis, Theory and Practice of Primary Aluminium Production, September 8-12 (2003), Quebec Canada
[2] R. J. Batterham and M. J. Hollitt, Erzmetall 56 (2003) 9, pp. 519-528
[3] R. P. Pawlek, "Inert anodes: an update", Light Metals 2004, ed. A. T. Tabereaux (TMS, Warrendale, Pa), pp. 283-287
[4] S. Karpel, Metal Bulletin Monthly 403 (July 2004), pp. 40-41
[5] Anonymous, "Alcoa to build new anode plant to serve Fjardaal in Iceland and Mosjoen in Norway", Alcoa press release 16 November 2004
[6] W. T. Choate, A. M. Aziz and R. Friesman, Light Metals 2005, ed. H. Kvande (TMS, Warrendale, Pa), pp. 495-500
[7] X. S. Zhang, "Current status of inert anodes study in aluminium electrolysis", J. Materials and Metallurgy 4 (2005) 1, pp.13-16
[8] W. Choate and J. Green, Light Metals 2006, ed. T. J. Galloway (TMS, Warrendale, Pa), pp. 445-450

[9] R. P. Pawlek, "Inert anode: a threat or challenge to the carbon industry?", 2nd International Carbon Conference, Kunming, September 17-19, 2006
[10] J. W. Evans, JOM 59 (2007) 2, pp. 30-38
[11] I. Galasiu, R. Galasiu and J. Thonstad, "Inert anodes for aluminium electrolysis", Aluminium-Verlag Düsseldorf (2007), 207 pp.
[12] A. M. Abakumov et al., X. Aluminium Siberia, Krasnoyarsk 7-10 September 2004, 12 pp.
[13] A. Krishnan, C. P. Manning and U. B. Pal, "Zirconia-based inert anodes for green sysnthesis of metals and alloys", Proceedings of Yazawa Int. Symp. on Metallurgical and Materials Processing: Principles and Technologies; Vol. 3, Aqueous and Electrochemical Processing (TMS, Warrendale, Pa) San Diego 2-6 March 2003, pp. 351-364
[14] W. Wen et al, Light Metals 2005, ed. H. Kvande (TMS, Warrendale, Pa), pp. 535-538
[15] X. Cao et al., Light Metals 2007, ed. M. Sørli (TMS, Warrendale, Pa), pp. 927-929
[16] Z. Shi et al., "300A bench scale aluminium electrolysis cell with Fe-Ni-$Al_2O_3$ composite anode", Light Metals 2005, ed. H. Kvande (TMS, Warrendale, Pa), pp. 1219-1223
[17] X. Cao et al., "Study on the conductivity of Fe-Ni-$Al_2O_3$ cermet inert anode", Light Metals 2007, ed. M. Sørlie (TMS, Warrendale, Pa), pp. 937-939
[18] Z. Qiu, Z. Shi and J.L. Xu, Can. Met. Q., 43 (2004) 2, pp. 259-264
[19] Z. Qiu, Z. Shi and J. Xu, "Metal-alumina cermet inert anode in a 100 A bench scale cell for aluminium electrolysis", Aluminium 80 (2004) 3, pp. 219-222
[20] N. Xie, Journal of the American Ceramic Society 88 (2005), pp. 2589-2593
[21] I. Galasiu et al., "Inert anodes for aluminium electrolysis; variation of nickel ferrite ceramics as a function of the way of preparation", Proceedings of The Int. Jomar Thonstad Symposium, ed. A. Solheim and G.M. Haarberg (October 16-20, 2002, Trondheim, Norway), pp. 171-176
[22] Z. Wang, Xiyou Jinshu Cailiao Gongcheng 34 (2005), pp. 158-161
[23] Y. H. Liu, "Study on grain composition of the biscuit for preparing inert anode based on $NiFe_2O_4$ spinel", EPD Congress 2004 (TMS, Warrendale, Pa), pp. 507-512
[24] X.Y. Yan, M.I. Pownceby and G. Brooks, Light Metals 2007, ed. M. Sørlie (TMS, Warrendale, Pa), pp. 909-913
[25] Y.-Q. Lai, "Solubility of composite oxide ceramics in cryolite alumina melts", J. Central South Univ. Technol. 34 (2003) 3, pp. 245-248
[26] T. Luo, Z. Wang and X. Yu, Light Metals 2007, ed. M. Sørlie (TMS, Warrendale, Pa), pp. 941-943
[27] L. Tao et al., "Preparation of cermet inert anode based on ferrous nickel and its use in an electrolysis study", Light Metals 2005, ed. H. Kvande (TMS, Warrendale, Pa), pp. 541-543
[28] X. Wu, Guizhou Gongye Daxue Xuebao 35 (2006) 3, pp. 22-24, 28
[29] Y. Lai et al., "An improved pyroconductivity test of spinel-containing cermet inert anodes in aluminium electrolysis cells", Light Metals 2004, ed. A. T. Tabereaux (TMS, Warrendale. Pa), pp. 339-344
[30] G. Zhang et al., Light Metals 2007, ed.M. Sørlie (TMS, Warrendale, Pa), pp. 931-936
[31] J. Wang et al., "Investigations of 5Cu-(10NiO-$NiFe_2O_4$) inert anode corrosion during low-temperature aluminium electrolysis", Light Metals 2007, ed. M. Sørlie (TMS, Warrendale, Pa), pp. 525-530
[32] Z. Tian et al., Light Metals 2007, ed. M. Sørlie (TMS, Warrendale, Pa), pp. 915-919
[33] Y. Lai et al., "On the corrosion behaviour of $NiFe_2O_4$-NiO based cermets as inert anodes in aluminium electrolysis", Light Metals 2006, ed. T. J. Galloway (TMS, Warrendale, Pa), pp. 495-500
[34] Y.Liu et al., Light Metals 2006, ed. T. J. Galloway (TMS, Warrendale, Pa), pp. 415-420
[35] V. Laurent and A. Gabriel, "Method for the manufacture of an inert anode for the production of aluminium by means of fusion electrolysis", EP patent 1,601,820 (12 March 2003)
[36] T. Luo et al., Light Metals 2006, ed. T. J. Galloway (TMS, Warrendale, Pa), pp. 491-493
[37] Y. Q. Lai, "Densification of Ni-$NiFe_2O_4$ cermets for aluminium electrolysis", Transactions of Nonferrous Metals Society of China 15 (2005) 3, pp. 666-670
[38] J. Li, Acta Metallurgica Sinica 18 (2005) 5, pp. 635-641
[39] Y. Q. Lai, "Corrosion analysis and corrosion rates prediction of $NiFe_2O_4$ cermet inert anodes", Zhongnan Daxue Xuebao 35 (2004) 6, pp. 896-901
[40] Q. W. Qin, Zhongnan Daxue Xuebao 35 (2004) 6, pp. 891-895
[41] Y. Lai et al., "On the corrosion behaviour of Ni-NiO-$NiFe_2O_4$ cermets as inert anodes in aluminium electrolysis", Light Metals 2005, ed. H. Kvande (TMS, Warrendale, Pa), pp. 529-534
[42] Y.-H. Liu et al., J. Mater.Metall. 2 (2003) 4, pp. 271-273
[43] S. T. Zhang, G. C. Yao and Y. H. Liu, "Study on sintering technique of $NiFe_2O_4/SiC_w$ used as matrix of inert anodes in aluminium electrolysis", Light Metals 2006, ed. T. J. Galloway (TMS, Warrendale, Pa), pp. 501-504
[44] J. H. Xi, G. C. Yao and Y. H. Liu, Light Metals 2006, ed. T. J. Galloway (TMS, Warrendale, Pa), pp. 479-483
[45] J. H. Xi et al., "Effect of $V_2O_5$ on electrical conductivity of cermet inert anodes", Light Metals 2006, ed. T. J. Galloway (TMS, Warrendale, Pa), pp. 485-489
[46] J. H. Xi et al., Light Metals 2007, ed. M. Sørlie (TMS, Warrendale, Pa), pp. 921-926
[47] S. P. Ray, X. Liu and D. A. Weirauch, "Cermet inert anode containing oxide and metal phases useful for the electrolytic production of metals", EP patent 1,666,640 (1 August 2000)
[48] S. P. Ray et al., US patent 6,821,312 (1 April 2002)
[49] R. K. Dawless et al., US patent 5,938,914 (19 September 1997)
[50] A. F. Lacamera et al., US patent 6,866,766 (5 August 2002)
[51] L. E. d'Astolfo and C. Bates, "Inert anode assembly", WO patent 2004/049467 (25 November 2002)
[52] S. Julsrud, WO patent 03/035,940 (25 September 2002)
[53] O.-J. Siljan and S. Julsrud, "A method and an electrowinning cell for production of metal", EP patent 1,364,077 (23 October 2001)
[54] O.-A. Lorentsen, O.-J. Siljan and S. Julsrud, EP patent 1,552,039 (23 August 2002)
[55] P. Tailhades et al., "Inert anode for producing aluminium by igneous electrolysis and method for producing said anode", EP patent 1,689,900 (7 October 2003)
[56] A.-P. Lamaze, WO patent 2005/033,368 (30 September 2003)
[57] L. Cassyre et al., "Gas evolution on graphite and oxygen-evolving anodes during aluminium electrolysis", Light Metals 2006, ed. T. J. Galloway (TMS, Warrendale, Pa), pp. 379-383

[58] B. Gaoet al., Light Metals 2006, ed. J. T. Galloway (TMS, Warrendale, Pa), pp. 467-470
[59] L.I. Kiss et al., "The bubble laden layer around inert anodes", Light Metals 2007, ed. M. Sørlie (TMS, Warrendale, Pa), pp. 495-500
[60] J. Thonstad, A. Kisza and J. Hives, Light Metals 2006, ed. T. J. Galloway (TMS, Warrendale, Pa), pp. 373-377
[61] V. de Nora, T. T. Nguyen and J.-J. Duruz, "Aluminium electrowinniung cells with metal-based anodes", WO patent 2004/074,549 (20 February 2003)
[62] T. T. Nguyen and V. de Nora, EP patent 1,763,595 (7 May 2004)
[63] V. de Nora, T. T. Nguyen and J.-J- Duruz, "Aluminium electrowinning cells with metal-based anodes", EP patent 1,554,416 (17 October 2003)
[64] V. de Nora et al., Aluminium 83 (2007) 1-2, pp. 48-53
[65] V. de Nora et al., "Semi-vertical de Nora inert metallic anode", Light Metals 2007, ed. M. Sørlie (TMS, Warrendale, Pa), pp. 501-505
[66] J.-J. Duruz, T. T. Nguyen and V. de Nora, EP 1,244,826 (6 December 2001)
[67] J. Xu et al., "Aluminium electrolysis in a low temperature heavy electrolyte system with Fe-Ni-$Al_2O_3$ composite anodes", Light Metals 2007, ed. M. Sørlie (TMS, Warrendale, Pa), pp. 507-511
[68] Z. Shi, J. Xu and Z. Qiu, Light Metals 2004, ed. A. T. Tabereaux (TMS, Warrendale, Pa), pp. 333-337
[69] S. Craig et al., "Cu-Ni-Fe anode for use in aluminium producing electrolytic cell", US patent 7,077,945 (8 May 2003)
[70] J. Yang et al, Light Metals 2004, ed. A. T. Tabereaux (TMS, Warrendale, Pa), pp. 321-326
[71] J. Yang, J. N. Hryn and G. K. Krumdick, "Aluminum electrolysis tests with inert anodes in KF-$AlF_3$-based electrolytes", Light Metals 2006, ed. T. J. Galloway (TMS, Warrendale, Pa), pp. 421-424
[72] V. A. Kryukovsky et al., Light Metals 2006, ed. T. J. Galloway (TMS, Warrendale, Pa), pp. 409-413
[73] D. R. Boyd et al., "Process for electrolytic production of aluminium", EP patent 1,689,913 (5 November 2003)
[74] E. V. Antipov et al., Light Metals 2006, ed. T. J. Galloway (TMS, Warrendale, Pa), pp. 403-408
[75] M. Glucina and M. Hyland, "Laboratory scale testing of aluminium bronze as an inert anode for aluminium electrolysis", Light Metals 2005, ed. H. Kvande (TMS, Warrendale, Pa), pp. 523-528
[76] D. Cao et al., Light Metals 2007, ed. M. Sørlie (TMS, Warrendale, Pa), pp. 483-486
[77] I. Galasiu et al., "Metallic inert anodes for aluminium electrolysis", 12th Int. Al Symposium September 7-10, 2003, Bratislava (CD-rom)
[78] Y. A. Lainer, Izv. V.U.Z. Tsvetn. Metall. (2004) 3, pp. 50-59
[79] N. P. Lyakishev et al., "Inert anodes synthesis and testing", ICSOBA, Travaux Vol. 31, No 35 (2004), pp. 29-34
[80] Yu. V. Kim, T. N. Drozda and D. A. Simakov, paper presented at XI Int. Conf. Aluminium Siberia 2005, Krasnoyarsk 13 September 2005
[81] D. A. Simakov et al., "Nickel and nickel alloys electrochemistry in cryolite-alumina melts", Light Metals 2007, ed. M. Sørlie (TMS, Warrendale, Pa), pp. 489-493
[82] O. Crottaz and J.-J. Duruz, EP patent 1,105,553 (30 July 1999)
[83] B. H. Boost, "Wettable cathodes and inert anodes", IX Int. Conf. September 9-11, 2003, Proc. Aluminium of Siberia, Krasnoyarsk 2003, pp. 153-159
[84] J.-J. Duruz and V. de Nora, EP patent 1,567,692 (3 December 2003)
[85] C- McMinn and V. de Nora, "Inert anodes for aluminium production", Aluminium 80 (2004) 3, pp. 135-140
[86] V. de NoraATP Aluminium 1 (2004) 2, pp. 38-41
[87] V. de Nora, "Inert anodes are ready for the use in aluminium production cells", Aluminium 81 (2005) 1-2, pp. 67-70
[88] T. Nguyen and V. de Nora, Light Metals 2006, ed. T. J. Galloway (TMS, Warrendale, Pa), pp. 385-390
[89] Z. Shi et al, "Copper-nickel superalloys as inert alloy anodes for aluminium electrolysis", JOM 55 (2003) 11, pp. 63-65
[90] T. T. Nguyen and V. de Nora, WO patent 03/078,695 (12 March 2003)
[91] R. von Kaenel et al., "The de Nora inert metallic anode for aluminium smelting", Aluminium 82 (2006) 3, pp. 162-166
[92] C. W. Brown and C. S. Bergsma, US patent 6,723,222 (22 April 2002)
[93] S. C. Bergsma, "Carbon containing Cu-Ni-Fe anodes for electrolysis of alumina", US patent 6,692,631 (15 February 2002)
[94] V. de Nora and T. T. Nguyen, WO patent 02/083,990 (12 April 2001)
[95] T. T. Nguyen and V. de Nora, "Non-carbon anodes for aluminium electrowinning and other oxidation resistant components with slurry-applied coatings", WO patent 03/087,435 (16 April 2002)
[96] T. T. Nguyen and V. de Nora, EP patent 1,495,160 (14 April 2003)
[97] V. de Nora and T. T. Nguyen, "High stability flow-through non-carbon anodes for aluminium electrowinning", EP patent 1,756,334 (3 June 2004)
[98] V. de Nora, EP patent 1,224,340 (26 October 1999)
[99] V. de Nora, "Aluminium electrowinning cells with oxygen-evolving anodes", EP patent 1,416,067 (10 January 2000)
[100] V. de Nora, WO patent 01/31,088 (25 October 1000)
[101] V. de Nora, "Alloy-based anode structures for aluminium production", WO patent 03/006,716 (13 July 2002)
[102] V. de Nora, WO patent 03/023,092 (29 August 2002)
[103] J.-J. Duruz, V. de Nora and G. Berclaz, "Aluminium electrowinning cell with sidewalls resistant to molten electrolyte", EP patent 1,230,436 (1 November 1999)
[104] V. de Nora, US patent 6,682,643 (17 April 2000)
[105] V. de Nora, "Aluminium electrowinning cells having a drained cathode bottom and an aluminium collector reservoir", WO patent 02/097,168 (28 May 2001)
[106] J. Antille et al., Light Metals 2006, ed. T. J. Galloway (TMS, Warrendale, Pa), pp. 391-396
[107] R. von Kaenel and V. de Nora, "Technical and economical evaluation of the de Nora inert metallic anode in aluminium reduction cells", Light Metals 2006, T. J. Galloway (TMS, Warrendale, Pa), pp. 397-402
[108] Anonymous, JOM 57 (2005) 8, p. 7
[109] R. J. Barnett, M. B. Mezner and D. R. Bradford, "Maintaining molten salt electrolyte concentration in aluminium-producing electrolytic cell", US patent 6,837,982 (25 January 2002)
[110] R. J. Barnett, M. B. Mezner and D. R. Bradford, US patent 6,800,191 (15 March 2002)

[111] D. R. Bradford, R. J. Barnett and M. B. Mezner, "Electrolytic cell for production of aluminium from alumina", US patent 6,811,676 (16 July 2002)
[112] D. R. Bradford, R. J. Barnett and M. B. Mezner, US patent 6,866,768 (17 December 2002)
[113] M. Kirko et al., "Aluminium electrolysis cells on bipolar electrodes with MHD control of process", paper presented at XI Int. Conf. Aluminium Siberia, Krasnoyarsk 13 September 2005

**Light Metals 2008** *Edited by: David H. DeYoung*
***TMS (The Minerals, Metals & Materials Society), 2008***

# Effects of the NaF to AlF3 Ratio on Fe-Ni-$Al_2O_3$ Anode Properties for Aluminum Electrolysis

Junli Xu[1], Zhongning Shi[2], Bingliang Gao[2], Zhaowen Wang[2]

1. College of Science, Northeastern University, Shenyang, China 110004
2. College of Materials and Metallurgy, Northeastern University, Shenyang, China 110004

Keywords: inert anode, low temperature aluminum electrolysis, light electrolyte, corrosion

## Abstract

Fe-Ni-$Al_2O_3$ anodes for aluminum electrolysis were prepared by powder metallurgy method and were tested in electrolytes with varying molar ratios. It was discovered that the anode corrosion rate increased with decreasing molar ratio of NaF to $AlF_3$. The appropriate ratio was between 1.6~1.8, and the produced aluminum purity was about 98%, while the corrosion rate of the anode was about 24mm/y. The effects of varying the ratio of NaF to $AlF_3$ on the corrosion rate of Fe-Ni-$Al_2O_3$ anode are explained.

## Introduction

Rapid corrosion rate is one of the application obstacles for inert anodes. The satisfactory corrosion rate of an inert anode in the operation cell should be less than 10 mm per year**[1]**. However, it is difficult to find an inert anode material..

The anode corrosion rate might be reduced by modifying the electrolyte composition [2-7]. One of the solutions to reduce the inert anode corrosion is to lower the electrolyte temperature. From the phase diagram of NaF-$AlF_3$, we can conclude that there is a eutectic point about 695 °C at the cryolite ratio (NaF/$AlF_3$) of 1.12.Using lower cryolite ratio can lower the melt point of the electrolyte, which may reduce the inert anode's solubility in the electrolyte. Good results were reported in a OIT report by using the electrolyte with a cryolite ratio (CR) of 0.56 [7].

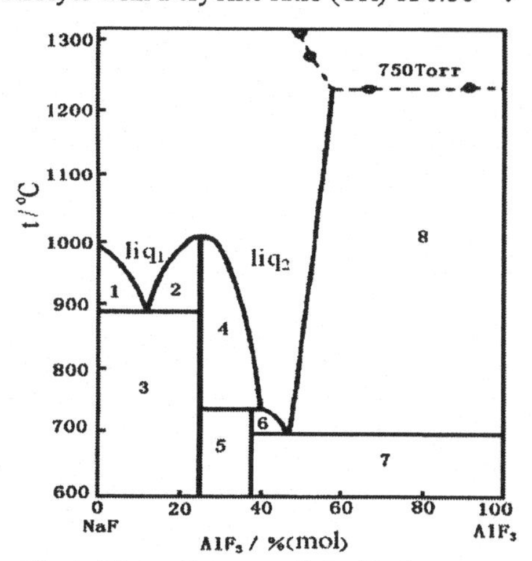


**Fig.1 Phase diagram of the NaF-$AlF_3$ system** [8]

1-NaF+liq1; 2-$Na_bAlF_{3+b}$+liq1; 3-NaF+$Na_3AlF_6$; 4-$Na_3AlF_6$+liq2; 5-$Na_3AlF_6$+$Na_5Al_3F_{14}$; 6-$Na_5Al_3F_{14}$+liq2; 7-$Na_5Al_3F_{14}$+$AlF_3$; 8-$AlF_3$+liq2

In the present work we tested the performance of Fe-Ni-$Al_2O_3$ anodes at different cryolite ratios CR=1.12, 1.25, 1.4, 1.5, 1.6, 1.8). The purpose was to lower the inert anode solubility in cryolite-based melts.

## Experimental

The fabrication process of the Fe-Ni-$Al_2O_3$ anodes is the same as given by reference **[9]**. The experimental cell is shown in Fig.2.The anode was weighed and the dimensions measured before each experiment. After an experiment was finished, the electrodes were taken out, leaving the electrolyte and aluminum metal in the two-compartments cell. The anode was put into a 30% $AlCl_3 \cdot 6H_2O$ solution at 80°C for several hours to remove the electrolyte from the anodes. Anodic weight loss and volume (by immersion) were determined.

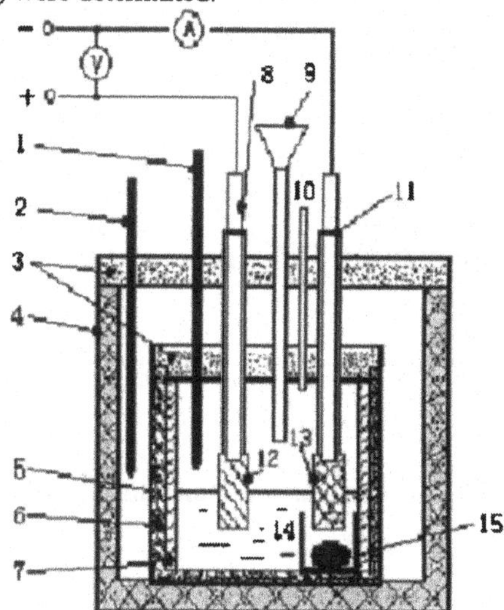


**Fig.2 Schematic diagram of the electrolytic cell used for aluminum electrolysis [9].**

1- Measuring thermocouple 2- Controlling thermocouple 3- lid 4- Furnace 5-Ferrite crucible 6- Graphite crucible 7- alumina insulation sleeve liner 8- anode rod & alumina sleeve 9- alumina feeder tube 10- vent-pipe 11-cathode rod & alumina sleeve.12- Inert anode 13- Carbon cathode 14- Electrolyte 15- Aluminum

Electrochemical tests were performed at 825 °C ( when CR=1.6 ) and 850 °C (when CR=1.8), respectively, using a three electrodes cell configuration and an Autolab instrument. The counter electrode was a carbon plate and the reference electrode was a molten Al electrode, which was contained in a sintered $Al_2O_3$ tube with a small hole. The tested electrode was impregnated with the electrolyte for 30 min before measurement. The dynamic polarization curves of the electrodes were measured by scanning the electrode potential from -1.5 to 2.5 V (versus open-circuit potential) at the scan rate of 10 mV/s. The EIS spectra of the electrodes were obtained in the frequency range from 2000 Hz to 0.5 Hz with an AC amplitude of 5 mV applied to the electrode under the open-circuit condition. In our experiments, we carried out cathodic polarization sweeps before each anodic polarization measurement in order to reduce the oxides on the electrode surface.

## Results and discussion

Cell Voltage

Cell voltage of the cell at different mole rate of $NaF/AlF_3$ is shown in Fig.3. It indicated that the electrolysis process was stable, although the cell voltage was larger than that in a typical cell configuration, which is mainly due to the alumina tube (between the anode and cathode compartment) increasing the electrolyte resistance. Most of the cell voltage ascended after several hours of electrolysis and the electrolysis had to stop. However, it was stable at CR=1.8. These indicated the electrolyte with CR=1.8 was more suitable for the used Fe-Ni-$Al_2O_3$ anodes.

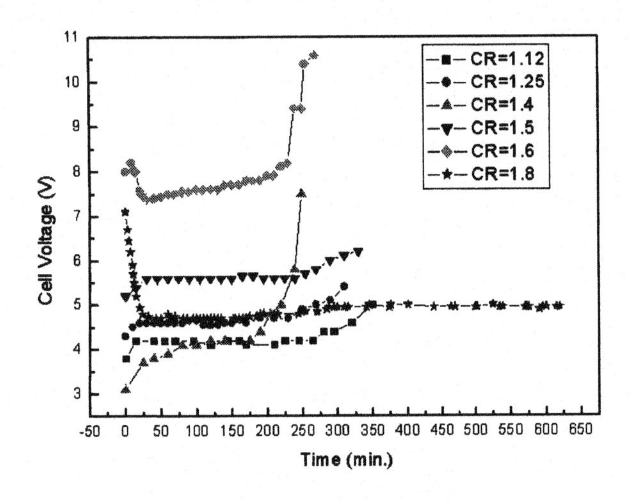


**Fig.3 Plot of cell voltage vs. time during electrolysis**

Anode Corrosion Rate

After electrolysis, It can be seen from anodes the appearance that the Fe-Ni-$Al_2O_3$ anodes corrosion rate is very severe when CR=1.12 and 1.25, while there was only small changes when CR=1.6 and 1.8. The anode corrosion rates were determined as described in reference [9], and the results are given for different cryolite ratios in Fig.4.

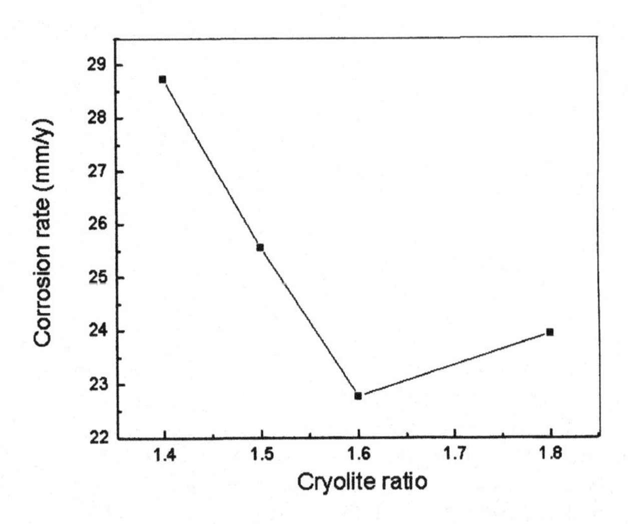


**Fig.4 Anodic corrosion rate as a function of cryolite ratio**

It can be seen From Fig.4 that the anodic corrosion rate increased with decreasing cryolite ratio. The lowest corrosion rate was about 23 mm/y at cryolite ratio 1.6.

Aluminum purity

A chemical analysis method was used to determine the aluminum purity. Since the impurities coming from the anode material are only Fe and Ni elements, only these elements were determined. The results are shown in Table II.

**Table II Impurities in aluminum metal produced**

| CR | 1.12 | 1.25 | 1.4 | 1.5 | 1.6 | 1.8 |
|---|---|---|---|---|---|---|
| Fe | 4.32 | 11.06 | 5.21 | 6.35 | 1.06 | 1.36 |
| Ni | 0.47 | 4.53 | 2.12 | 3.23 | 0.63 | 1.01 |

From Table II, we can see that the Fe contents in the produced aluminum were higher than the Ni contents. This indicates that Fe easier dissolves into the electrolyte.

Dynamic polarization analysis

Fig.5 shows the dynamic polarization curves of the working electrodes. It can be seen that the value of self-corrosion potential with CR=1.8 electrolyte is larger than that obtained at CR=1.6. This result indicates the inertness of the electrode is improved with increasing CR. However, the corrosion current in a CR=1.8 electrolyte is larger compared with that in a CR=1.6 electrolyte, which indicate the corrosion rate is slower in CR=1.6 electrolyte. This result was also confirmed by the anodic corrosion rate measurements and the purity of the produced aluminum.

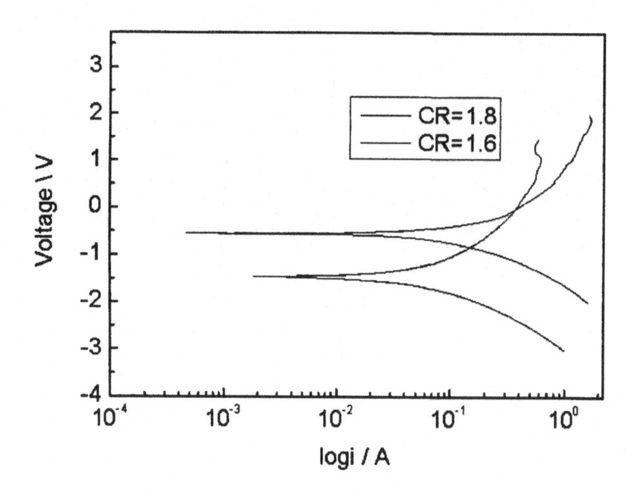


**Fig.5 Polarization curves of Fe-Ni-$Al_2O_3$ anodes in different CR electrolytes**

EIS measurements

The electrochemical impedance spectra obtained for the Fe-Ni-$Al_2O_3$ electrode at different CR electrolytes are shown in Fig.6 at open potential. It can be seen that the Cole-Cole plots have two obvious semicircles and no apparent linear response at low frequencies. This suggests that the rate-determining step is a charge-transfer reaction.

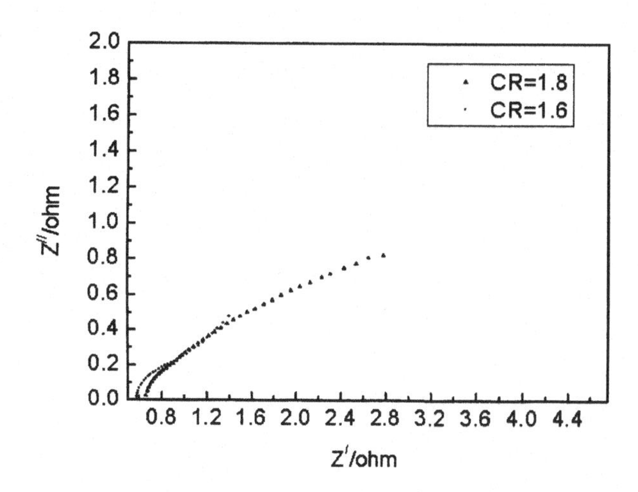


**Fig.6 Nyquist plots of Fe-Ni-$Al_2O_3$ anodes at different CR**

Corrosion Mechanism

During the electrolysis, Al-O-F complex anions and $F^-$ anions are transported to the anode, and oxygen atoms are discharged at the surface of the anode [10]. The oxygen reacts with the anode forming metal oxides on the anodic surface. At high alumina concentration, anode reaction may follow the equation:

$$[Al_2O_2F_4]^{2-}+2Me+2F^-=2MeO+2AlF_3+4e \qquad (1)$$

At low alumina concentrations, the anode reaction may be written:

$$2[Al_2OF_6]^{2-}+2Me=2MeO+4AlF_3+4e \qquad (2)$$

At high $Al_2O_3$ activity $Al_2O_3$ may react with MeO, forming a spinel according to the following equation:

$$Al_2O_3 + NiO=NiAl_2O_4 \qquad (3)$$

Some oxide films fall into the electrolyte because a loosely adherent anodic layer is detached by the flow of the electrolyte, and some of the metal oxide film dissolves in the molten cryolite since it reacts with cryolite as equation (4) shows[11]:

$$Me_xO_y+2y/3Na_3AlF_6=Me_xF_{2y}+y/3Al_2O_3+2yNaF \qquad (4)$$

or,

$$Me_xO_y+\frac{2y}{3}AlF_3 = xMeF_{\frac{2y}{x}}+\frac{y}{3}Al_2O_3 \qquad (5)$$

Metal oxide may also react with aluminum dissolved in the electrolyte by the following displacement reaction:

$$Me_xO_y+2y/3Al=y/3Al_2O_3+xMe \qquad (6)$$

When the metal oxide is thick enough, oxygen evolves from the anode surface. The anodic reaction may then be:

$$[Al_2O_2F_4]^{2-}+2F^- = 2AlF_3+O_2+4e \qquad (7)$$

or

$$2[Al_2OF_6]^{2-} = 4AlF_3+O_2+4e \qquad (8)$$

From the analysis above, we can conclude that one of the reason for the dissolution of the anode is that the metal components reacted with $AlF_3$. The dissolution accelerated with increasing concentration of $AlF_3$.

Moreover, as Fig.7 shows, the solubility of $Al_2O_3$ decreases with increasing CR.

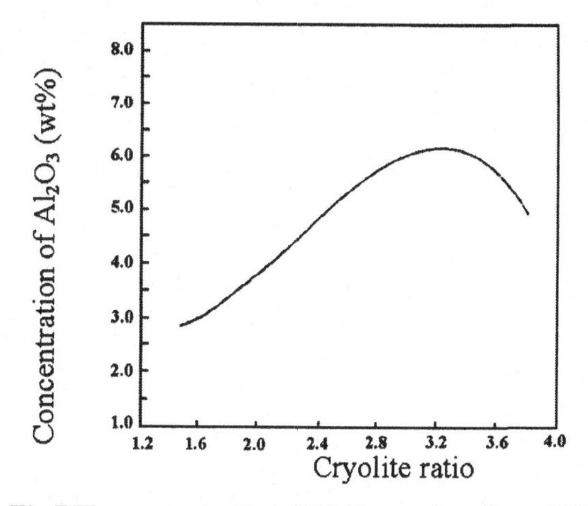


**Fig.7 The concentration of $Al_2O_3$ as a function of CR** [11]

The $Al_2O_3$ exists as Al-O-F complex anions after it is dissolves in the electrolyte. As mentioned above, Al-O-F complex anions and $F^-$ anions are transported to the anode, and oxygen atoms are discharged on the surface of the anode during electrolysis. With decreasing concentration of $Al_2O_3$, the discharged velocity of oxygen atom slowed down, and the electrons are absent at the surface of the anode. This makes the potential transfer to more positive, which can cause severe corrosion of the anode material. This can be seen from the cell voltage change during electrolysis ( as seen in Fig.3).

The corrosion rate is corresponding with electrolysis temperature. With decreasing CR ratio, the melting point of the electrolyte decreases. Generally, this cause the dissolution of Me component of inert anode to decrease, too. The corrosion rate is corresponding to the comprehensive effect of the change of melting point, $Al_2O_3$ solubility and concentration of $AlF_3$ with CR.

## Conclusions

The Fe-Ni-$Al_2O_3$ composite anode materials performs best and had a more stable electrolysis when CR=1.8, obtaining a 24 mm/y corrosion rate and with about 98% purity of the Al product. This is mainly due to the decreased melting point of the electrolyte, which makes it possible to perform the electrolysis at lower temperature, i.e. at about 840 $^0$C.
Although the electrolysis temperature can be decreased by using lower CR, it may be difficult to obtain a lower corrosion rate for the anode during operation at low CR. The lowest corrosion rates are obtained at unit activity of $Al_2O_3$, which might be difficult to maintain during operation at low CR since the $Al_2O_3$ solubility in the electrolyte is falling with decreasing CR.

## Acknowledgements

We greatly acknowledge the Ph.D Foundation of Liaoning Province for the financial supports granted in contract No.20041010.

## References

1. Halvor Kvande, Inert electrodes in aluminium electrolysis cell, Light Metals (TMS, Warrendale, PA), 1999: 369-376
2. V. de Nora and J. J Duruz, Low temperature operation cell for the electrowinning of aluminum, WO Patent 01/31,086 (26 Oct.1999)
3. T. R. Beck, "A non-consumable metal anode for production of aluminum with low-temperature fluoride melts", Light Metals 1995, Edited by J. Evans, (TMS, Warrendale, PA, 1995), 355-360
4. Craig W. Brown, "Laboratory Experiments with low-temperature slurry-electrolyte alumina reduction cells", Light Metals 2000, Edited by R. D. Peterson, (TMS, Warrendale, PA), 2000, 391-396
5. Alfred F, La Camera et al., Process and apparatus for low temperature electrolysis of Oxides, U.S. Patent 5415742, May 16,1995
6. A.M. Vecchio-Sadus, R.Dorin, E.J. Frazer, Evaluation of low-temperature cryolite-based electrolytes for aluminum smelting, Journal of Applied Electrochemistry, 1995, Vol.25: 1098-1104
7. Office of Industrial Technologies (U.S.DOE). Advanced anodes and cathodes utilized in energy-efficient aluminum production cells. Aluminum Project fact Sheet, 1999: 2-3
8. J.Thonstad, P.Fellner, G.M.Haarberg etal. Aluminium Electrolysis, 3rd edition.
9. Shi Zhong-ning, XU Jun-li and QIU Zhu-xian, Fabrication and testing of metal matrix composites as advanced anodes for aluminum electrolysis. Canadian Metallurgical Quarterly, 2004, Vol. 43(2): 259-264
10. Grjotheim K, Kvande H, LI Qingfeng, QIU Zhuxian. Metal production by molten salt electrolysis. China University of Mining and Technology Publishing House(Xuzhou).1998,79-81.
11. QIU Zhu xian. Theory and application of aluminium electrolysis[M]. Xuzhou: China University of Mining and Technology Press, 1998, 428-431.

**Light Metals 2008** *Edited by: David H. DeYoung*
***TMS (The Minerals, Metals & Materials Society), 2008***

# Anti-oxidation properties of Iron-Nickel Alloys at 800-900 °C

Zhonging Shi[1], Xingliang Zhao[1], Junli Xu[2], Xiaozhou Cao[1], Zhaowen Wang[1] and Bingliang Gao[1]

1. College of Materials and Metallurgy, Northeastern University, Shenyang, 110004 China;
2. College of Sciences, Northeastern University, Shenyang 110004 China

Keywords: Fe-Ni alloys, oxidation, metal-based inert anode, aluminum electrolysis

## Abstract

In order to find a suitable Fe-Ni base alloy as a kind of candidate metal inert anode materials for aluminum electrolysis, three kind of Fe-Ni (with different wt%) alloys with mass ratio of 50:50, 60:40 and 70:30 respectively were prepared by powder metallurgy process. Oxidation behaviors of the alloys were studied at 900 °C. The results indicated that the oxidation kinetic curves follow the parabolic law and anti-oxidation properties enhanced with increasing ratio of Fe to Ni in the alloy. X-Ray Diffraction showed that the oxide film formed on the anodic surface consisted of $Fe_2O_3$ and $Fe_3O_4$. 5wt% Al and 30wt% Co were added in different Fe-Ni alloy respectively. The results showed that anti-oxidation properties of Fe-Ni-Co is better than Fe-Ni-Al at 900 °C.

## Introduction

At present, the aluminum electrolysis inert anodes research work is a focus in aluminum electrolysis industry in China. Though there exit a lot of difficult, the huge profit and benefit promote the aluminum companies to development the technologies. In order to replace the traditional carbon anodes, the inventor of aluminum electrolysis Hall had tried to use copper as the anode for aluminum electrolysis more than 100 years ago. From then on, the research does not be interrupted ever. Now, because of the energy crisis and environmental issue, aluminum inert anode attracts us again. Ceramet base inert anodes, such as a representative Alcoa No.5324 $NiFe_2O_4$ have been studied for some time, but still not succeed commercially yet[1-5].

Because metallic anodes have good electrical conductivity, mechanical robust properties and easy to connect the bar-bus, metal inert anode for aluminum electrolysis was popular researched at these time. But the most obstacle for metal as aluminum electrolysis anode is its week anti-oxidation and corrosion resistance. There are two folds to solve the issue: (1) to find a suitable alloy system to replace the single metal element, which can improve the anti-oxidation and resistant corrosion during the course electrolysis. (2) to find out another suitable low temperature electrolyte to meet the demand of metal inert anode.

At 1997, J.A. Sekhar et. al[6] prepared a kind of Ni-6Al-10Cu-11Fe(wt%) alloys anode. By the results of some tests of oxidation and electrochemistry verified that the anode Ni-6Al-10Cu-11Fe-3Zn(wt%) was the suitable one for aluminum electrolysis. Submarian made a kind of Ni-Al-Cu-Fe-X, where the content of nickel is in an interval of 60-80wt%, while iron of 3-10wt%, aluminum of 5-20wt% and copper of 0-15wt%,X is the content of 0-5wt% one of some of the metals chromium, manganese, titanium, or molybdenum. At 1999, Duruz and Nora developed the Ni-Fe base inert anode for aluminum electrolysis[7-9]. At 2004, Zhuxian Qiu and Zhonging Shi[10] gave some results of the Ni-Fe base anode as aluminum anode, the purity of aluminum is 97%-98%. The properties of the surface layer are the key factors of the anode. In order to make the anode work soundly, the inertness function of the surface layer is necessary, which formed by the oxidation of the anode substrate.

## Experimental

Preparation of Ni-Fe base anode

The method of preparation anode is powder metallurgy process. The fine mixture of nickel, iron and additive were grinded by ball grind, and then cold press at the pressure of 500MPa. The green anode after cold press was sintered at the following temperature control drawing. See Fig.1.

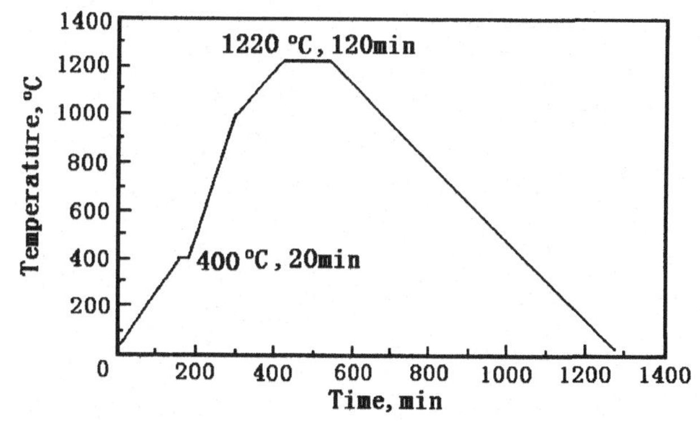


**Fig. 1 Sintering temperature of Fe-Ni base green anode**

In Fig.1, there were two stage of keeping temperature constant. The first stage is at 400 °C for 20 min to volatile the organic binder. The second one is the highest sintering temperature stage at 1220 °C for 2 hours, which would make the green anode structure compact and uniform.

Oxidation test

The sintered anodes were cut, and polished by No.600 sand paper, the samples was suspended in the furnace for oxidation test at 800 °C and 900 °C. The apparatus for oxidation test is shown in Fig.2. The constant temperature area of the furnace is 100 mm, the precision of the balance is $10^{-4}$g. The metallic wire, diameter of 0.5mm, for suspending sample is platinum.

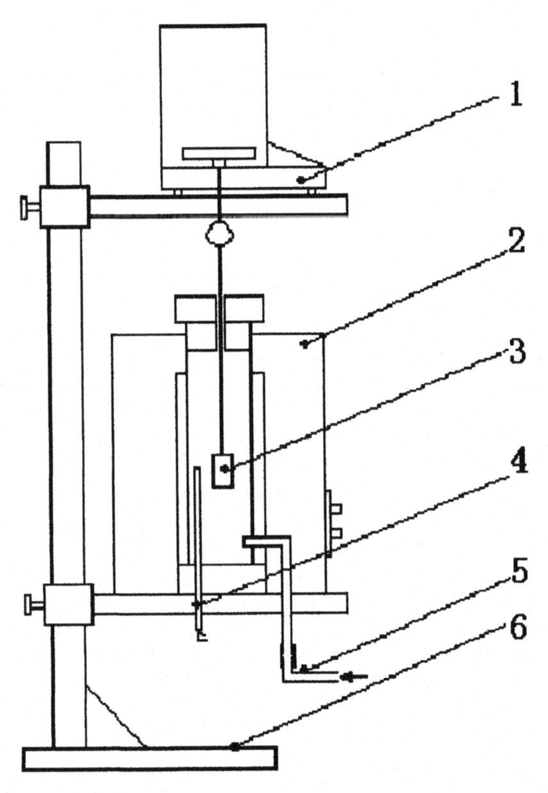


**Fig. 2 Equipment for anti-oxidation test**

1. Balance 2. Furnace 3.Fe-Ni Alloy sample 4. Thermocouple 5. Gas (oxygen or air) 6. Shelf base

According to the oxidation results, the oxidation kinetic curves of Fe-Ni base anode at 900 °C and 800 °C is plotted and shown in the Fig.3 and Fig.4, respectively.

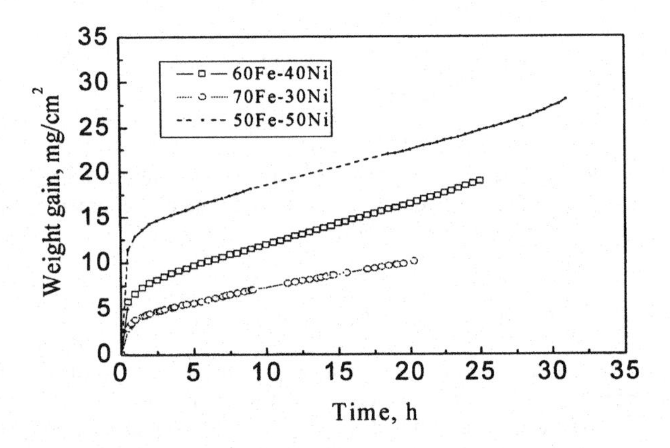


**Fig. 3 Oxidation kinetic curves of Fe-Ni alloy at 900 °C**

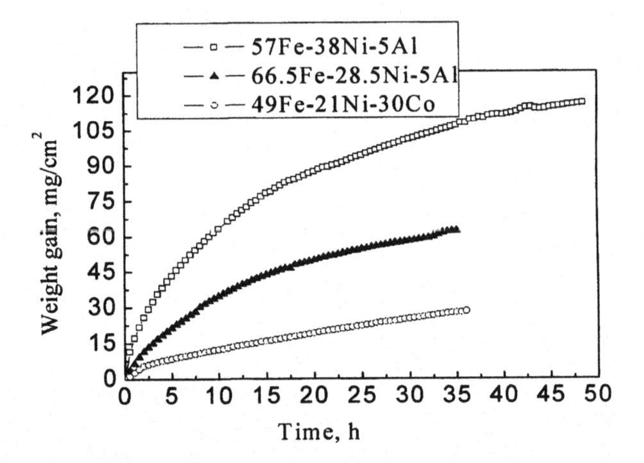


**Fig. 4 Oxidation kinetic curves of Fe-Ni base at 800 °C**

The oxidation rate of the Fe-Ni and Fe-Ni base alloy is lower obtained from the above figures. In the initial phase, the mass gain is increase great with the time increase, then the mass gain become slower, which is chemical reaction control step transferring to diffusion control step by the parabola principle. The phenomena indicate that the alloys have good anti-oxidation.

From the Fig.3, in a certain range of Fe/Ni ratio in Fe-Ni alloy the ratio of Fe to Ni increase, the oxidation rate of the Fe-Ni is increase at 900 °C.70Fe-30Ni is the best alloy for anti-oxidation, and its mass gain is 10mg/cm$^2$ after 25hours. The next to one is 60Fe-40Ni followed by the 50Fe-50Ni sample.

From the Fig.4, doping element of Al and Co can also make the oxidation kinetic curves of Fe-Ni base alloys follow the parabola principle. Though the doping aluminum and cobalt can not improve the anti-oxidation, they can make the surface layer stick to the substrate strengthen.

When comparing 57Fe-38Ni-5Al to 66.5Fe-28.5Ni-5Al samples to the same content of aluminum, the former sample has weaker anti-oxidation properties than the later one. When comparing the sample of 66.5Fe-28.5Ni-5Al to the sample of 49Fe-21Ni-30Co, 49Fe-21Ni-30Co shows better anti-oxidation than that of 66.5Fe-28.5Ni-5Al.

## Results and Discussion

Because the Gibbs free energy of $Fe_2O_3$ is more negative than that of NiO. $Fe_2O_3$ more stable than NiO thermodynamically, that is to say, when Fe and Ni coexist at the metal anode surface at high temperature, they will be oxidized by oxygen, and the metal Fe can also deoxidize Ni. So the oxides compound on the metal surface analysis by XRD is $Fe_2O_3$ and $Fe_3O_4$.

Fig.5 is SEM photograph of oxide film on Fe-Ni alloy surface (outer layer), the phase of the compound is $Fe_2O_3$ by EDXA. The structure of the oxide film is compact and dense, which is an obstacle to protect the alloy substrate from oxidation because the oxygen diffusion rata is reduced in this layer. From the Fig.5, the micro crystal structure of $Fe_2O_3$ is columnar crystal. As we known, the columnar has a characteristic of direction. Oxygen diffusion rate is decreased sharply, but oxygen can diffuse into the substrate along the crystal boundary of $Fe_2O_3$, and metal iron atom can migrate from the substrate to the alloy surface. Oxidation reaction will occur when oxygen atom and iron atom encounter at this layer. If the volume of the novel formation oxide expands too fast, the outer layer will crack by the stress.

Fig.6 SEM photograph of oxide film on Fe-Ni alloy surface(inner layer) consisted of $Fe_3O_4$ by SPS, the crystal structure the oxide of this layer is dendriform crystal, which is not dense enough to protect the underlining substrate. Oxygen pass through the outer layer will immigrate in the pore to the substrate easily in this layer.

Fig.7 SEM photograph of oxide film on Fe-Ni alloy substrate( covered by inner layer). Obviously, there exists a lot of pore since the Fe in the crystal migrate forward the inner layer, which will leave the pore or hollow alone. Iron diffuses toward out by mass transfer, which will form the cavity of iron. When the cavity of iron accumulates and swarms to over-saturation, which occur the pore on the interface between the substrate and inner layer. These interfacial pore or cavity will influence the oxide film sticking to the substrate. Further, the pore grows up gradually to the critical dimension. The connection of oxide film and metal substrate will be broken. As a result, the oxide film crack and fall off from the alloy surface. The function of protection substrate is terminative.

**Fig. 5 SEM photograph of oxide film on Fe-Ni alloy surface (outer layer)**

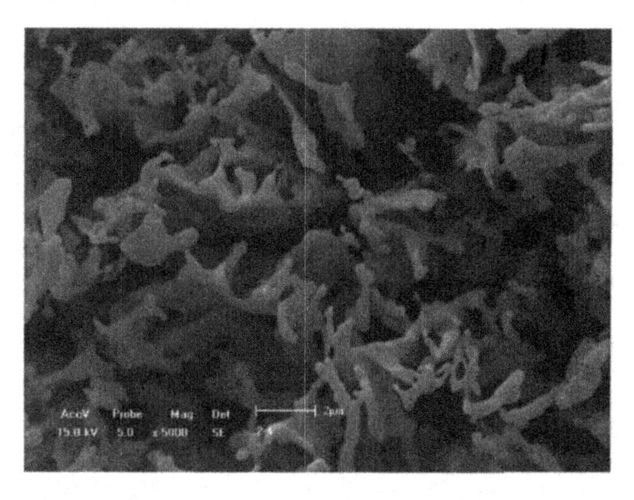

**Fig.6 SEM photograph of oxide film on Fe-Ni alloy surface (inner layer)**

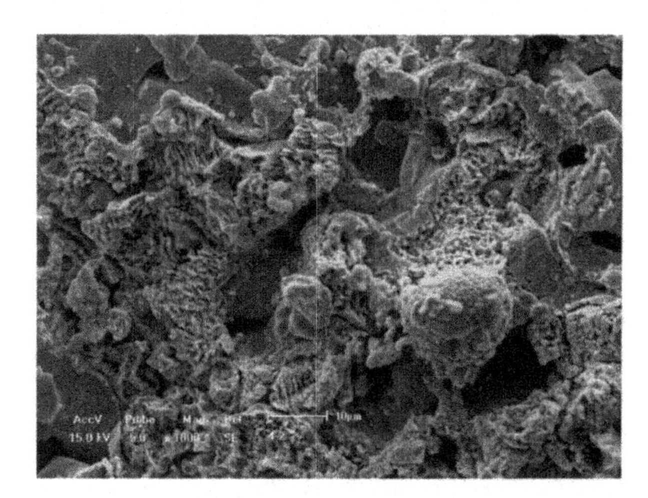

**Fig.7 SEM photograph of oxide film on Fe-Ni alloy substrate (covered by inner layer)**

### Conclusion

Fe-Ni and Fe-Ni base alloys oxidation kinetic curves follow the parabola principle at 800 °C and 900 °C, which indicates that the compact oxide film form on the alloys surface. Among 70Fe-30Ni, 50Fe-50Ni and 60Fe-40Ni alloys, 70Fe-30Niis the best one with anti-oxidation properties of weight gain $10mg/cm^2$ for 24 hours at 900 °C. In a certain range of Fe/Ni ratio in Fe-Ni alloy, the oxidation rate of the Fe-Ni increase at 900 °C with the increase of the ratio of Fe to Ni in the alloys. Doping aluminum and cobalt can also formation an oxide film on the alloys surface with more sticking to substrate. Oxidation rate of Fe-Ni-Al alloys with the same aluminum content increase with the decrease of iron content in the alloys. The effect of doping element cobalt is better than aluminum in respect to the anti-oxidation properties.

## Acknowledgements

The authors would like to appreciate the MOST of China for their financial support in the contract No. 2007CB210305 of "National Basic Research Program of China". We also really appreciate the financial support from Ph.D Foundation of Liaoning Province, China (No.20041010).

## References

[1] Issaeva L., et al. Electrochemical behavior of tin dissilved in cryolite-alumina melts. Eletrochem. Acta, 1997, 42(6), 1011-1018.

[2] Grjotheim Kai. Aluminium electrolysis. Dusseldolf: Aluminium-Verlag, 1982: 365.

[3] Vchida I, Laitinen H A. Proc. 1st international symposium on molten salts chemistry and technology. Japan, 1983:65.

[4] Billehaug K. and Φye H. A. Inert anodes for aluminum electrolysis in Hall-Héroult cell, Aluminum (Dusseldolf) 1980, 57(2): 146-150; 1980, 57(3): 228-231.

[5] Welch B., Hyland M. M. and James B.J., Future materials requirements for the hight-energy-intensity production of aluminum. JOM, 2001, 53(2): 13-18.

[6] J.A. Sekhar, H. Deng, J. Liu E.Sum, J.J.Duruz and V. de Nora, Micropyretically synthesized porous non-consumable anodes in the Ni-Al-Cu-Fe-X system, Light Metals, 1999, 347-354.

[7] Crottaz O. and Duruz J.J., Nickel-iron alloy-based anodes for aluminum electrowinning cells, WO patent, 2 Jan.199900/06, 804.

[8] Nora V. de and Duruz J. J., Slow consumable non-carbon metal-based anodes for aluminum production cells, WO patent, 2 Jan. 1999, 00/06,805.

[9] Duruz J. J. and Nora V. de, Metal-based anodes for aluminum electrowinning cells, WO patent, 9 Jan. 1999, 01/42, 534.

[10] Zhongning Shi, Junli Xu and Zhuxian Qiu, An iron-nickel metal anode for aluminum electrolysis, Light Metals 2004, Edited by Alton T. Tabereaux, 333-337.

**Light Metals 2008** *Edited by: David H. DeYoung*
***TMS (The Minerals, Metals & Materials Society), 2008***

# EFFECT OF RARE EARTH ELEMENT ON OXIDE BEHAVIOR OF Fe-Ni METAL ANODE FOR ALUMINIUM ELECTROLYSIS

Xiaozhou Cao[1], Zhongning Shi[1], Zhaowen Wang[1], Ting'an Zhang[1], Xianwei Hu[1], Xingliang Zhao[1], Dingxiong Lv[2], Youwei Wu[2]
1. Northeastern University, College of Materials and Metallurgy, Mail box 117, Shenyang, Liaoning 110004 China
2. Design & Research Institute of Northeastern University, Shenyang, Liaoning, 110004 China

Keywords: Metal anodes, Rare earth, Oxidation, Oxidation mechanism, Aluminum electrolysis

## Abstract

The additions of rare earth elements have significant effect on the oxidation behavior of Fe-Ni metal anode for aluminium electrolysis in air at 900℃ by thermogravimetry method. The microstructure and composition of oxide film have been analyzed by XRD and SEM. The results show that rare earth element can change the content of oxide film on the surface of metal anode. Rare earth element is inclined to distribute on grain interfaces and refined casting grains. The oxide film is compact and can prevent the matrix from oxidizing further. The oxidation mechanism was suggested by the component of oxide film of metal anode.

## Introduction

In the aluminum electrolysis production process, the temperature of molten salts reach to as high as 800~1000 ℃ in the aluminum reduction cell, above the molten salts anode exposes in the high temperature air, therefore researching the oxidation resistance performance of anode material has the important practical significance in the theory and the production practice[1]. At present, carbon anode because partially exposes in the air, the high-temperature oxidation loss approximately composes the anode consumption 17wt% in the aluminum electrolysis production, but using inert anode instead of carbon can reduce the anode material oxidation loss in high-temperature air. Metal anodes are idea electrode materials with good electrical conductivity, intensity and high thermal shock resistance. Protection layer theory was proposed by Sadoway[2], and they reported that the oxide film formed on the surface of metal anode could prevent high-temperature oxygen from attacking the interior metal body further. J.N.Hryn and Sadoway tested Cu-Al,Ni-Al,Fe-Cr-Al alloy materials[3]. J.A.Sekhar tested Ni-Al-Cu-Fe metal anode, the result show oxidation rate is high [4,5]. Duruz and De Nora tested Ni-Fe metal anode [6,7]. For oxidation resistance, it is crucial that the metal anode remain covered with compact, adherent oxide film.
Rare earth element is one kind of beneficial additives, it has been shown that addition of small amount of rare earth elements can usually improve the substrate compactness, reduces the oxidation rate, enhances the oxide film and the substrate binding force, thus remarkably improves the oxidation resistance of metal [8-10].
In this paper, rare earth element Nd was added in Fe-Ni metal anode. The oxidation kinetics of Fe-Ni metal anode with and without Nd were studied. Oxide film was analyzed by SEM, and discussing the effect Nd on oxidation mechanism.

## Experimental

### Metal Anode Preparation

Fe, Ni as the raw materials, 1% rare-earth element Nd was added, and smelting in the vacuum induction furnace. The column spindle was wire-cut into 30mm×10mm×10mm specimens which finally polished by 0.2um $Al_2O_3$ abrasive paste with the acetone solution clean degreasing, and dried in oven at 100℃.

### High Temperature Oxidation Test

Metal anodes with and without Nd were isothermally oxidized for 48h at 800℃in air. The oxidation kinetic curves were examined by thermogravimetry method. The schematic setting diagram for oxidation test was shown in Figure 1. The specimen was hung in the tubular electric furnace with Pt wire. The mass was reading by Daojin AY120 electronic balance accuracy for 0.1mg.

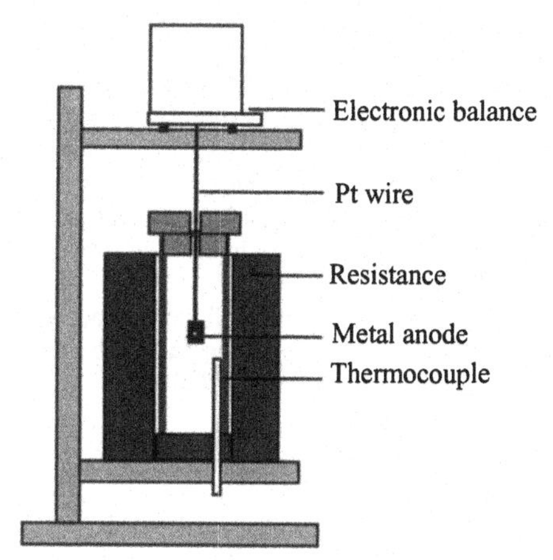


Fig. 1 Schematic setting diagram for oxidation test

Scanning electron microscopy (SEM) was used to investigate the oxide films.

## Result and Discussion

### Material Characterization

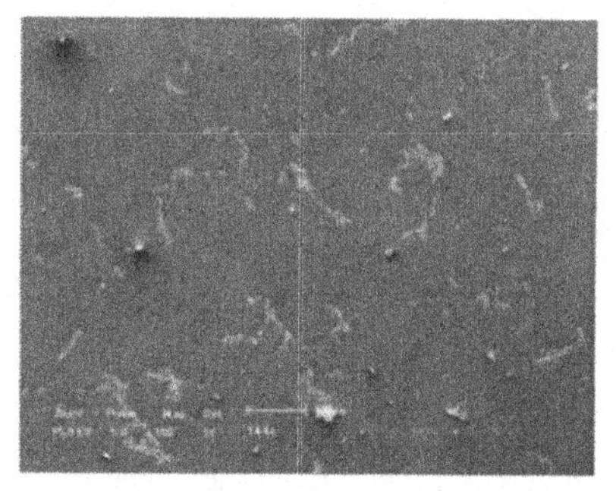

Fig.2 SEM photos of Fe-Ni-1%Nd metal anode

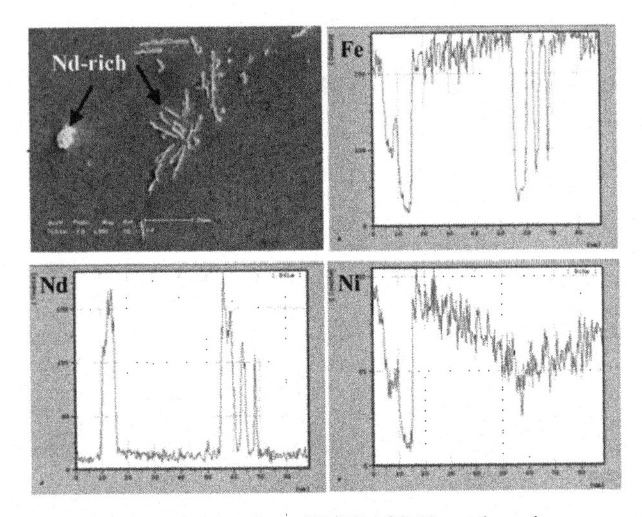


Fig.3 Line scaning of Fe-Ni-1%Nd metal anode

A SEM micrograph of polished Fe-Ni specimen with Nd is shown in figure 2. The results showed that the addition of Nd can refine the microstructure of Fe-Ni metal anode, some tiny granulated and branched Nd-rich granule distribute nearby the grain boundary. Because rare-earth element atomic radius is bigger than Fe and Ni, the distortion energy forming in the crystal bigger than that dissolving in the grain boundary, the rare-earth element distributed along the grain boundary and can strengthen grain boundary and suppress the impurity element to gather to the grain boundary. Grain boundary as the short distance diffused channel of metal cations, rare earth atom tends to enrichment on the grain boundary and obstructs this channel, lead to cation is difficult to diffuse along the grain boundary and can improve oxidation resistance of Fe-Ni metal anode in the high temperature oxidation test.

Figure 3 is the line scaning of Fe-Ni metal anode with Nd, it can be seen the distribution of Fe,Ni and Nd in the substrate. What he blue arrow point are the Nd-rich particles.

## Oxidation Kinetic Curves

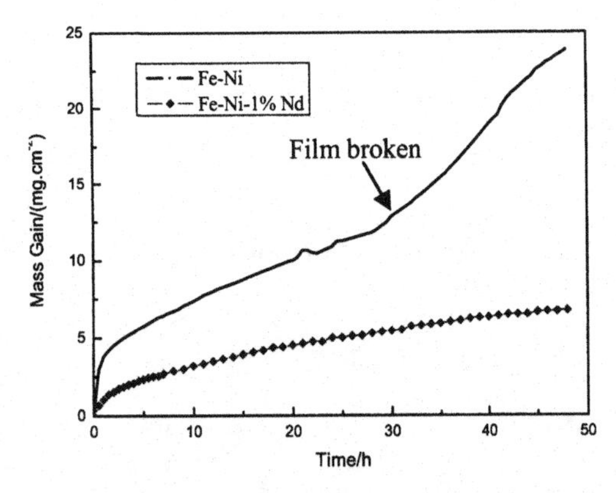


Fig.4 Oxidation kinetic curves of Fe-Ni with and without Nd at 800℃ in air

The oxidation kinetic curves of Fe-Ni metal anode with and without Nd after isothermal oxidation are shown in figure 4. It can be seen that addition of Nd reduced the oxidation rate of metal anode remarkably. The oxidation kinetic curve of Fe-Ni metal anode was divided into two parts, the oxidation rate increased suddenly at 28 hour which means the oxide film suffered from cracking and spalling and new oxide film formed on the metal anode surface. However Nd addition strengthened oxide film adhesion and decreased oxide film internal stress.

In order to seek the relation of mass gain and time, the data was fitted and discover the square of mass gain and time approach the linear relation. The fitted curves were shown in figure 5. That means the oxidation kinetic curves of metal anode with and without Nd are both follow parabolic law at 800℃. Oxidation rate constant k is 4.6174 and 0.9752. Oxidation kinetic equations are follow,

$$\Delta m^2=4.6174t \quad (1)$$

$$\Delta m_{Nd}^2=0.9752t \quad (2)$$

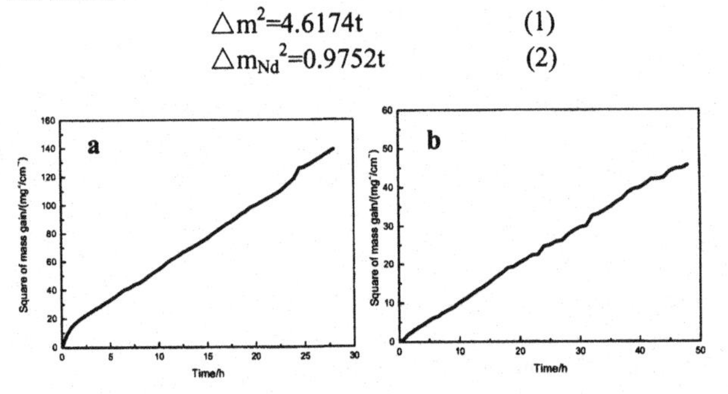


Fig.5 Square of mass gain of Fe-Ni vs oxidation hours at 800℃
(a)Fe-Ni (b)Fe-Ni-Nd

## Microstructure of the Oxide Film

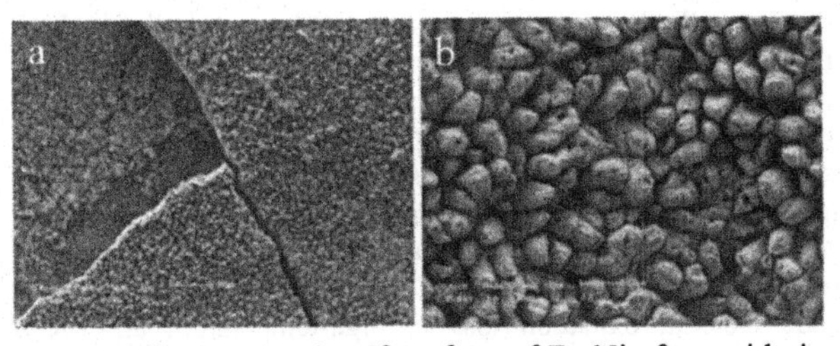


Fig.6 SEM micrography of surface of Fe-Ni after oxidation for 48h at 800℃ in air
(a) Fe-Ni 500× (b) Fe-Ni 1000×

Fig.7 SEM micrography of surface of Fe-Ni-Nd after oxidation for 48h at 800℃ in air
(a) Fe-Ni-Nd 500× (b) Fe-Ni-Nd 2000×

Figure 6 shows surface morphology of the oxides formed on Fe-Ni metal anode without Nd. It can be seen that the oxide film cracked at 28 hours correspond with figure 4. Figure 6 (b) is the SEM photography of oxide film at 1000 ×. Columnar crystal grows closely and has intense directivity, the compressive stress cause the oxide film broken. The single particle size is 5um. The oxide is mainly composed of $Fe_2O_3$ by XRD analyzed.

Figure 7 shows the SEM images of the oxide film with Nd addition, the oxide is needle and the grain size is 0.3um. Energy dispersive spectrum (EDS) analysis shows that the needle-shaped oxide film is only $Fe_2O_3$. Some small Nd-rich particles were found on the gas/oxide interface (fig.7 b), it was possibly $NdFeO_3$ by EDS analysis.

Oxidation Mechanism

Rare-earth element Nd has stronger affinity with O compared with Fe and Ni, at early oxidation stage Nd was oxidized rapidly to form $Nd_2O_3$ particle and as the nucleation center for $Fe_2O_3$, promoted the Fe selective oxidation and NiO proportion drop down which in the oxide film is easy to flake off. As seen from figure 7 b, Nd addition refines $Fe_2O_3$ oxide grain remarkably. In the intermediate stage, rare earth oxide and NiO gathered at the metal/oxide interface which lowered oxidation rate of Fe.
Nd addition in metal anode changed in the oxidation process and element mass transfer mechanism. No Nd adding, $Fe_2O_3$ in the high temperature oxidation process, $Fe^{3+}$ cation's outwards diffusion through oxide grain boundary is dominant, new $Fe_2O_3$ oxide formed in gas/oxide interface. In the long time oxidation process, lots of cations diffused to outside and leave vacancies, caused the internal oxide film to be loose, the internal compressive stress occur the oxide film broken and decreased oxide adhesive property.
Nd addition increased the $Fe_2O_3$ thermodynamics stability, partial pressure of oxygen was reduced because of Nd self oxidation lowered oxygen content in metal anode during the oxidation process[11,12]. Nd was found on the surface of oxide, new $Fe_2O_3$ oxide formed in the metal/oxide interface and growed slowly. So, Nd addition can inhibit $Fe^{3+}$ cation diffusion within $Fe_2O_3$ film, the predominant $Fe^{3+}$ cation outwards diffusion was instead of $O^{2-}$ anion inwards diffusion through oxide grain boundary.

## Conclusion

Nd addition can refine the grain size and segregate at grain boundaries in the forms of $Nd^{3+}$ ions.

Nd addition remarkably improve the oxidation resistance an 800 ℃ in air, oxidation kinetic curves follow parabolic law. Oxidation kinetics equation with and without are $\triangle m_{Nd}^2=0.9752t$, $\triangle m^2=4.6174t$ respectively.

Nd addition can promote $Fe_2O_3$ oxide nucleate rapidly in the early oxidation stage, and reduce the oxide growing rate. Adhesion of $Fe_2O_3$ film was greatly improved, it's due to the increase in creeping ability and the decrease in internal stress.

Nd addition improve the mass transfer mechanism in the oxidation process. The oxidation procedure was controlled by the outwards diffusion of Fe cations without Nd addition, however with Nd addition the oxidation procedure was controlled by the inward diffusion of $O^{2-}$ ion to grain boundary.

Nd mainly distributes nearby the oxide/gas interface, possibly exists in the $Fe_2O_3$ oxide grain boundary as the form of $NdFeO_3$.

## Acknowledgement

The authors appreciate the financial support from "National Natural Science Foundation of China (50674031)". We also sincerely acknowledge Mr. Yaluo Yu and Ms. Yaxin Yu for their valuable contribution to this study.

## Reference

[1] Zhuxian Qiu. Produce Aluminium in Prebake Cell. Metallurgical Industry Press, 2005,160.
[2] D.R.Sadoway, "Inert anodes for the Hall-Heroult cell: the ultimate materials challenge", *JOM*, 2001, 53 (5): 29-33
[3] John N. Hryn and Donald R. Sadoway, "Cell Testion of Metal Anode for Aluminum Electrolysis", *Light Metals,* 1993, 475-483.
[4] J.A.Sekhar, J.J.Liu and J.J.Duruz, "Stable anodes for aluminium production cells",US patent 5510008 (21 Oct,1994).
[5] J.A.Sekhar, J.Liu et al, "Graded Non-Consumable Anode Materials", Light Metals, 1998, 597-603.
[6] J.J.Duruz and V.de Nora, "Aluminium electrowinning cells operating with Ni-Fe alloy anode", WO patent 01/43,208 (9 Jan, 1999)
[7] J.J.Duruz and V.de Nora, "Metal-based anodes for aluminium electrowinning cells", WO patent 01/42,534(9 Jan 1999)
[8] R Cueff, H Buscail, et al. "Oxidation of alumina formers at 1173K: effect of yttrium ion implantation and yttrium alloying addition", *Corrosion Science*, 2003, 45(8): 1815.
[9] F.A. Golightl, F.H. Scott, et al, "The influence of yttrium additions on the oxide scale adhesion to an iron chromium aluminum alloy", *Oxid.Met,* 1976,10:163
[10] P Castello, F.H.Scott, et al, "Yttrium promoted selective oxidation of aluminium in the oxidation at 1100℃ of an eutectic Ni-Al-$Cr_3C_4$ alloy", *Corrosion Science,* 1999, 41(5): 901.
[11] Pieraggi B, Rapp R A, "Interface dynamics in diffusion driven phase transformations for metallic systems ", *Scripta Metallurgica et Materialia*, 1994, 30(11): 1491-1496..
[12] Przybylski K, "Observation of coherent perovskite particles in growing chromia films", *J Amer Ceram Soc*, 1986, 69(9): 264-266.

# Effect of additive CaO on corrosion resistance of 10NiO-$NiFe_2O_4$ ceramic inert anodes for aluminium electrolysis

Tian Zhong-liang, Huang Li-feng, Lai Yan-qing, Li Jie, Liu Ye-xiang
(School of Metallurgical Science and Engineering, Central South University, Changsha, 410083, China)

Keywords: inert anode; 10NiO-$NiFe_2O_4$ ceramic; additive CaO; corrosion resistance

## Abstract

10NiO-$NiFe_2O_4$ composite ceramic inert anodes with the additive CaO content of 0, 0.5, 1, 2 and 4%(by weight) were prepared and their corrosion resistance to $Na_3AlF_6$-$K_3AlF_6$-$Al_2O_3$ melts was studied in laboratory electrolysis tests. The results show that the content of additive CaO in ceramic inert anodes has great effects on the concentration of impurities in electrolyte. The addition of the oxide CaO to ceramic inert anode is adverse to improve its corrosion resistance. The additive CaO existed in the grain boundary accelerates the corrosion of ceramic inert anode.

## Introduction

The application of inert anode was always the target that the aluminum electrolysis industry is seeking for in the new technology field. As an inert anode material, it must meet the following requirements: good corrosion resistance in the molten salts and high stability with respect to oxidizing gases such as oxygen; good electrical conductivity and adequate resistance to thermal shock, especially at the practical temperature of Hall-Héroult cells; economically feasible[1,2]. Recently, the materials studied as inert anode for aluminum electrolysis mainly concentrated on the alloy[3] and the cermet[4]. Especially the nickel ferrite spinel was regarded as one of the most promising matrix of cermet inert anodes with respect to the corrosion resistance to melts[5].

The service condition for aluminum electrolysis, which generally happened at 950-970°C, is very rigorous for the performance of inert anodes. One of the most formidable challenges is the corrosion resistance to melts. The pilot scale of 6kA experiment, with the support by US Department of Energy, the $NiFe_2O_4$ based cermet inert anode exposed the weak corrosive resistance and unsatisfied thermal shock resistance[6]. The densification of material is the important factor to affect its corrosive performance. Generally, the electrolyte is easy to erode and destroy the electrode with low density. Hence, it is important to increase densification of material for improving its corrosive resistance and other properties.

In order to obtain desirable $NiFe_2O_4$ based cermet inert anode with high density and low porosity, the modifying agent or the sintering assistant (such as CaO, ZnO, $TiO_2$, $MnO_2$, and so on) is always used to improve the sintering property of ceramic material during preparing. The modifying agent CaO was used to restrain grain growth and improve the density of Mn-Zn ferrite[7,8].

In our previous works, CaO was used as the modifying agent to reform the sintering property and improve the density of $NiFe_2O_4$ based ceramic inert anode and showed good behavior[9]. In this paper, the 10NiO-$NiFe_2O_4$ composite ceramic inert anodes with additive CaO content of 0, 0.5, 1, 2 and 4%(mass fraction) were prepared and their corrosion resistance to $Na_3AlF_6$-$K_3AlF_6$-$Al_2O_3$ melts was studied in laboratory electrolysis tests. The purpose is to understand the effect of additive CaO on the corrosion resistance of composite ceramic inert anode and determine the content of additive CaO in ceramic material.

## Experimental

Preparation of ceramic and cermets

Proper amount of analytic grade NiO and $Fe_2O_3$ was mixed by ball milling and then calcined at 1200°C for 6 h in static air atmosphere to obtain the 10NiO-$NiFe_2O_4$ ceramic powder. Analytic grade CaO powder with different content of 0%, 0.5%, 1.0%, 2.0% and 4.0% was added to the 10NiO-$NiFe_2O_4$ ceramic powder prepared in advance. The calcined powders were mixed with dispersant and adhesive which were organic by ball milling again for 150min in stainless steel ball mixing pot and then compressed to form cylindrical blocks (Ø20mmx45mm) at the pressure of 200MPa. Then they were sintered in an atmosphere of efficaciously controlled oxygen partial pressure to get the desired samples [10]. The composition and correlative property of these inert anodes prepared are showed in table 1.

**Table 1** The composition and correlative property of inert anode samples

| Sample Nº | Content of CaO/wt% | Size in Diameter/mm | Relative Density/% | Electrical Conductivity (960°C)/S·cm$^{-1}$ |
|---|---|---|---|---|
| IA-1 | 0.0 | 17.3 | 98.15 | 1.03 |
| IA-2 | 0.5 | 17.4 | 97.30 | 1.22 |
| IA-3 | 1.0 | 17.3 | 97.47 | 1.36 |
| IA-4 | 2.0 | 17.3 | 98.75 | 16.29 |
| IA-5 | 4.0 | 17.5 | 95.92 | 13.57 |

Cell design

A sketch of the experimental cell is presented in Fig.1. A hole was drilled at the bottom of graphite crucible and 100g metal aluminum was added. Thus, a steady cathode surface could be obtained during electrolysis. Alumina sleeve was set in the graphite crucible and about 300g electrolyte was contained. Under the operating conditions of the laboratory test, the cell would not be thermally self-sustaining. It was necessary to provide extra heat by placing the cell with the anode in a vertical laboratory furnace and the furnace was heated to the desired temperature. The temperature was measured with a Pt/Pt-10%Rh thermocouple and controlled to ±1°C by a TCE-II programmable temperature control unit.

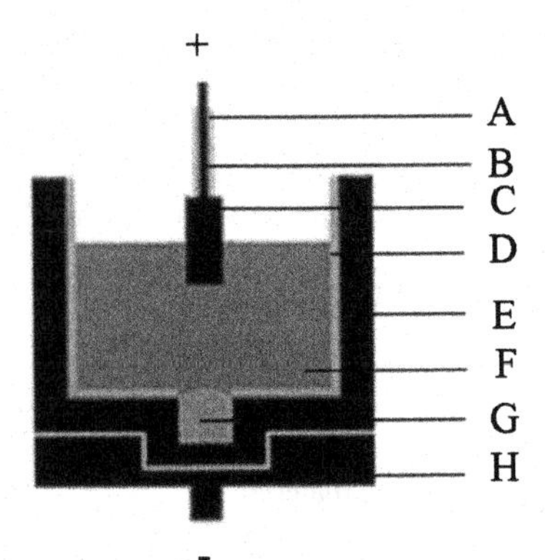


Fig.1. Experimental cell
A—$Al_2O_3$ sleeve; B—Stainless steel anode rod; C—Cermet inert anode; D—$Al_2O_3$ liner; E—Graphite crucible; F—Electrolyte; G—Metal aluminum; H—Graphite mechanical support.

Electrolysis tests
The electrolyte was made up of reagent grade cryolite ($Na_3AlF_6$, $K_3AlF_6$), $Al_2O_3$ and technical grade $AlF_3$, with the CR ((NaF+KF)/$AlF_3$ molar ratio) 1.7 and the concentration of $Al_2O_3$ 8wt%. Some certain amount of metal aluminum was added prior to electrolysis. All compositions were dried at 120°C for 48 h to remove the water before being used. The temperature was monitored during the experiment using a Pt/Pt+10%Rh thermocouple in a sintered alumina sleeve adjacent graphite crucible sidepiece and was maintained at 900°C, i.e. with a superheat of 10°C. The anode-bath contacting area was controlled by immersing the anode depth, which was 1cm. The current density of anode bottom was $1.0A\cdot cm^{-2}$. The current and the cell voltage were supplied and monitored by a Multi-Purpose Potentiostat/Galvanostat (model 273A/10, Perkin-Elmer Instruments). The current was kept constant throughout the experiments. During the testing, the alumina sleeve dissolves continuously into the electrolyte and contributes to keep the concentration of $Al_2O_3$. So, $Al_2O_3$ was not added throughout the testing.

After electrolysis the anode was raised out of the melt while maintaining polarization so as to prevent reduction of the anode material by dissolved metal. The cell was left to cool with the anode resting above the electrolyte. The anodes used were sectioned, mounted, polished, and analyzed by XRMA (JSM-5600LV) using a quantitative energy dispersive spectrometer (EDS) connected to the SEM. Some samples of bath were analyzed by X-rays fluorescence spectrum (Philips 8424 TW2424) (analytic error below 5%) for getting the concentration of Ni, Fe and Cu in bath and cathode aluminum.

## Results and discussion

Varieties of impurities concentration in bath during electrolysis
De Young [11] pointed out it took approximately 8 hours for stoichiometric $NiFe_2O_4$ in cryolite melts to reach steady-state concentrations, which were taken to be the solubilities. The work by Lai Yan-qing et al [5] showed that such a process would cost 4~6 hours. Therefore in present work all electrolysis experiments lasted 10 hours. As mentioned above, the electrolyte samples were taken during electrolysis to study the dissolution process of the anode material. The samples were analyzed for the concentration of anode components in electrolyte during electrolysis, and a typical set of results is plotted in Fig.1.

From Fig.1, it is clear that the different ceramic compositions showed a varying degree of dissolution of the anode components. For the 10NiO-$NiFe_2O_4$ inert anode, the steady state of impurities Ni, Fe and Ca is reached and the concentrations are 67ppm, 105ppm and 158ppm respectively when the electrolysis ends. Compared with the solubility of impurities Ni and Fe mentioned by DeYoung (bath ratio 1.1, melt with 6.5 wt% $Al_2O_3$, 1000°C)[11], the values are far below the solubility (90ppm for impurity Ni, 580ppm for impurity Fe).

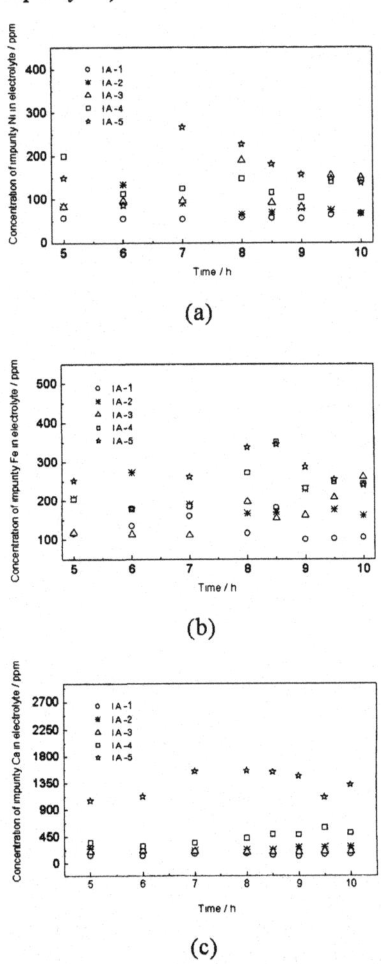


Fig.1 Elemental analyses of the electrolyte contamination of anode constituents versus time for the ceramic electrodes: (a) impurity Ni; (b) impurity Fe; (c) impurity Ca

However, for the ceramic inert anodes which are studied and contain the oxide CaO as modifying agent, the concentration of impurities in the electrolyte increases and the steady-state concentration cann't be reached even at the time of the electrolysis ending. For example, for the investigated ceramic inert anodes, which the content of the oxide CaO is 0.5%, 1.0%, 2.0% and 4.0%, the concentrations of impurity Ni in the bath are 67ppm, 150ppm, 144ppm and 138ppm, the concentrations of impurity Fe are 161ppm, 261ppm, 244ppm and 241ppm, the concentrations of impurity Ca are 282ppm, 203ppm, 517ppm and 1321ppm at the end of electrolysis testing.

So, from the concentrations of impurities in the bath during electrolysis, it indicates that the addition of the oxide CaO to ceramic inert anode as modifying agent is adverse to anti-corrosion to the molten cryolite. And the corrosion degree of ceramic inert anode deteriorates with the content of CaO increases.

Figure and microstructure of the inert anodes after electrolysis

The ceramic inert anodes after electrolysis are shown in the Fig.2. It can be seen that some parts of inert anodes besides the IA-1 swelled from the anode substrate and even there are cracks in the anodic surface of the IA-5. At the same time, from the Fig.2, the degree of swelling increases with the content of CaO in ceramic inert anodes increasing. It maybe implies that it is disadvantageous of addition of CaO to ceramic inert anode to improve the corrosion resistance; the corrosion resistance of inert anodes becomes bad with the content of CaO increasing.

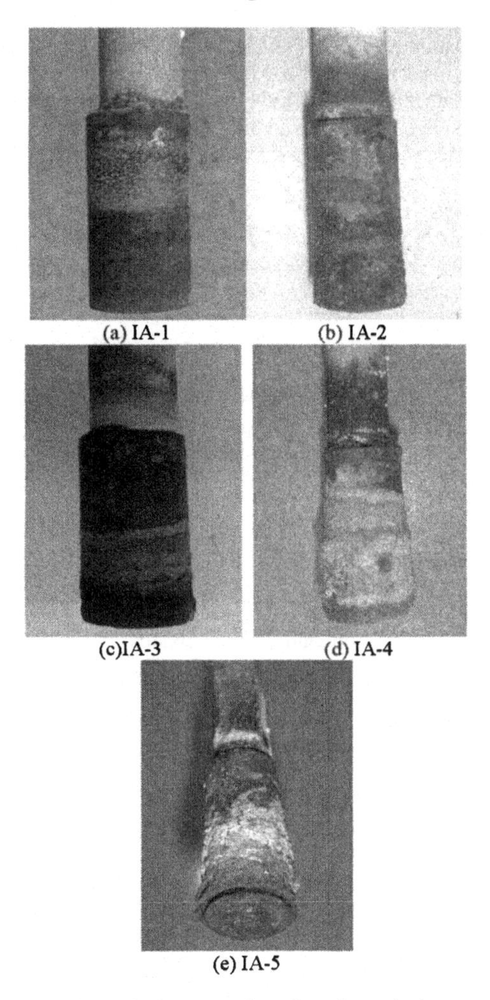

(a) IA-1 (b) IA-2 (c)IA-3 (d) IA-4 (e) IA-5

Fig.2 The inert anodes after electrolysis

To obtain more information about the corrosion process, the cermet inert anode IA-1 (composition $10NiO\text{-}90NiFe_2O_4$), IA-2 (composition $0.5\%CaO/(10NiO\text{-}90NiFe_2O_4)$) and IA-4 (composition $2\%CaO/(10NiO\text{-}90NiFe_2O_4)$) are sectioned, mounted and polished after electrolysis experiments. The back-scattered electron pictures of the cermet inert anodes are shown in Fig.3.

From Fig.3 (a), the porosity of inert anode is very low after electrolysis test, the anode could effectively hold back the bath penetration because of high relative density and shows good corrosion resistance to the molten cryolite during electrolysis. From fig.3 (c), it is very obvious that there are lots of holes and pores left on the left and the bath penetrates into these pores by capillary effects. The X-ray mapping analysis shows that there is a clear gradient across the surface for the aim elements Na, K, Ca (Fig.4). These elements scattered much more in the anode surface that in the bulk, which implies that ceramic inert anode is corroded badly by electrolyte penetration in spite of its high relative density at the beginning of electrolysis. However, for the aim elements Ni, Fe and O, the distribution is of uniform, indicating the reaction between bath and ceramic inert anode components($NiO$, $NiFe_2O_4$) is limited. The corrosion of inert anode is mainly affected by the reaction between bath and the additive CaO.

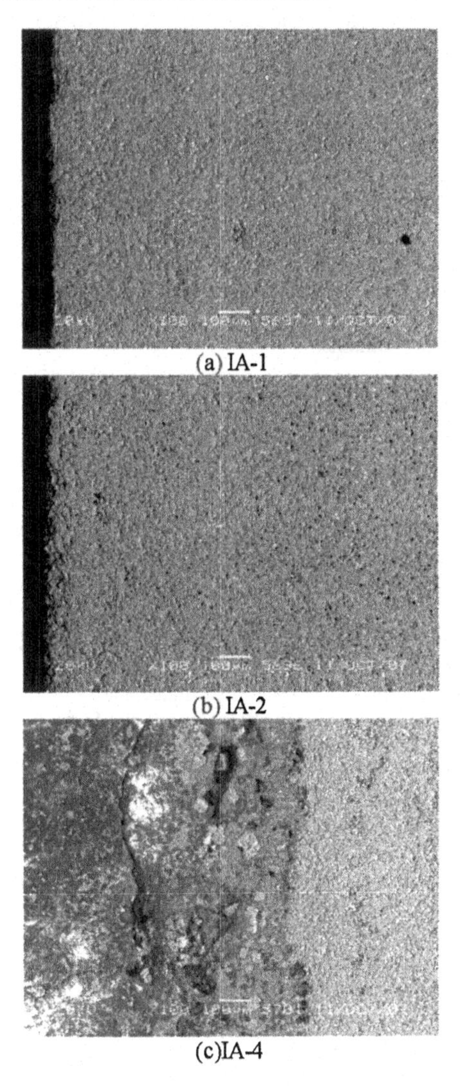

(a) IA-1 (b) IA-2 (c)IA-4

Fig.3 SEM backscattered images of the cermet inert anodes at anode bottom

Again compared with Fig.3 (a), (b) and (c), which are the SEM images for ceramic inert anodes containing different content of CaO, the corrosion degree is very different. The anode without CaO (IA-1) is hardly corroded. However, the corrosion of the anode containing 2%CaO (IA-4) is catastrophic.

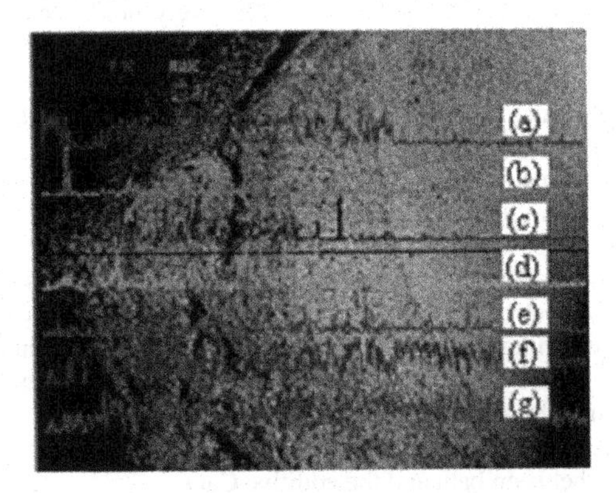


Fig.4 EDS images of the ceramic inert anode(IA-4) at anode bottom after electrolysis
(a)-F; (b)-Na; (c)-Al; (d)-K; (e)-Ca; (f)-Fe; (g)-Ni

Preliminary discussion on corrosion mechanism

Generally speaking, the ceramic inert anodes investigated may deteriorate in operating Hall cells by a number of possible mechanisms such as chemical dissolution, electrochemical dissolution, reduction by dissolved metal aluminum, electrolyte penetration, and grain boundary attack[12]. Under the current electrolysis conditions, chemical dissolution, electrolyte penetration and grain boundary attack serve as three major corrosion mechanisms.

For the former one, there were several possible reactions during electrolysis[13]:

$$2AlF_3\,(s) + 3\,FeO\,(s) = 3FeF_2\,(s) + 3/2O_2\,(g) + 2Al\,(l) \quad (1)$$

$$2AlF_3\,(s) + Fe_2O_3\,(s) = 2FeF_3\,(s) + 3/2O_2\,(g) + 2Al\,(l) \quad (2)$$

$$2AlF_3\,(s) + 3NiO\,(s) = 3NiF_2\,(s) + 3/2O_2\,(g) + 2Al\,(l) \quad (3)$$

In current electrolysis conditions, the probability of the reactions mentioned above is very low. So, for the ceramic inert anode without the additive CaO, its SEM image shows that the pores are minute and the bulk density of inert anode is high, and the corrosion mainly is affected by the chemical dissolution, which implies the inert anode is corroded at stoichiometric ratio and peeled off layer by layer. Thus, the stability of the inert anode can be maintained during electrolysis and it shows good corrosion resistance.

The results from our previous works[9] indicate that the additive CaO exists in the grain boundary if the content of doped CaO is high. During electrolysis, the chemical dissolution of the doped CaO is easier than that of other phase in ceramic inert anode. Thus, lots of holes and pores are come into being because of the loss of CaO, which implies that the corrosion resistance of ceramic inert anode to electrolyte penetration and grain boundary attack decreases. So, for the ceramic inert anode containing the additive CaO, its corrosion resistance is affected not by the chemical dissolution, but also by the electrolyte penetration and grain boundary attack. The stability of inert anode cannot be maintained during electrolysis and the anode shows bad corrosion resistance to molten cryolite.

## Conclusions

The content of additive CaO in ceramic inert anodes has great effects on the concentration of impurities in electrolyte at the same time of electrolysis. The concentrations are 67ppm ~ 150ppm for Ni, 105ppm ~ 261ppm for Fe and 158ppm ~ 1321ppm for Ca at the end of electrolysis with the content of CaO becomes from 0 to 4%.

According to the concentration of impurities Ni, Fe and Ca in electrolyte during electrolysis, the figures and microstructure of the inert anodes after electrolysis, the addition of the oxide CaO is adverse to improve the corrosion resistance of ceramic inert anode. The corrosion degree of ceramic inert anode deteriorates with the content of CaO increases.

The microstructure of inert anodes after electrolysis shows that the corrosion resistance of the ceramic inert anode containing the additive CaO is affected mainly by the electrolyte penetration and grain boundary attack. The additive CaO which exists in the grain boundary accelerates the corrosion of ceramic inert anode.

*Acknowledgement: The authors gratefully acknowledge financial support from National Basic Research Program of China (2005CB623703) and National Natural Science Foundation of China (No. 50474051).*

## References

[1] D. R. Sadoway. Inert Anodes for the Hall-Héroult Cell: the Ultimate Materials Challenge[J]. JOM 53(5)(2001): 34-35.
[2] R. P. Pawlek. Inert Anodes for the Primary Aluminium Industry: an Update[A]. W. R. Hale, eds. Light Metals 1996[C]. Warreudale PA, USA:TMS.243-248.
[3] M. Glucina and M. Hyland. Laboratory-Scale Performance of a Binary Cu–Al Alloy as an Anode for Aluminium Electrowinning[J]. Corrosion Science, 2006, 48: 2457-2469.
[4] T. E. Jentoftsen, O-A. Lorentsen, E. W. Dewing, et al. Solubility of Iron and Nickel Oxides in Cryolite-Alumina Melts[A]. J. L Anjier. Light melts 2001[C]. Warrendale, Pa: TMS, 2001. 455-460.
[5] Lai Yan-qing, Tian Zhong-liang, Qin Qing-wei, et al. Solubility of Composite Oxide Ceramics in the Melt of $Na_3AlF_6$-$Al_2O_3$ [J]. Journal of Central South University (Nature Science Edition), 2003. 34(3): 245-248.(In Chinese)
[6] H. Xiao, R. Hovland, S. Rolseth, et al. Studies on the Corrosion and the Behavior of Inert Anodes in Aluminum Electrolysis [J]. Metallurgical and Materials Transactions B: Process Metallurgy and Materials Processing Science, 1996, 27 (2): 185-193.
[7] A. B. D. Van Der Meer. Mechanical Strength of Magnesium Zinc Ferrites for Yokerings [A]. W. Hiroshi. Ferrite: Proceedings of International Conference [C]. Tokyo: D Reidel publishing company, 1980. 301-305.
[8] D. W. Johnson. Recent Progress on Mechanical Properties of Ferrites [A]. W. Hiroshi. Ferrite: Proceedings of International Conference [C]. Tokyo: D Reidel publishing company, 1980.

285-291.
[9] Lai Yan-qing, Zhang Yong, Zhang Gang, et al. Effect of CaO Doping on Densification 10NiO-$NiFe_2O_4$ Composite Ceramics[J]. The Chinese Journal of Nonferrous Metals, 2006, 16(8):1355-1360. (In Chinese)
[10] Lai Yan-qing, Zhang Gang, Li Jie. Effect of Adding Cu-Ni on Mechanical Capacity and Electrical Conductivity of $NiFe_2O_4$-Based Cermets [J]. Journal of Central South University (Science and Technology), 2004. 35(6):880-884.
[11] D. H. DeYoung. Solubilities of Oxides for Inert Anodes in Cryolite–Based Melts [A]. R. E. Miller. Light Metals [C]. Warrendale, Pa, USA: TMS, 1986: 299-307.
[12] S. R. Ray. Inert Anodes for Hall Cells [A]. R. E. Miller. Light Metals [C]. Warrendale, Pa, USA: TMS, 1986: 287-298.
[13] S. P. Ray. Effect of Cell Operating Parameters on Performance of Inert Anodes in Hall-Héroult Cells[A]. R. D. Zabreznik. Light Metals 1987[C]. Warreudale PA: TMS, 1987. 367-380.

# AUTHOR INDEX
# Light Metals 2008

## A



## B



## C



## D

# E



# F



# G



# H



# I

# J



# K



# L

## S



## T



## U



## V

## W



## X



## Y



## Z

# SUBJECT INDEX
# Light Metals 2008

## 1



## 5



## 8



## A



## B

# C



# D

# E



# F



# G

## H



## I



## K



## L



## M

# N



# O



# P



# Q



# R

# S



# T

# U



# V



# W



# X



# Y



# Z

9 780470 943397